MICROTECHNOLOGY AND MEMS

MICROTECHNOLOGY AND MEMS

Series Editor: H. Fujita D. Liepmann

The series Microtechnology and MEMS comprises text books, monographs, and state-of-the-art reports in the very active field of microsystems and microtechnology. Written by leading physicists and engineers, the books describe the basic science, device design, and applications. They will appeal to researchers, engineers, and advanced students.

Please view available titles in *Microtechnology and Mems* on series homepage http://www.springer.com/series/4526

Stephanus Büttgenbach
Arne Burisch
Jürgen Hesselbach
Editors

Design and Manufacturing of Active Microsystems

Springer

Editors
Prof. Dr. Stephanus Büttgenbach
Technische Universität Braunschweig
Institut für Mikrotechnik
Langer Kamp 8
38106 Braunschweig
Germany
s.buettgenbach@tu-bs.de

Dipl.-Ing. Arne Burisch
Technische Universität Braunschweig
Institut für Werkzeugmaschinen
und Fertigungstechnik
Langer Kamp 19 B
38106 Braunschweig
Germany
a.burisch@tu-bs.de

Prof. Dr.-Ing. Dr. h.c. Jürgen Hesselbach
Technische Universität Braunschweig
Institut für Werkzeugmaschinen
und Fertigungstechnik
Langer Kamp 19 B
38106 Braunschweig
Germany
j.hesselbach@tu-bs.de

ISSN 1615-8326
ISBN 978-3-642-26738-3 ISBN 978-3-642-12903-2 (eBook)
DOI 10.1007/978-3-642-12903-2
Springer Heidelberg Dordrecht London New York

Cover design: WMXDesign GmbH

Printed on acid-free paper

Springer is part of Springer Science+Business Media (www.springer.com)

Preface

Microsystems technology, which is considered to be a key technology for the 21st century, integrates signal processing with miniaturized sensors and actuators. This opens up a whole range of new applications going beyond purely microelectronics systems. For some time now, microsystem products have emerged as an important part of our everyday lives as witnessed by their successful application in various fields such as car-making, biomedical engineering and communications technology.

Whereas the field of microsensors is already highly advanced, microactuators are still in the basic development phase. Consequently, there is large demand for basic research and experimental development in the field of microactuators. This demand has been picked up at the Technische Universität Braunschweig in the late 1990s and has led to the establishment of the Collaborative Research Center "Design and Manufacturing of Active Micro Systems" in 1998. Five institutes belonging to the Department of Mechanical Engineering of the Technische Universität Braunschweig, institutes of the Leibniz Universität Hannover, laboratories of the Fraunhofer-Institut für Schicht-und Oberflächentechnik, the Physikalisch-Technische Bundesanstalt, and the Laserzentrum Hannover have jointly developed fundamentals for the design and manufacturing of active microsystems over a period of twelve years. The applicability of the methods and technologies developed has been verified on the basis of several prototypes of miniaturized stepper motors. This book summarizes the results obtained through the fruitful cooperation within the Collaborative Research Center (SFB 516).

Special thanks go to the German Research Foundation, which funded the Collaborative Research Center over twelve years. We are grateful to all authors for their participation and their contribution to this book. Last but not least we would like to thank all people who have helped to complete this book, especially Jan Torben Runte, Paul Frakes, and Robert John Ellwood for proofreading and many useful ideas.

Braunschweig (Germany),
December 2010

Stephanus Büttgenbach
Arne Burisch
Jürgen Hesselbach

Contents

1 Introduction ... 1
1.1 Initial Situation ... 1
1.2 Prototype Concepts for Active Microsystems ... 4
1.3 Scientific Development ... 6
1.4 Outline ... 7

Part I Design and Construction

2 Electromagnetic Design of Microactuators ... 11
2.1 Forms of Electromagnetic Microactuators ... 11
2.2 Design and Construction ... 12
2.3 Material Properties ... 13
2.4 Coil Forms ... 14
2.5 Functional Principles ... 14
2.6 Design of Micromotors ... 19
References ... 28

3 Drive Systems Based on Electromagnetic Microactuators . 29
3.1 Configuration of Drive Systems Based on Electromagnetic Microactuators ... 29
3.2 Power Supply ... 30
3.3 Current Command Generation ... 38
3.4 Position Control ... 40
References ... 47

4 Modular Computer Aided Design Environment for Active Microsystems ... 49
4.1 Introduction ... 49
4.2 Phase of Preliminary Design ... 52
4.3 3D-Model Synthesis and Optimization ... 56

4.4 Synthesis and Optimization of Processes and Process Sequences 58
References 66

Part II Guiding and Measuring in Active Microsystems

5 Wear Behavior in Microactuator Interfaces 69
5.1 Introduction 69
5.2 Test Systems 70
5.3 Fabrication Process of Tribological Surfaces 77
5.4 Experimental Investigations 79
5.5 Conclusion 86
References 86

6 Friction Behavior in Microsystems 89
6.1 Introduction 89
6.2 Preparation of Microguides and Bearings 90
6.3 Characterization 93
6.4 Results 97
6.5 Conclusion 106
References 107

7 Active Linear Guiding Concepts for Microsystems 109
7.1 Introduction 109
7.2 Active Aerostatic Guides for Microsystems 110
7.3 Active Magnetic Guides for Microsystems 117
7.4 Capacitive Displacement Sensors for Active Guides 121
7.5 Conclusion 125
References 126

8 Design of Sensors for Position Control of Microactuators . 127
8.1 Introduction, Requirements and Measurement Principle 127
8.2 Simulation of Integrated Laser Beam 3×3 Couplers 129
8.3 Production of Integrated 3×3 Couplers in Glass 132
8.4 Beam Guiding by Refractive and Diffractive Elements: Grating, Prism, Lenses 134
8.5 Signal Detection 137
8.6 Realtime Signal Processing 138
8.7 Alignment and Fixing of Optical Components 141
8.8 Summary 143
References 144

9 Tactile Dimensional Micrometrology 145
9.1 Introduction 145
9.2 3D Microprobes 146
9.3 Calibration of Probing Forces 157
9.4 Dimensional Standards for Micro Metrology 160
References 163

Part III Manufacturing and Fabrication

10 Fabrication of Magnetic Layers for Electromagnetic Microactuators 167
10.1 Introduction 167
10.2 Fabrication Technologies 169
10.3 Test Systems 171
10.4 Experimental Investigations on Soft Magnetic Materials 173
10.5 Experimental Investigation on Hard Magnetic Materials 179
10.6 System Integration Aspects 184
10.7 Conclusion 186
References 186

11 Fabrication of Excitation Coils for Electromagnetic Microactuators 189
11.1 Introduction 189
11.2 Photoresist Pattern Creation 191
11.3 Electroplating of Cu Microcoils 196
11.4 Insulation 197
11.5 Integration of Coils into Microactuators 198
11.6 Conclusion 205
References 205

12 Development and Fabrication of Electromagnetic Microactuators 207
12.1 Introduction 207
12.2 Linear VR Stepper Motor 209
12.3 Rotating VR Stepper Motor 215
12.4 Rotating Synchronous Micro Motors 219
12.5 Conclusion and Outlook 223
References 224

13 Development and Fabrication of Linear and Multi-Axis Microactuators 225
13.1 Introduction 225
13.2 Linear VR Microstep Motor 226
13.3 Linear Hybrid Microstep Motor 227
13.4 Linear Synchronous Micromotor 230
13.5 xy-Actuator Investigations 234

13.6 VR Micro- and Nanopositionier for xy-Actuators 235
13.7 Experimental 241
13.8 Conclusion 242
References 243

14 Micromachining of Parts for Microsystems 245
14.1 Introduction 245
14.2 Microgrinding 246
14.3 Microgrinding of Boreholes 260
14.4 Conclusions 262
References 264

Part IV Microassembly

15 Size-Adapted Manipulation Robots for Microassembly 269
15.1 Introduction 269
15.2 Size-Adapted Robot for Microassembly 272
15.3 Miniaturized Robot for Desktop Factories 277
15.4 Conclusion 285
References 286

16 Tools for Handling and Assembling of Microparts 287
16.1 Introduction 287
16.2 Electrostatic Forces in Microhandling Processes 289
16.3 Mechanical Microgrippers with Integrated Actuators .. 297
16.4 Pneumatically Driven Auxiliary Microtools 305
References 307

17 Stereophotogrammetry in Microassembly 309
17.1 Introduction 309
17.2 Photogrammetry 310
17.3 Conclusions 325
References 326

18 Use of Hot Melt Adhesives for the Assembly of Microsystems ... 327
18.1 Adhesive Bonding as Micro Joining Technology 327
18.2 Properties of Hot Melt Adhesives 328
18.3 Adhesive System Selection Criteria 329
18.4 Different Particle Shapes of Micro-Scale Hot Melt Adhesives . 331
18.5 Application Methods for Micro-Scale Hot Melt Adhesives 333
18.6 Properties of Hot Melt Adhesive Bonds 339
18.7 Conclusion 342
References 343

19 Design of a Microassembly Process Based on Hot Melt Adhesives 345
19.1 Introduction 345
19.2 Process Design for the Joining with Hot Melt Adhesives 348
19.3 Implementation of a Passive Heat Managment 352
19.4 Implementation of Active Heat Managment 358
19.5 Conclusion 362
References 363

20 Design of an Automated Assembly for Micro and Nano Actuators 365
20.1 Introduction 365
20.2 Assembly Concept for a Linear Microactuator with Levitation System 367
20.3 Assembly Concept for a xy-Micro- and Nanopositioner 369
20.4 Conclusion 371

Part V Industrial Applications

21 Bistable Microvalve for Biomedical Usage 375
21.1 Introduction 375
21.2 Concept 377
21.3 Transfer of SFB Knowledge 379
21.4 Realization 379
21.5 Combination of the Valve Layers 387
21.6 Pre-Evaluation of the Intermediates 388
21.7 Summary and Prototype 390
21.8 Outlook 391
References 392

22 Microassembly Following the Desktop Factory Concept 393
22.1 Introduction 393
22.2 Miniaturized Components 397
22.3 New Prototype of the *Parvus* Robot 399
22.4 Experimental Verification 402
22.5 Conclusion 407
References 408

23 Automated Optical BGA-Inspection – AUTOBIN 411
23.1 Introduction 411
23.2 Specification of the Target System 416
References 422

24 Slider with Integrated Microactuator (SLIM) for Second Stage Actuation in Hard Disk Drives 423
24.1 Introduction: 423
24.2 Concept 424
24.3 Micromagnetics Design and Fabrication 426
24.4 Micromechanics Design and Fabrication 430
24.5 SLIM System Integration 432
24.6 Experimental Investigations 436
24.7 Conclusion and Outlook 438
References 439

Index 441

Chapter 1
Introduction

S. Büttgenbach[1], A. Burisch[2]

[1] *Institute for Microtechnology*
Technische Universität Braunschweig
s.buettgenbach@tu-braunschweig.de

[2] *Institute of Machine Tools and Production Technology*
Technische Universität Braunschweig
a.burisch@tu-braunschweig.de

1.1 Initial Situation

Microsystem technology has progressed rapidly over the past few decades and plays an ever-growing role in the development of innovative technical products. In microsystems, sensors and actuators are integrated with information processing components resulting in compact and lightweight devices which offer further benefits, such as low energy consumption, high reliability, adaptivity and improvement of the cost-benefit ratio. Microsystems cover a broad range of application areas such as car manufacturing, biomedical engineering, communications technology and environmental protection. A major factor here is the fact that microsystems are used to add value much higher than the value of the microsystems themselves.

Whereas the field of microsensors is already highly advanced, microactuators are still in a more basic development phase, although microactuation opens up a multitude of important new opportunities to microsystems. Initially favored microactuation principles were electrostatic and thermal actuation because all processes for the fabrication of these microactuators were available from currently existing microelectronics technology. Principles technologically more difficult to realize have developed more slowly, for example piezoelectricity, shape memory alloys, pneumatics and hydraulics.

Magnetic microactuators exhibit considerable advantages such as high forces, large deflections, low input impedances and thus, the involvement of only low voltages. However, the basic structure of magnetic microactuators has imposed limitations for their broad adaptation. Key elements include three-dimensional microcoils and complex hard and soft magnetic microstructures, which correspond to wound coils and magnets in bulk actuators, respectively. In addition, in order to achieve high forces, both the electric conductors of the microcoils and the flux guiding structures of the magnetic circuits have to be fabricated in such a way that they allow for sufficiently high current and magnetic flux, respectively. These constraints require tech-

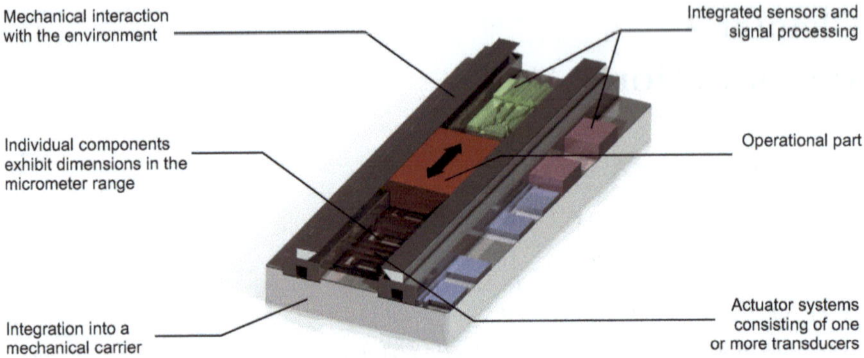

Fig. 1.1 Schematic of an active microsystem as defined in the context of the Collaborative Research Center

nologies to facilitate the fabrication of hybrid microsystems made of complex three-dimensional microstructures with high aspect ratios.

In particular, microactuators are an essential component of active microsystems, which can be characterized by the following features (see Fig. 1.1): sensors, signal processing functions and actuator systems consisting of one or more transducers are integrated into a mechanical carrier. The total construction size of the system is typically in the range of several centimeters whereas essential structures are in the micrometer range. Such a typical hybrid microsystem is made of several different components and types of materials (see Fig. 1.2). For the successful development of systems based on magnetic actuators, further technologies and processes are required for the fabrication of microcoils and magnetic microdevices. These include handling and bon-

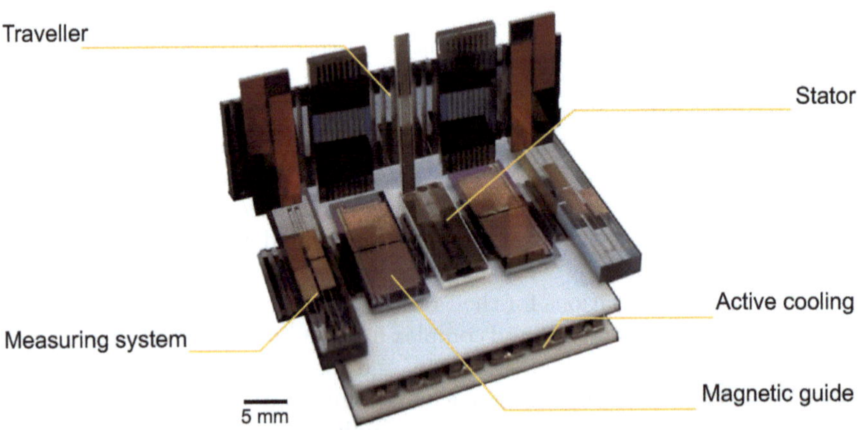

Fig. 1.2 Test setup with components of a linear microhybrid stepper motor

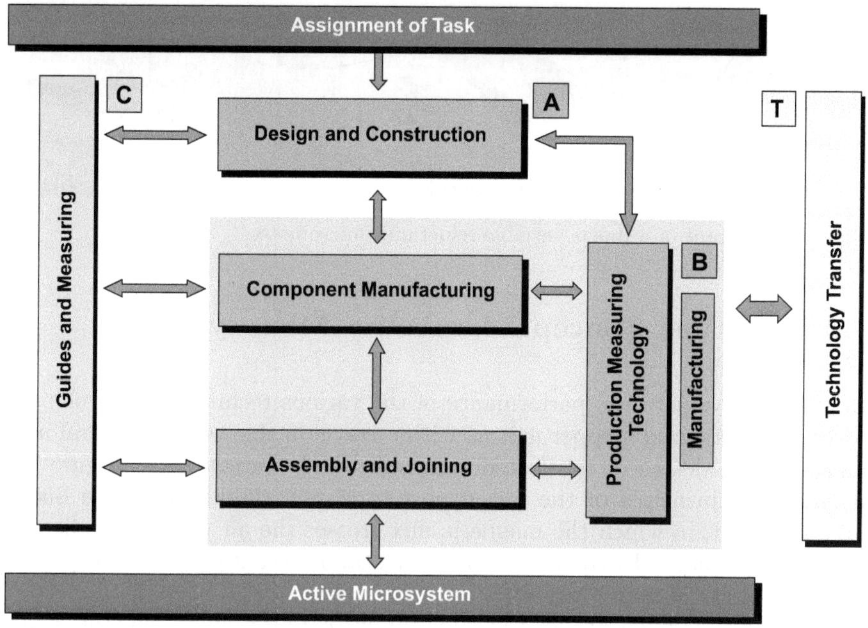

Fig. 1.3 Structure of the Collaborative Research Center

ding techniques, production measurement technology, guides and measuring techniques, and design engineering methods.

In order to tackle these challenges and to stimulate the development of magnetic active microsystems for a variety of applications, the Collaborative Research Center "Design and Manufacturing of Active Micro Systems" (Sonderforschungsbereich SFB 516) has been established at the Technische Universität Braunschweig. Five institutes belonging to the Department of Mechanical Engineering of the Technische Universität Braunschweig as well as institutes of the Leibniz Universität Hannover and laboratories of the Fraunhofer-Institut für Schicht- und Oberflächentechnik, the Physikalisch-Technische Bundesanstalt, and the Laserzentrum Hannover are involved in the Callaborative Research Center. It consists of 17 subprojects and is subdivided into four main project groups (see Fig. 1.3). Project group A deals with design engineering methods, project group B centers on microproduction technologies and project group C is dedicated to guides and measuring techniques. Project group T was established during the last phase of the research center and focuses on the transfer of selected technologies into industrial applications.

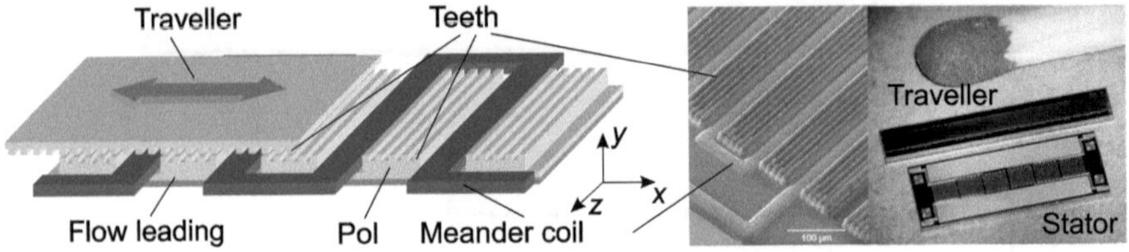

Fig. 1.4 Concept of a linear variable reluctance micromotor

1.2 Prototype Concepts for Active Microsystems

In order to evaluate the performance of the various technologies developed, several miniaturized stepper motors with a travel in the order of several millimeters were designed, fabricated, assembled and characterized. Figure 1.4 presents the principle of the initial prototype, a variable reluctance linear microactuator, in which the magnetic flux crosses the air gap perpendicular to the wafer plane. In the course of the research activities this concept has been expanded in many respects.

In order to compensate for the Maxwell normal forces between stator and traveler, which are common to any electromagnetic motor which uses soft magnetic material, the so-called "horizontal reluctance stepping motor concept" (Fig. 1.5.b) was developed. In this design, the horizontal magnetic flux generated by 3D meander coils (see Fig. 1.5.a) attracts the traveler poles from both sides, compensating for the normal forces. This results in very low friction forces, allowing for the implementation of a purely tribological or a

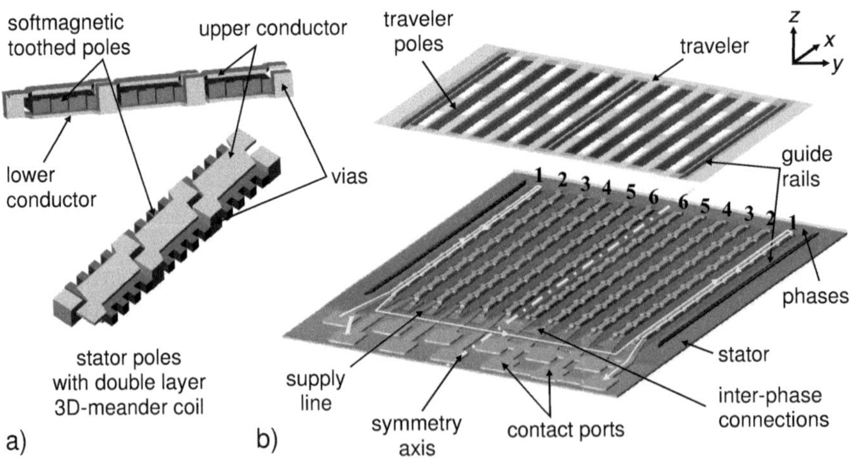

Fig. 1.5 Concept of a linear variable reluctance micromotor with horizontal flux guidance: (**a**) Meander coils; (**b**) Complete system design

Fig. 1.6 Concept of a linear micro hybrid stepper motor

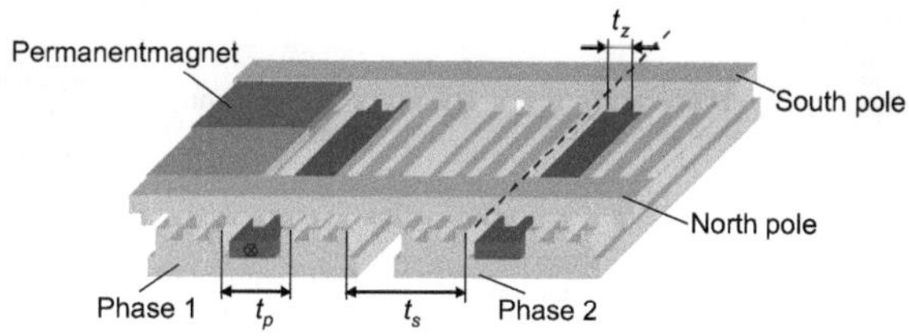

passive magnetic guidance. This basic idea has also been successfully applied to a rotating variable reluctance microstepper motor.

Another concept is the linear micro hybrid stepper motor (see Fig. 1.6). This motor is based on a variable reluctance (VR) stepper motor but equipped with permanentmagnets to increase the driving force.

As the simple VR motor is easier to fabricate than the hybrid one, additional prototypes have been created based on this concept. The VR motor requires the fabrication of microcoils and soft magnetic circuits. To overcome the normal forces between stator and traveler, an active magnetic guidance was developed. Based on this concept, the final prototype comprising contributions from all subprojects has been defined; the prototype includes an xy-nanopositioning stage comprising of four linear VR stepper motors controlled in the microstepping mode, magnetic and tribological guidance and xy-position sensors (Fig. 1.7).

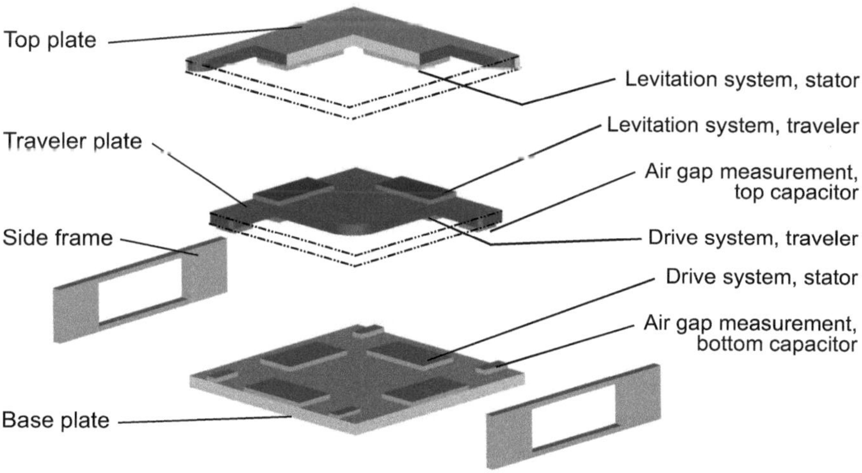

Fig. 1.7 Concept of an xy-nanopositioning stage

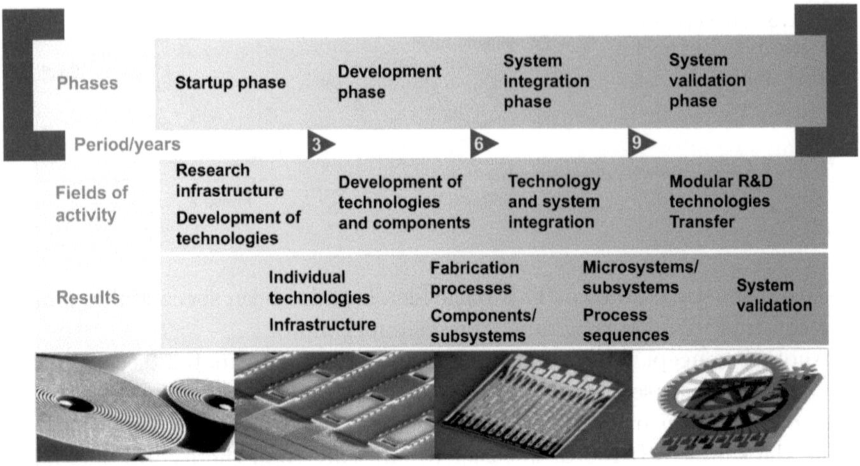

Fig. 1.8 Scientific development of the Collaborative Research Center

1.3 Scientific Development

The activities of the Collaborative Research Center comprised four three-year phases (see Fig. 1.8). During the startup phase, the necessary research infrastructure was established. Experimental equipment was installed and several technologies, such as UV depth lithography for the fabrication of microcoils and magnetic flux guiding structures, were developed and optimized.

In the development phase, the research activities were focused on the adaptation of the individual technologies developed during the startup phase to the fabrication of components of the magnetic micromotors. The actuators of the variable reluctance prototype micromotors have been manufactured, and the feasibility of the concepts has been demonstrated. By the end of the second phase, methods and processes for the design and manufacturing of components and subsystems of active microsystems have been made available.

During the system integration phase, the process technologies, including handling devices, assembly processes and the development of tribological layers, were combined to process sequences for the fabrication of magnetic micromotors. To assist with the design of active microsystems, simulation models were improved and extended to further prototype concepts, and a construction kit for microactuators and a software tool for the design of microparts were developed. Furthermore, control circuits, which accomplish continuous stepping of the micromotors, were developed and successfully tested.

The main purposes of the system validation phase were the integration of active guidance and measuring systems into the prototypes, the development of an xy-nanopositioning system, and the application of the technologies to other kinds of magnetic microactuators. Three additional projects were devoted to the transfer of selected technologies into industrial applications in close cooperation with industrial partners. This covers the development of a microvalve, a miniaturized precision assembly robot, and automated inspection methods for electronic production.

1.4 Outline

This book describes in detail the results developed within the Collaborative Research Center. First, the essential design tools and theory for magnetic microactuators are presented. The simulation of drive systems and concepts for position control of microactuators is investigated in Part I. The open and closed loop control algorithms that are discussed take into account the unique dynamic behavior and aspects of miniaturized systems.

Necessary methods for guides and measuring systems within active microsystems are discussed in Part II. This includes an investigation of the frictional and abrasive behavior as well as their reduction by applying appropriate protective layers. In conjunction with this, the integration of measuring systems into the actuator is presented. Production measuring techniques are used to measure the dimensions of the microsystems and their characteristics.

The development of fundamental manufacturing techniques for active microsystems is presented in Part III. This covers thin-film technologies such as coating, etching and lithography for the manufacturing of the functional components: guides for the magnetic flux, coil systems and isolation layers. Other micromachining processes such as drilling, milling, and grinding are explored in order to machine the conventional components.

Part IV presents solutions for the automated assembly process using new handling devices, sensor guidance, and joining technologies. Here, the influences of the tolerances of three-dimensional microparts and the behavior of these components on the assembly process are investigated.

Based on results from research models, Part V presents industrial solutions using microsystem technology for biomedical applications, new assembly devices, electronic production, and consumer electronics.

Part I
Design and Construction

Chapter 2
Electromagnetic Design of Microactuators

G. Janssen, R. Gehrking, J. Edler, B. Ponick, H.-D. Stölting

Institute for Drive Systems and Power Electronics
Leibniz Universität Hannover
ponick@ial.uni-hannover.de

Abstract The electromagnetic design of microactuators differs significantly from that of conventional motors in at least two respects: The use of microfabrication allows the design of completely new topologies of the active part that could not – or not economically – be manufactured from currently used iron and copper wire. However, the physical properties of micro-manufactured magnetic materials can differ significantly from those of conventional iron lamination or permanent magnets. The laws of growth have certain consequences to be taken into account when designing microactuators: For example, the influences of mechanical friction and the ohmic resistance of the windings are of increasing importance. This chapter thus focuses on the special design rules that apply for microactuators as well as on the advantages and disadvantages of different electromagnetic designs.

2.1 Forms of Electromagnetic Microactuators

Two out of the three classical functional principles for electromagnetic motors and actuators have strong disadvantages when being applied to microactuators: DC motors will require a mechanical commutator which introdues additional friction into the system – and friction is a major problem for all microactuators due to the laws of growth. Induction motors on the other hand, will have a high slip due to the high resistance / inductance ratio – another result of the laws of growth – leading to high losses in the rotor and a very bad control of speed. Thus, synchronous motors are the best choice for electromagnetic microactuators. There are three variations of this motor principle, the variable reluctance (VR) step motor, the permanent magnet and a combination of both, the hybrid synchronous motor. These functional principles are suited for both, rotating and linear motors. Material characteristics of microactuators are dependent on the alloys used, the geometry of the active components and the lithographic manufacturing process. These

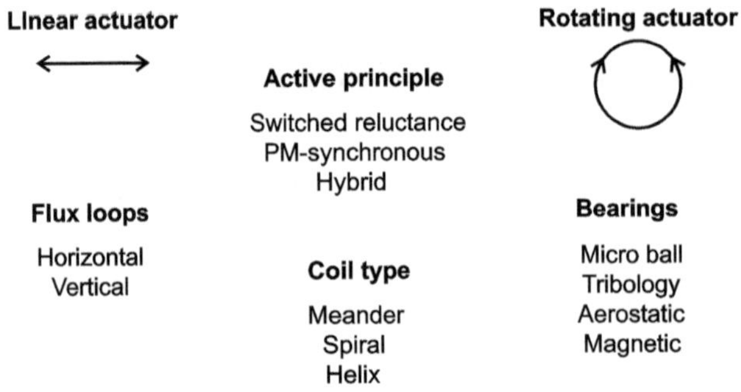

Fig. 2.1 Forms of electromagnetic microactuators

factors have to be taken into account carefully during the conceptional phase of micromotor construction. Further differences in the basic model form of a microactuator are the horizontal or vertical flux loops in the active components, the form of the coils, and the type of bearings and guidance used between the fixed and moving structures of the system. The different forms of electromagnetic microactuators are shown in Fig. 2.1

2.2 Design and Construction

The special material characteristics and processes used for the microactuator thin-film fabrication technology do not allow any simple projection of macroscopic motor construction forms to the microscopic levels [7]. The continuous development of process technologies such as electroplating or gas flow sputtering and the findings gained, serve as base for the electromagnetic concept. Technological limits imposed by the process-dependant values set limits to the possible geometries of microactuators. The main components of the electromagnetic conceptual design are the soft and hard magnetic materials and the conductor itself in form of different coil geometries. Knowledge of magnetic material characteristics is essential for the conceptual design, as optimization of the actuator unit is only possible with these data.

A general target of the electromagnetic conceptual design is to have a greatest possible motive force generated within the smallest possible space, taking into account the available technical fabrication capabilities. However, positioning accuracies or a preferably simple actuator fabrication process have also to be taken into account to meet user specific requirements. The investigation into electromagnetic, thermal and structural mechanical effects supplies a broad analysis which serves as constructional base even in the conceptual phase.

A further challenge in the development and successful implementation of microactuators is the reduction of friction. Not only the reduction of weight and Maxwell prime forces, but also the choice of the bearing or guidance system is important. Electromagnetic guides may be preferable instead of mechanical guides.

The analytical conception and calculation of microactuators using magnetic and thermal symbol elements should be supported by numerical simulations using the Finite Element Method (FEM). Static and transient calculations, e.g. the determination of the effects of harmonics in the supply current, serve to model microactuators in combination with the necessary power stage electronics.

The soft magnetic components of a microactuator show geometry-influencing material characteristics. As the permeability may e.g. depend on the geometry of the magnetic core, a special pole design can lead to a maximization of the motional force while simplifying pole construction. The verification of simulation results on electromagnetic actuators through force measurements lead to improvements in the simulation model, calculated values comparing well to the actual measured results.

A classical conceptual design process for electromagnetic microactuators consists of the following steps:

1. Defining the requirements for the microactuator.
2. Ascertaining the basic form that meets the requirements defined.
3. Validation with respect to the manufacturing technology used.
4. Creation of a parametric model.
5. Implementation within a simulation environment.
6. Optimisation by parameter analysis within set constraints.
7. Checking the results.
8. Construction of the microactuator based on the optimised model.

2.3 Material Properties

Calculation of microactuator properties requires solid knowledge of the electromagnetic material properties, especially the B-H curves of the magnetic materials that are cased to determine the differential permeability at the operating point and the saturation.

Using Vibrating Sample Magnetometer (VSM) measurements, permeability and anisotropies were determined, which played an important role in the conception phase of the active part. Material properties which serve as knowledge base for microactuator construction are treated in detail in Chap. 10. Important in this respect is a dependence on geometry. Simple ring core structures have shown a better permeability as complex pole and tooth structures. Linear VR micromotors allow various complexities of pole and tooth structure. To avoid mechanical stress, which leads to deterioration of permeability, e.g. linear micromotors with split tooth construction have been manufactured.

2.4 Coil Forms

The different coil forms used in microactuators are shown in Fig. 2.2 as single layer models. The continuous development of manufacturing technology has allowed to produce double layer coils as well. The out-of-plane meander and

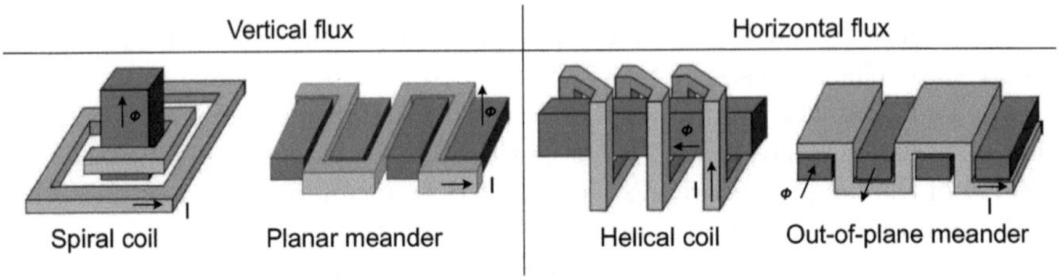

Fig. 2.2 Coil forms

the helix coil are typical forms of actuators with vertical flux loops. The planar meander and the spiral coil are consequently used in horizontal actuators with flux loops. The coil forms are treated in detail in Chap. 11

2.5 Functional Principles

The different functional principles governing microactuators are covered as follows:

2.5.1 Variable Reluctance (VR) Principle

The VR principle is the generation of force caused by the reduction of magnetic resistance and the increase of inductivity. The greater the change in inductivity, the greater the resulting force. The absolute value of inductivity, however, for the force of a linear actuator or the torque of a rotating motor is insignificant. Microactuators according to the VR principle are constructed and manufactured in many different forms as linear and rotating step motors.

The linear model depicted in Fig. 2.3 shows the principle construction of a symmetrical functional part of a three phase VR actuator ($m = 3$).

The traveller forms the magnetic back iron of the construction [2]. Its surface adjacent to the air gap is evenly toothed along its whole length. The stator is also similarly toothed along its air gap face side. The number of teeth per pole $n_{z,p}$ is preliminarily taken to be arbitrary.

$$n_{z,p} \in \mathbf{N} \tag{2.1}$$

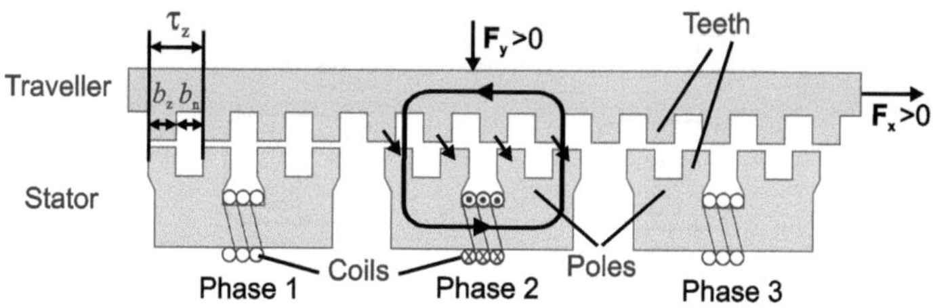

Fig. 2.3 Sketch of the principle of a linear three phase VR step motor.

The "Pole-based VR step micromotor" is a special case with $n_{z,p} = 1$, where one tooth represents one pole formation (e.g. Fig. 2.15). The displacement of poles belonging to the same phase is an integer multiple of the tooth pitch. The pole pitch is therefore given as:

$$\tau_p = g \cdot \tau_z \qquad g = 1, 2, 3, ... \tag{2.2}$$

The number of poles per phase are in turn arbitrary.

$$n_p \in \mathbf{N} \tag{2.3}$$

The distribution τ_{Str} determines the relationship of the phases to each other:

$$\tau_{Str} = n_p \cdot \tau_p + \left(k + \frac{1}{m}\right) \cdot \tau_z \qquad k \in \mathbf{N} \tag{2.4}$$

In an electrically unexcited condition $I = 0$, no force results between traveller and stator. If a constant current I_1 is assumed, the resulting force has a component in the direction of movement $\mathbf{F_x}$ and a normal component $\mathbf{F_y}$, both dependent on traveller position x as follows:

In aligned condition, i.e. tooth to tooth position, $\mathbf{F_x}$ is zero (Fig. 2.4.b) and $\mathbf{F_y}$ is at maximum. This rest position is stable. As the traveller is moved out of this stable position, force $\mathbf{F_x}$ increases and works against the movement. At increasing displacement, force $\mathbf{F_x}$ increases to a maximum and then decreases (Fig. 2.4.c). When the traveller reaches full tooth displacement (teeth disaligned Fig. 2.4.d), the force $\mathbf{F_x}$ is again zero. The rest position is unstable, i.e. a minor displacement leads an increase of $\mathbf{F_x}$, which amplifies the displacement forces working on the traveller.

The force $\mathbf{F_y}$, however, decreases steadily between teeth aligned and teeth disaligned, reaching minimum at the latter. Further displacement of the traveller leads to force $\mathbf{F_x}$ now supporting the movement, Fig. 2.4.e, passing a maximum and decreasing to zero and regaining tooth aligned position. Force $\mathbf{F_y}$ normally increases continuously again to the maximal value. The traveller position as in Fig. 2.4.f is identical to the start position as in Fig. 2.4.b. A complete cycle has been executed. The quantitatively described force-way

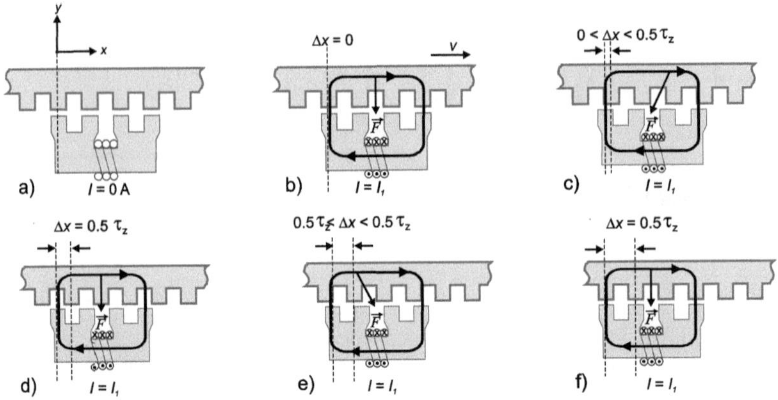

Fig. 2.4 Variable reluctance (VR) principle

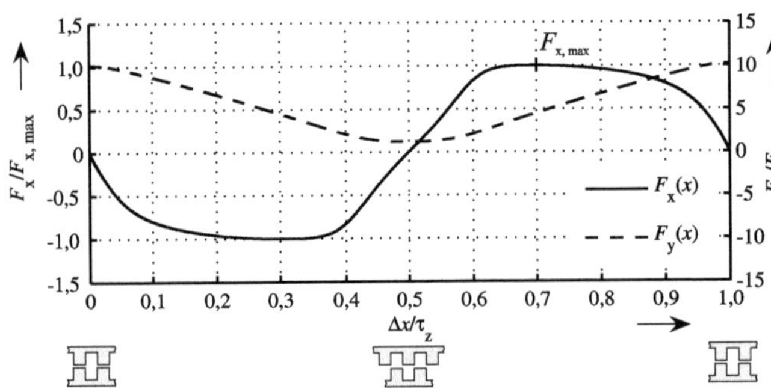

Fig. 2.5 Force-way characteristic curves of a VR step motor

characteristic curves are shown in Fig. 2.5. This motor principle is found in the macroscopic world in so-called "Switched reluctance motors".

2.5.2 PM Synchronous Principle

Synchronous motors are classical motors with m-phases (usually $m = 3$) in the stator, which generate a rotating field in the air gap. In the traveller or rotor, there is either a direct current exciter winding or permanent magnets, which generate a constant magnetic field. There is a large number of different designs, differing in the form of stator winding (groove laid, interlaced windings, pole-wound, concentric coils (concentrated winding), self-supporting iron-less air gap windings), in the method of generation of the magnetic field (electrical, permanent magnet), in the rotor design (cylindrical rotor, salient pole rotor, surface mounted magnets) and in the number of phases and poles.

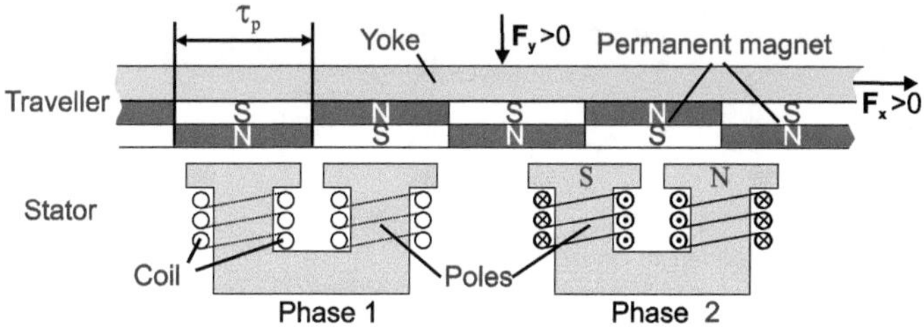

Fig. 2.6 Principle of a PM synchronous motor

Permanent magnet excited systems are especially suitable for synchronous micromotor manufacturing, these being either conventional magnets or microtechnically manufactured polymer magnets. Synchronous micromotors composed of concentric or overlapping windings without soft-magnetic back iron can be simply produced in many different sizes and forms. Synchronous micromotors can in principle be driven as step motors when the stator phases are fed by cyclical current pulses. Instead of a continuously rotating air gap field, the magnetic field will then move stepwise. Fig. 2.6 shows the principle of a PM synchronous motor.

2.5.3 Hybrid Principle

Being a hybrid of the VR and the PM synchronous principle, the hybrid principle contains permanent magnets either in the stator or in the rotor and soft magnetic teeth on the surfaces of stator and rotor [1]. Fig. 2.7 shows a typical assembly of a two phase hybrid motor. Hybrid step motors, unlike VR step motors, require a minimum of only two phases to be able to move in both directions. This requires, however, a bipolar servo control drive system. If the hybrid step motor has to be supplied from a unipolar servo system, at least three phases are necessary like for a VR step motor. In the following study, it will be assumed that the hybrid step motor is supplied by a bipolar servo system, which is usual for motors.

In the example shown in Fig. 2.7, the permanent magnets are placed in the stator between the soft-magnetic poles and coils of both phases. The traveller consists of a toothed soft-magnetic back iron. Analogue to the VR step motor, the sum of tooth width and slot is referred to as tooth pitch. While the pole distribution of a VR step motor is an integer multiple of the tooth pitch Eq. (2.2), the geometrical displacement of poles belonging to the same phase in a hybrid motor is:

$$\tau_p = k \cdot \tau_z + \frac{1}{2} \cdot \tau_z \qquad k \in \mathbf{N} \qquad (2.5)$$

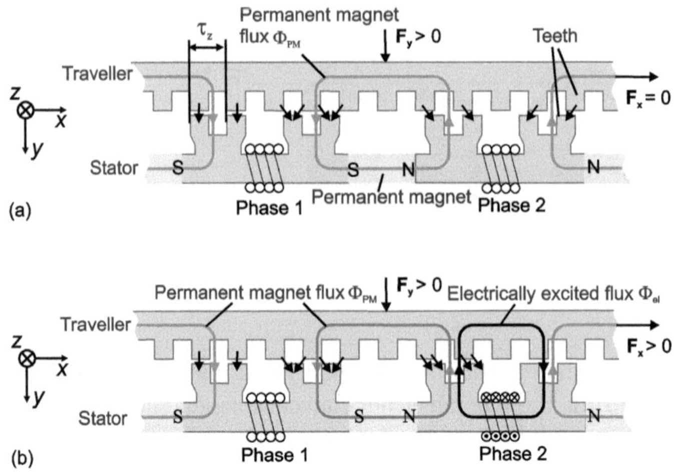

Fig. 2.7 Hybrid princple

The geometric displacement between different phases of bipolar driven hybrid step motors is:

$$\tau_p = k \cdot \tau_z + \frac{1}{2 \cdot m} \cdot \tau_z \qquad k \in \mathbf{N} \tag{2.6}$$

A comparison of Eq. (2.4) and (2.6) shows the necessary displacement between phases for hybrid step motors, which is different from VR step motors.

Due to the integrated permanent magnets, a magnetic field exists also in the electrically unexcited condition, leading to a detent torque or force, which does not exist in case of VR step motors. In Fig. 2.7.a, the traveller is shown at magnetic rest, i.e. all horizontal forces acting on the traveller sum to zero. If phase 2 is excited with the traveller position as in Fig. 2.7.a, the permanent magnet generated flux is superimposed by the electrically excited flux and is either weakened or strengthened in the pole areas. The direction of the electrical current in a hybrid step motor has influence on the force generated, which is not the case for VR step motors. The current direction determines which poles and teeth are magnetically weakened or strengthened. In Fig. 2.7.b, the direction of current is chosen in order to strengthen the magnetic flux in the left pole and weaken it in the right pole of phase 2. The horizontal force at the left pole of phases 2 is larger than at the right pole and, as a result of this inequality, the resulting total force leads to a mechanical movement of the traveller. The quantitatively described force-way characteristic of a hybrid step motor is shown in Fig. 2.8.

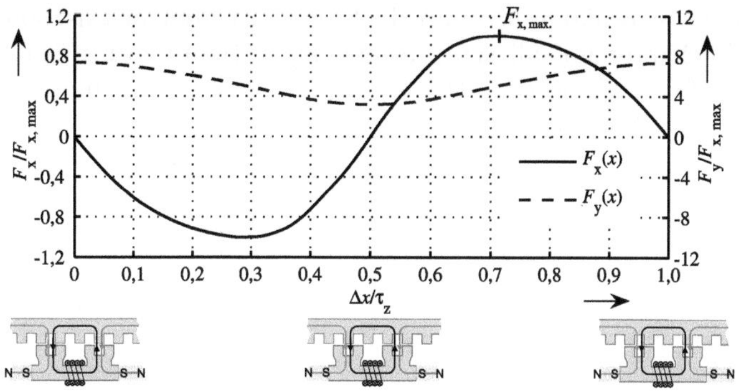

Fig. 2.8 Force-way characteristic curves of a hybrid step motor

2.5.4 Bearing and Guidance Concepts

The different types of bearings and guidance concepts are split into active and passive systems. The micro-ball guidance, friction guide and some designs of magnetic guidances belong to the passive forms. Active guidance systems include aerostatic pressure and active electromagnetic systems. Especially magnetic and active electromagnetic guidance systems have been simulated and calculated. A special and very attractive form of a passive magnetic guidance case can be realized by an integrated levitation function in VR step micromotors with horizontal magnetic flux, as shown in Fig. 2.12.

Magnetic bearing systems allow an almost frictionless, wear-free operation of motors and actuators magnetic forces [8]. E.g. one form of passive magnetic guidance provides only supports to another guidance system [9]. The different types of bearings utilize magnetic forces between two permanent magnets, two electromagnets, one permanent and one electromagnet or a magnetic and a soft magnetic core. In general, the force per PM volume or PM is higher for systems, where a north pole on one side of the air gap faces a south pole on the other side of the air gap, than for systems whith identical magnetic poles on both air gap sides. Active systems require an electromagnet either on the rotor for one stator, so that the force developed is controllable via the excitation current. The disadvantage of active magnetic bearings is that a fast and powerful control of the exciter current is required, as well as assumed sensors that measure the width for the air gap to be controlled.

2.6 Design of Micromotors

In recent years, many different types of electromagnetic microactuators have been designed, calculated, manufactured and measured. The great diversity of usable micromotors available has already been illustrated in this book. Therefore, in this chapter not all types of existing actuator forms can be mentioned. In the following section, the characters of the most important design alternatives are discussed.

2.6.1 Linear Synchronous Micromotors

The principle design of a permanent magnet excited linear synchronous micromotor with three phases is shown in Fig. 2.9 [6].

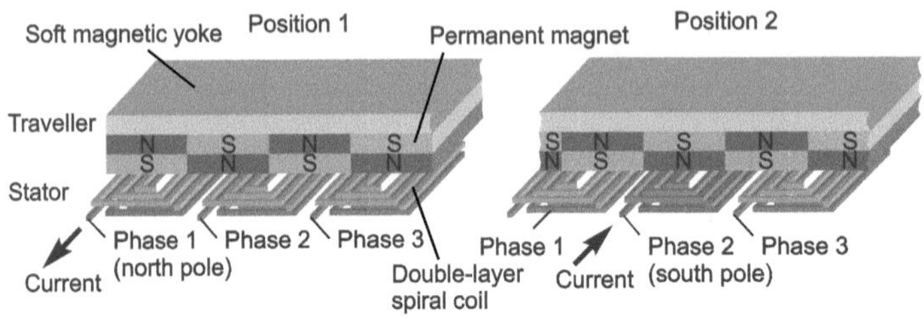

Fig. 2.9 Principle of linear synchronous micromotors

For easier manufacturing, it is preferable to use an air gap winding embedded in electrically isolating material. The two stator models, with and without soft magnetic yoke underneath the coils, have been investigated. In the traveller, permanent magnets with alternating poles are arranged on a soft magnetic yoke. For the permanent magnets, both commercially available magnet foils and SmCo layers in thin film technology, were used. The step-width is given as τ_p/m, with τ_p being the pole pitch and m being the number of phases. This represents one third of the width of the permanent magnets, as these are placed directly next to each other. Fig. 2.9 depicts two different positions for a current flow in phase 1 or in phase 2, respectively.

The permanent magnets have a width of 2.0 mm, leading to a step width of 0.667 mm. Force measurements with the magnet foil motor showed a maximal motive force of 5.4 mN at 2 A phase current. Thin-film technology produced permanent magnets lead to a motive force of 1.65 mN. Related to the effective air gap surfaces, a surface area of 72 mm^2 provided force densities of 75 $\mu\mathrm{N\,mm}^{-2}$ and 22.9 $\mu\mathrm{N\,mm}^{-2}$, respectively.

2.6.2 Synchronous Micromotor with Radial Flux

A two pole rotating PM-synchronous step motor with two phases and a NiFe45/55 stator is shown in Fig. 2.10.a. The pole windings are copper helix

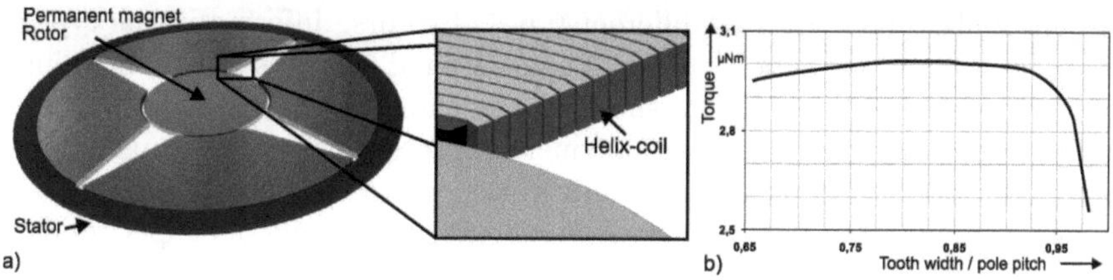

Fig. 2.10 (**a**) Design of synchronous micromotor with radial flux; (**b**) Influence of tooth width on generated torque

coils. The rotor consists of a single NdFeB-permanent magnet with radial magnetisation mounted on a central ball bearing [11].

An increase in motor torque was attained by optimizing the stator geometry. By raising the geometry of the stator teeth, an optimization of the motor torque can be achieved without the need for change of coil characteristics. The influence of the tooth width on the torque is shown in Fig. 2.10.b.

2.6.3 Synchronous Micromotor with Axial Flux

Rotating PM-excited synchronous motors can also be manufactured with an axial flux in the air gap, thus allowing to eliminate all soft magnetic materials. The manufacturing of such motors is simple as the stator consists of just an air gap winding and the rotor of permanent magnets [12]. Two different stator winding types are considered. The concentrated winding type can be used for rotating and linear synchronous micromotors. Supplementary, synchronous micromotors can also be designed with an overlapping winding. The overlapping winding is the most common form in conventional three phase motors and is generally placed in suitable slots in the stator.

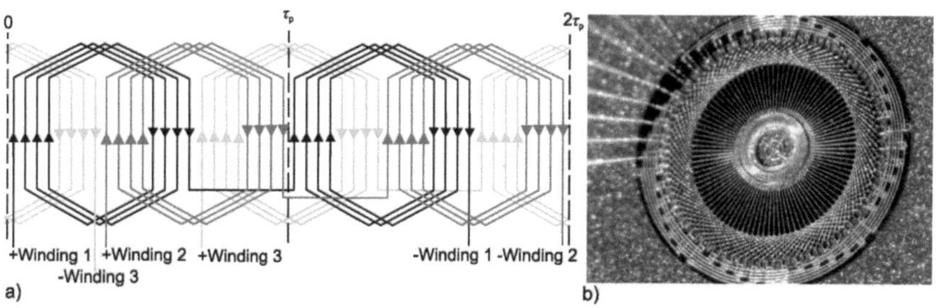

Fig. 2.11 (**a**) Principle of a three phase overlapping winding model; (**b**) Manufactured stator

Fig. 2.11 shows the design principle of a three phase winding with overlapping coils. There are two different types of rotors, differing in the type of magnet used. In one type, the use of conventional magnets limits the miniaturization of the motor. Very small motor diameters of e.g. 1 mm require micro-technically manufactured polymer magnets. PM-synchronous motors, especially those without soft magnetic yokes, proved to be very reliable and robust drive systems, as the active components consist only of copper conductor and permanent magnets. Further, no special bearing system is required, a simple friction system is sufficient. A thermally optimized operation allows continuous duty drive without difficulty. For precision positioning functions, however, a servo-controlled drive with a position sensor output is required.

2.6.4 Linear VR Step Micromotors

VR step motors allow exact positioning and smallest stepwidths to be realized without requiring positioning sensor systems. Step motors are usually controlled in open loop. Various types of VR step micromotors have been designed and manufactured in the last years. One of the earliest forms is depicted in Fig. 2.12 [3, 4, 5, 6].

The stator consists of toothed poles wound with a planar meander coil. A number of poles bound with such meander coil form one of three or six phases.

The magnetic field generated by the stator winding closes over the air gap and the toothed core of the traveller. The Maxwell forces, which provide the motive forces of the motor, are accompanied by normal forces attracting stator and traveller. If these normal forces are not compensated by a special motor design or bearing forms, they can have a severe impact on the functionality of the micro-drive servo system due to the increasing friction.

As the magnetic characteristics of thin-film technology produced layers depend on their thickness, an increase in the thickness of the active component will not necessarily lead to an increase in force, but may even lead to a decreasing force. In Fig. 2.13, three possible variations of the stator geometry of a linear VR step motor are shown.The first geometry has narrow pole cores and wide pole shoes, as it is usual for macroscopic motors as well. The space

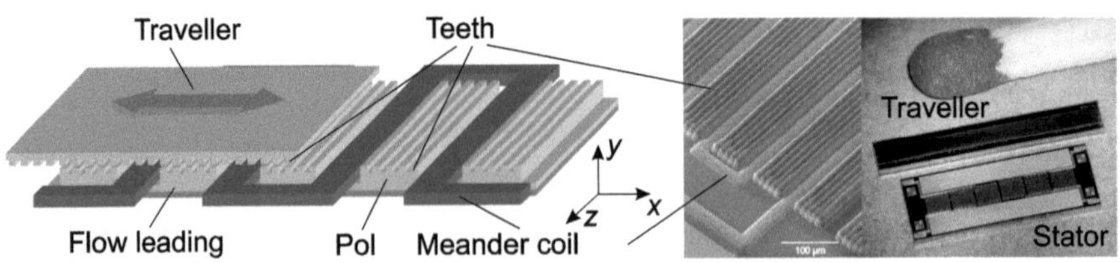

Fig. 2.12 Early forms of linear VR step micromotor

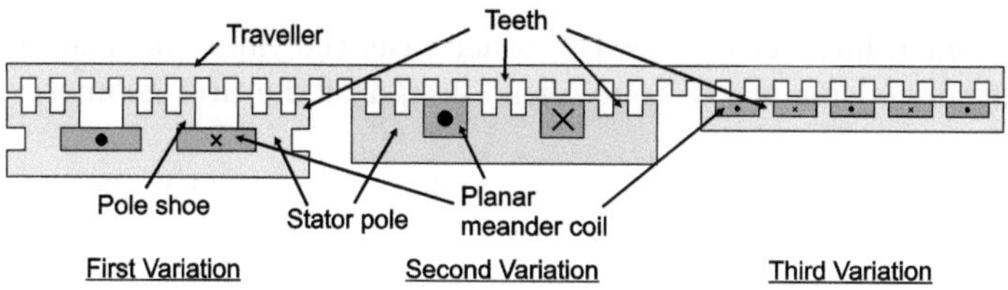

Fig. 2.13 Possible variations of the geometry of VR step micromotors

available for the winding is therefore increased. This form of stator requires, however, a greater stator thickness.

The second geometry reduces the space for the winding in order to reduce the height of the stator, and the third geometry is optimized for minimum stator height, as this will lead to maximum permeability of the soft magnetic yoke. Disadvantage of the second and third geometry is the comparatively small amount of space available for the winding.

The following simulation results have been gained using the same boundary conditions with respect to active component dimensions, tooth pitch, air gap, permissible power dissipation per unit area and material characteristics. The rotor tooth pitch is chosen to be $\tau_z = 68\,\mu\text{m}$. In order to have more space available for the conductor, the stator tooth pitch is chosen to toe a multiple of the rotor tooth pitch:

$$u_{z,S} = k \cdot \tau_z \qquad k \in \mathbf{N} \tag{2.7}$$

However, the width of the teeth is identical at least close in both, stator and rotor. Permissible power dissipation per unit area is chosen to be $p_{v,cu,p} = 272\,\text{mW}\,\text{mm}^{-2}$, being a value determined by thermographic measurement.

Fig. 2.14 depicts the calculated forces for the third geometry dependant on the stator tooth pitch. For the motive forces, the results from two calculations assuming different permeability of the soft magnetic parts are taken. In one case, the increase in permeability due to reduction of the active compo-

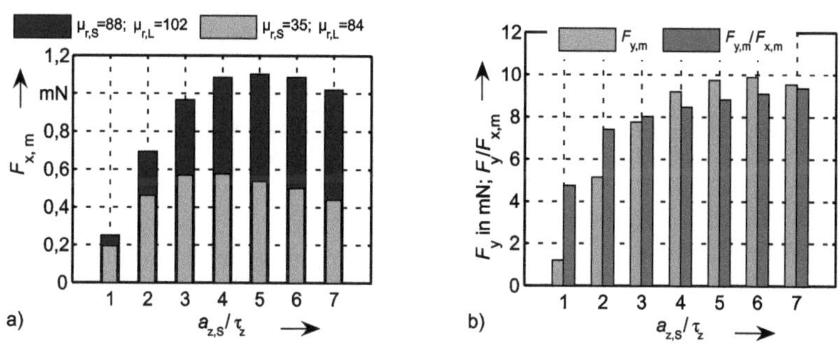

Fig. 2.14 Calculated forces dependant on stator tooth pitch

nent thickness is taken into account; in the other case, the values of relative permeability determined for active component design of the first geometry (see Fig. 2.13 left) are taken.

The reduction of active component thickness leads to a relative permeability of $\mu_{r,S} = 88$ and for the traveller of the stator of $\mu_{r,A} = 102$. In comparison, the first geometry shows values of $\mu_{r,S} = 35$ for the stator and $\mu_{r,A} = 84$ for the traveller. For the normal forces and power ratio, the results are calculated using the higher permeability values only.

Generated forces and the ratio of normal to motive force increase with increasing tooth pitch up to a maximum. If the stator tooth pitch is increased, motive forces up to $\mathbf{F}_{x,m} = 1.1\,\mathrm{mN}$ are obtained, being an increase of 55% compared to the first geometry.

2.6.5 Modular Linear VR Step Micromotor

Possibilities for the reduction of step width are either the increase of the number of phases or the reduction of the tooth pitch. To reduce manufacturing costs, a linear VR step micromotor was designed that can be produced in two variations with two different full step widths. There are two different stator designs, one with poles only (Fig. 2.15.a), the other with teeth being applied to the poles in a further manufacturing step (Fig. 2.15.b) [10]. Each of the three phases consists of two groups with four poles (Fig. 2.15.c). In the second geometry, each pole is equipped with three teeth. The traveller tooth width is the same as the stator tooth width. The calculated motive forces of

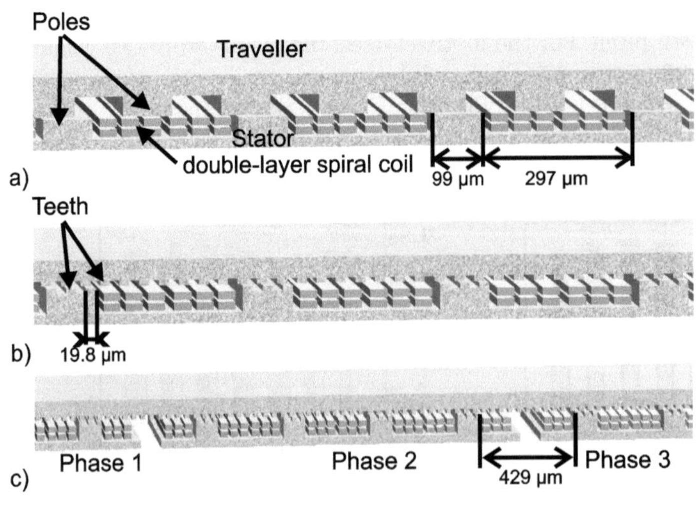

Fig. 2.15 Modular linear VR step micromotor

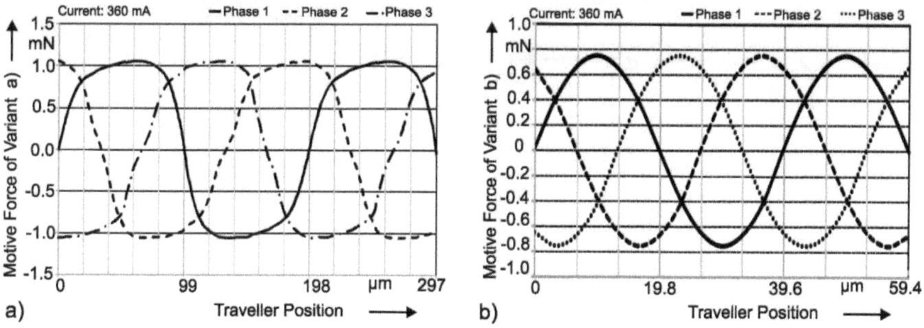

Fig. 2.16 Motive force of the modular linear VR step micromotor

the first geometry a) and the second geometry b) is shown in Fig. 2.16.a and b. Even though the step width of variant b) is only 20% of variant a) the motive force generated by variant b) is only 25% lower. The linear VR step micromotor can be equipped with an active magnetic guidance for the compensation of normal forces. Using a position sensor and closed-loop control, positioning accuracies of less than a micrometer are possible. The linear VR micromotor is therefore also suitable for planar motors by using four motors mounted in a square shape.

2.6.6 Linear VR Step Micromotor with Integrated Magnetic Guidance

Fig. 2.17 shows a linear VR step micromotor design that generates a horizontal magnetic field which is used not only for the production of motive force, but also can be used for passive magnetic levitation of the traveller. The normal forces between stator and traveller compensate each other. Without current, the vertical position of the traveller core is lower than the stator core. When the stator current creates a magnetic flux, part of it will enter the traveller from the top, thus leading to a levitating flux well force that is easily able to compensate the weight of the traveller. The vertical forces are calculated using a FEM model which includes the traveller displacement in vertical direction.

The reduction of the moving force due to the vertical displacement of the traveller is only low. Calculations show that the rotor's gravitational force of 0.13 mN is already compensated at a vertical displacement of 1.1 µm. An additional load can also be compensated by the integrated magnetic guidance forces.

A high harmonic frequency content of the VR step motor stator current, when supplied by a direct current chopper source, was investigated as well. Additional losses due to eddy currents causing increased resistance in the stator windings or eddy currents in the magnetic cores were considered. Fig. 2.18

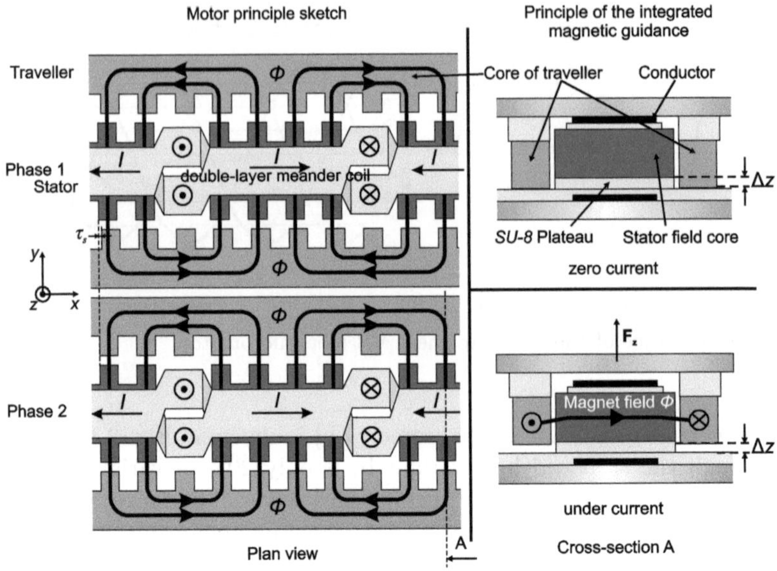

Fig. 2.17 Linear VR step micromotor with integrated magnetic guidance

shows the distribution of eddy currents and copper losses in the active components of the VR step micromotor. Fig. 2.18.a shows the distribution of the eddy current losses on the surfaces of the soft magnetic yoke, and Fig. 2.18.b in a stator pole cross-section along the broken line. The associated current density vectors are shown in Fig. 2.18.d. As the eddy currents flow mainly in the outer area of the pole cross-sections, the highest eddy current losses are found here. Fig. 2.18.c shows the distribution of copper losses in the stator conductors. The highest copper losses are found in the area between the upper and lower stator conductors and the vias, where upper and lower conductor pieces are joined. Here, the conductor cross-section is smallest. Investigation results with respect to harmonic frequency content of the VR step motor stator current, when supplied from a direct current chopper source, showed that operation with chopper frequencies up to 10 MHz caused no significant increase in the overall losses of the motor. Main motor losses are the I^2R losses in the stator windings, these being almost constant over an observed frequency range from 1–10 MHz. Operation on chopper power supplies is therefore not restricted with respect to the motor temperature.

2.6.7 Rotating VR Step Micromotors

The rotating VR step micromotor shown in Fig. 2.19 was designed in different layouts, consisting basically of a stator with concentrical soft magnetic pole structures wound with helix coils. Dependent on the system configuration,

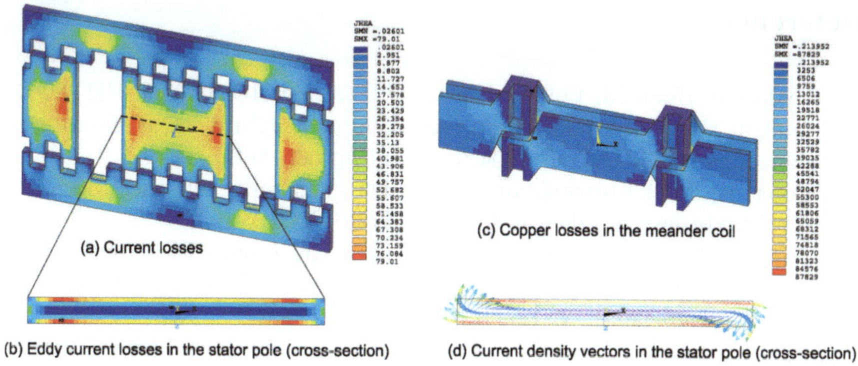

Fig. 2.18 Harmonic frequency

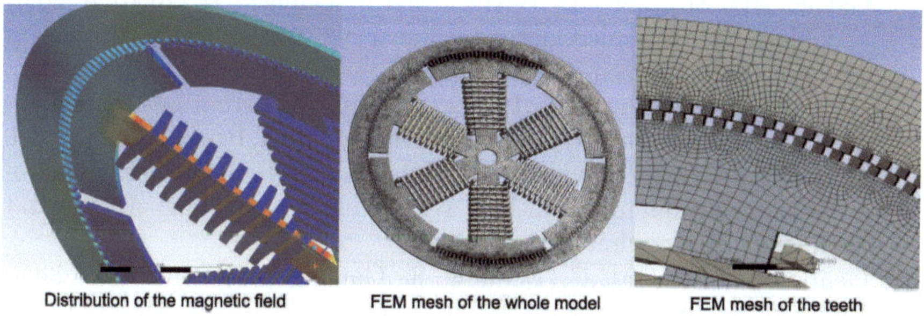

Fig. 2.19 FEM model of a VR step micromotor

three or six phases were designed with coils of 7 to 25 turns and pole diameters of 6 mm and 8 mm. The opposing coils are connected via so-called system connections to form a phase. The rotor consists of a soft magnetic toothed structure mounted on the stator using a guide pin. In full-step operation, step angles from 0.24° to 0.64° can be attained according to system configuration. In half-step and micro-step operation, higher position accuracies can be reached. The external stators consist of a simple ring structure with a high relative permeability. The verified simulation results using a 3D-FEM model met the actual measured results.

Acknowledgements The authors gratefully acknowledge the funding of the reported work by the German Research Foundation within the Collaborative Research Center 516 "Design and Manufacturing of Active Micro Systems".

References

[1] Budde T, Hahn M, Föhse M, Edler J, Stölting HD, Gatzen HH (2004) Design, fabrication and characterization of a hybrid linear micro step motor. In: ACTUATOR 2004, Bremen, pp 665–668
[2] Edler J (2006) Entwurf von Aktivteilen elektromagnetischer Mikrolinearaktuatoren nach dem Reluktanzprinzip. VDI Verlag
[3] Edler J, Föhse M, Stölting HD, Gatzen HH (2001) Elektromagnetischer Mikrolinearmotor. In: GMM-Fachbericht 33, Innovative Klein- und Mikroantriebe 2001, Mainz, pp 117–121
[4] Edler J, Föhse M, Stölting HD, Gatzen HH (2001) A linear microactuator with enhanced design. In: Micro System Technologies 2001, Düsseldorf
[5] Gatzen HH, Stölting HD, Büttgenbach S, Dimigen H (2000) A novel variable reluctance micromotor for linear actuation. In: Proceedings of actuator 2000, Bremen, pp 363–366
[6] Gatzen HH, Stölting HD, Ponick B (2004) Alternatives for micromachined linear actuators. In: Proc. Actuator 2004, Bremen, pp 317–320
[7] Gehrking R (2009) Entwurf von Aktivteilen elektromagnetischer Mikrolinearmotoren unterschiedlicher Wirkprinzipien. Leibniz Universität Hannover, Dissertation
[8] Gehrking R, Ruffert C, Demmig S, Gatzen HH, Ponick B (2005) Entwicklung einer magnetischen Führung für Mikroaktoren. In: Mikrosystemtechnik-Kongress, VDE Verlag, Berlin, Freiburg, pp 383–386
[9] Gehrking R, Demmig S, Ponick B, Feldmann M, Büttgenbach S (2006) A micro linear motor with integrated passive magnetic guidance. In: 32nd Annual Conference on IEEE Industrial Electronics, IECON 2006, Paris, France, pp 1245–1250
[10] Hansen S, Janssen G, Ganesan V, Ponick B, Mertens A, Gatzen HH (2009) Linearer skalierbarer Mikro-Reluktanzschrittmotor. In: 4. Kolloquium Mikroproduktion, Bremen
[11] Janssen G, Traisigkhachol O, Gatzen HH, Ponick B (2008) Design and fabrication of a rotary micro synchronous motor. In: ACTUATOR 08 International Conference and Exhibition on New Actuators and Drive Systems, Bremen, pp 457–490
[12] Waldschik A, Feldmann M, Büttgenbach S (2007) Entwicklung von Synchron-Mikromotoren mit speziellen Rotoren basierend auf Polymermagneten. In: Mikrosystemtechnik-Kongress, Dresden

Chapter 3
Drive Systems Based on Electromagnetic Microactuators

V. Ganesan, K. Wiedmann, A. Beradinelli, S. Demmig, H.-D. Stölting, A. Mertens

Institute for Drive Systems and Power Electronics
Leibniz Universität Hannover
mertens@ial.uni-hannover.de

Abstract This chapter deals with concept and design of complete drive systems based on electromagnetic microactuation. This includes the simulation of the drive system, various approaches for the electronic power supply, and concepts for the control of the step motors as well as for closed-loop control of microactuators with position control. For all the components of the drive system, it is important to consider the special requirements of miniaturization. For the electronic power supply, this means that passive components need to be minimized. This requires very high switching frequencies and switched capacitor topologies are of special interest. For the position control, it is important to find algorithms that are effective while remaining simple. This issue is resolved by a new kind of modified, combined linear and sliding-mode controller.

3.1 Configuration of Drive Systems Based on Electromagnetic Microactuators

Fig. 3.1 depicts the general structure of a drive system based on electromagnetic microactuators. The drive structure consists of five main parts: the microactuator, the power electronics, the current control, the current command, and the position control.

The power electronics excites a current in the phases of the motor, which actuates the moving part. The power electronics is driven by the current controller. The position of the moving part may be controlled using either a feed forward or feed back control. The feed back control is represented by the dashed lines in Fig. 3.1 and will be discussed in detail in Sect. 3.4. The current setpoint for the current controller is generated by the current command block. The setpoint for the current command is fed forward by the position controller. Here it has to be distinguished between the kind of motor to be driven. The two main categories are step motors and permanent magnet

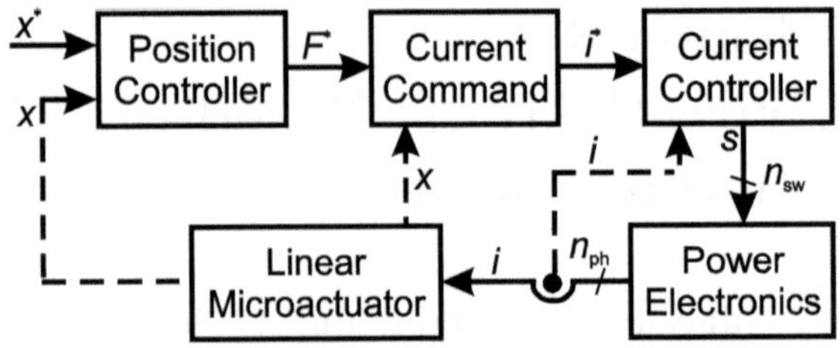

Fig. 3.1 Drive system model

synchronous motors (PMSM). Both types of motors can be designed as rotational or linear motors. The different types of electromagnetic microactuators are discussed in detail in Chap. 2. For linear motors, an additional position measurement can be implemented [1]. If a position feedback is available, the motor can be controlled in closed loop. Step motors can be driven either in closed loop or open loop position control, while the PMSM is generally controlled in closed loop. The current command generation and position control are discussed in detail in Sects. 3.3 and 3.4 respectively.

3.2 Power Supply

This section deals with the power converters used to excite the electromagnetic linear microactuators for achieving precise position control of their travelers. A power converter is used to supply the actuators with the desired currents from a constant voltage source. The power required to drive the linear microactuators is in the range of some watts and can be even smaller in PMSMs. Due to the unique characteristics of linear microactuators, the following requirements are placed on power converters:

- higher power density – compact design
- low current ripple
- a wide adjustable output range
- good dynamic response.

Apart from these, the power converters should generally have a high efficiency. Depending on their working principle, the miroactuators require power converters capable of supplying unipolar or bipolar current. For example the actuators based on synchronous and hybrid reluctance principle require converters capable of working at least in two quadrants, as they need bipolar current to control the moving part. Contrary to this, the variable reluctance (VR) linear microactuators discussed in Chap. 2 require only a unipolar current. In this section, the power converters for such (VR) linear microactuators are presented. An overview over the existing power converters, with

[1] Different approaches of position measurement are presented in Chap. 8

Table 3.1 Electrical parameters of linear microactuators

Type	n_{ph}	R_{m}	L_{m}	$\hat{i}_{\mathrm{max}}$	$\tau = \frac{L_{\mathrm{m}}}{R_{\mathrm{m}}}$
Permanent magnet synchronous	3 or 6	22 Ω	3.1 μH	200 mA	140.9 ns
Axial reluctance	3 or 6	0.4 Ω	12 nH	2 A	30 ns
Hybrid reluctance	3 or 6	0.8 Ω	36 nH	2.5 A	45 ns

their advantages and disadvantages when employed with linear microactuators, is given. DC-DC converters based on Switched Capacitor (SC DC-DC converters) topologies, which are more suitable for low power applications are explained in detail. Here, only a single quadrant power converter is explained. The proposed concepts and their analysis form the basics for other multi quadrant converters.

3.2.1 Electrical Parameters of the Linear Microactuators

Linear microactuators, that are fabricated using microfabrication techniques (electroplating in conjunction with photolithographic sputtering) exhibit different electrical parameters opposed to the electromagnetic actuators manufactured through conventional methods. The most important parameters required for designing the power supply are number of phases n_{ph}, stator coil resistance per phase R_{m}, stator coil inductance per phase L_{m} and maximum current required per phase $\hat{i}_{\mathrm{max}}$. In contrast to conventional drive converters, the induced voltage due to the movement of the traveler (often called back EMF) can usually be neglected. Table 3.1 summarizes the electrical parameters of different linear microactuators [1]. As can be seen, the electrical time constant of the microactuators, especially those of VR and hybrid step motors are very small and in the range of nano seconds. This means that their inductance can be neglected.

3.2.2 Linear Series Regulator

Due to the small amplitude of the current required to drive the linear microactuators, the linear series regulator is one opportunity for a power supply. Fig. 3.2 shows the principle of a series voltage regulator [14]. An operational amplifier (Op-Amp) is working as an error amplifier. It senses the voltage across the shunt resistance R_{shunt}, which is a measure of the actual current flowing through the motor winding and compares it with the command voltage v^*_{control}. The output of the error amplifier proportional to the difference of the input signals drives the gate of the MOSFET M. The gate voltage of the MOSFET M is adjusted until the desired current flows through the

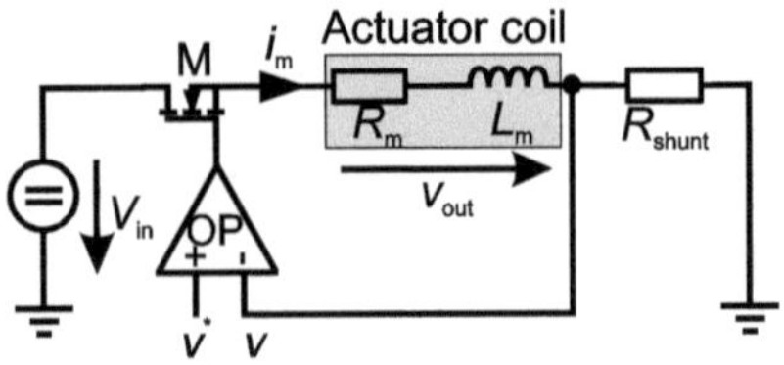

Fig. 3.2 Linear series regulator

motor coils. The MOSFET M in its linear region operates as a gate controlled variable resistance. This means that the voltage difference between input and output drops across the MOSFET M. The efficiency of a series voltage regulator for a constant output current $I_m = \frac{V_{out}}{R_m}$ can be calculated as follows:

$$\eta = \frac{P_{out}}{P_{in}} = \frac{V_{out} \cdot I_m}{V_{in} \cdot I_m} = \frac{V_{out}}{V_{in}}. \tag{3.1}$$

This shows that at lower operating points, the power loss is very high. This is a negative aspect in terms of energy efficiency and heating of the system. The output current follows the command signal irrespectively of the shape of the current command waveform and load, which eliminates the necessity of additional filter components.

3.2.3 Switching DC-DC Converter – Inductive

Switching DC-DC converters consist of semiconductor devices like MOSFETs operating as switches and passive components to regulate the output voltage. Due to the low power requirement of the linear microactuators, only non-isolated converter topologies are considered. The detailed analysis and design of such converters can be found in several text books [5].To keep the analysis simple, a single quadrant step-down DC-DC (buck) converter is explained. This buck converter topology is suitable for driving a VR based actuator, as it needs only a unipolar current. Fig. 3.3 shows the basic structure of a current controlled step-down converter.

The average output voltage $\bar{v}_{out}$ in a step-down converter can be set between zero and the input voltage V_{in} by adjusting the duty cycle $d = \frac{T_{ON}}{T}$ of the MOSFET M, where T_{ON} is the on time and T the period of the switching signal.

Fig. 3.4 shows the characteristic voltage and current waveforms in a stationary operation.

Current ripple. For a given operating point, the output current ripple is determined by the switching frequency f_s and the total inductance $L_m + L_F$, where L_F is the additional filter inductance connected between the converter and the actuator winding. The current ripple may lead to copper and iron losses in the actuator. They also cause force ripple and consequently interfere the positioning of the moving part. Hence to minimize the current ripple, ei-

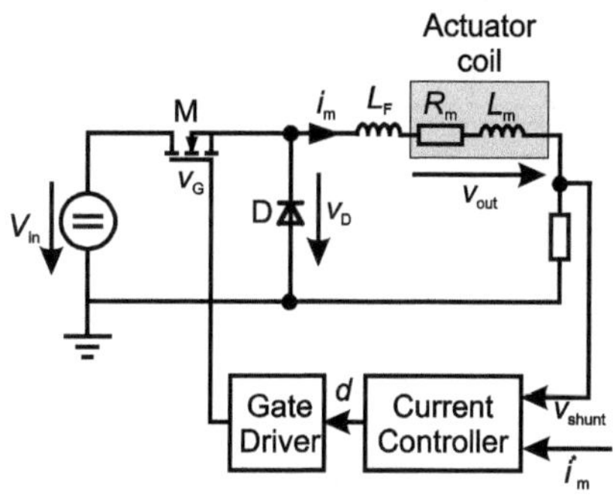

Fig. 3.3 Switching step-down converter

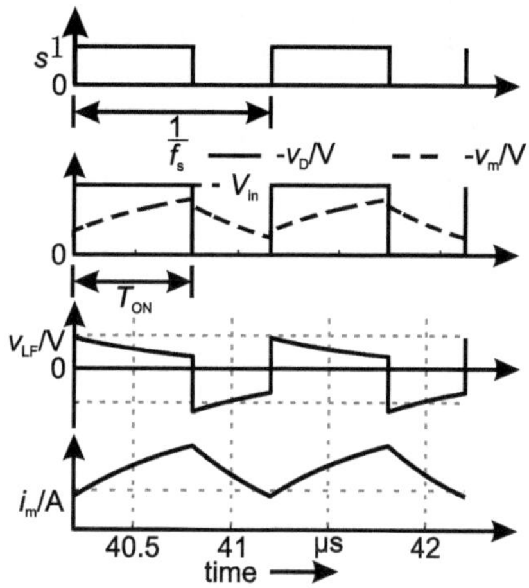

Fig. 3.4 Stationary time response of a step-down converter

ther a very high switching frequency or an additional filter inductance has to be selected. The maximal switching frequency is determined by the semiconductor technology. The additional inductance increases the volume of the power electronic circuit. Moreover, considering the integration of the power electronic circuit in an integrated chip, inductance is the component, which is difficult to integrate.

Efficiency. There are three loss mechanisms associated with a buck converter, (1) conduction losses, (2) switching losses, and (3) driver losses. A detailed analysis of these losses can be found in several publications [11]. The efficiency of such switching converters is generally in the range of 60% to 90% [8], considering all the three losses mentioned above.

3.2.4 Switched Capacitor DC-DC Converter

Similar to conventional switching DC-DC-converters with coil, the Switched Capacitor DC-DC converter (SC DC-DC converter) is used to step-down or step-up the input voltage to the desired value. It consists of semiconductor devices operated as switches and capacitors as energy storage devices, as opposed to the inductors in the buck converter. Due to the elimination of magnetic components, the converters offer the option for monolithic integration at low power levels.

Working principle. The working principle of the SC DC-DC converter can be explained taking a two stage series parallel topology shown in Fig. 3.5 as an example. Fig. 3.6 shows the switching signals and voltage signals in a

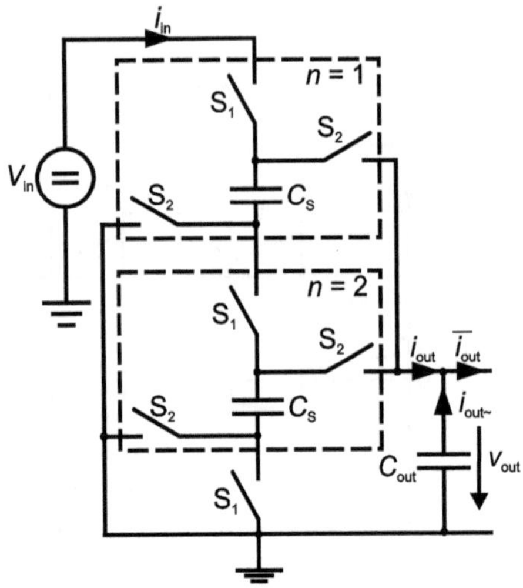

Fig. 3.5 Two stage switched capacitor DC-DC converter

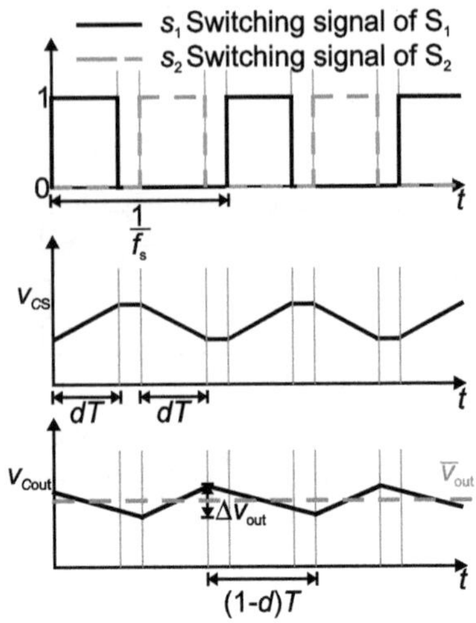

Fig. 3.6 Voltage response of two stage SC DC-DC converter in stationary operation

stationary condition. In each switching period, the SC DC-DC converter is switched through three different states. When s_1 is high and s_2 is low, the switches S_1 are closed, and the switching capacitors C_S are connected in series with the input voltage and each of them is charged to half the input voltage. When s_2 is high and s_1 is low, the switches S_2 are closed, and each of the charged switching capacitors are connected to the output in parallel. This results in an output voltage of $\frac{V_{in}}{2}$. When both s_1 and s_2 are low, all the switches are switched off, and the circuit remains in an inert state. The voltages over the switching capacitors remain unchanged. During the discharging state and the inert state, the load is supplied from the filter capacitor C_{out}. Thus, based on the voltage divider principle, the output voltage is reduced by a factor of two. By controlling the charging and discharging times of the switching capacitors, the output voltage can be set to desired values between zero and $\frac{V_{in}}{2}$. This can be achieved by varying either the duty cycle d defined as

$$d = \frac{T_{\text{charge}}}{T} \tag{3.2}$$

at a constant switching frequency f_s, or the switching frequency f_s with a constant duty cycle d or both. The other possibilities of output voltage regulations will be explained in later sections.

Analysis of SC DC-DC converters. In this section a general method for determining the output characteristics and efficiency of a simple SC DC-DC converter is discussed. A simple two stage SC DC-DC converter shown in Fig. 3.5 is considered for this purpose. Several other topologies have been

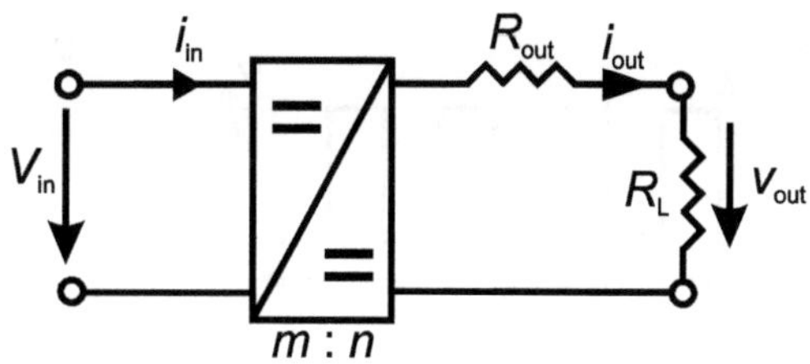

Fig. 3.7 Two port model representing a SC DC-DC converter

proposed in past works [10]. The analysis explained here forms the basis for all other topologies. This analysis takes into account only the losses due to resistive charging and discharging of the capacitors and the $I^2 \cdot R$ losses due to parasitic resistances in the circuit. The switching losses in the semiconductor devices and the driver losses have to be calculated separately, depending on the specific topology of the SC DC-DC converter.

For performing voltage regulation (step-down or step-up), the SC DC-DC converter should transfer energy between input and output. To accomplish this, the capacitors are charged and discharged through switches with parasitic resistance. This results in a certain amount of energy loss in charging and discharging the capacitors. These losses result in a voltage drop across the converter proportional to the output current.

A simple two port SC DC-DC converter model shown in Fig. 3.7 can be used as an idealized model for further analysis. The ideal DC-DC converter in the model represents the no-load conversion ratio of $\frac{n}{m}$, which corresponds to the number of stages in the SC DC-DC converter. The output resistance R_{out} models the voltage regulation, the load-dependent voltage drop and the losses. At a given operating point, the output resistance R_{out} is a function of

$$R_{\text{out}} = f(f_{\text{s}}, d, \frac{m}{n}, R_{\text{para}}, R_{\text{L}}). \tag{3.3}$$

Depending on the switching frequency, two limits have been defined for the output resistance R_{out}. When the switching frequency is less than the natural frequency of the RC circuit ($\frac{1}{f_{\text{s}}} >> R_{\text{para}} \cdot C_{\text{S}}$), the currents are impulsive as shown in Fig. 3.8.a.

At such switching frequencies, the equivalent output resistance R_{out} is independent of the duty cycle d. This limit is termed Slow Switching Limit (SSL) [10]. When the switching frequency is greater than the natural frequency of the RC circuit ($\frac{1}{f_{\text{s}}} << R_{\text{para}} \cdot C_{\text{S}}$), the capacitor charging and discharging currents are quasi constant, and the voltage over them changes linearly as shown in Fig. 3.8.b. This limit of the equivalent output resistance is termed Fast Switching Limit (FSL) [10]. The output resistance can be calculated separately for these two limits. Such a modeling has the advantage of having a unified model for the entire switching frequency range, as opposed to the average state space analysis method, which is valid for high switching frequencies only [7]. Table 3.2 shows the analytically calculated [10] fast and

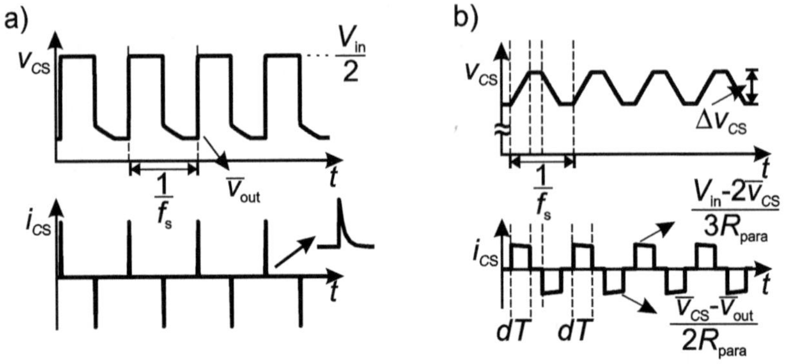

Fig. 3.8 Switching capacitor voltage and current waveform for a two stage SC DC-DC converter: (**a**) $f_s = 1$ kHz; (**b**) $f_s = 1$ MHz

slow switching limits of the equivalent output resistance for a SC DC-DC converter with different number of stages N.

Table 3.2 Fast switching and slow switching limits of the equivalent output resistance for SC DC-DC converter with different number of stages

Stages	1	2	3	4	5
R_{SSL}	$\frac{1}{C_S \cdot f_s}$	$\frac{1}{2 \cdot C_S \cdot f_s}$	$\frac{1}{3 \cdot C_S \cdot f_s}$	$\frac{1}{4 \cdot C_S \cdot f_s}$	$\frac{1}{5 \cdot C_S \cdot f_s}$
R_{FSL}	$\frac{4 \cdot R_{DS(on)}}{\alpha}$	$\frac{7}{4} \cdot \frac{R_{DS(on)}}{\alpha}$	$\frac{10}{9} \cdot \frac{R_{DS(on)}}{\alpha}$	$\frac{13}{16} \cdot \frac{R_{DS(on)}}{\alpha}$	$0.64 \cdot \frac{R_{DS(on)}}{\alpha}$

Output resistance R_{out}. The total output resistance can be approximated as [10]:

$$R_{out} = \sqrt{R_{SSL}^2 + R_{FSL}^2}. \tag{3.4}$$

Fig. 3.9 shows the comparison between the analytically calculated R_{out} and the simulated values.

Output characteristics and efficiency. From Fig. 3.7, the ratio of output voltage to input voltage is calculated as

$$\frac{\bar{v}_{out}}{V_{in}} = \frac{m}{n} \cdot \frac{1}{(1 + \frac{R_{out}}{R_L})}. \tag{3.5}$$

Fig. 3.10 shows the comparison of analytical and simulation results of the ratio of average output voltage to input voltage $\frac{\bar{v}_{out}}{V_{in}}$, as a function of duty cycle d for the two stage SC DC-DC converter shown in Fig. 3.5.

From Fig. 3.7, the efficiency of the SC DC-DC converter is calculated as

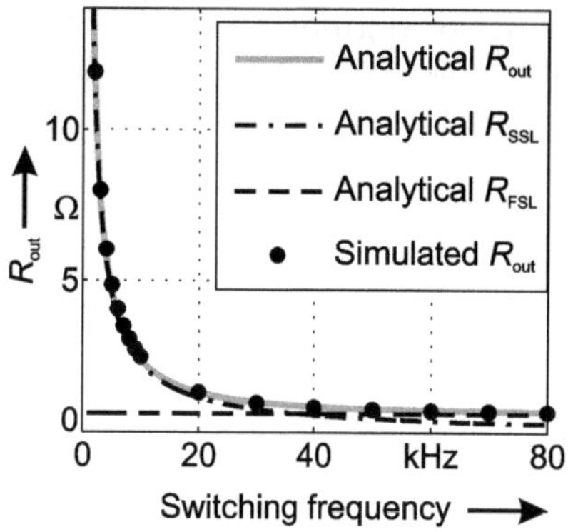

Fig. 3.9 Comparison of analytical and simulated equivalent output resistance

$$\eta = \frac{P_{\text{out}}}{P_{\text{out}} + P_{\text{loss}}} = \frac{1}{(1 + \frac{R_{\text{out}}}{R_{\text{L}}})}$$
$$= \frac{n}{m} \cdot \frac{\bar{v}_{\text{out}}}{V_{\text{in}}}. \tag{3.6}$$

The efficiency in (3.6) is linearly proportional to the output voltage. This means that lower output voltages result in smaller efficiencies. Increasing the number of stages in the converter circuit and selecting the suitable number of stages depending on load results in an overall better efficiency, as shown in Fig. 3.11. The switching and driver losses are not included in the efficiency calculations.

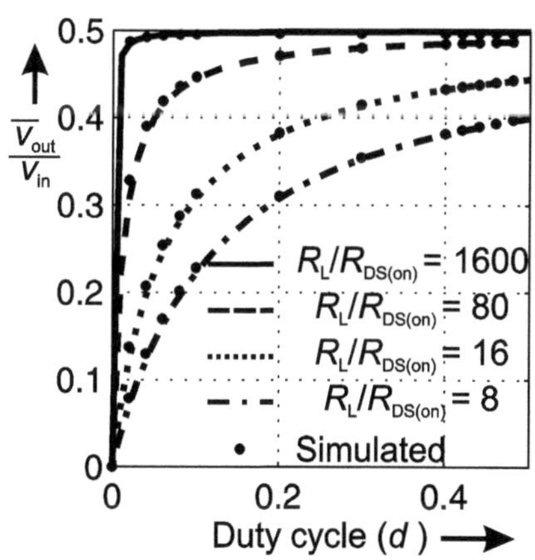

Fig. 3.10 Relative average output voltage of a two stage SC DC-DC converter with $f_s = 1$ MHz, $C_s = 22\ \mu$F and $C_F = 100\ \mu$F

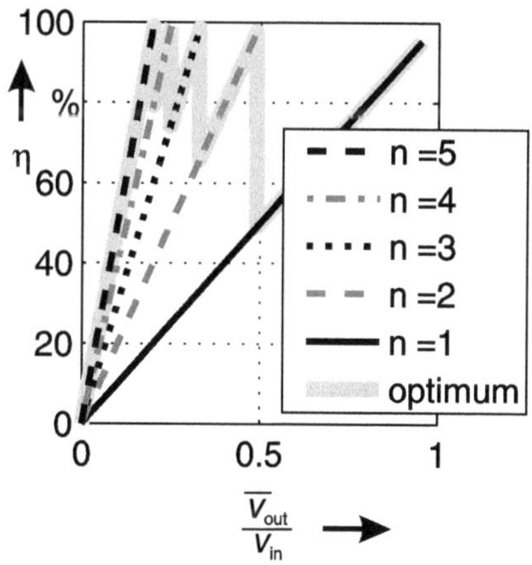

Fig. 3.11 Efficiency of SC DC-DC converter operating with different stages and the optimum efficiency over the output voltage

Voltage regulation of SC DC-DC converters. Unlike the conventional step-down converter, where the duty cycle is the most predominant way to regulate the output voltage, in SC DC-DC converters the voltage regulation can be achieved in more than one way. Equation (3.5) shows that the output voltage is a function of the number of stages $\frac{m}{n}$, duty cycle d, and switching frequency f_s. In addition, the input current i_{in} can be adjusted to regulate the output voltage. The different voltage regulation methods do not influence the efficiency of the converter, as η is always linearly proportional to the output voltage (3.6). But they have an influence on the output voltage ripple of the converter. A suitable regulation method can be chosen depending on the limitations of the hardware and the requirements of the application.

This section has presented different concepts of power supply for microactuators. It has been outlined that phase current ripple, power efficiency, and overall size are the most important criteria for evaluating the power supply. The linear series regulator offers the most simple power supply with better current ripple. But its efficiency over a wide operating region is low. A buck converter provides better efficiency, but requires additional inductors to limit the current ripple. The SC DC-DC converter has better efficiencies for certain voltage conversion ratios and at the same time does not require additional inductors to filter current ripple. But it requires a complex current control strategy and more semiconductor switches.

An optimal power supply for microactuators may be a combination of a SC DC-DC converter and a linear series regulator. The series regulator can be used as a current controller whose input voltage is adjusted using the SC DC-DC converter in discrete steps depending on the load. This increases the efficiency of the linear controller.

3.3 Current Command Generation

As shown in Fig. 3.1, the drive structure contains a Current Command block that provides the current setpoint for the power electronics. Here it has to be distinguished, if a closed loop or open loop position control is applied. For closed loop position control, those phase currents are calculated that apply a certain thrust which is demanded by the position control. In the open loop control mode, those phase currents are calculated that force the moving part of the motor to move to the desired position. The motor types which are discussed here are the permanent magnet synchronous motor (PMSM) and the VR step motor. These motors are described in detail in Chap. 2.

3.3.1 Permanent Magnet Synchronous Motor

The configuration of micromanufactured PMSMs is the same as for macromanufactured ones. Most of the conventional drive concepts for macroscopic PMSM require the detection of the permanent magnet flux [9]. In this respect, rotor position transducers (resolvers, etc.) or magnetic field sensors (Hall-effect sensors e.g.) can be used. Here it has to be considered, that for the discussed rotational micromanufactuered motors, no position feed back or flux detection is provided, as described in Chap. 8. Thus only an open loop control can be applied by exciting a three phase current system in the stator phases. Hereby the permanent magnet flux tries to align with the stator flux, causing a rotation of the rotor. A detailed description of the principle of torque production for a PMSM can be found in [12].

In contrast to rotational motors, for micromanufactured linear motors a position feedback can be provided, as optical position measurement principles can be utilized as described in Chap. 8. Thus drive concepts, as they are known from macromanufactured motors, can be applied. Here the Field Orientated Control (FOC) is an adequate control strategy for the discussed linear PMSM, as it allows to minimize the copper losses [9]. Applying the FOC, the flux producing d-component (i_d) and the force producing q-component (i_q) of the stator current can be decoupled. Through this, the active and reactive power of the drive system can be controlled by FOC. In general, the microactuators considered here are controlled to produce a maximal thrust per ampere. Thus the stator current only contains a q-component ($i_\mathrm{d} = 0$). In terms of thermal strains and copper loss minimization, this is the optimal solution. As FOC is a standard control method in drive system control theory, it is not discussed in detail at this point.

3.3.2 Variable Reluctance Step Motor

The current setpoint generation of VR step motors fabricated using microtechnology is explained on the basis of the horizontal VR step motor described in Chap. 2. The physical configuration of this special type of VR step motor differs from conventional step motors. Nevertheless, the basic control strategy is the same and can also be applied to rotational VR step motors. Generally, step motors are driven in an open loop stepping mode which is explained in the following section. In a full step operating mode, only one stator phase is excited at a time in the sequence shown in Fig. 3.13. The resulting thrust is shown as a bold gray line, and it causes a travel in a positive moving direction. Fig. 3.12 shows the thrust over position characteristics of the microactuator obtained by exciting one phase of the stator winding with a constant current of 1 A over the entire position of the moving part. With this excitation strategy, the moving part is forced to move to the position where the force characteristic of the excited phase crosses zero. The

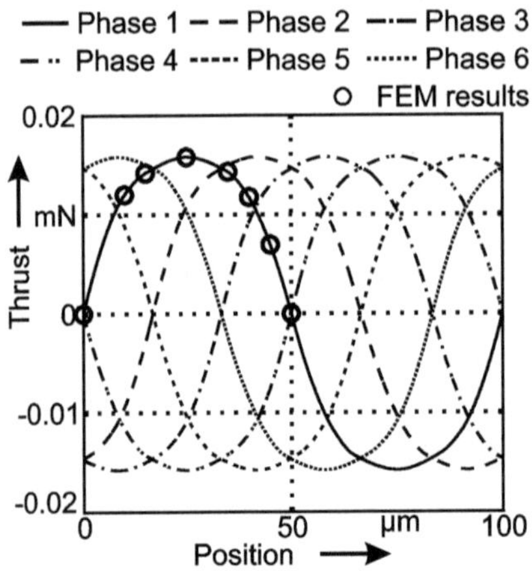

Fig. 3.12 Thrust over position characteristics of each string of the horizontal VR step motor for an excitation of 1 A

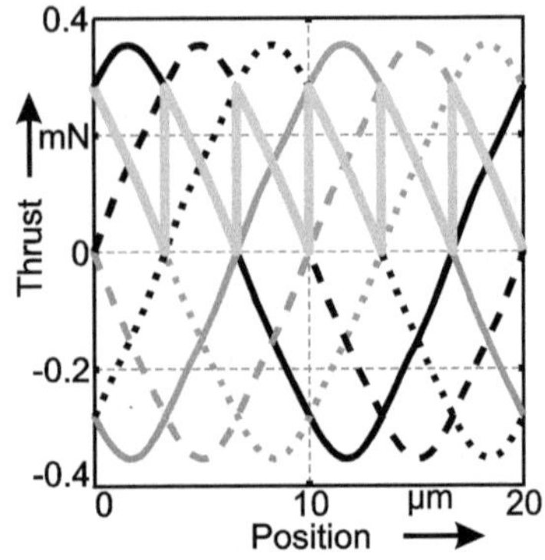

Fig. 3.13 Resulting thrust over position characteristics of the horizontal VR step motor operating in full step stepping mode for an excitation of 2.5 A. Step size is $\frac{100\mu m}{6}$

step size in full step operating mode is equal to the pole pitch τ_P, divided by the number of phases n_{ph}. The step size can be further reduced by operating the microactuator in the micro stepping mode, i.e. varying the motor currents in an analogue fashion, as explained in [13]. By knowing the thrust over position characteristic of the motor, it is also possible to implement a closed loop position control[2]. In this, those phase currents are excited that generate the thrust which is demanded by the position control.

3.4 Position Control

Beside the stepping mode which is discussed in Sect. 3.3, it is possible to apply a closed loop position control. As mentioned before, if position control for micromanufactured motors is compared with the control of macromanufactured ones, the conditions are quite different. On the one hand, effects of disturbance forces can be much bigger. (For example, even dust can affect the friction coefficient of a micromanufactured motor.) On the other hand, manufacturing tolerances are much bigger than in macromanufactured motors. These operation conditions make the design of a robust position control a challenging task.

[2] In this case, it has to be distinguished, if rotational or linear micromanufactured motors are considered, as the rotational motors discussed do not provide a position measurement.

3.4.1 Principle of Sliding Mode Control

The controller which is discussed in this section is based on the Sliding Mode Controller (SMC) which is known for its robust control characteristics. Due to its poor dynamic behavior, the SMC has been modified to achieve a time-optimal position control. In general, the SMC approach is not limited to systems of a certain order. Nevertheless, the inertia system considered here has two system state variables (position and velocity). Thus it is possible to illustrate the system behavior with the help of the phase portrait. But first, the mathematical basis of the SMC is given. The equation of motion is defined as follows:

$$\ddot{x} = f(\dot{x}, x) + u; \tag{3.7}$$

where u represents the manipulated variable. It does not matter, if $f()$ is a linear or nonlinear function. The so called sliding surface (s)is defined next.

$$s = (\dot{x} - \dot{x}_{\text{set}}) + \lambda \cdot (x - x_{\text{set}}) = \dot{x}_{\text{diff}} + \lambda \cdot x_{\text{diff}} = 0 \tag{3.8}$$

x_{set} stands for the position setpoint, while x_{diff} represents the difference between the actual position and the setpoint. Equation (3.8) describes the dynamic behavior of the position error. Here, λ represents the eigenvalue and therefore defines the dynamics. The SMC should guarantee that the system dynamics equals the desired one, which is defined by the sliding surface. Therefore, it has to be guaranteed, that the system state vector approaches the sliding surface from every possible state. This operating condition when approaching the sliding surface is called reaching mode. In reaching mode, the following control law is applied:

$$u = \begin{cases} -U_{\text{const}} & \text{for} \quad s > 0 \\ +U_{\text{const}} & \text{for} \quad s < 0. \end{cases} \tag{3.9}$$

U_{const} represents a constant manipulated variable and has usually its maximal possible value[3]. After the state vector has reached the sliding surface, it should remain on the surface. This operating condition is called sliding mode. As long as the derivative of s is zero, the system remains in sliding mode.

$$\begin{aligned} \dot{s} &= \ddot{x} - \ddot{x}_{\text{set}} + \lambda \cdot (\dot{x} - \dot{x}_{\text{set}}) = 0 \\ \text{with } \ddot{x} &= f(\dot{x}, x) + u \end{aligned} \tag{3.10}$$

The so called equivalent control fulfills this requirement:

$$u_{\text{eq}} = -f(\dot{x}, x) + \ddot{x}_{\text{set}} - \lambda \cdot \dot{x}_{\text{diff}} \tag{3.11}$$

As mentioned before, an absolutely accurate estimation of the real system behavior is not possible, and therefore the control law defined by (3.9) is

[3] Note that U and u here have the dimension of acceleration.

applied predominantly. This results in a two-level controller switching characteristic, that causes chattering of the moving part during the sliding mode. To avoid this undesired effect, a saturation area is defined around the sliding surface. The modified control law is then defined as follows:

$$u = \begin{cases} -U_{\text{const}} \cdot \text{sign}(s) & \text{for} \quad |s| > s_{\text{sat}} \\ u_{\text{eq}} - u_{\text{add}} \cdot \frac{s}{s_{\text{sat}}} & \text{for} \quad |s| \leq s_{\text{sat}}. \end{cases} \tag{3.12}$$

To guarantee that the state vector remains inside the saturation area, the manipulated variable u_{add} is added. In terms of stability, it is important that the amount of u_{const} or u_{add} respectively is bigger than an unknown disturbance. In this way, stability according to Ljapunov is fulfilled [1].
As mentioned at the beginning, the behavior of the controlled system can be illustrated with the help of the phase portrait. The phase portrait can be derived from the equation of motion (3.7) applying separation of the variables [3]. This leads to the following equation if no disturbance and only a constant acceleration is considered ($f(\dot{x}, x) = 0; u = u_{\text{const}}$):

$$\frac{1}{2} \cdot \dot{x}^2 = U_{\text{const}} \cdot x + \text{const}. \tag{3.13}$$

The initial condition of the system is defined by const. The phase portrait which illustrates the state vector characteristics defined by (3.13) is depicted in Fig. 3.14. The characteristics vary with the sign of U_{const} and the initial conditions. Furthermore, the sliding surface is shown. Equation (3.8) represents a straight line in the phase portrait, where λ is the slope. Fig. 3.15 illustrates the behavior of the controlled system, when the SMC is applied. Here again, no disturbances are considered for a better understanding. The initial position and velocity are zero. At the beginning, the moving part of the motor is accelerated with constant thrust, until it reaches the saturation area. Then the equivalent control and the additional manipulated variable are applied (3.12), so that the state vector remains inside the saturation area. Within this boundary, it "slides" into the desired setpoint state vector which is represented by the setpoint position in this case.

3.4.2 Modified Sliding Mode Control

The main advantage of the SMC is its capability of achieving a defined system dynamic even for nonlinear systems with nonlinear system disturbances. The disadvantage is the poor dynamic behavior. The dynamic of the position error in sliding mode is defined by the sliding surface which represents a first order delay system[4]. A time optimal control for an inertia system is a two-level or "bang bang" control with optimal switching [3]. The goal of the modified sliding mode controller (MSMC) presented here is to incorporate the optimal

[4] The time constant is defined by $\frac{1}{\lambda}$.

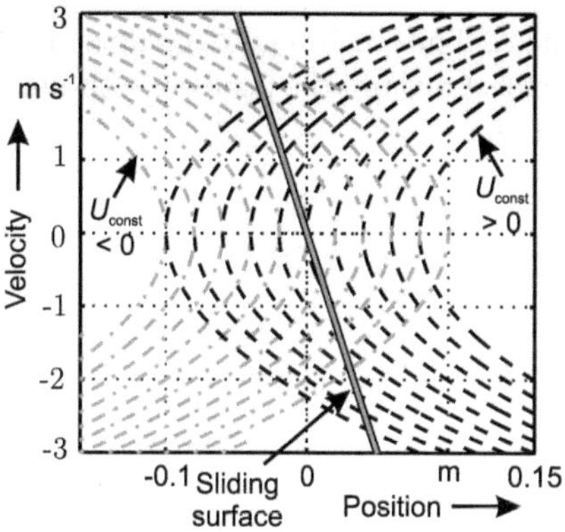

Fig. 3.14 Phase portrait

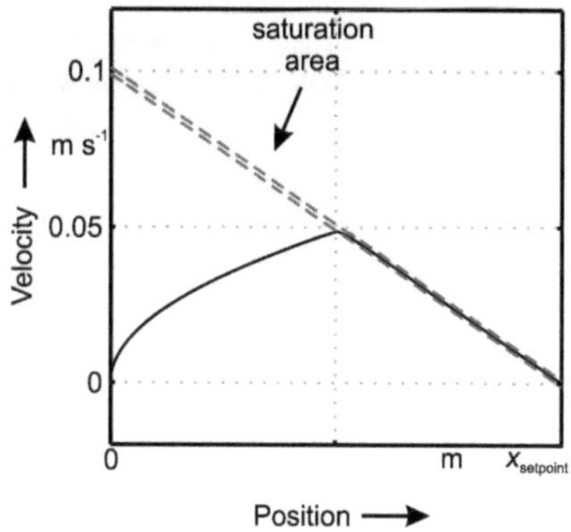

Fig. 3.15 Phase portrait of the controlled system

dynamic behavior of the two-level controller and the robustness of the classic SMC. For achieving this, the variable λ is calculated adaptively, as explained in the following section.

First, the time optimal switching trajectory of a two-level controller is derived. Eq. (3.13) is considered for the state vector difference $(\dot{x}_{\text{diff}}; x_{\text{diff}})$ and for maximal acceleration or deceleration respectively.

$$\dot{x}^2_{\text{diff}} = 2 \cdot u_{\max} \cdot x_{\text{diff}} \Rightarrow |\dot{x}_{\text{diff}}| = \sqrt{|2 \cdot u_{max} \cdot x_{\text{diff}}|}. \tag{3.14}$$

This expression is inserted in the equation of the sliding surface:

$$s = \dot{x}_{\text{diff}} + \lambda \cdot x_{\text{diff}} = \sqrt{|2 \cdot u_{\max} \cdot x_{\text{diff}}|} + \lambda \cdot x_{\text{diff}} = 0 \tag{3.15}$$

In this way, a “time optimal” λ can be calculated, depending on the moving part’s postion.

$$\lambda_{\text{adapt}} = \sqrt{\left|\frac{2 \cdot u_{max}}{x_{\text{diff}}}\right|} \tag{3.16}$$

Next, it has to be considered that the derivation of the equivalent control in Eq. 3.11 assumes a constant λ. To receive the correct value comprising an adaptive λ, the calculation of the equivalent control has to be modified as follows:

$$u_{\text{eq}} = -f(\dot{x}, x) + \ddot{x}_{set} - \frac{1}{2} \cdot \lambda_{\text{adapt}} \cdot \dot{x}_{\text{diff}} \tag{3.17}$$

As a consequence of the fact that the moving part is accelerated or decelerated with $u_{\max}$ during sliding mode, the state vector remains within the saturation area. As the numerator of the adaptively calculated λ contains the position difference, the λ_{adapt} has to be limited, as soon as the position reaches the setpoint. In this context, it has to be considered that λ defines the stiffness of the controller during the sliding mode (3.11). This means that sensitivity to

measurement noise increases with the value λ. Thus, it is a good compromise to choose a relatively small, constant value for λ around the position setpoint. This leads to the overall control algorithm as follows:
Calculation of λ:

$$|x_{\text{diff}}| > |x_{\text{diff,min}}| \Rightarrow \lambda_s = \lambda_{\text{adapt}}; \qquad \lambda_u = \frac{\lambda_{\text{adapt}}}{2} \tag{3.18}$$

$$\tag{3.19}$$

$$|x_{\text{diff}}| \leq |x_{\text{diff,min}}| \Rightarrow \lambda_s = \lambda_{\text{const}}; \quad \lambda_u = \lambda_{\text{const}}. \tag{3.20}$$

Calculation of s and u_{eq}:

$$s = v_{\text{diff}} + \lambda_s \cdot x_{\text{diff}} \tag{3.21}$$

$$u_{\text{eq}} = -f(\dot{x}, x) + \ddot{x}_{\text{set}} - \lambda_u \cdot \dot{x}_{\text{diff}}. \tag{3.22}$$

Calculation of the manipulated variable:

$$|s| > s_{\text{sat}} \Rightarrow u = -U_{\text{const}} \cdot \text{sign}(s) \tag{3.23}$$

$$|s| \leq s_{\text{sat}} \Rightarrow u = u_{\text{eq}} - u_{\text{add}} \cdot \frac{s}{s_{\text{sat}}}. \tag{3.24}$$

The variable λ_s represents the λ which is used for the calculation of the sliding surface. The variable λ_u is used for the calculation of the equivalent control. Both variables are introduced, as they do not equal each other, if λ is calculated adaptively and the standard equation of the equivalent control (3.11, 3.21) is used. For a time optimal control, U_{const} has to equal u_{max}. A detailed description of the MSMC can be found in [1, 2, 15].

System identification. A time optimal, energy efficient control of the motor depends on a good knowledge of the system. In this context, the calculation of the adaptive λ needs the actual value of the maximum thrust that is available to decelerate the moving part of the motor (3.16). Therefore, disturbance forces and system parameters have to be identified. For this purpose, an approximation of the general description of the equation of motion (3.7) is done as follows :

$$\ddot{x} = p_1 \cdot F_{\text{set}} - p_2 \cdot \dot{x} - p_3 \cdot \text{sign}(\dot{\text{x}}) \tag{3.25}$$

$$\ddot{x} = \mathbf{p}^T \cdot \mathbf{x} \tag{3.26}$$

$$\text{with } \mathbf{x} = (F_{\text{set}}; -\dot{x}; -\text{sign}(\dot{\text{x}}))^{\text{T}} \text{and } \mathbf{p} = (\text{p}_1; \text{p}_2; \text{p}_3)^{\text{T}} \tag{3.27}$$

This equation assumes that the power electronics can be modeled as a proportional gain (p_1). Furthermore, the resulting disturbance force has a component that increases linearly with velocity (p_2) and a constant component (p_3), whose sign depends on the velocity. This is a general approach, where the reason of the disturbance force is not distinguished. For example, (p_2) can be effected by friction, but also by eddy currents. As (3.25) represents a linear combination of the system parameters, linear estimation methods can be employed for identification. Here, estimation methods based on the Least Mean Square (LMS) algorithm represent an adequate approach, as the LMS algorithm is known for its simplicity of implementation, computational efficiency, and robust performance with respect to process and measurement noise. As the identification can run online or offline, advantages or disadvantages have to be weighted up. At this point, the different estimation methods shall not be discussed in detail, as they are explained in standard literature of control theory [6, 4].

Measurement results. To validate the time optimal control in combination with the system identification, a macroscopic test bench was used, consisting of a linear hybrid step motor part[5]. The control algorithm and current setpoint generation was implemented in a rapid prototyping system. The current setpoint is fed forward to a commercial power electronics that drives the motor. Measurements have shown that the acceleration can be modeled as described by (3.25). For parameter estimation, a Gauß-Markov estimator was used, as described in [1]. The Gauß-Markov estimator is an offline estimator reducing the online computational effort to a minimum. The identified maximal deceleration ($\ddot{x}$ or u_{max}, respectively) as a function of the current and velocity is comprised in (3.16), where the adaptive λ is calculated.

Fig. 3.16 shows the state vector characteristic of the real motor compared to a simulated state vector, where the estimated parameters are used. For this test, the motor was excited with a constant current. Both state vectors have almost the same characteristics indicating a good estimation of the system parameters. Furthermore, the time optimal MSMC was implemented. The resulting state vector characteristic is depicted in Fig. 3.17. It can be seen that the moving part is accelerated much longer than it is decelerated and does not overshoot at the setpoint. This is possible, because the "intrinsic" decelerating thrust of the system[6] is comprised in the adaptive calculation of λ. The resulting state vector trajectory has a time optimal characteristic. Fig. 3.18 depicts the current amplitude which is applied to the three phase system of the stator. This amplitude is proportional to the applied thrust. Hereby, the current is superimposed by a ripple. This ripple is caused by the additional manipulated variable listed in (3.23), that compensates unknown disturbances and thus forces the state vector to stay within the saturation area.

Finally the MSMC also was implemented to a micromanufactured linear PMSM. Here, the position was measured by a commercial laser interferometer. The control algorithm was implemented in a rapid prototyping system, and the current setpoint was fed forward to a commercial analog amplifier. The good result of the MSMC, that was achieved is shown in Fig. 3.19, where the characteristic of the position over time is depicted. This charac-

[5] A macroscopic test bench was used in the first place, because side conditions can be controlled much better, as it is possible for micromanufactured motors.

[6] This thrust is represented by $p_2 \cdot \dot{x} - p_3 \cdot \mathrm{sign}(\dot{x})$ in the equation of motion.

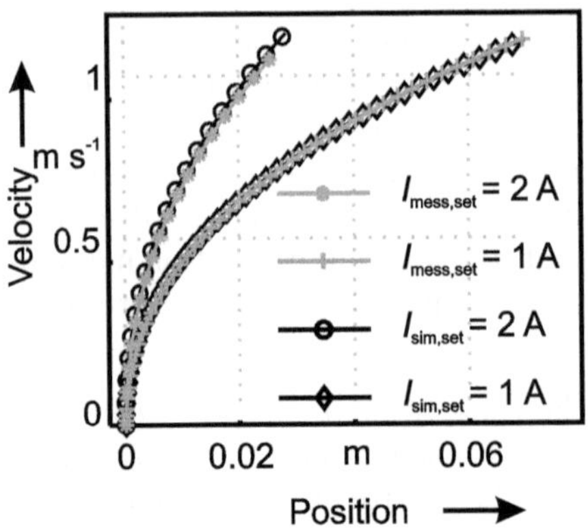

Fig. 3.16 Comparison of the measured phase portrait and the simulated phase portrait with identified system parameters

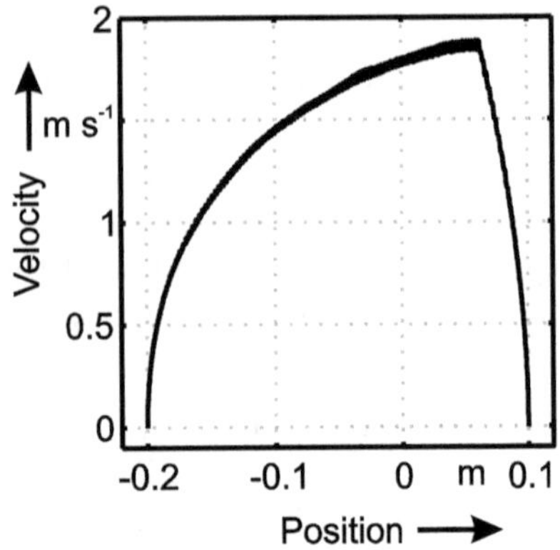

Fig. 3.17 Resulting phase portrait, if the MSMC is applied

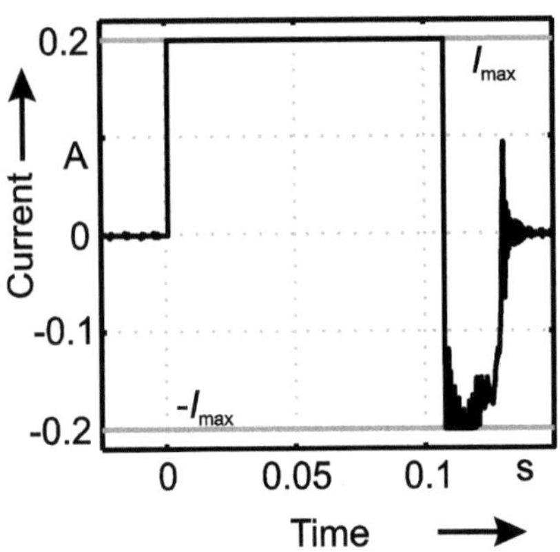

Fig. 3.18 Resulting current, if the MSMC is applied

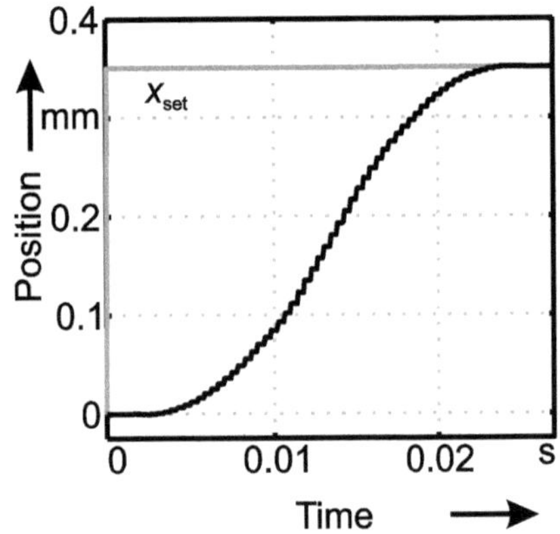

Fig. 3.19 Position characteristics of a micro linear PMSM, if the MSMC is applied

teristic reflects a time optimal control without overshooting. The relatively low sampling rate of the laser interferometer (2.5 kHz) is reflected by the discretized characteristics of the measured position.

Acknowledgements The authors gratefully acknowledge the funding of the reported work by the German Research Foundation within the Collaborative Research Center 516 "Design and Manufacturing of Active Micro Systems".

References

[1] Demmig S (2008) Ansteuerung und Regelung von Elektromagnetischen Mikromotoren

[2] Demmig S, Gehrking R, Hahn M, Wiedmann K, Ponick B, Gatzen HH, Mertens A (2008) Technical progress in designing and controlling of active microsystems. Microsystem Technologies Volume 14:12

[3] Föllinger O (1993) Nichtlineare Regelung 1. R. Oldenbourg Verlag, München

[4] Haykin S (1996) Adaptive Filter Theory. Prentice-Hall, Inc

[5] Mohan N, Undeland M, Robbins P (2003) Power Electronics. John Wiley & Sons

[6] Nelles O (2001) Nonlinear System Identification. Springer Verlag Berlin Heidelberg

[7] Ngo K, Webster R (1992) Steady-state analysis and design of a switched-capacitor dc-dc converter. In: Power Electronics Specialists Conference, 1992. PESC '92 Record., 23rd Annual IEEE, pp 378–385 vol.1, DOI 10.1109/PESC.1992.254649

[8] Pressman AI (1998) Switching Power Supply Design. McGraw-Hill

[9] Schröder D (2009) Elektrische Antriebssysteme – Regelung von Antriebssystemen. Springer-Verlag Berlin Heidelberg

[10] Seeman MD (2009) A design methodology for switched-capacitor dc-dc converters. PhD thesis, Electrical Engineering and Computer Sciences University of California at Berkeley

[11] Sodhi R, Brown S S, Kinzer D (1999) Integrated design environment for dc/dc converter fet optimization. In: The 11th International Symposium on Power Semiconductor Devices and ICs, 1999. ISPSD '99. Proceedings., pp 241 –244, DOI 10.1109/ISPSD.1999.764108

[12] Stölting HD, Beisse A (1987) Elektrische Kleinmaschinen – Eine Einführung. B. G. Teubner Suttgart

[13] Stölting HD, Kallenbach E (eds) (2006) Handbuch Elektrische Kleinmaschinen. Carl Hanser Verlag München Wien

[14] Tietze U, Schenk C (2009) Halbleiter-Schaltungstechnik. Springer-Verlag Berlin Heidelberg

[15] Wiedmann K, Demmig S, Mertens A (2007) Modified sliding mode controller for positioning of micro linear motors. In: Proc. European Conf. Power Electronics and Applications, pp 1–10, DOI 10.1109/EPE.2007.4417430

Chapter 4
Modular Computer Aided Design Environment for Active Microsystems

U. Triltsch, U. Hansen, C. Boese, J. R. Ziebart, H.-J. Franke[2], T. Vietor[2], S. Büttgenbach[1]

[1] *Institute for Microtechnology*
Technische Universität Braunschweig
s.buettgenbach@tu-braunschweig.de

[2] *Institute for Engineering Design*
Technische Universität Braunschweig
h-j.franke@tu-braunschweig.de; t.vietor@tu-braunschweig.de

Abstract With the ambition to increase the yield and the ability of microsystems as well as to speed up development, a computer aided design environment for active microsystems has been developed. Multiple software modules have been combined to form a complex, object-oriented development framework that is able to support designers throughout the entire design process. Besides 3D-CAD-, development- and process simulation modules the tool comprises a workflow manager and the ability to analyze process parameters. Furthermore, it is connected to databases acting as a knowledge-, a product-, a process-, and a tool storage.

4.1 Introduction

The design process for microelectromechanical systems (MEMS) needs to solve three basic tasks: a functional and a structural model of the system has to be conceived and optimized and, as manufacturing dominates the design-layout, a physical design corresponding to a feasible process-chain must be developed during the design process.

For microelectronic devices, there already exist highly sophisticated tools for design. The EDA (Electronic Design Automation) frameworks allow a high degree of automation. However, these EDA processes rely heavily on similar production steps for each variant of the basic elements regardless of dimensions, a low aspect ratio and a limited choice of materials. Most MEMS do not fit these criteria.

The three-dimensional geometry of MEMS imposes the lateral shape created by 2D-Design with the third dimension depending on a sequence of the 2D-Steps, the process sequence. Therefore, a design process model and corresponding design software is needed [3].

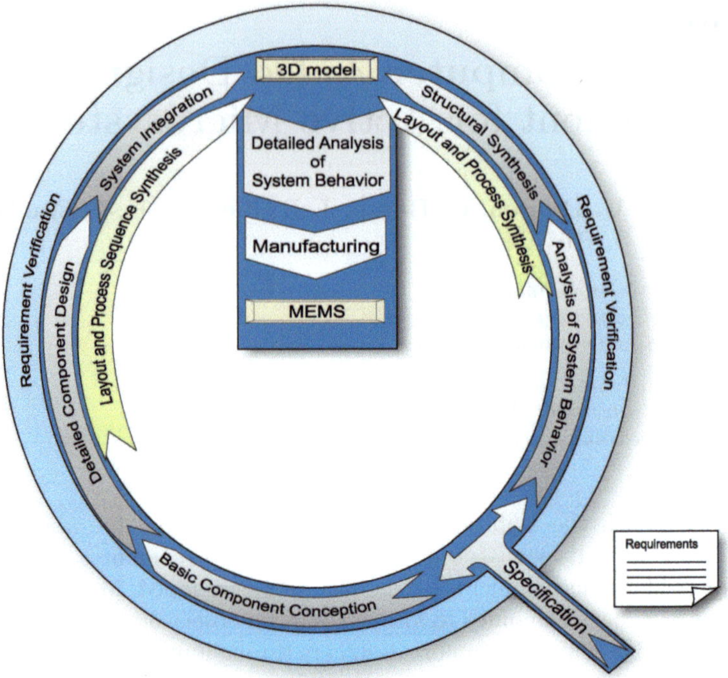

Fig. 4.1 Q-Model [10]

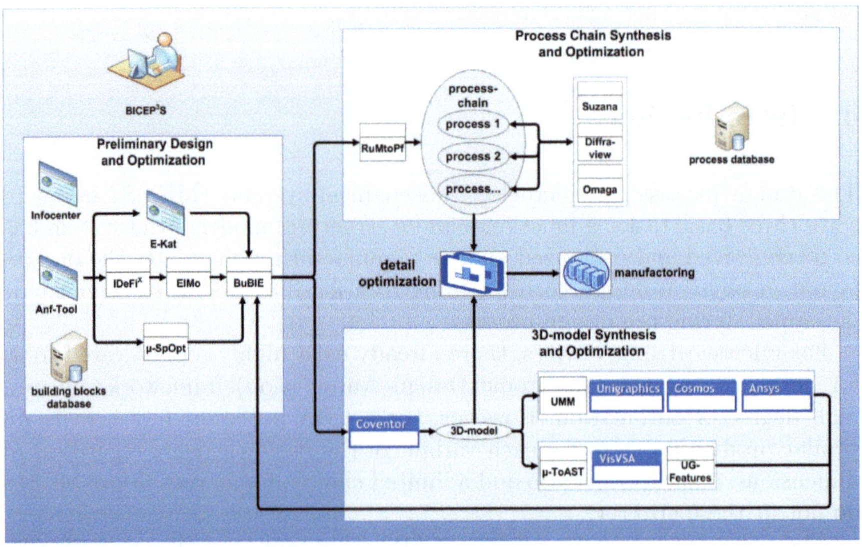

Fig. 4.2 Software overview

The Q-Model (Fig. 4.1) allows both ways of MEMS design: the functional design based process in accordance to microelectronic design (right fork in the figure) and a MEMS-specific process relying on basic components and their integration into a complete system (left fork in the picture) [10]. While commercial software was available for the first process, the second process lacked software. This gap is closed by the pesented modular software environment .

Both development processes start with a specification of the desired system. Based on this, a first concept of the system is conceived. The function based process continues with the analysis of the system function and its optimization. Then the layout-, process chain-, and structural synthesis results in a three-dimensional model of the device. The MEMS specific process starts with the conception of each of the basic components leading up to a detailed design. Then these components are integrated into a complete system represented by a 3D-model. These last two steps are accompanied by a parallel development of the needed process steps and their integration into a process chain.

These results serve as a source for the detailed analysis of the complete system function in order to optimize the system functionality and the process chain. The MEMS device is a product of the subsequent manufacturing process.

As mentioned before, the available TCAD (**T**echnology **C**omputer **A**ided **D**esign) tools are focused on the functional simulation of complex systems and do not consider the fabrication process.

Our system couples commercial system-level tools and several component level tools to form a closed design environment called "**B**raunschweigs **I**ntegrated **C**AD-**E**nvironment for **P**roduct **P**lanning **P**rocess **S**imulation" – $BICEP^3S$. These tools can be categorized according to their function in the design process [10].

In the phase of preliminary design, starting with a system specification, a first concept is developed. It consists of a set of basic elements, the "building blocks" and a structural layout (topology). It serves two parallel optimization cycles: the optimization of the 3D-model on the one hand and the process chain optimization on the other.

Each of these three sub-processes contains a series of design steps supported by corresponding software modules with interfaces to adjacent modules. To ensure the consistent information flow throughout the software environment, a novel flexible data model has been developed (Fig. 4.2) [7]. It has the capability of storing expert knowledge e.g. analytical, heuristical, fuzzy values and rules of thumb. This expert knowledge is kept in central databases for use by all modules in the environment.

4.2 Phase of Preliminary Design

The phase of preliminary design allows for flexible strategies to get from the system specification to the first concept. The specific strategy chosen is determined by the product to be developed.

4.2.1 Intraweb Knowledge Storage Service

To provide all subprojects an overview of active and discarded approaches and to gather and provide all relevant data for the manufacturing of a concept, an intraweb based solution called "Infocenter" has been developed. Each concept is presented in the form of a flow-chart, each block is hyperlinked to a list of required parts for the system. Corresponding manufacturing processes are stored along with critical parameters inside the system and linked to these parts [6].

4.2.2 Interactive Design of Multi-Dimensional Function Structures

To select the necessary building blocks which make up the first concept, the desired main function must be partitioned into sub-functions. The main problem is to find the appropriate level of detail to which the desired main function should be partitioned.

To support the MEMS designer working on this task, the software module *Idefi*^{X} (Interaktives Werkzeug zur Definition mehrdimensionaler Funktionsstrukturen) has been developed. *Idefi*^{X} is a *Microsoft Visio* Add-In utilizing intelligent *Visio* Shapes to represent the function structure elements [12].

The design of *Idefi*^{X} allows scalability, depending on the desired features. The implemented *Visio* Shapes on the first level assure correctness of the function structure depiction. On a second level, an interactive thesaurus and interactive function tree representing the third dimension of the structure are available (Fig. 4.3).

The highest level of interaction can be achieved by activating the integrated SQL Connection to the building-blocks database. For each selected sub-function representing shape, the number of possible realizations in the database is displayed. Thus, the designer gets information about the actual level of detail such as whether or not the achieved detail-level is correct (Fig. 4.3).

After the specification of the desired function structure has been carried out satisfactorily, it can be exported to an XML-file. This file serves as input to *ElMo*, the software module for the creation and evaluation of morphologies.

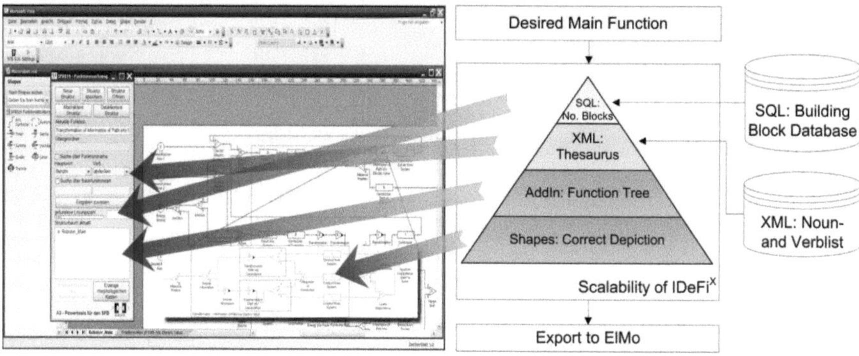

Fig. 4.3 Example use of *Idefi*^{X} for the synthesis of a function structure

4.2.3 Evaluation of Morpholgies

The "Erzeugungs- und Evaluationswerkzeug für Morphologien" – *ElMo*, a *Microsoft Excel* Add-In, is assigned to support the designer to develop morphological variants. It helps to reduce the amount of variants to a controllable amount and to evaluate the variants in order to determine the most promising.

Starting with a collection of sub-functions generated by *Idefi*^{X}, the *ElMo*-internal workflow leads the designer through the main steps, starting with a visualisation and selection of the sub-functions to be treated. It is followed by a global filter which can be applied to the connected building-block database. Then, the first stage of a morphological field based on the remaining sub-functions and the found solutions in the database is presented to the user.

This morphological field still lacks three aspects. Local incompatibilities between solutions and "extra" sub functions required to compensate negative effects are not considered. Also, the potential of functional integration is not taken into account, which arises from the multiple purposes that most solution elements can serve.

Interactively, these considerations can be applied to the first stage of the morphological field in combination with a local-optimum filter selecting the two best solution elements per sub-function in regards to a desired optimization goal. Each update of the morphological field is accompanied with an internal estimation of the amount of variants to be evaluated.

If a satisfactory morphological field is reached, the designer can start the automatic evaluation process, which leads to a ranking and visualisation representation of the three best performing variants (Fig. 4.4). The selection of the most promising variant leads to the "**Bu**ilding **Bl**ocks **E**ditor" – *BuBlE* through an export interface (see Sect. 4.2.6).

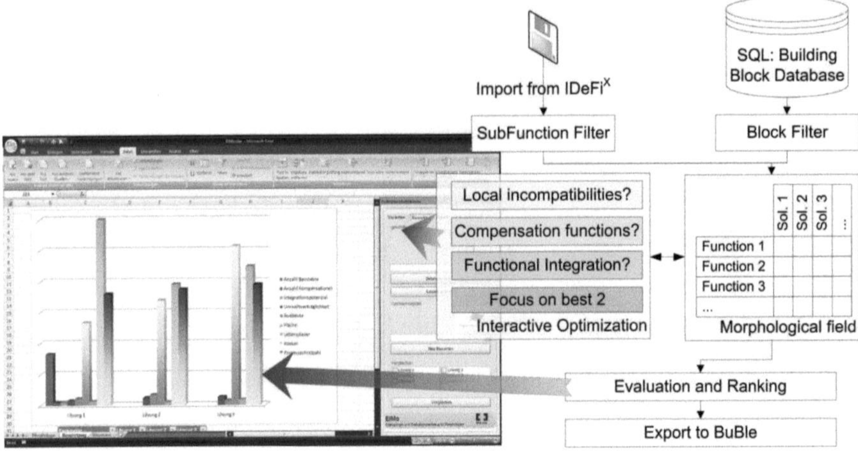

Fig. 4.4 Screenshot of *ElMo*

4.2.4 Web-based Design Catalogue System

Using an existing web-based design catalogue system developed in the Federal Ministry of Education and Research funded Joint-Project "GINA", a design catalogue has been created for MEMS design. Design catalogues serve as structured data storage for common design elements. In order to implement the catalogue into the software environment, a direct interface to *BuBLE*, *Unigraphics* and *ANSYS* has been added. By taking advantage of the implemented systematic and specific element properties, the designer is now able to find a design element, e.g. a micro coil, and to easily integrate it into the design workflow [4].

4.2.5 Optimization of Microcoils

One of the most important elements of magnetic MEMS is a microcoil because of its application in sensor and actuator systems. The main aim in coil optimization is to obtain a large value for the inductance of the coil system while keeping its electrical resistance at a minimum.

Constraints exist due to thermal considerations and space availability. In order to achieve a high inductance, a large number of windings is preferred but leads to either a larger space requirement or a higher resistance. A common approach is to use a multilayered design which leads to a complicated process-chain on the other hand.

μ-SpOpt (**Mikro-Spulen Opt**imierung) has been developed enabling the designer to estimate the inductance and the resistance of disc-shaped coils dependending on the design parameters. Also, an automated optimization based on parameter ranges is possible [8].

4.2.6 Building Blocks for Active Microsystems

The MEMS-specific design process following the left fork in the Q-Model (Fig. 4.1) generates an entire system made up from sub-functional elements, called building blocks. The major innovation regarding these building blocks is the connection of a list of possible fabrication sequences to a three dimensional parametric geometric representation.

The designer chooses the needed elements from a database through *ElMo* (Sect. 4.2.3) or the web-based design catalogue system (Sect. 4.2.4). The blocks are arranged in *BuBlE* and geometrical parameters and materials are selected. Afterwards material interfaces and layers that can be fabricated in a single process step are found by an iterative algorithm. A process sequence for the system is generated and validated using the validation tool [10].

If incompatibilities are detected, alternative sequences are generated and validated. The possible solutions are proposed to the designer for a final review. Once a feasible process sequence has been identified, the layout data is merged to derive a suitable mask set. A three-dimensional model can be generated and transferred to *Coventor Designer* or other proprietary software tools in order to visualize the result and do further computational analysis. The workflow in *BuBlE* is shown in Fig. 4.5.

New building blocks can be defined by extended TCL-Scripts and then introduced into the System. For this purpose a special TCL-based description language has been developed [10].

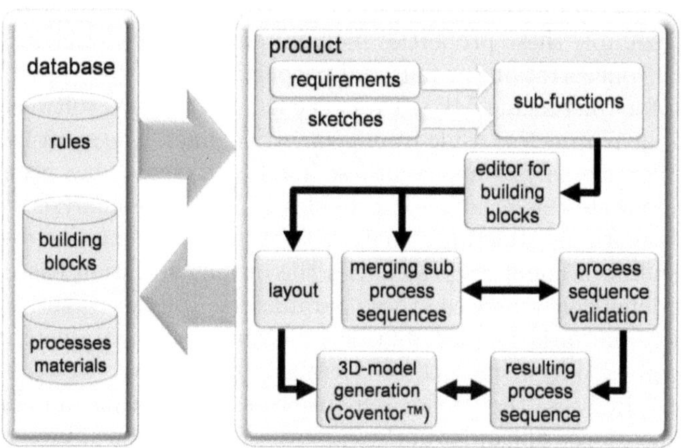

Fig. 4.5 Workflow in the building blocks editor [11]

4.3 3D-Model Synthesis and Optimization

The geometric information coming out of *BuBlE* has to be synthesized to a 3D-Model for further assessment and optimization. This is possible by taking advantage of proprietary software like *Coventor Designer*. Depending on the design task, a variety of simulations and optimizations can be carried out until a satisfactory geometry is achieved. This geometry, in combination with the process sequence, serves as the basis for production documentation

4.3.1 Linking CAD and FEA Models for Optimization

For the optimization and analysis of MEMS, e.g. the electromechanical part of an active micro system, an FEA model in *ANSYS* is useful. In addition, a CAD model is necessary for the visualization of the system, e.g. in *Unigraphics*. Both models have the same parametric options regarding the geometry, but unfortunately cannot be linked to each other.

Therefore a new tool has been created that connects and controls both models with central parameters, the "**U**nigraphics **M**odel **M**anager" – *UMM*. With this module, it is possible to link the parameters of both models and to access the associated calculation instructions. In addition to the CAD model it is possible to define limits to guarantee reasonable parameter values [5].

4.3.2 Tolerance Analysis and Synthesis in MEMS-Design

In mechanical engineering the tolerances are typically of geometrical nature. Conversely, in MEMS-Design the process steps and the corresponding deviation of non-geometrical process parameters influence the third dimension of the system. For example these properties include temperatures, electrical and magnetic fields, concentrations or material properties.

A new approach for tolerancing MEMS was needed as standard software did not allow for input parameters aside from geometrical ones. The software module "**Mikro To**leranz**a**nalyse- und **S**ynthese **T**ool" – μ-*ToaST* has been developed to accommodate this need [5]. It provides for tolerance analysis as well as for tolerance synthesis. Tolerance analysis takes the input parameters and their deviations into account and calculates the output deviation based on a mathematical dependency between these parameters. Tolerance synthesis is used to quantify the variation distributions of certain input parameters with respect to the given output parameter tolerance.

For the tolerance analysis, the Taylor expansion is used because of its independence of dimensions. The order of the expansion is determined by the differential, but in most cases the first order is sufficient. Disadvantageous to the Taylor expansion is the necessity to know an appropriate mathematical description of the problem. In order to reduce this hindrance the software tool

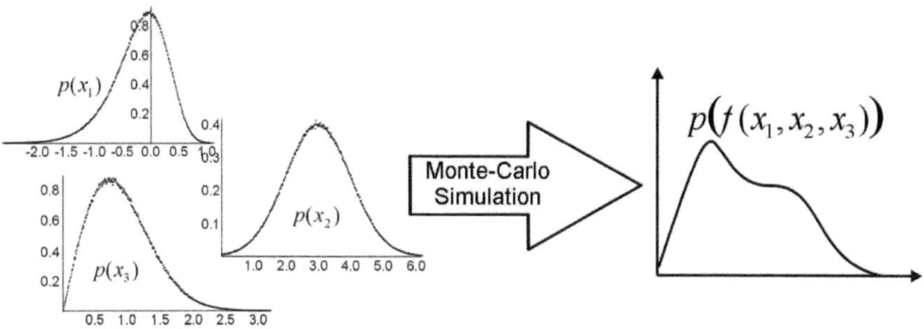

Fig. 4.6 Graphical representation of Monte Carlo method

allows two methods, each of which operates on sets of experimental data. In the first approach, the Taylor expansion is applied on a derived analytical fitting function. The second method is to perform a multidimensional linear data-interpolation. The partial derivatives of the Taylor expansion mentioned above are called the tolerance sensitivities. They are necessary to perform a tolerance synthesis.

The software tool is capable of coping with statistically distributed tolerances in addition to simple worst-case-studies. The Gaussian distributions of the input parameters can be specified directly by naming the limits of the tolerance or it is possible to transform an input data set into a distribution automatically. A Monte-Carlo simulation is performed using these input tolerances. Fig. 4.6 shows schematically how the Monte-Carlo Simulation process works. For each pass of the Monte Carlo Simulation, a set of stochastically selected input parameters is created and combined to form a single output result. This pass being executed many times leads to the output distribution. The Monte-Carlo simulation results are not necessarily normally distributed as shown in Fig. 4.6 [5].

The program architecture is illustrated in Fig. 4.7. The graphical user interface (GUI) contains the main module and is connected to a symbolic mathematics software, an object database and a relational database through JavaAPIs, and a JDBC-Interface [5].

4.3.3 Rule Based Verification of Assembly

Since a monolithical layout is not always achievable, it is not possible to avoid MEMS assembly in every case. Two aspects of MEMS assembly are a major challenge: first, the handling of the systems throughout the assembly process puts restrictions on the design. Second, the inevitable tolerances of geometric dimensions and positioning have to be considered.

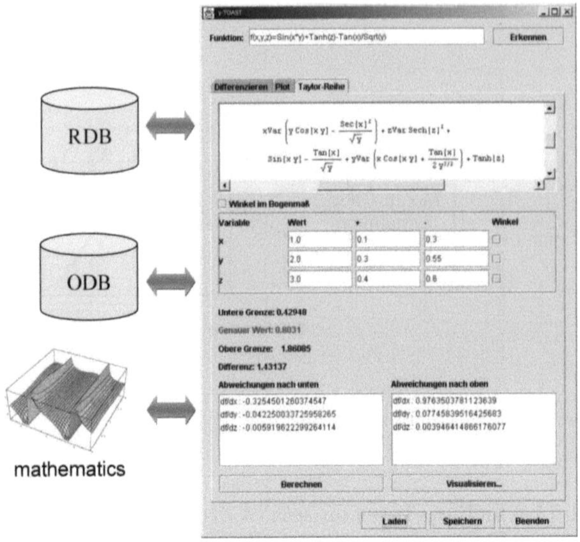

Fig. 4.7 System architecture: GUI, relational DB, object DB and mathematics software

3D-CAD models of the MEMS created with *Unigraphics* and *VisVSA* are used to create assembly scenarios and to validate the assembly process. The scenarios allow for an analysis of single assembly steps or geometric tolerances in context with the resulting tolerance distributions on critical output parameters [5]. To validate the handling, positioning and the joining of non-monolithical MEMS, so called "assembly features" have been introduced into the 3D-CAD System *Unigraphics.* Based on design-rules, the geometric information from the model, the assigned materials and the centrally stored grippers, these features provide an evaluation of the assembly process. In the event the assembly is not feasible, alternatives are suggested.

4.4 Synthesis and Optimization of Processes and Process Sequences

The third dimension of the MEMS and the yield is highly dependent on the process sequence. Therefore a feasible process sequence needs to be determined and then optimized. In context with the modular software environment, the process sequence alternatives attached to the selected building blocks are used to generate a process sequence for the complete system. In order to optimize the critical process steps, simulation modules have been conceived and implemented.

4.4.1 Rule Based Validation of Processing Sequences

Usually MEMS are fabricated using a combination of bulk and surface orientated structuring technologies. The single steps are processed successively and can be represented in process sequences. The sequences have to be checked for consistency in order to assure feasibility. Therefore several constraints have to be considered while defining a process sequence. In practice, errors are often identified while iteratively trying to build the first prototypes. This slows down the overall development process.

The modeling software *RuMtoPf* (**R**echner**u**nterstützte **M**odellierung **t**echnologie-**o**rientierter **P**rozess**f**olgen) was developed to assist the MEMS designer concerning the process sequences. Using the tool makes it possible to define complex process sequences and to detect process incompatibilities before the fabrication process is carried out. Examples of incompatibilities include: a certain resist may not serve as a masking layer in a wet chemical etching process or a thermal oxidation should always be preceded by a cleaning step [7].

Aside from just determining the feasibility of process sequences corresponding to no incompatibilities, the attractiveness has to be evaluated. This attractiveness depends on company-specific parameters like the potential of rationalization and also on general values like environmental impact. In order to evaluate process sequences and fabrication steps, detailed information and in depth knowledge on the processing is essential. To enable a computer based validation, the expert knowledge needs to be coded. In *RuMtoPf* this is achieved by setting up rules, which the validating algorithm uses for check. These rules are directly derived from the expert knowledge as rules of thumb, or better, as analytical descriptions. An execution strategy makes sure that only applicable rules are taken into account. Additionally, the user may define default presets, e.g. the clean room temperature or minimum acceptable etch selectivity. Fig. 4.8 depicts the general architecture of the software.

The specified data for each fabrication step can be obtained from a database, as mentioned in Sect. 4.1, whereby the data model contains the technology, geometric properties, materials and fuzzy parameters that qualify attractiveness.

The composition of the media and changes to the material properties due to processing are considered. So-called "dependent parameters" are also available, allowing for the modeling of dependencies between process, medium and/or material parameters. They can be used for determining e.g. the layer thickness of a resist in a spin-on process (using spin-on constants from a database table) or the depth of an etched hole (processing time · etch rate).

In order to calculate the attractiveness of two concurring process sequences, the fuzzy parameters are subjected to a rule based evaluation process (inference). The results are defuzzified and serve as a criterion for selection by the designer.

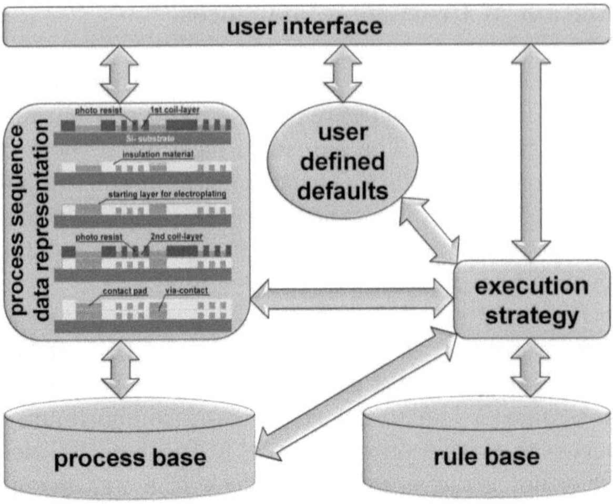

Fig. 4.8 General architecture of *RuMtoPf* (according to [7])

4.4.2 Simulation of the Silicon Wet-Etching Process

The primary technique for the fabrication of three-dimensional MEMS devices is anisotropic chemical wet-etching of crystalline silicon. Simulation tools are essential to make full use of the potential of the process. The simulation tool *SUZANA* uses a cellular automaton model to simulate a wide range of three-dimensional geometries, including etch-stop layers and even locally disordered crystal zones. This makes *SUZANA* one of the most accurate and flexible tools in this area available today [9].

Two main approaches for the simulation of anisotropic chemical wet-etching of silicon are known. Geometric models using a more abstract level describing the etch process. They do not focus on atoms but regard changes of geometric forms like planes, edges and vertices. Geometric models can be solved very quickly but lack precision simulating more complex shapes.

Atomistic models directly represent the atomic structure of the silicon crystal. Modifications of the shape are depending on the binding energy of atoms within the crystal structure. The evaluation of these energies leads to an etching probability of individual atoms or cells of atoms. The advantage of atomistic models is a high accuracy and the consideration of undefined, broken crystal planes. Thus, arbitrarily complex geometric shapes may be simulated.

The etch simulator *SUZANA* is based on an atomistic model using a cellular automaton. Cellular automata may be considered as discrete systems. They consist of a large number of simple and identical components. These "cells", in *SUZANA* a cluster of silicon atoms, are arranged in a lattice corresponding the crystal structure of silicon. They interact locally whereby each cell has a finite number of possible states. The cells evolve synchronously and

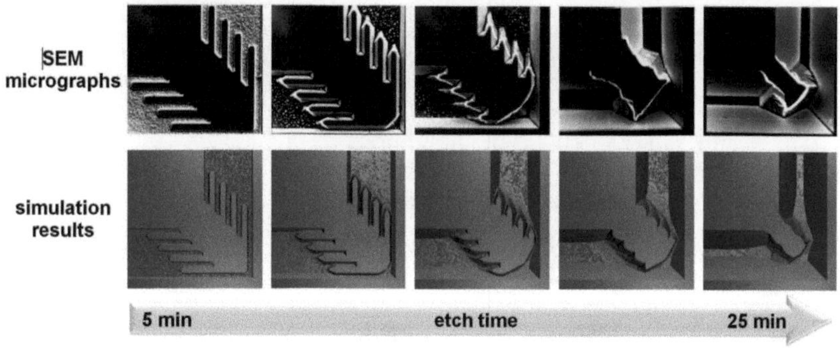

Fig. 4.9 A comparison between simulation and experimental results shows the high accuracy of the algorithm

in discrete time steps in accordance with transition rules. The rules can be regarded as functions composed of the state of a considered cell and the states of the next two adjacent cells. The result is the new state of the considered cell.

Fig. 4.9 demonstrates the removal of a comb-like compensation structure along the etch limiting planes ⟨111⟩. At the top, SEM micrographs are shown. The pictures below illustrate the simulation results of the same geometry. In both cases KOH (30%) at 80°C is used. Comparing the SEM micrographs and the simulation shows a high degree of matching [1].

Commonly, a simulation task is defined by several geometric and process parameters and the mask layer. The most important parameters are model dimensions, resolution, etch time, intended etch depth as well as concentration and temperature of the etchant. Furthermore etch rates for the three main plates of silicon need to be provided, because the etch probability is calculated based on these values. The mask layer can be defined via a graphical interface or can be loaded as graphic file. In addition, the definition of etch stop layers and the number of simulation steps is possible. Also, double sided and multilevel etch processes of both ⟨100⟩- and ⟨110⟩-oriented silicon can be simulated (Fig. 4.10). For visualization of the 3D geometry of the etched microstructure, the 3D solid model is translated into a 3D surface model.

Furthermore the simulation data are stored in voxel files according to the specified model resolution. Voxels are volume elements representing a value on the regular grid in three dimensional space. Each saved simulation step or rather etch state can be translated into a surface data model that can be displayed in a built-in 3D model viewer. Furthermore, it can be exported in ACIS format for further use in proprietary software, for example, in FEM analysis.

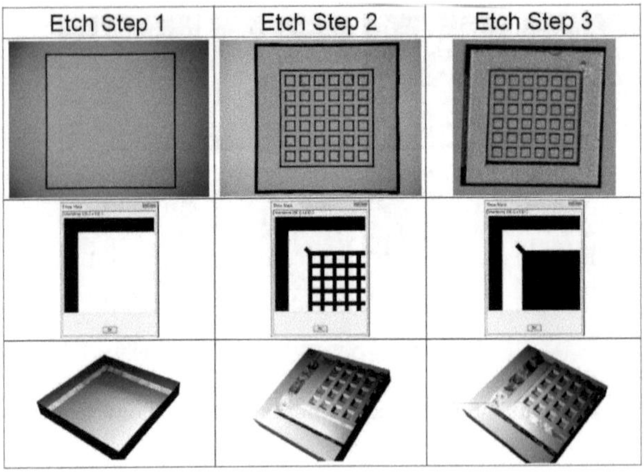

Fig. 4.10 Multilevel etch process

4.4.3 Simulation of Diffraction Effects During UV Depth Lithography

The main difficulty with lithographic pattern transfer is that the different levels of the lower substrate and the upper microstructures lead to a proximity distance. Due to this, diffraction effects distort the mask structure (see Fig. 4.11).

The diffraction effects mainly depend on the structure geometries, the proximity distance, the wavelength and the exposure dose used.

Besides the Optical Proximity Correction (OPC), simulation tools from the semiconductor industry are known compensating such diffraction effects. Unfortunately, these techniques are only suitable for fixed proximity distances or work only in the near field within a proximity distance of up to some microns. The level difference during a MEMS fabrication process can be up to 150 μm and more (see Fig. 4.11) which makes these tools unusable.

To estimate diffraction effects for the UV-depth lithography, where arbitrary mask geometries and several different proximity distances have to

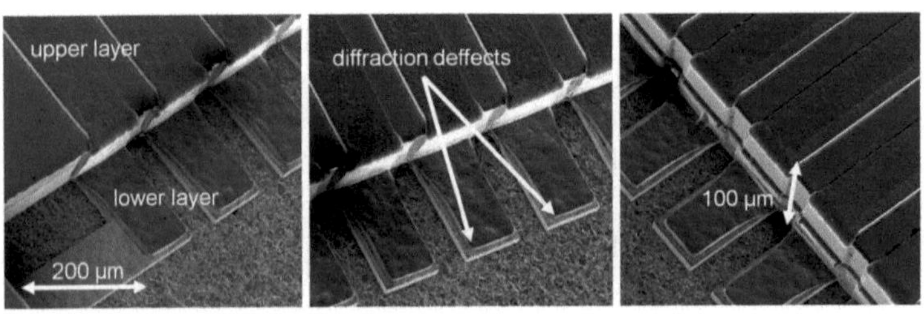

Fig. 4.11 SEM showing fabricated microstructure with diffraction effects

be considered a special simulation tool called *DiffraView* has been developed [11].

The simulation is based on the adapted Fresnel diffraction analysis whereby the intensity of each point on the surface is determined. For the calculation of the intensity the numerical solution of the Fresnel-Sine and Fresnel-Cosine integrals is used. If more than one wavelength is dominant during the exposure, intensities for every relevant wavelength are calculated. The superposition of all intensities leads to an intensity profile of the surface. Together with the resisting sensitivity on a certain wavelength the intensity profile can be used to calculate the absorbed dose [11].

Afterwards an example will depict the accuracy of the developed simulation tool. Fig. 4.12.a shows a calculated intensity plot for the structure presented in Fig. 4.11. Four windings of the upper copper conductor were simulated. The distance between the structures is 40 µm and their width is set to 60 µm. The calculation was done at a single wavelength of 365 nm. Regions where the calculated intensity is larger than 60% of the incoming intensity are painted as white areas. This proofed to be very close to the limiting dose where the resist still reacts and will be completely developed. The thin grey line marks the position of the 100 µm deep step. The white line qualitatively states the boundary of the resulting resist structure. In comparison to the technological result the simulation shows a high degree of matching: a sharp image on top of the ferromagnetic core and a distorted outline in the lower regions of the microstructure.

Fig. 4.12.b shows the mask layout of the same shape with additional compensation structures, while Fig. 4.12.c image shows the corresponding calculation result. Different approaches to compensate diffraction are shown. The first approach is to simply make the masked areas a little bit wider in the distorted areas and add additional structures to the corners. This compensates the diminution of the structure width but still does result in rounded corners. The second approach shows how additional structures, which are detached from the actual mask structure, can compensate diffraction effects

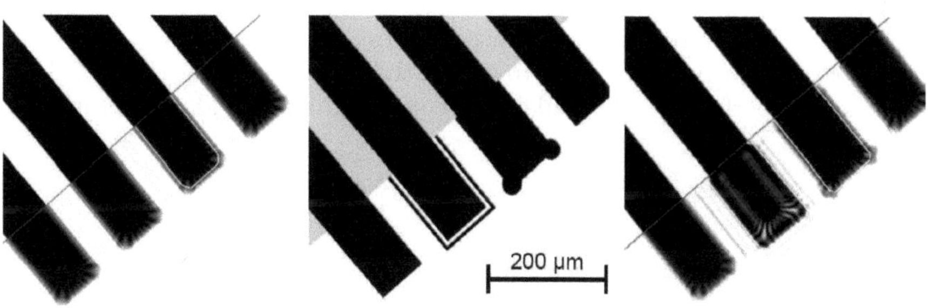

Fig. 4.12 (**a**) Intensity plot for the structures shown in Fig. 4.12; (**b**) Mask layout with additional compensation structures; (**c**) Intensity plot with compensation structres

(see Fig. 4.12.c). However, this approach leads to further diffraction under the masking layer. Several iterative steps would be necessary to reach an optimum design. It becomes clear that a manual optimization would be a challenging task and an automated optimization method should be derived.

4.4.4 Optimization of Lithographic Masks

As stated before, the anisotropic behavior of the silicon crystal limits the variety of feasible geometries. Moreover, it complicates the design of suitable lithographic masks. Commonly, the problem is solved by adding specially calculated compensation structures to the mask design. Several approaches calculating these structures are known but they fail whenever space is limited, the structures are very complex, or undercutting and etching from both sides is required.

To avoid these problems, the software tool *OMaGA* (**O**ptimization of **Ma**sks using **G**enetic **A**lgorithms) was developed. It uses genetic algorithms for the generation of suitable mask layouts within a predefined design space [10, 2].

The use of a genetic algorithm is a stochastic, iterative working optimization technique based on the evolution theory whereby a population of solution candidates is generated. The initial population is generated randomly. Afterwards, all individuals are evaluated using a problem specific function and a performance value for each individual is calculated. Then, genetic operations, such as selection, crossover, etc. are applied to the population forming the individuals of the next generation. The iteration continues until an individual reaches a specified performance value.

Fig. 4.13 shows the iterative procedure the algorithm is based on. During the iteration the solutions are divided in optimal and suboptimal etch processes which can be evaluated with the etch simulator (see Sect. 4.4.2) via a software interface. Furthermore, the number of generations can be specified whereby multiple generations are represented as a DNA sequence.

Also, *OMAGA* is connected to the object-orientated database mentioned in Sect. 4.1. Therefore, former optimization results can be loaded and used as input for other projects. In order to reduce the calculation time, the *OMAGA* tasks can be performed in parallel on a computer cluster. The integration of

For each Generation:
{
- Selection of good solutions for new population
- Generation of new solutions from the parent individuals
- Evaluate new solutions relating to their fitness
- Replace old generation with new generation

}

Fig. 4.13 General structure of the genetic algorithm

OMAGA in *BICEP³S* makes it possible using the results from the other tools, especially *SUZANA* and *DiffraView*.

4.4.5 Conclusion and Outlook

As soon as an optimized physical layout, represented by the 3D-Model, and a corresponding process sequence have been generated, the manufacturing process of the MEMS can be carried out, closing the left fork of the Q-Model.

This MEMS-specific design process allows for a vast selection of basic elements with great liberty in the choice of materials, high aspect ratios and even hybrid designs to be integrated into a complete system. The presented software modules support the designer throughout every design step, accelerating the design process, reducing iterations and elevating the design quality.

The software environment still shows some flaws: Being a “grown” modular system, a higher integration of all the modules into one global software system for MEMS design is desirable. Furthermore, the sophisticated simulation tools in late steps of the process should be coupled to modules in the phase of preliminary design. A possible approach would be to design more interfaces between the corresponding modules. But as some of the information needed for simulation still lack in early steps, condensing the gained knowledge of late steps into heuristic design rules seems more promising. This would include factory specific knowledge about process stability or throughout rates to assure not only feasible devices but also inexpensive ones [6].

With the software environment being a relatively new approach, the stored building blocks are yet limited to those entered by the Collaborative Research Center. To increase the number of available basic elements in a short time, standard libraries should be integrated by a specific interface. In order to raise acceptance of the system in industrial practice, an integrated protection for intellectual property seems promising. Concerning the output of the environment, an interface to manufacturing workflow-systems or an integrated export of a final process sequence in the form of a process-plan would make the transition to manufacturing easier. Nonetheless, an already very handy software environment has been achieved, of which the benefit in everyday use has been proven by the Collaborative Research Center.

Acknowledgements Many thanks go to Christoph Germer who developed the software module µToaST, and to Lars Steffensen who introduced a previous version of the design environment. The authors gratefully acknowledge the funding of the reported work by the German Research Foundation within the Collaborative Research Center 516 “Design and Manufacturing of Active Micro Systems”.

References

[1] Büttgenbach S, Than O (1996) SUZANA: A 3D CAD tool for anisotropically etched silicon microstructures. In: Proceedings of the 1996 European conference on Design and Test, IEEE Computer Society, p 454
[2] Büttgenbach S, Hansen U, Steffensen L (2001) Computational synthesis of lithographic mask layouts for silicon microcomponents. In: Proceedings of SPIE, vol 4407, p 126
[3] Ehrfeld W (2002) Handbuch Mikrotechnik. Hanser Verlag
[4] Franke HJ, Löffler S, Deimel M (2004) Increasing the efficiency of design catalogues by using modern data processing technologies. Dubrovnik
[5] Germer C (2004) Interdisziplinäres Toleranzmanagement. Logos Verlag, Dissertation
[6] Germer C, Franke HJ, Hansen U, Büttgenbach S (2001) Der rechnerunterstützte Entwurf in der Mikrotechnik. Konstruktion 12
[7] Hansen U, Büttgenbach S (2004) T-CAD for the analysis and verification of processing sequences. Microsystem Technologies 10(3):193–198
[8] Hansen U, Güttler J, Seidemann V, Büttgenbach S (2001) An optimization software for the design of micro coils. Proc 12th Micromechanics Europe Workshop pp 241–244
[9] Hansen U, Germer C, Büttgenbach S, Franke HJ (2003) Mixed system and component level T-CAD for micro fabrication. In: Proceedings of the Symposium on Design, Test, Integration and Packaging of MEMS/MOEMS, IEEE Computer Society, pp 4–9
[10] Triltsch U, Büttgenbach S (2008) TCAD tool for innovative MEMS and MOEMS: an all-in-one solution. In: Proceedings of SPIE, vol 6882, p 68820G
[11] Triltsch U, Feldmann M, Boese C, Büttgenbach S (2008) Simulation tool for proximity effects in high aspect ratio UV-lithographic patterning. Sensors and Actuators A: Physical 142(1):429–433
[12] Vietor T, Franke HJ, Ziebart J, Büttgenbach S, Boese C (2010) IDeFi^{X} – Interaktives Werkzeug zur Definition mehrdimensionaler Funktionsstrukturen. Konstruktion 4 / 2010:73–74

Part II
Guiding and Measuring in Active Microsystems

Chapter 5
Wear Behavior in Microactuator Interfaces

R. Bandorf, F. Pape, H. H. Gatzen[2], G. Bräuer[1]

[1] *Fraunhofer Institute for Surface Engineering and Thin Films*
Fraunhofer-Gesellschaft
braeuer@ist.fhg.de

[2] *Institute for Microtechnology (now: IMPT)*
Leibniz Universität Hannover
gatzen@impt.uni-hannover.de

Abstract The growing number of active microsystems applications leads to a demand for enhanced reliability and durability of the moving microparts. The lifetime of the microsystems is closely related to protective coatings on the mating surfaces. This chapter deals with the improvement of the microwear resistance of various thin films, the deposition techniques, and the analysis tools on the microscale. The thin films were prepared using Physical Vapor Deposition (PVD) and Plasma Enhanced Chemical Vapor Deposition (PECVD). From a variety of coatings, the most promising results were gained for carbon based films. The typical film thickness was in a range of a few hundred nanometer. The micro-/nanowear was investigated by an Atomic Force Microscope (AFM) based under single asperity contact by methods of nanoindentation and nanoscratch testing. For an investigation of multi asperity contact specific test equipment was set up for characterization under rotating and oscillating motion. Along with the improvement by the coatings it was found that additionally the substrate material plays an important role for the microtribological system.

5.1 Introduction

The growing number Micro Electromechanical Systems (MEMS) with microactuator applications featuring moving parts in contact [12] leads to a demand for a great mechanical reliability and durability of the moving microparts. The lifetime of such microsystems is closely related to the quality of protective coatings on the interface surfaces [21]. For the tests, sliders were fabricated which were used as counterpart for the AFM based methods. For the wea tests, the sliders or chips themself were used as test body. A classification of the applied tests and the corresponding test bodies and counter parts is given in Table 5.1. Along with the improvement accomplished by the coatings, it was found that additionally the substrate material plays an important role in a microtribological system. Fig. 5.1 depicts the schematic of the different test methods.

Table 5.1 Classification of the applied wear tests

Method	Test body	Counter part
AFM/Hysitron	Diamond tip	Slider/plate
Tape abrasive test	Chip/plate	Tape
Rotational test	Slider	Disc
Oscillating test	Slider	Plate

5.2 Test Systems

The characterisation of the coated samples was performed by AFM based methods for investigation under single asperity contact. For the determination of the micro-/nanowear, single and oscillating scratch tests were carried out. The micromechanical properties were determined by nanoindentation experiments. For the investigation of the microwear under micro flat-flat contact, specific testers were set up. The investigations were performed under rotating or oscillating motion. Wear investigations on multi asperity contact surfaces mating each other are well known from macroscopic testing. Re-

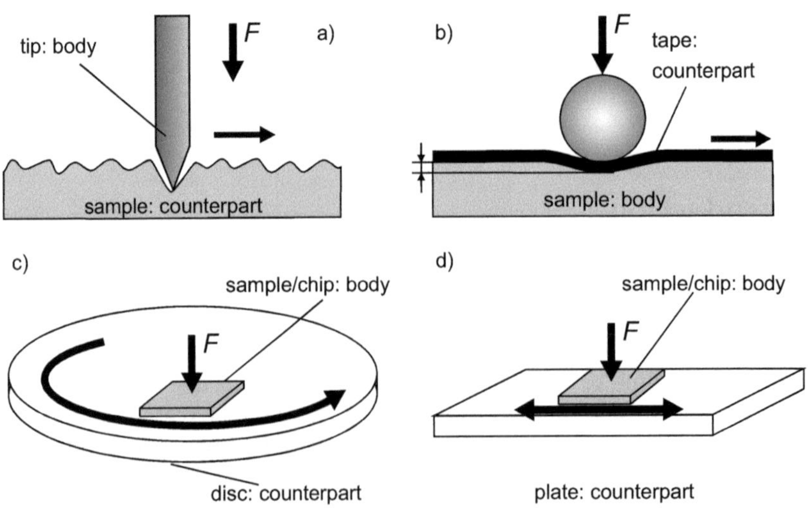

Fig. 5.1 Schematics of the used wear tests: (**a**) AFM methods; (**b**) Tape abrasive test; (**c**) rotational tester; (**d**) oscillating tester

ducing the lateral size of the samples, the test equipment needs also to be adapted. Using polished silicon surfaces, it is not clear whether a full face microcontact occurs or if similar to macroscopic results the contact area will be just a small fraction of the apparent contact area. Therefore, specific test equipment was designed and built allowing areal contact conditions.

5.2.1 Nanoindentation and Scratch Test

Test Method. The micro mechanical properties in point-contact of the prepared films were determined using a *Hysitron TriboScope* [24, 28, 29] in conjunction with a *Veeco Di3100 AFM* and a *Hysitron TriboIndenter*. The nanoindenters consists of a diamond tip that can be moved along two axes. A well defined normal load can be applied to the tip causing a vertical (z-direction) displacement, additionally, the tip can be moved along a line (x-direction). The resulting vertical and horizontal displacement, as well as the resulting friction forces can be measured as a function of time during the whole experiment. The typical loads for testing at single asperity contact are 10 µN to 15 mN. For the determination of hardness of the films by indentation, a Berkovich indenter with a tip radius of about 300 nm was used. For micro wear and micro friction experiments, conical diamond tips with 90° opening angle and tip radii between 0.6 µm and 6.5 µm were used. The experiments consist of either single scratch with increasing load or an oscillating wear test under constant load condition with a predetermined number of cycles.

For the investigation of the microwear, single and oscillating scratch tests were applied. In a single scratch, a diamond tip with typically a conical shape and a defined tip radius first scans the surface under minimal load to map the original topography (pre-scan). In a second step, the tip is moved over the surface with continuously increasing load. This offers the chance

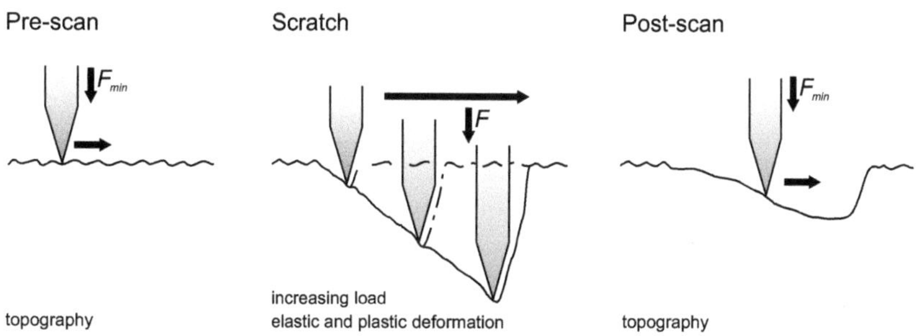

Fig. 5.2 Schematic of single scratch. Pre-scan: mapping of the topography under minimal load; Scratch: lateral tip movement with increasing normal load resulting in elastic and inelastic deformation, and material loss; Post-scan: second mapping under minimal load after scratching

to monitor the onset of deformation, define critical loads, etc. The resulting depth profiling is the sum of elastic and inelastic deformations, as well as occuring material loss, e.g. due to ploughing. After completing the scratch, the surface is once again mapped under minimal load (post-scan). Comparing pre-scan and post-scan, the densification and material loss of the investigated system can be determined as function of the applied load [18]. The test cylce is presented in Fig. 5.2.

Similarly to the single scratch, the oscillating scratch also maps the topography before and after applying a load. In contrast to the single scratch, the applied load under oscillating conditions is kept constant for a defined number of cycles (in our case typically 25). Correlating the pre-scan and post-scan data provides information on the time dependent plastic and elastic deformation of the surface, as well as the material loss and critical loads or cylces, respectively [18]. As a standard of the micro wear, the remaining wear depth after 25 cycles of oscillating scratch test was used. The applied load with a conical indenter typically was 2 mN.

To investigate multilayer coatings, the nanoindenter allows for obtaining acoustic emission signals from samples during nanoindentation. An acoustic emission sensor is integrated into the Berkovich tip. Crack-induced release of elastic strain energy achieved due to indentation of the tip is measured accoustically.

Test Body and Counter Part. For the investigations, the test body was always the diamond tip of the *AFM/Hysitron* system. The counterpart was varied using different substrate materials with coatings. In general, typical materials relevant for microtechnology applications were employed without and with protective coating. Table 5.2 summarizes the counterparts used.

Table 5.2 Materials used as counterpart for nanoindentation and scratch test

Bulk material	Surface/Interface material	Tribocoating
Si	bulk material	DLC (single layer)
Glass	NiFe	DLC multilayer
NiZn-ferrite	SmCo	
	SU-8	
	Cu	

5.2.2 Tape Abrasive Wear Test

Test Method. The abrasive wear behavior of the thin films was determined by a specifically developed abrasive wear tester (Fig. 5.3) [30]. To realize a two body contact, a videotape as abrasive medium was used, since commercial video tape has alumina particle on their backside to clean the recorder head during its employment and to minimize wear of the oxide layer. The tape was turned and the abrasive side was used for grinding. The tape was moved under a loaded Al_2O_3 sphere with 10 mm diameter, pressing the tape with 20 mN at a sliding speed of 22 mm s^{-1} against the substrate. After grinding, the resulting crater depth was measured by profilometry. As measured for the abrasive wear, the tape length required to grind a 100 nm deep crater is used. By this method, highly abrasive films as well as highly wear resistant ones can be quantified.

Fig. 5.3 Abrasive wear tester

Test Body and Counter Part. The counterpart in an abrasive tape tester was the tape with embedded abrasive particles. During the investigation, no abrasive particles except the tape remain in the created crater since the moving tape removes any debris. The test body was a plate, both coated and uncoated for an abrasive classification of a huge variety of materials. For the investigation of the different coatings, typically polished silicon chips or plates were used as substrates (Table 5.3). Keeping the substrate for the coatings constant is beneficial for a reliable classification, since the substrate can significantly influence the resulting abrasive wear resistance [3].

Table 5.3 Materials used as test body for abrasive wear test

Bulk material	Coating	
Si	none	CN_x
Glass (BK7)	DLC	TiN
	Me-DLC	Cr
	cBN	MoS_2

5.2.3 Rotational Test System

Test Method. For the wear measurement in a flat contact at low sliding speeds, a test stand, which is based on the pin-on-disk principle, was set up and used. The standard pin-on-disk test is an experiment to measure friction and wear of a pin or a ball in sliding contact against a rotating disk [14]. In our case, a gimbaled slider replaced the pin, thus substantially increasing the tester's flexibility. Fig. 5.4 depicts the rotating wear tester used for the tribological tests. The tester has a friction drive free of backlash and an integrated angel transmitter *Heidenhain RON 905*. It is an ideal tool for conducting wear tests for analyzing the influence of a protective layer on the life expectance of a micro tribological interface [6].

Test Body and Counterpart. The slider was developed based on the idea that a quasi-areal contact can only be achieved under three-point contact conditions. Choosing three protruding contact areas arranged in a triangle achieved these conditions. The contact surfaces are ultra-smooth and highly coplanar. To allow these contact areas to comply with the counterpart, the slider was gimbaled.

For such a contact, the precondition for the slider and its counterpart are optimal flatness and surface finish. To accomplish both, the slider pad areas are machined by nanogrinding [13], followed by Chemical-mechanical Polishing (CMP) [15]. The fabrication was conducted on a *Peter Wolters 3R40* lapping machine, equipped with nanogrinding plates or CMP pads, respectively. The load was applied through weights. The rotational speed of both plate and wafer was $18\,\mathrm{U min^{-1}}$. In all experiments, a slurry flow rate of $15\,\mathrm{ml\,min^{-1}}$ was used.

The three pad areas were created by recessing the slider material outside the pads by ion beam etching (IBE). To evaluate different materials at the interface, the sliders were made of a base material and coated with various

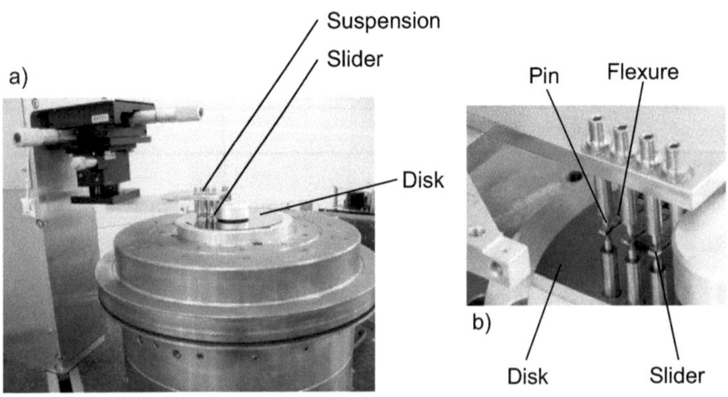

Fig. 5.4 Rotating wear tester: (**a**) setup; (**b**) close view of slider and counterpart

materials. Coating techniques typically used were sputter deposition or electroplating. These sliders were used with and without a wear-resistant DLC coat.

Three slider bulk materials were used: Si, NiZn-ferrite, and glass. Used without additional coating, a bulk material also represented the interface material. For alternative interface materials, as well as for applying a protective DLC layer, the pads areas were coated. Sliders with an anorganic coat like SmCo were prepared by sputter deposition on the protruding tripads. In case electroplating was applied like for forming NiFe or Cu surfaces, the tri-pad pattern was created directly by a micromold. Organic coatings like *SU-8* or *SU-8* filled with Fullerene particles were fabricated by spin-on technique and patterned photolithography.

The sliders were fabricated in different sizes. The first design featured a $2.4 \times 2\,\mathrm{mm}^2$ slider footprint and three pads, each with a diameter of 705 µm. For the second generation of sliders, the size of the pads was reduced to avoid an overlap of the three wear-tracks. In this case the diameter of the pads was 500 µm. A smaller slider was also introduced. It had a footprint of $1.24 \times 0.99\mathrm{mm}^2$ and a pad diameter of 250 µm. The latter one's dimensions agree with a Hard Disk Drive (HDD) "pico"slider.

For one series of tests, the pad surface was not flat but patterned. This was accomplished by structuring the surface with a Focused Ion Beam (FIB) [20]. As mentioned before, the interface materials were used "naked" or in combination with a wear-resistance coat. For such a coat, a single layer or multilayer DLC was chosen. For the test, the tri-pad sliders were bonded on flexures, which were mounted on a suspension creating the required load. The preload on the slider was in the range of 20 mN to 30 mN. The suspension also allowed to add a load on the slider through a small pin. Table 1 provides an overview over the interface materials used for the slider.

The counterparts for the rotational Tribotester were 4 in wafers with 30 mm location holes. Each wafer was protected by an AZ-resin coat against scratches during the drilling of the hole. For the tests, the wafer was used without any coat and with a base material deposited on the wafer like sputter deposited SmCo, electroplated NiFe and Cu, or spin coated *SU-8*. The counterparts were used without any wear-resistance coat. Table 5.4 provides an overview of the materials used.

Table 5.4 Materials used as body and counterpart for rotating test

Bulk material	Surface/Interface material	Tribocoating (body only)
Si	bulk material	DLC (single layer)
Glass	NiFe	DLC multilayer
NiZn-ferrite	SmCo	
	SU-8	
	Cu (counterpart only)	

5.2.4 Oscillating Test

Test Method. For oscillating micro wear investigations, samples were tested in flat-to-flat micro contact using a specifically developed oscillation testing machine (Fig. 5.5) [26]. The samples ($10 \times 10\,\mathrm{mm}^2$) were mounted on a carrier performing oscillations with a maximal lateral displacement of $80\,\mu\mathrm{m}$. DLC coated Si ⟨100⟩ served as a flat counterpart to the micro patterned test samples. The normal load was applied by dead-weights in the range of 0.05 N to 1 N. Using oscillation frequencies of up to 100 Hz, sliding speeds of up to $5\,\mathrm{mm\,s^{-1}}$ were achieved. Additionally, the piezo drive could be exchanged by an excenter motor for wider strokes.

Test Body and Counter Part. For the oscillating tests, Si ⟨100⟩ served as substrate material, both for the body and the counter part. The body was both polished and microstructured. The coatings applied were alternative modifications of DLC. The counterpart was usually polished, since the resulting coefficient of friction was minimized for a pairing of patterend and polished samples [5]. Typically, the counterpart was DLC-coated while the body surface varied (Table 5.5).

Table 5.5 Materials used as test body and counterpart for oscillating test

Bulk material	Coating
Si	none
	DLC
	Me-DLC
	Si-DLC

Fig. 5.5 Oscillating wear tester with excenter drive

5.3 Fabrication Process of Tribological Surfaces

The investigations of the microtribological interfaces were performed using bulk material, as well as coated systems. Therefore, various coatings were applied on different substrate materials (brittle, ductile, organic). As standard reference, polished silicon with the different coating systems was used. The choice of silicon was based on the importance of this material in MEMS as one of the standard materials. Else, typical materials relevant for electromagnetic microactuators were used.

For the fabrication of the microtribological coatings, vacuum deposition was applied. The basic principle is the use of a technical plasma for the generation of the film forming species either from a precursor or from a solid state source, the target [8, 10, 16]. The focus of the results presented was on diamond-like carbon films (DLC). The films can be subdivided into hydrogen-free films (a-C) and hydrogenated films (a-C:H). The a-C:H-films can be doped with a number of elements (a-C:H:X, with X: Si, W, Au) altering the resulting coating properties. Table 5.6 provides an overview of the plasma gases and target materials used [1, 2]. All coatings were prepared in industrial sputtering tools. The base pressure was well below $1.0 \cdot 10^{-5}$ mbar, while the working pressure was typically $0.5 \cdot 10^{-3}$ mbar.

Table 5.6 Preparation parameters for DLC coatings

Coating	Process	Process Type	Gases	Target
a-C	PVD	DC-sputtering	Ar	Graphite
a-C:H	PECVD	reactive plasma CVD	Ar, C_2H_2	
a-C:H:W	PVD + PECVD	reactive DC-sputtering	Ar, C_2H_2	W
a-C:H:Au	PVD + PECVD	reactive DC-sputtering	Ar, C_2H_2	Au
a-C:H:Si	PECVD	reactive plasma CVD	$C_4H_{12}Si$	

5.3.1 Physical Vapor Deposition (PVD)

For the preparation of the a-C-films, DC sputtering of a graphite target in an argon atmosphere was used. Fig. 5.6.a shows the schematics of DC sputtering. Due to its negative voltage, the target mounted on the cathode serves as a source for the film forming material. The gas ions are accelerated towards the cathode and sputter the target material. The sputtered material is deposited on the substrate surface, thus forming the film. Additionally, a bias voltage can be applied to the sample holder. In this case, the growing film is additionally bombarded by positive ions. This bombardement can lead to a densification, sputtering of loosely bonded material, and improvement of adhesion [22, 27].

5.3.2 Plasma Enhanced Chemical Vapor Deposition (PECVD)

Alternatively to PVD processes, such based on chemical vapor deposition (CVD) were applied. Using a reactive gas, typically acetylene with RF-bias at the substrate, hydrocarbons are formed and a film is grown by plasma enhanced chemical vapor deposition (PECVD) as shown shematically in Fig. 5.6.b. The a-C:H, and a-C:H:Si-films finally were prepared using plasma enhanced chemical vapor deposition in an argon/acetylene or argon/TMS environment, respectively. In this case no target is used and the process gas serves as precursor for the film forming species. The CVD process is enhanced by a plasma, generated at the substrate due to an applied bias voltage. The hydrocarbons condensate in the chamber and on the substrate.

5.3.3 Hybrid Processes (PVD + PECVD)

The third type of processes were hybrid, combining a PVD process with CVD. Metal containing diamond-like carbon DLC films a-C:H:W and a-C:H:Au were prepared by reactive DC-Sputtering using argon/acetylene gas. To modify the coating properties, an RF-Bias was applied to the substrate. Furthermore, the growing film is bombarded by positive ions due to the applied negative RF-bias at the substrate.

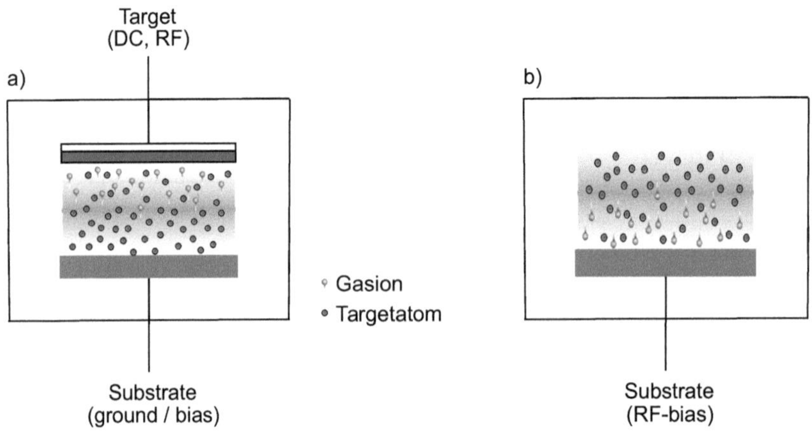

Fig. 5.6 Schematic of used vacuum coating technologies: (**a**) PVD: DC sputtering; (**b**) PECVD

5.4 Experimental Investigations

The base materials that were investigated, both uncoated and coated. Typical materials used showing brittle behavior were silicon (Si), glass, and NiZn-ferrite. Ductile materials were copper (Cu), nickel-iron (NiFe), and samarium-cobalt (SmCo). Finally, organic materials were photo resins, mainly *SU-8* with and without filler. The body for the wear investigations was mostly consisting of brittle material.

The experimental investigation focused on the behavior under point contact using nanoindantation and scratch tests. For the classification of the abrasive wear, the tape abrasive test was used. The coatings with the highest potential for effective protection were DLC coatings. The investigation of the microwear close to an application was performed using rotating and oscillating tests.

5.4.1 Investigations with AFM Based Methods

The investigation under point contact were carried out using nanoindentation for the determination of the micromechanical properties (hardness, Young's modulus). The microwear behavior was examined by single and oscillating scratch tests. The crack behavior under load can be analyzed by conducting acoustic emission measurements during the nanoindentation process. Such a test allows measuring the integrated acoustic energy released during the crack propagation.

Penetration. Investigating the micromechanical properties of a-C:H films, the measured micro hardness was depending on the applied bias voltage during deposition. For determining microhardness and Young's modulus, a Berkovich indenter (tetrahedral shape) using the approach of Oliver and Pharr [23] was used. Table 5.7 summarizes the micromechanical properties of relevant materials [1, 18]. The hardness of the a-C:H films is increased from 10 to 20 GPa by rising the working pressure from $3 \cdot 10^{-2}$ mbar to $5 \cdot 10^{-2}$ mbar. A further increase was realized by varying the bias voltage during deposition. The hardness of the deposited a-C:H films varied as a function of the applied bias voltage. The hardest films with 28 GPa were prepared at −650 V. A change in hardness typically is also an indication for a modification of the materials microstructure. The SEM top view and cross section shows a transition from a cauliflower like surface and shell like fracture to a smooth surface with tiny holes and a glassy structure [4].

Scratch Tests. Depending on the hydrogen content in the films, the micro wear behavior differed. While the hydrogenated a-C:H films show nearly no wear under low load, the hydrogen free (more tough) a-C films exhibit an initial wear due to a running-in behavior. At higher load, the response changed and the a-C films showed no significant further modification after running-in.

Table 5.7 Micromechanical properties

Material	Hardness	Young's modulus
Si	11.8±0.5 GPa	155±5 GPa
NiFe	5.5±0.75 GPa	136±15 GPa
Cu	1.7±0.1 GPa	162±40 GPa
Floatglass	7.1±0.25 GPa	83.9±5.5 GPa
Pyrex	7.3±0.3 GPa	65.8±3.9 GPa
SU-8	0.55 GPa	5.2±0.2 GPa
a-C:H	26±2 GPa	185±15 GPa

The a-C:H films showed continuing wear. Therefore, the used load regime was essential for the right selection of the films [4].

Besides glass, NiFe, *SU-8*, and NiZn-ferrite sliders, DLC multilayer coatings were applied to the Si slider. The DLC multilayer coatings were deposited in a sequence of soft and hard layers achieved by varying both the process pressure and the power applied during the deposition [24, 25]. The total multilayer thickness was 600 nm, each multilayer was 75 nm thick. To benchmark the multilayer, soft and hard monolayers were also created. On a glass substrate, a hard DLC coatings of such a thickness showed delamination caused by intrinsic stress of the coat. The hardness values were as follows: 10.5 GPa for the Si substrate, 13 GPa for a soft monolayer, 18 GPa for a hard one, and approx. the same value for the DLC multilayer with a hard top layer, as presented in Table 5.8. Scratch test results of these combinations are shown in Fig. 5.7.

Table 5.8 Properties for soft, hard, and DLC multilayer compared to Si substrate

Material	Hardness [GPa]	Indentation depth [nm]	Plastic deformation [nm]
Si substrate	10.5	140	40
600 nm soft DLC deposited at 0.023 Torr	13	120	30
600 nm hard DLC deposited at 0.038 Torr	18	110	10
600 nm thick eight layer DLC coating	18	110	10

Influence of Substrate Material and Coating Thickness. When investigating the micromechanical properties of the thin-films, it turned out that coating thickness plays an essential role for the proper determination of the properties. If the coating is very thin, an influence of the substrate on the results of this measurement seems unavoidable. Furthermore, keeping the 10% rule for the indentation depth sometimes was difficult.

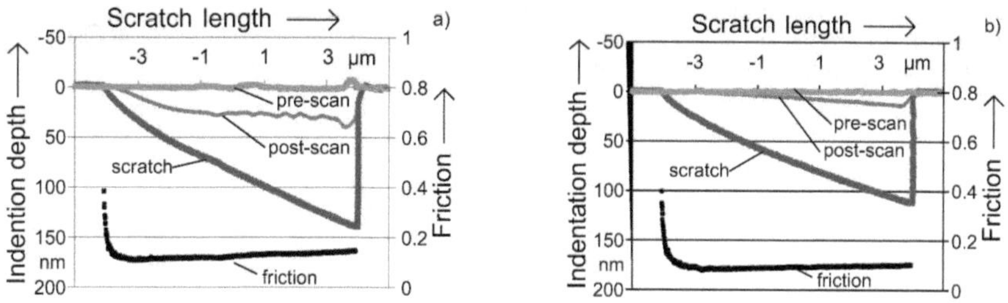

Fig. 5.7 Scratch tests on (**a**) Si substrate and (**b**) DLC multilayer

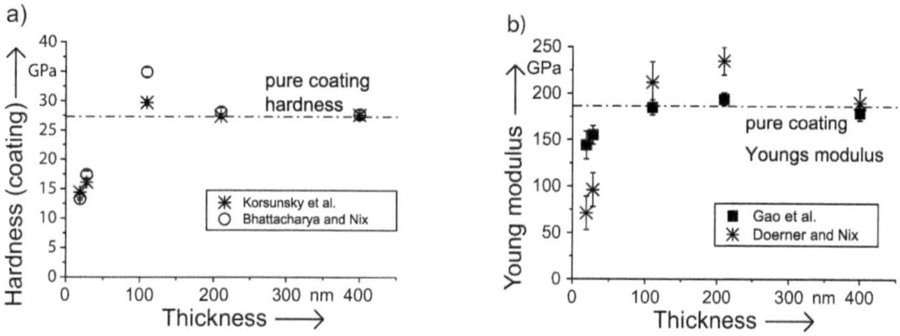

Fig. 5.8 Influence of a-C:H coatings as a function of the film thickness: (**a**) Calculated hardness of a-C:H coatings in dependence of the coating thickness; (**b**) Calculated Young's Modulus of a-C:H coatings in dependence of the coating thickness

Using nanoindentation experiments to determine the micromechanical properties, the final load was in the range of 500 μN to 10 mN. The interpretation of the results is based on the use of two theoretical models. For the evaluation of the hardness in dependence of the coating thickness, the model of Korsunsky et al. [17] as well as the theory of Bhattacharya and Nix [7] were used. Fig. 5.8 shows the influence of the coating thickness on the calculated hardness (a) and the Youngs modulus (b) of a-C:H coatings. For films with a thickness below 250 nm, the resulting hardness varies. Exceeding a thickness of 250 nm, the determined hardness reaches a constant level. This is attributed to the pure coating properties whereas thinner films always include contributions from the substrate material to the measured values. For the thin-films, there are some differences in the results depending on the applied models. In the case of thicker coatings, the discrepancy is within the expected error limit. For investigations of the Young's modulus of the coatings, the models of Gao et al. [11] as well as of Doerner and Nix [9] were used. Fig. 5.8.b shows the calculated Young's Modulus of the a-C:H films in dependence on the thickness. Analogous to the hardness, the values scatter for thinner films and reach "bulk properties" for thicker coatings exceeding 250 nm thickness.

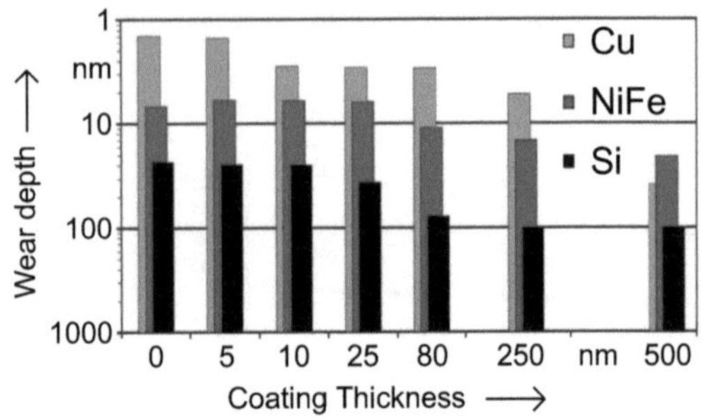

Fig. 5.9 Remaining wear depth after 25 cycles of oscillating scratch test; applied load: 2 mN

Scratch tests on 250 nm a-C:H coatings deposited on NiFe and Si show the influence of the substrate material on resulting depth under load, elastic and plastic deformation, and material loss. While a-C:H on Si shows a hardness of approximately 25 GPa, the hardness of NiFe on Si with 5 μm NiFe thickness is 7 GPa. For NiFe the hardness is the pure coating hardness due to the high thickness. The indentation depth depends on the substrate coating system. For a softer substrate, the indentation depth increases. Depending on the materials involved, the elastic and plastic deformation, as well as the material loss differs. Scratching pure NiFe coatings, the film is deformed plastically by approx. 40%. Another 30% are lost due to material wear out. Coating the NiFe surface with a-C:H the wear depth is reduced by approx. 45%. Additionally, the highest fraction ($> 80\%$) of the deformation is due to elastic effects [19]. Therefore, an effective protection is ensured by applying thin hard coatings, like here of 250 nm a-C:H.

Fig. 5.9 shows the influence of a-C:H coatings as a function of the film thickness for various substrates (Cu, NiFe, Si). For the coatings on Si, (the hardest substrate material), it can be seen that an increasing coating thickness results in a reduction of the remaining wear depth after scratch testing. The values for 250 nm and 500 nm differ only within the error limit. Therefore, for thicknesses exceeding 250 nm a-C:H on Si substrate, the remaining wear properties are attributed only to the coating. For the softer substrates Cu and NiFe, the required coating thickness to reach "quasi-bulk" properties of the applied coating is higher. Also for those substrates, a reduction of the remaining wear depth was found with increasing coating thickness.

Acoustic Emission. The most common mode of a wear layer's catastrophic failure is delamination due to crack propagation. Therefore, compared to DLC monolayers, multilayers consisting of alternative harder and softer DLC layers are less prone to delamination since the softer layers are acting as crack stops. This crack behavior under load can be analyzed by conducting acoustic emission measurements during nanoindentation process. Such a test allows measuring the integrated acoustic energy released during the crack propagation.

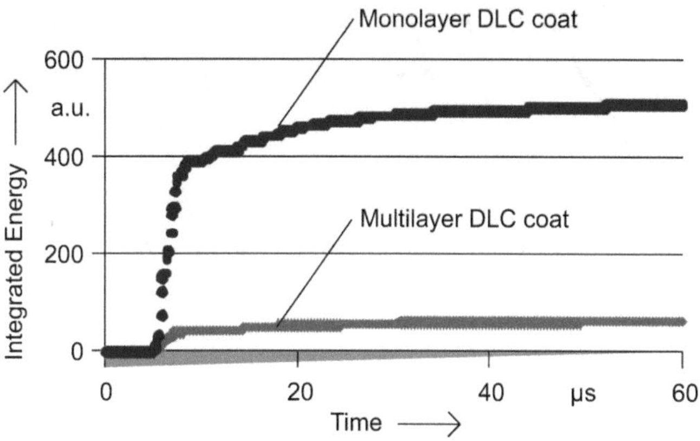

Fig. 5.10 Integrated energy of an induced crack for a monolayer DLC coat and a multilayer DLC coat on *SU-8* coating

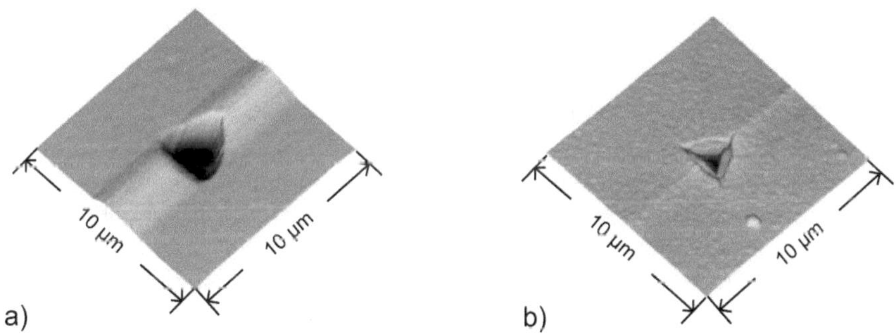

Fig. 5.11 Scan of the surface after indent on (**a**) monolayer DLC coat; (**b**) multilayer DLC coat

For these measurements, the crack propagations of a hard single DLC coat with a thickness of 600 nm and a multilayer DLC coat consisting of eight individual layers (each of a thickness of approx. 75 nm) were compared. In both cases, a maximal load of 5 mN was used on a Berkovich tip. Fig. 5.10 depicts the integrated energy for the fracture of a monolayer DLC coat as well as for a DLC multilayer coat. For both tests, *SU-8* was used as a substrate material, since its softness fosters a large DLC deformation. The diagram shows, that the fracture energy for the monolayer is more than seven times greater as for the multilayer. A crack is stopped on the following soft coat, therefore, the released energy correlates to the coating thickness of one eighth of the monolayer coat. Fig. 5.11 presents AFM scans of the surface after indent for the monolayer DLC coat and for the multilayer. The interface of the monolayer coat is destroyed, while the crack for the multilayer is stopped in the coat.

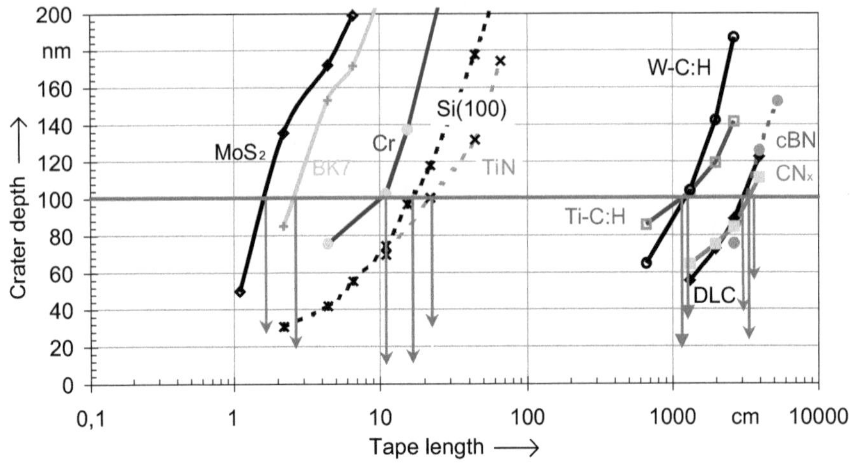

Fig. 5.12 Abrasive wear resistance of different coatings; the tape length required to grind a 100 nm deep crater serves as an evidence for the abrasive wear resistance

5.4.2 Tape Abrasive Wear Tests

For a first classification of the alternative coating systems, the abrasive wear was investigated using a tape based abrasive wear tester. Fig. 5.12 shows the abrasive wear resistance of the coatings investigated with the tape length required to grind a 100 nm crater as an evidence. On one hand, it is obvious that soft materials like MoS_2 show a very low abrasive wear resistance. On the other hand the carbon based films a-C, a-C:H, and a-C:H:X, with X being a dopand like Si or Ti, exhibit abrasive wear resistance orders of magnitute greater [30]. As results of the abrasive wear tests, the carbon based systems turn out to show the highest potential as wear resistant or protective coatings.

5.4.3 Rotational Tests

For the rotational wear tests, an average velocity of $7\,\mathrm{mm\,s^{-1}}$ was applied. After the completion of the wear tests, debris at the sliders could be detected. Particularly interesting results were achieved with Si sliders (large size footprint, large size contact pads). For all tests, a flat Si surface served as counterpart. Even for a low load, debris was found at the edges of the pad, lending proof to the theory that any sliding contact resutls in wear [2]. Fig. 5.13 depicts both slider (Fig. 5.13.a), as well as one of its pads after a test (Fig. 5.13.b).

Sliders with contact pads were used for evaluating multilayer DLC coats. For the tests, uncoated and coated small size sliders with small size contact pads were used. As before, a flat Si surface served as counterpart and the

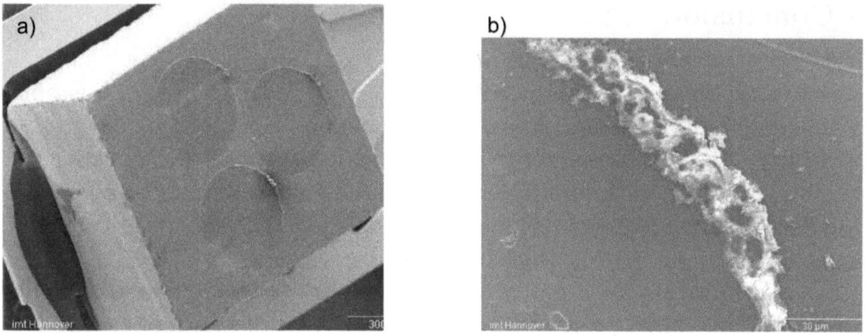

Fig. 5.13 Slider after test: (**a**) Test specimen; (**b**) Debris at one edge of the specimen [2]

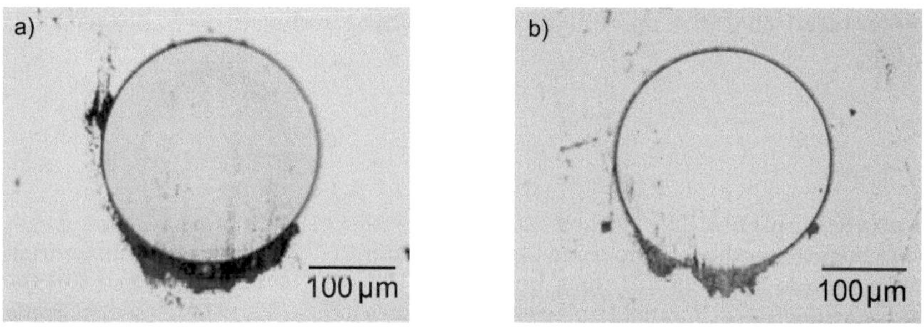

Fig. 5.14 Micrograph of Si-slider after test against Si-counterpart: (**a**) Si-slider after test; (**b**) Si-slider with DLC multilayer

average velocity was 7 mm s^{-1}. For these tests, the vertical preload force was 23 mN. After the tests, for both types of sliders, wear particles were observed. For the multilayer DLC coat, the amount of wear particles was lower. Fig. 5.14 depicts contact pads after the conclusion of the tests for uncoated and coated sliders. The particles around the pad's rim are clearly visible. For neither uncoated nor coated pads, an height reduction was observed.

5.4.4 Oscillating Tests

With the oscillating wear tester, similar to the rotating investigations, debris at the sliders could be detected after the wear test. At low loads only slight wear was observed. The formation of debris increased with a growing load [25].

5.5 Conclusion

For active MEMS comprising a tribological system, there are various optimization possibilities. Both wear and friction of micro contacts can be reduced substantially by submicrometer thick coatings on the surface. Superior behavior was obtained with interfaces featuring DLC coatings. The nanoindentation and stress tests proved, that in dependence on the substrate hardness, there is an optimal coating thickness. The wear behavior could be further improved by applying DLC multilayer coatings due to their resistance to delamination. For evaluate contact surfaces of slide bearings for microactuators, rotating wear tester were found to be particularly well suited. The tests conducted allowed an analysis of the wear behavior for areal contact. Even for very low loads, wear occured on the contact pads. As previously, it could be demonstrated that the amount of wear could be reduced by applying DLC coatings.

Acknowledgements The authors acknowledge the contributions of several colleagues involved in the Collaborative Research Center 516 "Design and Manufacturing of Active Micro Systems": Michael Beck, Cord Henke, Rolf Küster, Holger Lüthje, Claudia Neumeister, Dominik Paulkowski, Jan-Hinrich Sick, Thorsten Staedler, Sonja Wiendl, and Andreas Wortmann. Additional the authors gratefully aknowledge the funding of the reported work by the German Research Foundation within the Collaborative Research Center 516.

References

[1] Bandorf R, Lüthje H, Schiffmann K, Staedler T, Wortmann A (2002) Sub-micron coatings with low friction and wear for micro actuators. Microsystem Technologies 8(1):51–54

[2] Bandorf R, Lüthje H, Schiffmann K, Beck M, Gatzen H, Schmidt M, Büttgenbach S, Bräuer G (2004) Sub-micron coatings with low friction and wear for micro actuators. Microsystem Technologies 10(3):223–226

[3] Bandorf R, Lüthje H, Staedler T (2004) Influencing factors on microtribology of dlc films for MEMS and microactuators. Diamond and Related Materials 13(4-8):1491–1493

[4] Bandorf R, Lüthje H, Henke C, Wiebe J, Sick J, Küster R (2005) Different carbon based thin films and their microtribological behaviour in mems applications. Surface and Coatings Technology 200(5-6):1777–1782

[5] Bandorf R, Luthje H, Kuster R, Sick J (2006) Microtribology: Carbon based coatings on micro structured surfaces. In: Proceedings of the An-

nual Technical Conference-Society of Vacuum Coaters, vol 49, p 554
[6] Beck M (2004) Increase of the wear resistance of micro-components. Dissertation, Leibniz Universität Hannover
[7] Bhattacharya A, Nix W (1988) Analysis of elastic and plastic deformation associated with indentation testing of thin films on substrates. International Journal of Solids and Structures 24(12):1287–1298
[8] Chapman B (1980) Glow discharge processes: sputtering and plasma etching. Wiley New York
[9] Doerner M, Nix W (1986) A method for interpreting the data from depth-sensing indentation instruments. J Mater Res 1(4):601–609
[10] Frey H, Döbele H (1995) Vakuumbeschichtung. 1. Plasmaphysik-Plasmadiagnostik-Analytik. VDI-Verl.
[11] Gao H, Chiu C, Lee J (1992) Elastic contact versus indentation modeling of multi-layered materials. International Journal of Solids and Structures 29(20):2471–2492
[12] Gatzen HH (2009) Magnetic micro electro-mechanical systems for sensor and actuator applications. ECS
[13] Gatzen HH, Beck M (2003) Nano-roughness standards with a very low roughness value. In: Proceedings euspen 4th Int. Top. Conf., vol 2, pp 495–498
[14] Gatzen HH, Beck M (2003) Wear of single crystal silicon as a function of surface roughness. Wear 254(9):907–910
[15] Gatzen HH, Cvetkovic S (2007) Modellierung des CMP-Prozesses zum Planarisieren von Mikrosystemen. In: 63. Jahrbuch Schleifen. Honen, Läppen und Polieren, Vulkan Verlag
[16] Haefer R (1987) Oberflächen-und Dünnschicht-Technologie. Springer Berlin
[17] Korsunsky A, McGurk M, Bull S, Page T (1998) On the hardness of coated systems. Surface and Coatings Technology 99(1):171 182
[18] Küster R (2006) Mikrotribologische Untersuchungen dünner Schichten unter Anwendung rastersondenbasierter Verfahren, Dissertation, vol 1. Fraunhofer IRB Verlag, Stuttgart
[19] Kuster R, Schiffmann K (2004) Nano-scratch testing on thin diamond-like carbon coatings for microactuators: friction, wear and elastic-plastic deformation. Zeitschrift für Metallkunde 95(5):306–310
[20] Küster R, Bandorf R, Lüthje H, Bräuer G, Gatzen HH, Neumeister C (2005) Verschleißarme Beschichtungen für Mikroaktoren
[21] Maboudian R, Carraro C (2004) Surface chemistry and tribology of mems. Annual review of physical chemistry 55:35–54
[22] Mattox D (1989) Particle bombardment effects on thin-film deposition: A review. Journal of Vacuum Science & Technology A: Vacuum, Surfaces, and Films 7:1105
[23] Oliver W, Pharr G (1992) Improved technique for determining hardness and elastic modulus using load and displacement sensing indentation experiments. Journal of Materials Research 7(6):1564–1583

[24] Pape F, Küster R, Bandorf R, Bräuer G, Gatzen HH (2007) Nano-scratch studies on diamond-like carbon coatings on various materials applied in micro actuators. In: GFT Tribologie-Fachtagung, GFT, vol 1, pp 28/1–28/10
[25] Paulkowski D, Bandorf R, Achilles S, Pape F, Gatzen HH, Bräuer G (2008) Studies on diamond-like carbon coatings for the application in micro actuators. Advanced engineering materials 10(7):644–647
[26] Schmidt M, Wortmann A, Lüthje H, Büttgenbach S (2001) Novel equipment for friction force measurement on mems and microcomponents. In: Proceedings of SPIE, vol 4407, p 158
[27] Smidt FA (1990) Use of ion beam assisted deposition to modify the microstructure and properties of thin films. International Materials Reviews 35(1):61–128
[28] Staedler T (2001) Mechanische und tribologische Charakterisierung dünner Schichten mit Hilfe rastersondenbasierter Verfahren, Dissertation, vol 1. Fraunhofer IRB Verlag, Stuttgart
[29] Staedler T, Schiffmann K (2001) Correlation of nanomechanical and nanotribological behaviour of thin dlc coatings on different substrates. Surface Science 482–485(2):1125–1129
[30] Wortmann A (2002) Entwicklung einer Prüfmethode zur messtechnischen Evaluation abrasiver Verschleißresistenz auf der Mikroskala, Dissertation, vol 1. Fraunhofer IRB Verlag, Stuttgart

Chapter 6
Friction Behavior in Microsystems

A. Phataralaoha, R. Bandorf, G. Bräuer[2], S. Büttgenbach[1]

[1] *Institute for Microtechnology*
Technische Universität Braunschweig
s.buettgenbach@tu-braunschweig.de

[2] *Fraunhofer Institute for Surface Engineering and Thin Films*
Fraunhofer-Gesellschaft
braeuer@ist.fhg.de

Abstract In active microsystems, low lateral propulsive forces act under high contact pressures, thereby requiring a low friction coefficient μ for successful operation. The maximum acceptable friction coefficient is determined by the available driving forces. This chapter focuses on the reduction of microfriction at the contacting interfaces, the use of micropatterned surfaces, and the realization of microguidance. For an investigation on flat-flat-microcontact, specific test equipment was built and used. Physical Vapor Deposition (PVD) and Plasma Enhanced Chemical Vapor Deposition (PECVD) were used for applying thin films. In contrast to macroscopic examples, microfriction under single asperity contact correlates to the applied normal force. At a low load range from 50–100 μN, the friction coefficient decreases with increasing load. Above a critical value the friction coefficient increases with increasing load due to inelastic effects. Under multi asperity microcontact, the friction coefficient is dependent on the contact pressure. To tailor the resulting friction coefficients, the influence of different micropatterns are studied. For carbon based films the friction coefficient μ can be decreased below 0.1.

6.1 Introduction

Friction is an unavoidable characteristic of systems comprising movable components, which are in contact with each other. The effects of friction increase significantly as the system size decreases. Therefore, friction becomes critical on the microscale, and is one of the fundamental limitations in the design and implementation of reliable, efficient microsystems.

In the last decade various types of actuating principles used in micromachines have been reported [8, 12]. Great efforts have been made to improve the reliability of these devices, but the understanding of friction, wear and other related phenomena is still insufficient [4]. Several guides and support structures based on two categories have been demonstrated for the application in micromachines: contact and non-contact type bearings. In contact type bearings, support structures are in direct contact with the moving parts. Due to the tribological contact, the reliability of such devices drastically depends on friction and wear. Non-contact type structures use more complicated mechanisms such as electrostatic [11], electromagnetic [5], or aerostatic bearings [6]. These systems show much less friction and wear compared to contact type systems, however, as major drawbacks, their fabrication is rather complex, so that reliabilty of the systems can be negatively affected. Furthermore, the bearing systems have to be designed and properly integrated into a particular micromachine to assure their function. Thus, tribological guides seem to be well suited for MEMS applications, due to the fact that the guide mechanism is relatively simple and the structures can be fabricated within a sequence of photolithographic and etching steps. In addition, the guide structures provide a stable and robust support for the moving structures.

Most commonly used substrate materials show relatively high friction, especially on the microscale. Several approaches have been introduced to improve the tribological behavior of sliding motion, such as coating of wet or dry lubricating films [24, 3]. Wet lubricating films are widely applied in mechanical engineering. Since on the microscale the contact surfaces are extremely smooth and the gap between the surfaces is very close, capillary force and viscosity of wet lubricating films can negatively affect the friction behavior. Alternatively, dry lubrication films with very low friction and high wear resistance seem more suitable for microscale applications. Nowadays dry lubricating coatings are established in many macroscopic applications like tools or aerospace.

Many coatings are deposited using vacuum technology. These processes are also compatible with typical MEMS processing. Using vacuum technology nearly any material combination can be deposited. Fundamentals of the friction behavior of microparts and improvement by use of coatings and micropatterned surfaces will be described in the following sections. Process parameters for film deposition and fabrication of microcomponents are chosen according to the tribological demands. Furthermore, effects of testing parameters and topography of contact surface are investigated.

6.2 Preparation of Microguides and Bearings

Microguides are inevitable parts of active microsystems. Due to the size of microsystems, the achievable driving force of microactuators is very low. For this reason friction of the guiding parts has to be reduced as much as possible. As a demonstrator, micro linear motors using the electromagnetic principle have been developed and fabricated (see chapter 11). Various microguides and coating processes were designed and developed in order to study the tribological behavior and to reduce the static as well as the kinetic friction between the moving parts.

6.2.1 Vacuum Coating

For the improvement of the friction coefficient of sliding microparts vacuum coating processes were used. The general description of the processes of the different coatings used for microtribological applications is found in chapter 5.

6.2.2 Tribological Microguide

According to fabrication processes, the microguides can be created using various micromachining techniques, such as wet chemical etching, plasma dry etching and electroplating. The tribological guide constructions are categorized into sliding guide and ball bearing.

Sliding Guide

Due to its mechanical properties and micromachinability, single crystalline (100)-silicon has been proved as an appropriate material for linear microguides [4]. The linear guide sample consists of two main parts, namely a base plate and a sliding plate. In the first design, the sliding as well as the base plate were fabricated by wet chemical etching using KOH. The guide structures were etched into the form of linear tracks. Due to the anisotropic properties of silicon, typical V-grooves with 150 µm heights and extremely smooth surface on (111)-planes were generated. In this design, the (111)-planes were used as contact surface. Furthermore, the etched V-grooves allowed a self-alignment between the base and sliding plates, which assured minimum lateral clearance between the plates and allowed the slider to move only in one direction (Fig. 6.1).

Tribological testing on nanoscale shows that the friction force varies with the size of the contact area [18]. To investigate the effect of the contact area on microscale, the contact area of track structures was reduced to 50% and 33% of the entire contact surface. Fig. 6.2 shows a linear guide with 50% of contact surface. After fabrication, the sliding plate and the base plate were coated with a 500 nm thick dry lubricating layer (a-C). The tribological characterizations are described in section 6.3.

However, the reduction of the contact surface by wet chemical etching is limited. Structures with a very small contact area cannot be properly fabricated, because the surface may be completely etched away. Therefore, the form of track structures fabricated with KOH is restricted to rectangular shapes. To fabricate the various shapes of the guides, an Inductively Coupled

Plasma (ICP) etching process was applied, whereby the guide structures can be fabricated independently from the crystal orientation (Fig. 6.3). Cylindrical support structures with a thickness of 20 µm were etched into the base plate using ICP. The top surface of the support structures is used as contact surface. Circular micropatterns with a diameter of 5 µm and thickness of 2 µm fabricated with plasma etching techniques were applied to the reduce contact area (Fig. 6.4).

A similar guide construction can also be made of SU-8. A thick layer of 20 µm of SU-8 is used for support structures. A thin SU-8 layer of 4 µm is applied on the top of the support structures for the micropatterns. The fabricated guides are coated with 500 nm thick a-C dry lubricating film to improve the tribological property. The characterization of the linear guide will be described in section 6.4.

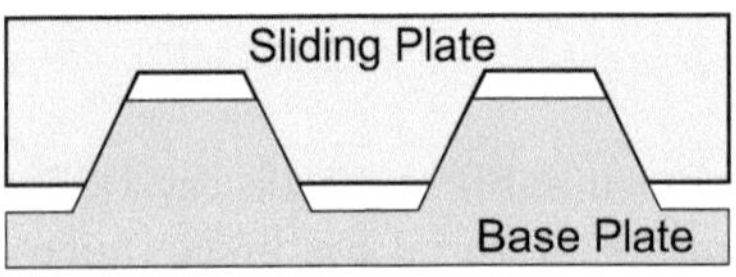

Fig. 6.1 Construction of self-aligned linear guide structure fabricated by KOH etching

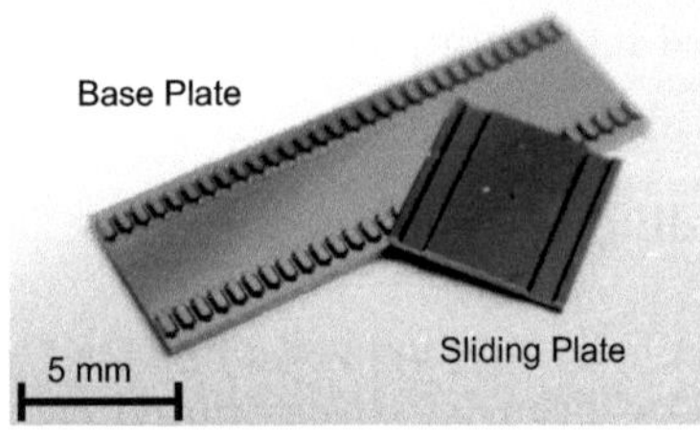

Fig. 6.2 Track structure of a linear guide with 50% contact area

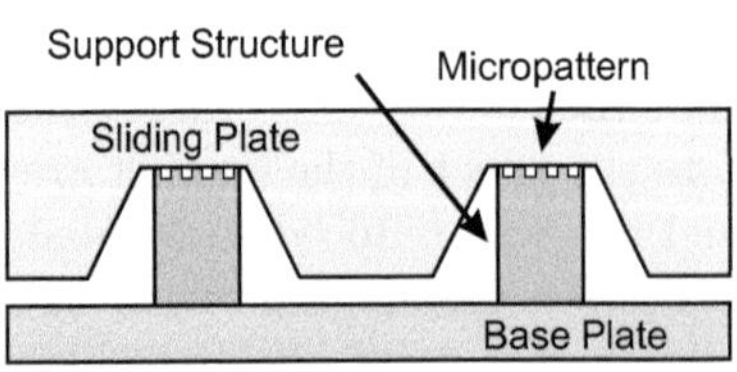

Fig. 6.3 Silicon linear guide structure fabricated using ICP

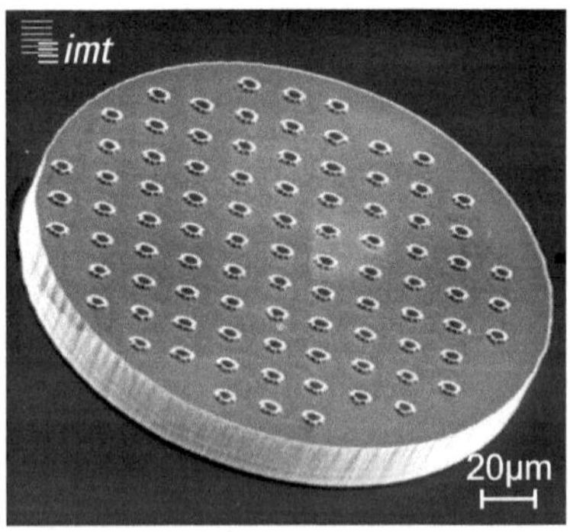

Fig. 6.4 Cylindrical support structure with a diameter of 200 µm and 5 µm circular micropatterns to reduce contact area

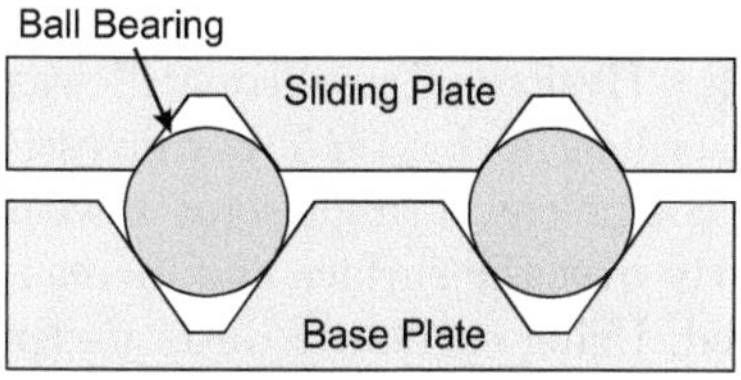

Fig. 6.5 Construction of a silicon linear ball bearing

Fig. 6.6 Rotatory ball bearing with silicon substrate and SU-8 microgear

Ball Bearings

Guide samples with ball bearings were developed to compare their tribological properties with the sliding principle. Alumina balls with a diameter of 0.5 mm were used as support elements. For the linear guide systems, the two parallel V-grooves on the sliding plate, as well as on the base plate were fabricated by KOH etching. The balls were then placed into the V-groove tracks on the base plate, so that at least four balls were below and supporting the sliding plate (Fig. 6.5).

Additionally, rotatory ball bearings were designed and fabricated. Circular grooves for the ball bearing can be made of silicon or SU-8. The groove dimensions have to be accurately calculated, so that the balls exactly fit into the grooves with an appropriate amount of clearance. For example, nine 0.5 mm balls are required to fit into a circular groove with a middle diameter of 1.02 mm and with an overall clearance of 20 µm. An oversized or undersized groove size can lead to a stick-slip and thus a high friction force. For the fabrication of the silicon circular grooves, plasma dry etching of a silicon wafer was applied. On the other hand, a thick layer of SU-8 can also be used to fabricate ball grooves. Fig. 6.6 shows a silicon ball bearing with a diameter of 1.02 mm and a SU-8 microgear as a rotor.

6.3 Characterization

Various measuring systems have been developed to investigate friction forces and friction coefficients. Due to the measurement principle, resolution, and driving range, a measuring system is only appropriate for a particular testing configuration. Therefore, different measuring systems are required for characterizing diverse microguides and coating systems.

6.3.1 Microfriction under Single Asperity Contact

For the investigation of the friction under single asperity contact an atomic force microscope (AFM) in combination with a Hysitron TriboScope^{TM} was used. The principle of the scratch tests is described in chapter 5. For investigation of the load dependence of the friction coefficient, a single scratch with increasing load was used. While moving the tip over the surface and increasing the load, the lateral forces were measured. Using oscillating tests under constant load, the time evolution can be studied by varying the number of cycles as well as the applied load.

6.3.2 Oscillating Friction Tester

The oscillating friction tester described in chapter 5 was additionally equipped with a quartz sensor for detection of the lateral forces, i.e. the resulting friction forces in oscillating motion. The application of the normal load was realized either through magnets, or after modification by using deadweights. The counterpart is moving and the friction force is measured on the sample [21].

6.3.3 Flat-Flat-Contact Friction Force Measuring System

Macroscopic friction measurements with oscillating or rotating driving systems require a large sample surface with a normal load of several newtons. A drawback of these systems, especially for microscopic measurements, is that the coating film with a thickness of several nanometers can be damaged after a couple of measurement cycles. The investigations of microfriction using macroscopic measuring systems are therefore very limited. AFM are widely used for investigations of mechanical and tribological properties of materials. Due to the high resolution in the lateral direction, the friction measurements can be performed on small areas with a submicrometer measurement range. Furthermore, there are no standardized measuring systems for the investigations of the tribological properties of microsystems which cover the range between nano and macroscale. Because of these facts, friction force microscopes have been developed for fundamental investigations of the tribology of microparts.

A Friction Force Measuring System (FFMS) has been constructed to measure friction force under flat-flat-contact particularly in linear oscillating movement. Due to the fact that the size of microparts can vary from the submillimeter range to some micrometers, the FFMS has to be flexible and to allow various parameters, e.g. normal load, moving speed, contact area and moving distance, to be set. Particularly for the tribological investigation of linear microsystems, the FFMS can perform friction testing with a driving

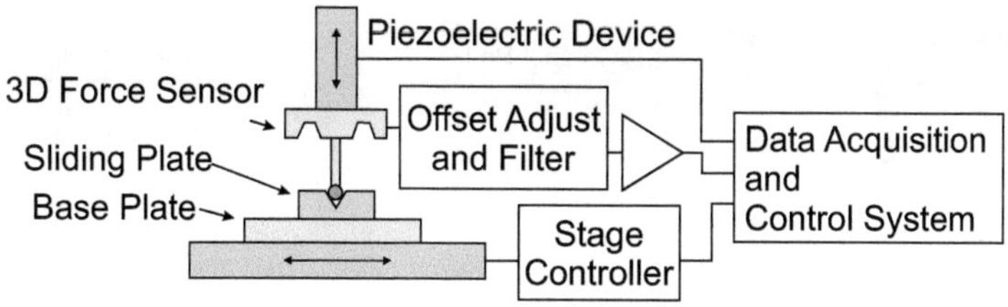

Fig. 6.7 Schematic construction of the Friction Force Measuring System (FFMS)

speed of some $\mu m\, s^{-1}$, and detect static as well as kinetic friction force with the accuracy of several micronewton.

The main part of the FFMS consists of a motorized linear stage (IntelliStage M-531, *Physic Instruments*), a piezoelectric device (MT60, *Owis*), a 3D piezoresistive micro force sensor and a data acquisition system (Fig. 6.7). The motorized linear stage is used for oscillation. For the measurements, a base plate is mounted on the linear stage, and then an opposite sliding plate is placed on top of the plate. The sliding plate is held by the probe's tip of the force sensor, while the stage and the base plate are moving. The normal force can be controlled by pushing the sliding plate with the piezoelectric device via the probe's tip. The resulting lateral friction forces are detected. By using the 3D force sensor, the normal load and friction forces can be simultaneously measured [15].

The key component of FFMS is the 3D micro force sensor, which allows force measurements with high accuracy (see also chapter 9). The sensor is based on a boss membrane structure fabricated using bulk micromachining The overall size of the sensor is 6.5 mm x 6.5 mm. Piezoresistors are fabricated by diffusion into the membrane's backside and then are connected to Wheatstone's bridges. A 5 mm long stylus with a 300 µm ruby ball at its end is fixed in the center of the boss. An external force exerted on the probing ball will cause a membrane deflection and induce a change in resistance, which depends on the magnitude and direction of the force [16].

The first prototype of the force sensor comprised 24 n-piezoresistors and the bridge connections were arranged on an external circuit board (Fig. 6.8 a). Each bridge corresponded to the x, y and z direction, respectively. The construction of the entire sensor was therefore relatively large and heavy, so that it was not appropriate for the characterization of microparts. Furthermore, due to the large probe size, the monitoring of the measured positions and samples was limited.

A further development of the sensor was the integration of the bridge circuits on the membrane structure. The number of contact pads were reduced from 48 to 12, and the size and weight of the sensor probe could also be drastically reduced [19]. However, due to the layout and the number of the piezoresistors, the crossing of metal conductors could not be avoided, so that two metal layers were required which caused a complication in the fabrication

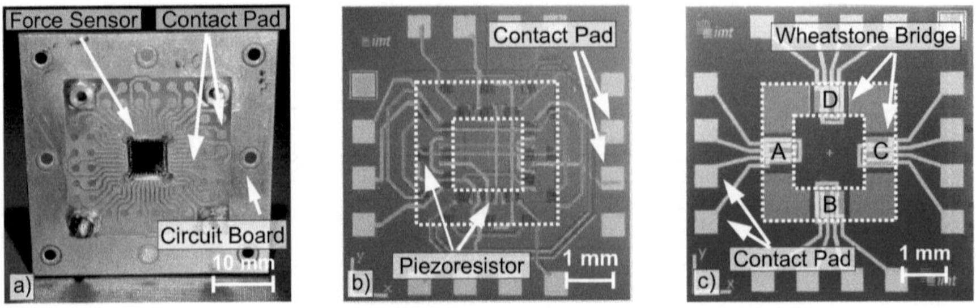

Fig. 6.8 Different designs of three dimensional force sensors with: (**a**) an external bridge connection; (**b**) three bridges on-chip connection and **c** four bridges on-chip connection (A, B, C, and D refer to piezoresistor bridge circuits on each side of sensor's membrane)

(Fig. 6.8 b). To simplify the layout and increase the production yield, force sensors with four bridges have been developed. Full Wheatstone's bridges were placed on four edges of the membrane. To obtain higher sensitivities, p-piezoresistors were used instead of n-resistors (Fig. 6.8 c). Table 6.1 shows the sensitivities of the force sensors with full boss membranes for vertical (S_z) and horizontal (S_x and S_y) directions. Compared to the previous sensor designs, the optimized sensor sensitivities can be drastically increased about nine times and 50 times for the horizontal and vertical probing directions, respectively.

Table 6.1 Sensitivities (mV (V mN)$^{-1}$) of the force sensor with different circuit layout

Layout	Vertical sensitivity (S_z)	Horizontal sensitivity (S_x and S_y)
Three bridge design	0.016	0.28
Four bridge design	0.79	2.57

A further development of the force sensor was the improvement of the isotropic properties in all probing directions. The typical boss membrane structure shows a difference of stiffness between vertical and lateral probing directions by a factor of 40. To balance the stiffness of both probing directions, the edges of full membrane was through etched to create a cross membrane structure. Subsequently, the cross membrane structures were extended and collateral stiffening structures were added in the membrane structures (Fig. 6.9). The membrane stiffening causes an increase in the torsion stiffness and consequently the horizontal stiffness, but has no influence on the vertical stiffness. As a result, the ratio of the vertical to the horizontal stiffness can be reduced to a factor of 6 [17].

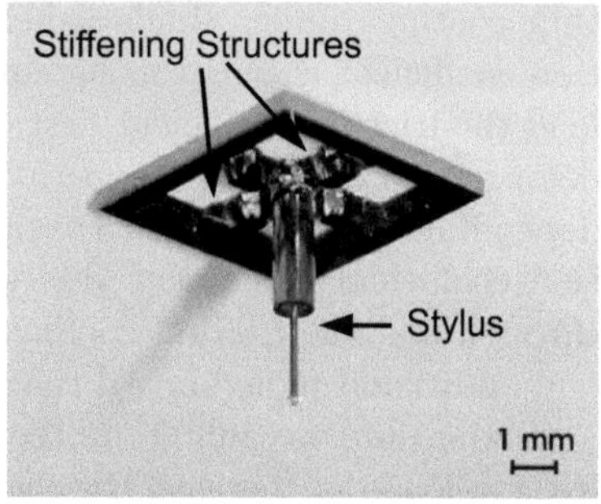

Fig. 6.9 Force sensor with modified boss membrane for an improvement of the isotropic characteristics

For application as a force sensor in FFMS, the sensors have to be characterized and calibrated, and the probing sensitivities have to be determined. The characterization and calibration procedures are described in chapter 9.

6.4 Results

Using the different described characterization methods and tools general trends in microfriction as well as the behavior in different applications is described. The first part of this section focuses on the effect of coating systems and surface modification regarding microfriction. The second part presents the friction behavior of various microguides as well as microactuators.

6.4.1 Microfriction

Models for Microfriction

For macroscopic applications the coefficient of friction is commonly defined as a dimensionless scalar, describing the ratio of the friction force between two bodies under a defined applied load. The friction coefficient is usually correlated to the mating materials. For dry friction Amontons' law is used. The law states that the friction force is directly proportional to the applied load (1st law) and that the friction force is independent of the contact area (2nd law).

$$\mu = \frac{F_F}{F_N} = const. \tag{6.1}$$

F_F is the friction force and F_N is the applied normal force. Moving from multi to single asperity contact and using the Hertzian elastic theory, a non-linear friction-load dependence is found, contradicting Amontons' law. For thin coatings under single asperity contact a load dependence of the friction coefficient is found using single scratch tests. Depending on the applied load the friction coefficient first decreases with increasing load due to elastic deformation of the coating-substrate-system. After reaching a critical load, depending on the applied coating system, the slope changes and the friction coefficient rises again due to inelastic effects. Finally, the film is worn through and the friction coefficient is determined by the substrate material and additional ploughing of the probe tip [23, 22].

Taking into account that the single asperity contacts show a non-linear friction-load dependence, the theory of Greenwood and Williamson describes the direct proportionality of the friction force to the applied force by a modification of the contact area [9]. Since contact in technical surfaces is defined by the number of asperities under contact, the number of asperities will increase with increasing load confirming Amontons' law.

Friction Reduction by Different Coating Systems

Besides the definition of the friction coefficient by the mating materials, the surfaces were coated with various thin film materials. The friction coefficient of the films were determined for multi asperity contact by a macroscopic pin-on-disc test using different counterparts, and under single asperity contact using an Hysitron nanoindentation system with diamond tips as counterparts by scratch tests (table 6.2) [2].

Table 6.2 Friction coefficient of coating systems for different counterparts using pin-on-disc (Ø: 4.76 mm, 3 N) and AFM based methods (Hysitron) (Ø: 1 μm, 1 mN)

Coating	Counterpart (pin-on-disc)			Hysitron
	steel	Al_2O_3	DLC	Diamond
Al_2O_3	0.89	0.54	0.15	0.11
cBN	0.46	0.45	0.12	0.10
CN_x	0.26	0.17	0.11	0.12
DLC	0.15	0.10	0.09	0.07
MoS_2(Ti)	0.15	0.13	0.06	0.22

Load Dependence for Microcontacts

Describing friction under microtribological conditions, the resulting theory must be chosen depending on the contact situation. To study the influence of the applied normal force on the friction force, a flat punch, i.e. a diamond indentor with a flat tip, and an oscillating friction tester were used [1]. In the case of the flat punch a flattened tip with a diameter of 46 μm was used

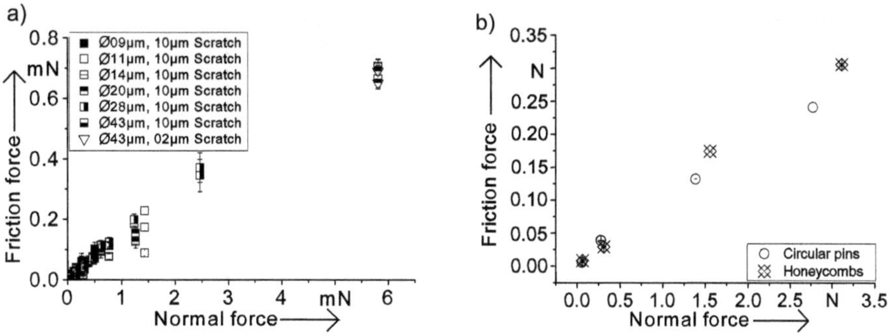

Fig. 6.10 Friction force over normal force; counterpart: microstructured silicon with 130 nm a-C: (**a**) flat punch with Ø 46 µm; (**b**) oscillating friction tester

(Fig. 6.10 a). The counterpart was a microstructured silicon sample with pin structure and varying pin diameter, coated with 130 nm a-C. For the oscillating tests the samples also consisted of microstructured silicon with pin and honeycomb structures and 130 nm a-C, tested on polished silicon with 130 nm a-C (Fig. 6.10 b) [14]. In both cases, a linear dependence of the friction force on the normal force was observed, in contrast to an expected load dependence for the flat punch. Therefore, in accordance to the contact theory the real contact area was calculated [10]. For real contact areas significantly smaller than the nominal contact area, the models from Greenwood and Williamson were applied [9]. After calculation, a ratio of $3 \cdot 10^{-6}$ of $A/A_{nominal}$ was found, even for the flat punch [14].

Influence of the Resulting Contact Area

By the different microstructure pattern, the nominal contact area was varied. While a linear dependence of the friction force on the applied normal force occurs for a given structure under oscillating friction tests, the variation of the nominal contact area, i.e. the pattern size, shows a different behavior. Since the load dependence of the friction force shows a linear dependence on the resulting friction force, a varied contact area under given normal force should be constant. In contrast to the expectation, the oscillating friction tester shows a dependence of the friction force on the nominal contact area. It was previously mentioned that the friction coefficient is constant for both, AFM based method and oscillating test. Thus, the reason for the dependence on the contact area must be different.

The deviation from ideal flat-flat contact of the microstructured silicon wafer is in the range of a few tens of micrometers. Therefore, the contact to the surface is realized only by a small fraction of the pins on the surface instead of the nominal contact area available. Involving only a fraction, e.g. 5% of the

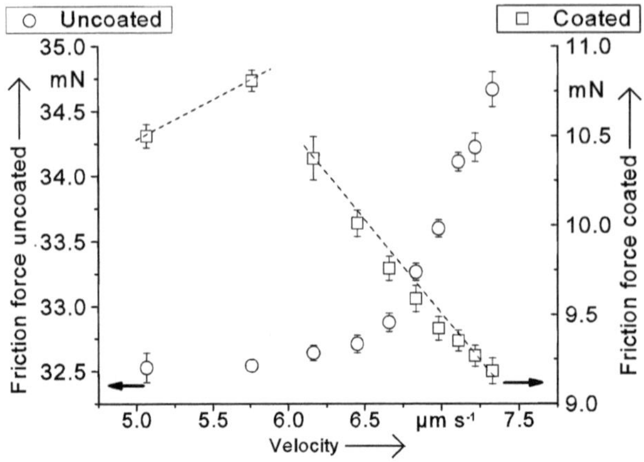

Fig. 6.11 Velocity dependence of microstructured samples with 20% nominal contact area; uncoated silicon, coated silicon with 450 nm a-C, 10.0 kPa contact pressure

nominal area, will lead to a significant increase of the local pressure by orders of magnitude. In that range the contact area is no longer compensating the increase of the local pressure, but resulting in nearly full face contact [13].

Stick-Slip and Adhesion Phenomena

Comparing uncoated and coated samples in oscillating motion the resulting friction behavior differs significantly. Figure 6.11 shows the velocity dependence of microstructured samples with 20% nominal contact area. For the uncoated samples, i.e. silicon running against silicon, the measured friction force was increasing with increasing load. Under dry conditions, after short running in time, stick-slip occurred. With increasing velocity also the stick-slip phenomena increased drastically in combination with increased wear and debris formation.

For coated samples the resulting friction force was reduced by a factor of three, compared to pure silicon. For slow motion the friction force increased with increasing velocity due to stick-slip. Further increase of the velocity led to a continuous reduction of the resulting friction force due to influence of the wetting behavior of the films. The reason for the decreasing friction force with increasing velocity were meniscus forces. For hydrophile surfaces the friction force will decrease with increasing velocity [20]. Measurement of the water contact angle confirmed that the deposited carbon films were hydrophile.

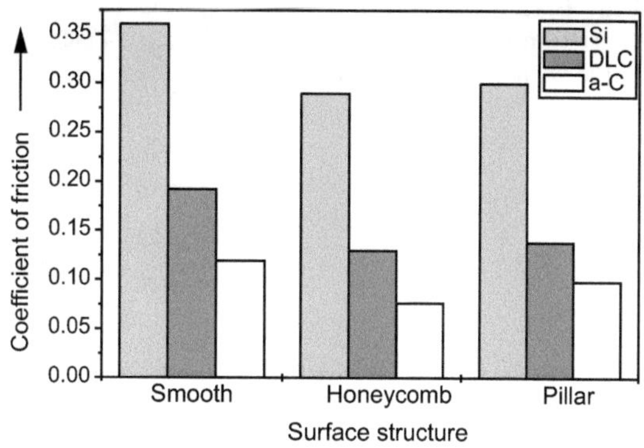

Fig. 6.12 Average coefficients of friction of silicon samples with and without DLC or a-C coating layer including various surface structures

6.4.2 Friction of Microguides

Microguide

The tribological investigation of microparts was performed using FFMS, while the influences of contact area, form of the micropatterns and carbon based dry lubricating films were considered. To study the effect of the lubricating film and contact surface, smooth silicon samples with a contact area of 1 mm^2 were tested on a smooth and on a patterned silicon plate. Their contact surfaces were coated with the same coating material, either a-C or DLC. The thickness of the coating film was 500 nm. The sliding speed was set to 0.1 $mm\,s^{-1}$ with a normal load of 30 mN. Fig. 6.12 shows the test results, whereas the coefficient of friction was calculated from the resulting average friction force and the applied normal load on the samples. The tests on micropatterns – honeycomb as well as pins – showed less friction, compared to uncoated samples. The DLC dry lubricating coating layer can significantly reduce the coefficient of friction by about 50% compared to uncoated samples. A further reduction of the coefficient of friction by more than 60% was obtained by using an a-C coating. In addition, a high static friction force can be observed for uncoated samples. With the dry lubricating layer, the static friction can be considerably reduced to the same level of kinetic friction force.

Fig. 6.13 shows the coefficient of friction of silicon linear guide samples fabricated by KOH etching with different contact areas, as described in section 6.2.2. The sign of the coefficient of friction, as shown in the graph, corresponds to the driving direction of the samples. All samples were coated with 500 nm thick a-C dry lubrication layer. The sliding speed was set to 100 $\mu m\,s^{-1}$ with a normal load of 30 mN. The coefficient of friction was decreased by 30% by reducing the contact area from 100% to 33%.

Due to technological constraints, the reduction of the contact surface with wet chemical etching is very limited. For this reason, a plasma dry etching process was applied to fabricate the guide structure and micropatterns. Fig. 6.14 shows the tribological characteristic of the microguides for different

Fig. 6.13 Coefficients of friction of silicon linear guide with various contact surface

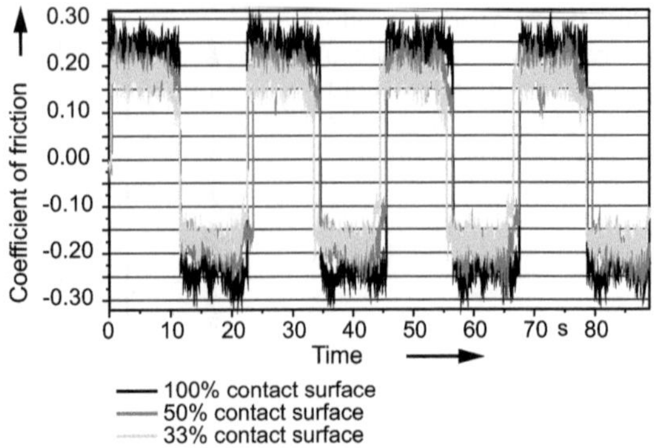

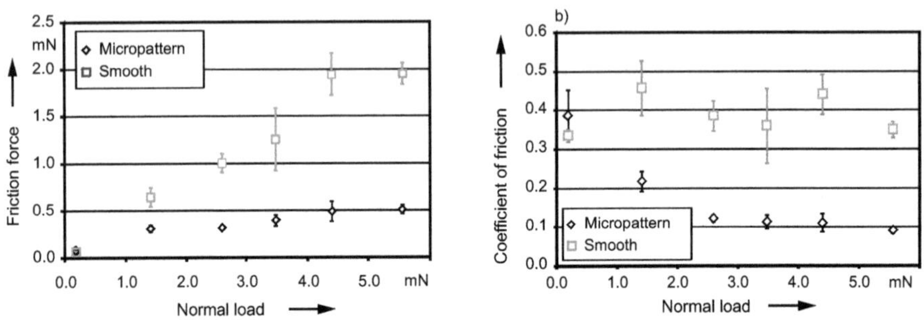

Fig. 6.14 Tribological characteristic of silicon linear guide with 200 μm diameter support structures with micropattern and smooth contact surface: (**a**) friction force; (**b**) coefficient of friction

normal loads. The friction force of the guides with smooth contact surfaces increases considerably faster than the guides with micropatterns when loads rise. The average coefficient of friction of the samples with smooth surfaces are nearly constant and stay in a range between 0.3 to 0.5. On the contrary, the coefficient of friction of the sample with micropatterns declines gradually at the beginning and remains constant at 0.1, when the load is higher than 3 mN. Theoretically, the micropatterns afford the removal of the particle out of the contact surface, and as a result, the coefficient of friction can be reduced. Nevertheless, the results show that the removal of particle can be found, when the normal load is high enough.

There are various applications of SU-8 in the production of microsystems, e.g. for galvanic molding, or electrical isolation [7]. SU-8 guide samples with circular support structures and micropatterns were fabricated. Depending on the configurations of the structure, the coefficient of friction of the samples can be varied (Fig. 6.15). The uncoated samples feature a very high coefficient of friction of 0.7. Similar to the results of the silicon guide samples, the friction decreases by using micropatterns. A 50 nm thick Ti-interlayer was applied before the a-C layer was deposited on the samples. SU-8 has a very high elasticity, but the Ti-interlayer can prevent the deformation of the SU-8,

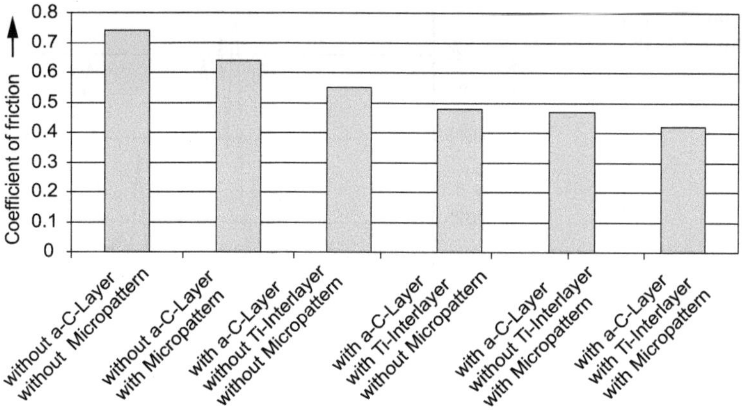

Fig. 6.15 Average coefficients of friction of SU-8 linear guides

hence the samples with Ti-interlayers show a lower coefficient of friction than the other samples.

The lowest coefficient of friction of SU-8 samples was determined to be 0.4, which is still too high for applications in microsystems. Due to the high shear stress at the contact surface, the SU-8 micropatterns are strongly damaged after a few driving cycles, which causes the rapid increase in the friction force.

Friction of Ball Bearings

The tribological behavior of linear ball bearings was also investigated. The silicon substrate and sliding plates were etched using KOH in order to create V-grooves. Ruby balls with a diameter of 0.5 mm were placed into the grooves, as described in section 6.2.2. In spite of a very low friction force, a high stick-slip effect of the sliding plate was observed, which was a result of the topography of the V-grooves, the ball's roundness, and the surface interactions between the balls and the grooves (Fig. 6.16).

The comparison of the maximum and average friction force of linear ball bearings is shown in Fig. 6.17. The maximum force is the break-through force in a sticking phase, before the sliding plate slips over. For friction tests, the normal load from 0.2 to 3.4 mN is applied. The maximum friction force is approximately five times higher than the average friction force for the applied loads. Furthermore, the average, as well as the maximum forces, are proportional to the normal load.

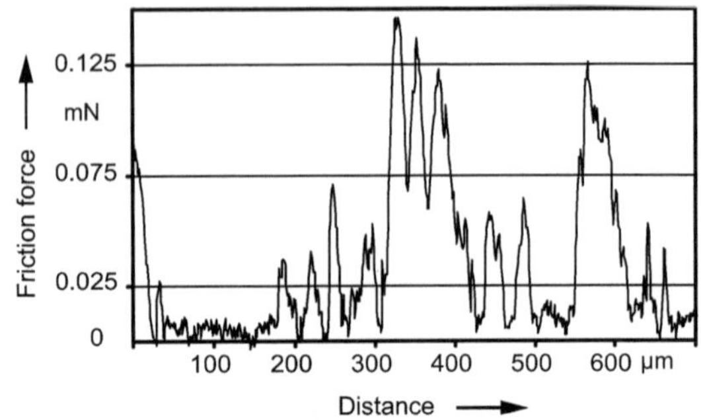

Fig. 6.16 Friction force of silicon a linear ball bearing

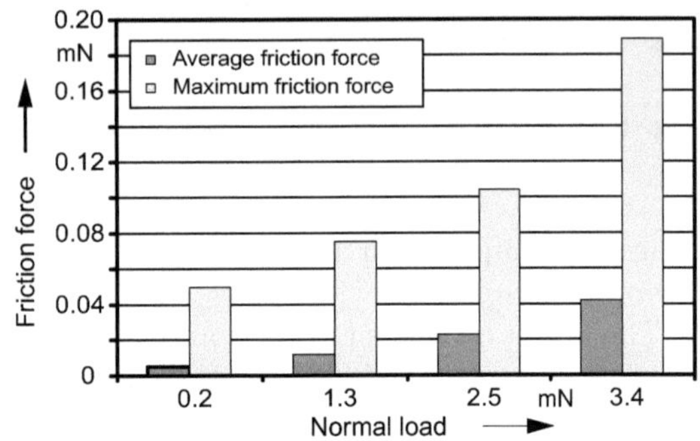

Fig. 6.17 Average and maximum friction force of silicon linear ball bearing with various normal load

Friction of Integrated Guides of Microsystems

Micro electromagnetic linear actuators using a horizontal magnetic field have been developed as described in chapter 12. To reduce the assembly steps and increase the process yield, tribological guide structures were integrated into the systems. For friction testing, the stator was mounted on the driving stage, and the probe tip was placed at one edge of the slider. The driving speed was set to $100\,\mu\mathrm{m\,s}^{-1}$ and no additional normal load was exerted onto the slider. Fig. 6.18 shows tribological characterizations of the actuator with and without a 500 nm thick dry lubricating film. A significant reduction of the friction force, by more than 50%, can be determined.

Tribological Properties of Rotary Actuators

In comparison to the micro linear actuators, tribological properties of rotary systems were also determined. The friction of the rotary systems is defined by a friction moment. SU-8 rotors were fabricated in form of a microgear (Fig. 6.19). To measure the friction force, the FFMS probe tip was attached to a gear tooth. According to the attachment point and measured friction force, the friction moment can be calculated. Fig. 6.20 shows the friction moment of the SU-8 bush bearing with a shaft diameter of 1 mm. An average friction

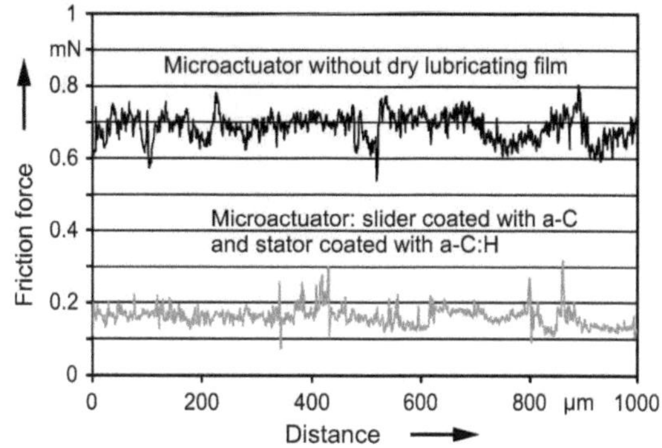

Fig. 6.18 Friction force of linear motor with and without dry lubrication coating

Fig. 6.19 SU-8 microgear with bush bearing

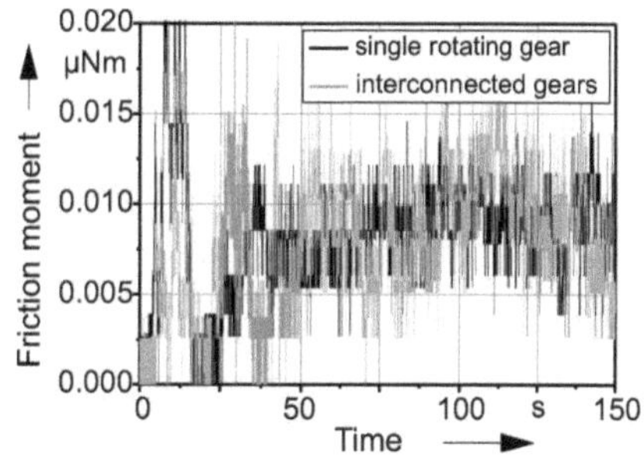

Fig. 6.20 Friction moment of single and interconnected gear

moment of 0.07 μN m for the bush bearing was determined. Furthermore, a high stick-slip was observed, approximately twice the average moment. In the case of interconnection of two gears, a drastic increase of the average friction moment was found. The resulting friction moment of interconnected gears was approximately three times higher than the single rotating gear.

Similar to the measurements with linear ball bearings, as described in section 6.4.2, the rotary ball bearing features a very low friction (Fig. 6.21). With the same measuring parameters as used for the bush bearing, the ball bearing shows a friction moment lower than 0.02 μN m and no stick-slip was observed. In case of interconnection of two microgears, the friction moment was nearly equal to that of the single rotating gear (Fig. 6.22).

Various types of micro rotary motors were designed and fabricated as described in chapter 11, and their tribological properties were subsequently determined. In general, the friction moment of micro rotary motors depends on the rotor weight. The friction moment can be as low as 0.1 μN m, but apart from the interaction of contact surfaces and rotor weight, the fitting clearance also plays a significant role in the frictional responses. Fig. 6.23 a shows the surface profile of three identical rotors on a identical stator. Due to the geometric deviation of some micrometers, rotor A and B have surface contact not only on the shaft in the middle, but also at the outer circumfer-

Fig. 6.21 SU-8 microgear with ball bearing

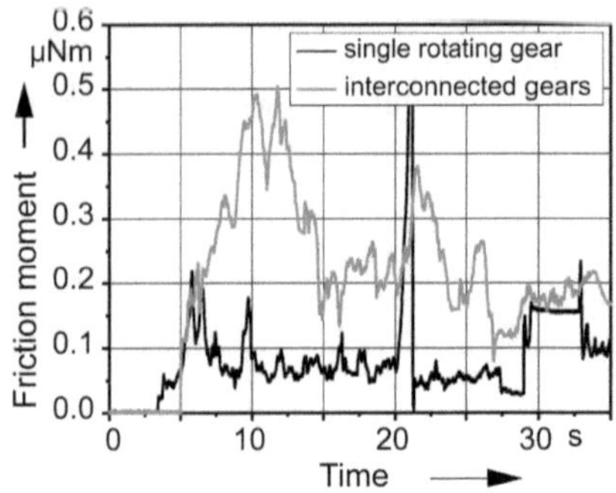

Fig. 6.22 Friction moment of single and interconnected gear

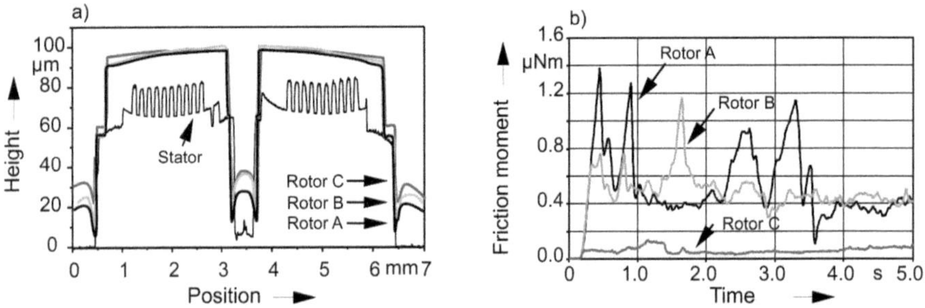

Fig. 6.23 (**a**) Comparison of cross section profile of various rotors on a stator; (**b**) Friction moment of the rotors

ence of the stator. In contrast, rotor C touches the stator only in the middle. Thus, rotor A and B show a high friction with average friction moment of approximately 0.4 µN m. stick-slip behavior, around three times of the average friction moment, were also observed (Fig. 6.23 b).

6.5 Conclusion

For microparts in relative motion the resulting friction coefficient plays a crucial role for the improvement of reliability and durability of the active microsystems. The reduction of the resulting friction coefficient and the minimization of stick-slip phenomena can be realized by a combination of thin film coatings and micropatterning of the sample surface. According to the driving force of micromotors, the increase of the friction force and stick-slip directly affect the function and the reliability. Thus, surface modifications as well as geometrical optimizations regarding the tolerance of microparts are the key aspects to optimize the friction behavior of the active microsystems.

Acknowledgements The authors acknowledge the contributions of several colleagues involved in the Collaborative Research Center 516 "Design and Manufacturing of Active Micro Systems": Cord Henke, Rolf Küster, Holger Lüthje, Dominik Paulkowski, Martin Schmidt, Jan-Hinrich Sick, Thorsten Staedler and Andreas Wortmann. Additionaly the authors gratefully acknowledge the funding of the reported work by the German Research Foundation within the Collaborative Research Center 516.

References

[1] Arfsten J, Kampen I, Kwade A (2009) Mechanical testing of single yeast cells in liquid environment: Effect of the extracellular osmotic conditions on the failure behavior. International journal of materials research 100(7):978–983

[2] Bandorf R, Lüthje H, Schiffmann K, Staedler T, Wortmann A (2002) Sub-micron coatings with low friction and wear for micro actuators. Microsystem Technologies 8(1):51–54

[3] Bandorf R, Lüthje H, Henke C, Wiebe J, Sick JH, Küster R (2005) Different carbon based thin films and their microtribological behaviour in MEMS applications. Surface and Coatings Technology 200

[4] Bhushan B (1999) Handbook of Micro/Nanotribology. CRC Press LLC

[5] Denkena B, Li J (2005) Untersuchung einer magnetischen Mikroführung – Modellierung und Simulation. wt Werkstattstechnik online 5

[6] Denkena B, Li J, Kopp D (2004) An aerostatic linear guidance for microsystems. Annuals of the German Academic Society for Production Engineering (WGP) XI/2

[7] Feldmann M (2007) Technologien und Applikationen der UV-Tieflithographie: Mikroaktorik, Mikrosensorik und Mikrofluidik. Shaker

[8] Feldmann M, Ruffert C, Gatzen HH, Büttgenbach S (2005) Fertigung von Funktionskomponenten für elektromagnetische Mikroaktoren. In: Kolloquium Mikroproduktion – Fortschritte, Verfahren, Anwendungen

[9] Greenwood J, Williamson J (1966) Contact of nominally flat surfaces. Proceedings of the Royal Society of London Series A, Mathematical and Physical Sciences pp 300–319

[10] Johnson K (1987) Contact mechanics. Cambridge Univ Pr

[11] Kumar S, Cho D (1990) A proposal for electrically levitating micromotors. Sensors and Actuators A 24

[12] Mehregany M, Nagarkar P, Senturia S, Lang J (1990) Operation of microfabricated harmonic and ordinary side-drive motors. In: Proceeding 3rd Annual IEEE Microelectromechanical Systems Workshop

[13] Paulkowski D (2009) Strukturelle Eigenschaften von dünnen amorphen Kohlenstoffschichten und ihre Auswirkungen auf Mikrotribologie und Deformationsverhalten. Fraunhofer Verlag, Dissertation

[14] Paulkowski D, Bandorf R, Schiffmann K, Bräuer G (2009) Friction on flat-flat micro contacts coated with amorphous carbon. In: Proceedings of GfT Tribologie Fachtagung, vol 2, p 51

[15] Phataralaoha A, Büttgenbach S (2004) Microscopic friction force measuring system for the investigation of micro components. In: 4th Euspen Conference

[16] Phataralaoha A, Büttgenbach S (2005) A novel design and characterization of micro probe based on a silicon membrane for dimensional metrology. In: Eurosensors XIX

[17] Phataralaoha A, Büttgenbach S (2006) A novel design and characterization of a monolithic three axial micro probe for dimensional metrology. In: Proceedings of APCOT06

[18] Phataralaoha A, Büttgenbach S, Schiffmann K, Sick JH, Bandorf R, Küster R (2005) Tribologische lineare Mikroführungen. In: Mikrosystemtechnik Kongress

[19] Pornnoppadol P (2004) 3D-Mikrotaster mit piezoresistiven Elementen. Shaker

[20] Riedo E, Lévy F, Brune H (2002) Kinetics of capillary condensation in nanoscopic sliding friction. Physical review letters 88(18):185,505

[21] Schmidt M, Wortmann A, Lüthje H, Büttgenbach S (2001) Novel equipment for friction force measurement on MEMS and microcomponents. In: Proceedings of SPIE, vol 4407, p 158

[22] Staedler T (2001) Mechanische und tribologische Charakterisierung dünner Schichten mit Hilfe rastersondenbasierter Verfahren. Fraunhofer-IRB-Verl.

[23] Staedler T, Schiffmann K (2001) Micromechanical and microtribological properties of thin CN_x and DLC coatings. Advanced engineering materials(Print) 3(5):333–337

[24] Sundararajan S (2001) Micro/nanoscale tribology and mechanics of components and coatings for MEMS. The Ohio State University

Chapter 7
Active Linear Guiding Concepts for Microsystems

H. Kayapinar, H.-C. Möhring, B. Denkena

Institute of Production Engineering and Machine Tools
Leibniz Universität Hannover
denkena@ifw.uni-hannover.de

Abstract High accuracy positioning of machines, below one micrometer, is often limited by conventional rolling or sliding guides due to friction that causes stick-slip effects. To ensure high-precision, frictionless and failsafe motion for microsystems, the application of active aerostatic and electromagnetic guides has been investigated and is presented in this chapter. In both cases, physical contact between moving microparts is avoided. Aerostatic guides offer high bearing capacity and stability and work free of self-heating. Electromagnetic guides can be designed more compactly and the air gap can be controlled more dynamically. Both concepts need an accurate displacement measurement system to control the air gap of several micrometers under variable loads. Due to the high degree of miniaturization, novel design, manufacturing, and calibration concepts were investigated.

7.1 Introduction

In contrast to macrosystems, friction can cause a complete restriction of movement in microsystems. One of the special characteristics of the systems that has been investigated in the Collaborative Research Center 516 is the small drive forces of about 1 mN, which can be 10 times lower than their normal forces caused by the weight of the moving slide and the reluctance forces of the electromagnetic actuator. Thus, handling the friction becomes very important to enable motion. Unlike tribological guides, active guides provide a frictionless operation of the actuator by levitating the slide. In the first part of this chapter, the design of active aerostatic guides for microsystems is presented. In the second part, the focus is on the design of magnetic guides based on reluctance forces. In the third part, the design of a displacement sensor is presented which is required by both the active aerostatic and the magnetic guide.

7.2 Active Aerostatic Guides for Microsystems

The application of aerostatic bearings and guides in precision engineering machines is well-proven. Thus, it is obvious to apply this technology to microsystems as well.

An aerostatic guide consists of multiple bearing sections. A simple aerostatic thrust bearing with central feed hole and pocket is shown in Fig. 7.1.a. Initially, the compressed air flows through a restricting nozzle into a pocket. It flows further through the air gap into the environment. The pressure built up under the slide acting on the bottom area represents the carrying bearing force. As the pressure distribution within the pocket is statically constant, the bearing force can be increased by increasing the pocket's diameter. But as the pocket's volume represents dead volume that can lead to the unstable "air hammer effect", the possibilities of increasing the bearing force are limited with this type of bearing. The flow resistance of the nozzle is independent from the gap width, whereas the flow resistance of the gap is inversely proportional to the cube of its width. As a result, the bearing force becomes dependent on the gap width. Thus, static stiffness is achieved. During dynamic movements the pocket acts like a gas spring stimulating vibrations. Stability is given as long as the dynamic stiffness (spring rate) is higher than the static stiffness. The basics of these types of aerostatic bearings can be found in [3].

7.2.1 Requirements

In addition to general requirements for guides like high carrying force, stiffness and damping, specific requirements result from the application in microsystems. While the air is supplied via the slide in macroscopic aerostatic

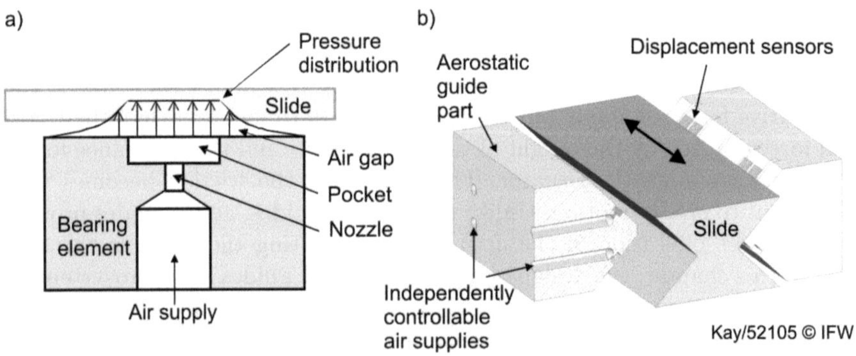

Fig. 7.1 (**a**) Simple aerostatic thrust bearing with central feed hole and pocket; (**b**) Principle of a miniaturized active aerostatic guide with air supply through the guide

guides, this approach cannot be applied to microsystems due to missing installation space on the slide and the resulting disturbances of pneumatic connections on the slide's movement. Due to the low weight of the miniaturized slide, the pressure has to be applied at opposing areas of the slide to reach the operating point at the maximum stiffness. In addition, the air gap should be kept adjustable. Furthermore, the guidance characteristics must not depend on the slide's travel and the air consumption must be kept low for economical operation.

To meet all the above mentioned requirements, a guidance concept according to Fig. 7.1.b is proposed. The air supply is installed in the stationary guide parts. It is possible to apply pressure on both sides of the x-shaped slide for prestressing. One of the stationary guide parts is fixed while the other one can be adjusted to obtain the air gap at the operating point. To avoid unnecessary air consumption, overlap of the slide's guiding surfaces and the bearing's surface has to be assured over the whole travel range. Therefore the slide is longer than the guides by the length of the travel range. To keep the air gap constant while the slide is moving or the load is changing, a closed-loop gap control is applied. Displacement sensors sense the current gap at eight positions while a digital control regulates the pressures of the independent air supplies feeding the aerostatic bearing sections.

7.2.2 Design and Properties of Aerostatic Microbearings

The dynamic behaviour of air bearings can be improved by reducing dead volumes. A well-known technique to obtain perfect properties for macroscopic air bearings is the application of porous ceramics. The microstructure of porous ceramics behaves like multiple nozzles with nearly no dead volume.

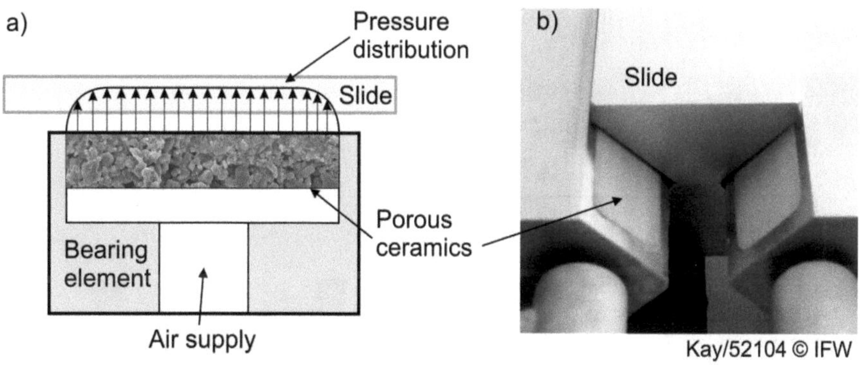

Fig. 7.2 (**a**) Principle of an aerostatic bearing with porous ceramics; (**b**) Manufactured aerostatic guide with sol-gel ceramics

Borosilicate glass provides a uniform porosity, with pore sizes down to 1–1.6 µm. By using double layered borosilicate glass, the throttling effect caused by the different pore sizes of the coarse layer (20–100 µm) and the sintered fine layer can be used for building aerostatic bearings (Fig. 7.2.a). With a preliminary up-scaled model of the guide, the aforementioned properties have been demonstrated, wherein the ceramic has been glued into metal holders (Fig. 7.2.b). Borosilicate glass is of limited use for cutting manufacturing. In machining, transcrystalline separation mechanisms have not been observed, but outbreaks of grains, plastic deformation and clogging of the pores.

The substitution of borosilicate glass by sol-gel ceramics was also analyzed. Due to the necessary grinding of the surface when using metal holders, all the fine-pored layers have been removed. Attempts to coat the sol-gel ceramics with finer layers subsequently, led to inhomogeneous permeability due to variable layer thicknesses. This effect has been further supported by absorption of adhesive that has been applied to seal different guide segments together.

By using discrete micronozzles, homogeneous air permeability can be ensured. One way to manufacture air bearings with micronozzles is shown in Fig. 7.3. Grooves that are ground into a solid ceramic carrier serve as an air intake. A thin layer of *Foturan* glass is patterned with micronozzles with a diameter of 20 µm. It is then glued to the surface of the carrier. The air flows through the grooves and the micronozzles to the guide's surface.

The cross sections of the grooves must be at least greater than the total sum of the cross sections of the nozzles, so that throttling takes place in the nozzles.

Another possibility to obtain a well-defined pressure distribution and to have good utilization of the bearing's surface area and good dynamics due to less dead volume at the same time is the application of the bypass technology [3]. Instead of using one large pocket (see Fig. 7.1.a), ideally, only small grooves are layed enclosing the whole bearing's surface. In order to adapt

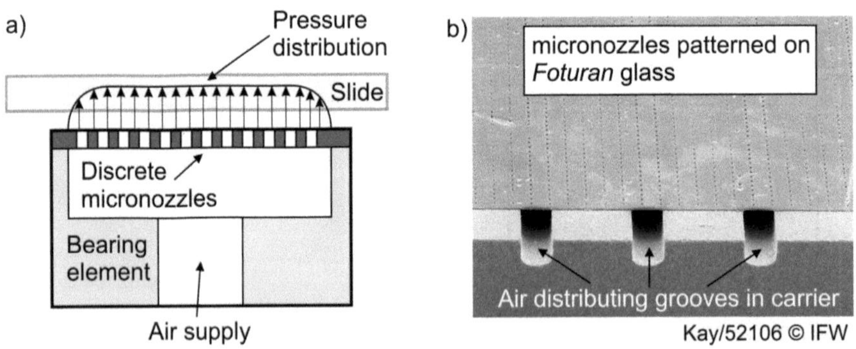

Fig. 7.3 (**a**) Aerostatic bearing with micronozzles; (**b**) Air bearing with micronozzles patterned on *Foturan* glass

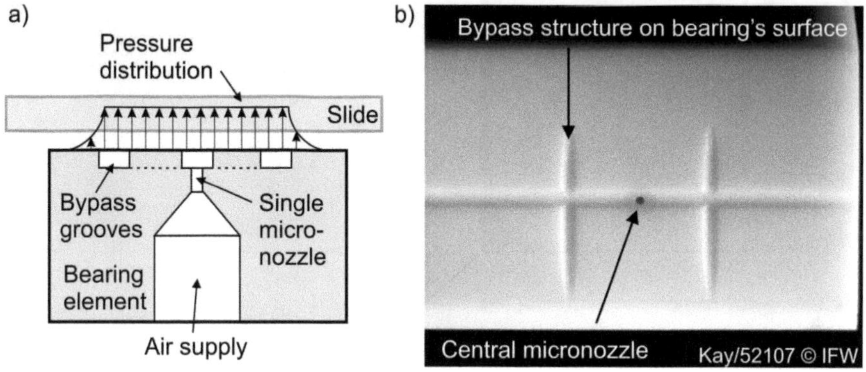

Fig. 7.4 (**a**) Aerostatic bearing using bypass grooves; (**b**) Manufactured bearing with branched grooves

to constructive and manufacturing requirements, branched bypass structures have been chosen for a first macroscopic guide design (Fig. 7.4.b). When designing an active aerostatic micro-guidance system, the entire system must be viewed holistically at an early stage. In the selection of the material and the design of the bearing structures, the ease of integration of a measurement system has been especially considered. Al_2O_3 has been chosen as a construction material. On the one hand, this electrically non-conductive ceramic material allows the direct deposition of sensor electrodes, which is to be further elaborated in Sect. 7.2.3. On the other hand, due to the high hardness, there is less wear during settling of the slide onto the guide.

The design process has been supported by Computational Fluid Dynamics (CFD) calculations. Due to the extreme differences in size of the simulated air gaps of 0–15 µm and the bearing dimension of several millimeters, the meshing process has been controlled manually to obtain exact results.

To study the properties of the designed air bearing, in particular compared to a pocket bearing (Fig. 7.1.a), Finite Elements Method (FEM) models of both bearings have been used. In order to have a direct comparison, the bearings' surfaces, nozzle geometries and pocket and groove volumes respectively were chosen equal. As expected, the simulation results showed 42% more load capacity and stiffness with the H-structure bearing compared to the pocket bearing at an air gap of 5 µm. The simulation results were furthermore used for parameter optimizations. With respect to the bearing capacity, optimal positions of the two transverse grooves could be found.

As explained in Sect. 7.2.1, a closed-loop controller is needed to keep the gap constant. The controller should be able to control all degrees of freedom of the slide except for the direction of movement. To obtain a controllability of the system, the individual bearing sections must be controlled independently. This can either be achieved by segmenting the guide [2], leading to a higher assembly and adjustment effort. Particular attention should be paid

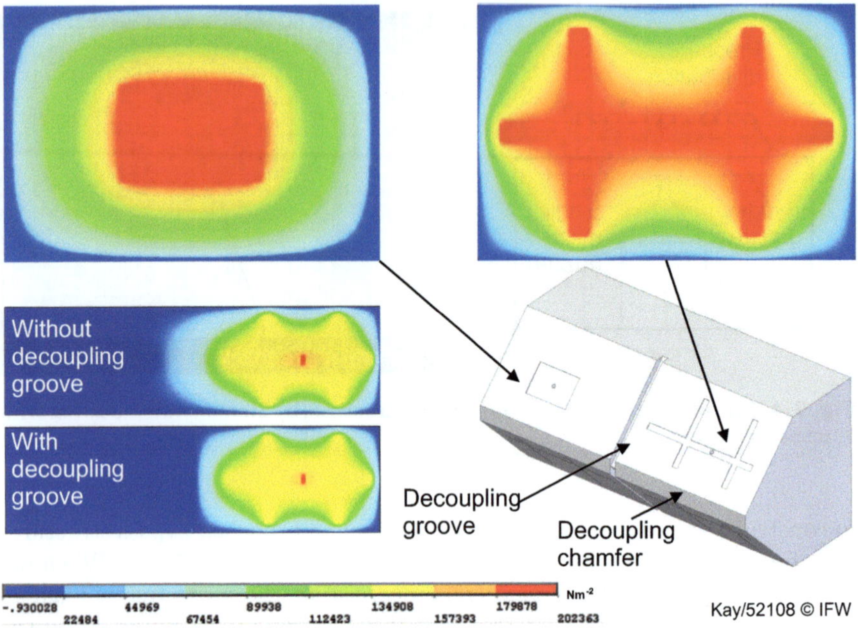

Fig. 7.5 CFD simulation results: (Top) Comparison of pressure distributions of rectangular pocket bearing and bearing with branched structures; (Bottom) Effect of the decoupling groove

to the decoupling groove in Fig. 7.5. It ensures that the pressure profile remains local, thus preventing interference of adjacent bearing areas. In the lower simulation-visualization of the pressure distribution, the effectiveness of decoupling can be seen. Segmentation can be avoided by the decoupling approach. s A macroscopic guide was designed to allow more complex structures and thus to examine the boundary conditions of the challenging machining of ceramics. As a minimum, a nozzle diameter of 250 µm could be achieved (Fig. 7.6). As with decreasing nozzle diameter the stiffness increases, it is limited here to 90 mN µm^{-1}. To keep the dead volume small, the grooves' depth should be kept small. But if it is too small, pressure drops in the grooves, so that the desired pressure distribution is not achieved. Leaning on the experience and simulation results, the grooves' depth and width have been defined to 40 µm and to 400 µm respectively. According to CFD-simulation results, a bearing capacity of 4.951 N was achieved at an air gap of 5 µm and a supply pressure of 2 bar. The measurement results were 520 mN below the simulation results with comparable stiffnesses. This discrepancy can be explained in particular by the unconsidered pressure drops through the supply lines in the simulation. An important aspect is also the position of the center of pressure. It was designed far eccentrically, so that higher stability is given against tilting.

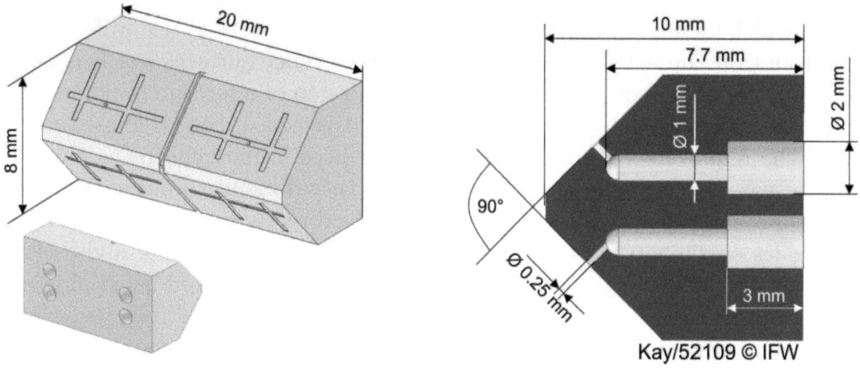

Fig. 7.6 Design of a macroscopic guide with branched groove stuctures

In the smaller version, the complex groove structures could not be maintained due to the selected machining technology and available space. In this case, the principle of the simple thrust bearing according to Fig. 7.1.a has been applied. The width and depth dimensions for the pockets correspond to the groove dimensions of the large version. Unlike the large guide, the supply channels have to be arranged side by side (Fig. 7.7). This inevitably results in asymmetrical arrangements of the nozzles. The height of the guide is limited by the minimum diameter of the supply channels. In order to ensure that the main throttling takes place in the micronozzle, the flow resistance of the supply channel must be smaller. A stage is provided in the supply hole, which serves as a limit stop for the supply hoses. The bearing capacity of this version is 0.6612 N at an air gap of 5 µm and a supply pressure of 2 bar.

7.2.3 Construction and Alignment

The integration of the displacement sensors in the aerostatic microguides requires an unconventional approach. In case of the magnetic guide, the sensor electrodes are fabricated on a wafer and mounted on a common plane base

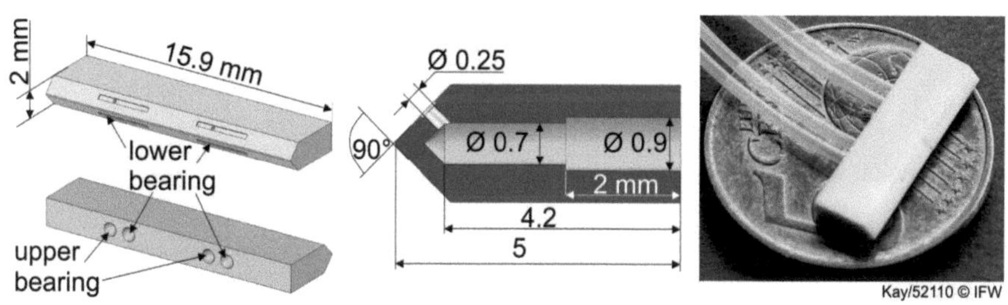

Fig. 7.7 Design of a microscopic guide with rectangular pockets

plate, which facilitates the alignment (Fig. 7.10). In order to use wafer-based electrodes on the aerostatic guides, precise reference surfaces are necessary to assemble the sensor parts. In view of the air gap of a few micrometers and the strong decrease in sensitivity with capacitive sensors at large measuring gaps, the requirements on the position tolerances in manufacturing the reference surfaces and microassembly would be very difficult to meet. With the new approach presented here, it is possible to directly apply the measuring electrodes on the guide surfaces in order to avoid manufacturing and assembly steps. For this purpose, the sensor electrodes are deposited using physical vapour deposition directly on the ceramic bearing surface. First, the whole guide body (Fig. 7.8.a) is covered with a mask, which has been structured by laser cutting before (Fig. 7.8.b). In this step, it is very important that there is no gap between mask and guide surface. This can be ensured by gluing the mask. Care must be taken that fine structures are not blocked by adhesive particles. After the deposition of gold with a thickness of 400 µm (Fig. 7.8.c) the mask can be removed. This results in electrodes on a 3D surface (Fig. 7.8.d). In the case of a short circuit, the electrodes can be electrically isolated by controlled scratching with a diamond. Since a metallic slide is used, the electrodes, except those on the soldering pads, must be covered with electrically isolating Si_3N_4. The pads must be kept electrically accessible for later soldering.

Figure 7.9 shows the experimental setup for the aerostatic guides. To adjust the guides to each other, an automated 5-axis adjustment system is used. As a small shift in slide direction is tolerable, the sixth axis has been omitted. The adjustment is done in two steps. In the first step, the adjustable guide is unmounted from the collet and both guide parts are clamped against each other with the slide in between. There will be a passive self-alignment. This target position of the adjustable guide is determined by a coordinate measuring system by measuring at least two 3D-points on the guide. In the

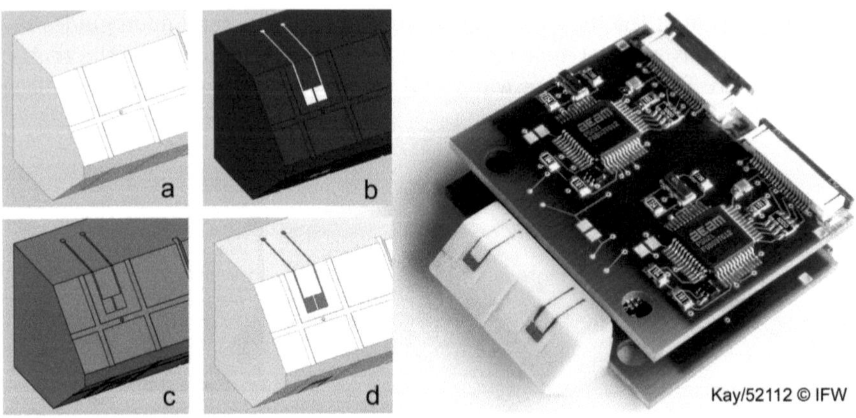

Fig. 7.8 Deposition of sensor electrodes on the guide surface

Fig. 7.9 Test rig for active aerostatic guides

second step, the adjustable guide is mounted into the collet and another measurement is taken of the corresponding points. These points represent the actual position. With the application of the spatial Helmert transformation and using the measured points, three translational and three rotational transformation parameters are determined. With the exact knowledge of the positions of the axis of the adjustment system, the driving commands are obtained by inverse kinematics.

The gap measurement data is read out digitally by a microcontroller (*AT91SAM7A3*) via Serial Peripheral Interface and transferred to the control implemented on a *dSpace 1103* real-time controller board. The pressure-regulating valves are located as close as possible to the guides to prevent lag times.

Controller performance tests in a manual test rig have shown that the decoupled PID control is more robust compared to state control, but the latter offers a faster step response [1]. Due to the huge similarity of the controlled aerostatic guide plant to the plant of the magnetic guide, further details can be found in Sect. 7.3.2.

The manufacturing and metrological investigations of aerostatic guides have been supported in Chaps. 14, 9, 11 and 18.

7.3 Active Magnetic Guides for Microsystems

Whereas a magnetic guide based on repulsive forces has been developed in [5], a magnetic guide with attractive forces is presented in this chapter.

7.3.1 Requirements and Design

The design of the electromagnetic guide takes into account constraints resulting from the micromanufacturing technology as well as demands on guiding qualities, such as locking of degrees of motion or the controllability of disturbances that result from varying payloads and actuator forces. Both a horizontal drift and a yaw of the traveller are prevented by the actuator's forces F_a of the nanopositioners built in the Collaborative Research Centre 516 (Fig. 7.10). Therefore, additional lateral guidance by the electromagnetic guide is unnecessary. According to electromagnetic force simulations verified by observations on the prototype (see Sect. 7.3.2), passive self-centering of the traveller occurs if the width of the primary part corresponds to the width of the secondary part in the linear nanopositioner.

In the case of the "XY-nanopositioner" with two degrees of freedom (Chap. 20), the feed forces of the actuators that are actually not used for drive are due to lateral guiding. In both cases, the remaining three degrees of freedom (z, φ, ϑ) are locked by the guide or controlled respectively, Where z is the altitude of the center of mass of the slide, φ is the tilt angle about the x-axis and ϑ is the tilt angle about the y-axis.

7.3.2 Prototypical Magnetic Guide

The aim of the prototype of the electromagnetic guide is to achieve the possibility of studying the control behaviour together with hardware. By using a mechanically rigid rack, it becomes possible to adjust the carrier of the electromagnets and the holder of the distance sensors in the three translational degrees of freedom. Complex precision assembly as it is presented in [4] is avoided in this way. The traveler is an epoxy resin board with 1.6 mm thickness that is coated with copper on the bottom to provide a target surface for the four capacitive sensors (Fig. 7.11). Four iron parts are arranged

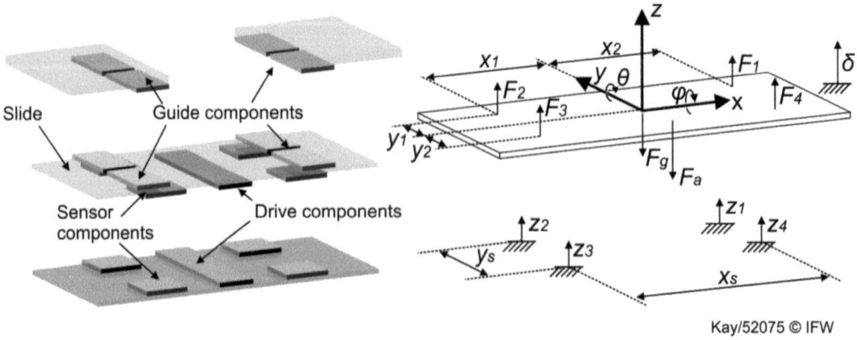

Fig. 7.10 Principle of the magnetically guided linear nanopositioner

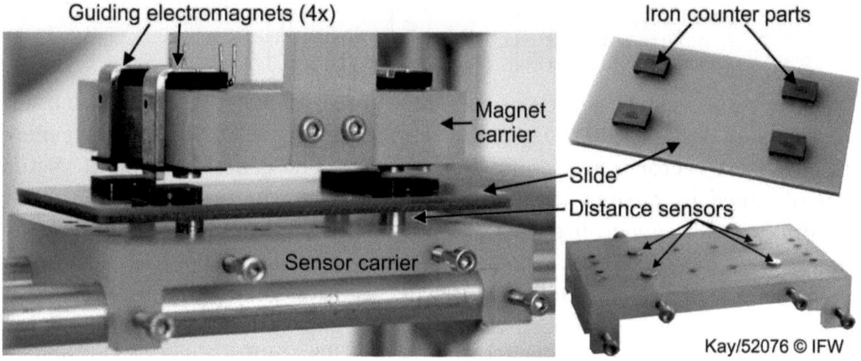

Fig. 7.11 Prototypical magnetic guide

on the traveller, which cause reluctance forces below the guiding magnets, which have previously been surface ground to obtain equal air gaps. The sensor signals are sampled with the control cycle and then conditioned by signal processing. The controller structure is decoupled referred to the controlled degrees of freedom. The nonlinear model of the electromagnets is linearized at the operating point. Due to appropriate control parameters, the controller is able to levitate the traveller autonomously from the ground to the operating height. In the controller design, the weight force F_g and the driving force F_a are assumed to be known and are compensated. In the system modeling, these quantities are taken into account and thus a linear system model is obtained. The equations of motion of the traveller in the local coordinate system in the center of gravity can be specified as follows:

$$\begin{bmatrix} m & 0 & 0 \\ 0 & J_\varphi & 0 \\ 0 & 0 & J_\vartheta \end{bmatrix} \begin{bmatrix} \ddot{z} \\ \ddot{\varphi} \\ \ddot{\vartheta} \end{bmatrix} = \begin{bmatrix} 1 & 1 & 1 & 1 \\ y_1 & y_1 & -y_2 & -y_2 \\ -x_2 & x_1 & x_1 & -x_2 \end{bmatrix} \begin{bmatrix} F_1 \\ F_2 \\ F_3 \\ F_4 \end{bmatrix} \tag{7.1}$$

The actuator forces F_1 to F_4 are in a nonlinear relation both to the excitation current vector of the coils $\mathbf{i}$ and the air gap vector $\boldsymbol{\delta}$ between the active and passive guide components ($F_1..F_4 = f(\mathbf{i}, \boldsymbol{\delta})$). This is described by the characteristic diagram whose partial knowledge is required for controller synthesis (Fig. 7.13.a).

By the application of SISO (Single Input Single Output) controllers for each degree of freedom, the complexity of the control can be reduced for real-time computing. Therefore it is necessary to transform the coupled MIMO (Multiple Input Multiple Output) plant to multiple decoupled SISO plants. The state vectors that are controlled by the independent full state feedback controls are given as:

$$\boldsymbol{\zeta} = \begin{bmatrix} z \\ \dot{z} \end{bmatrix}, \boldsymbol{\varphi} = \begin{bmatrix} \varphi \\ \dot{\varphi} \end{bmatrix}, \boldsymbol{\vartheta} = \begin{bmatrix} \vartheta \\ \dot{\vartheta} \end{bmatrix} \tag{7.2}$$

Because the velocities are not measured directly, state observers are used. For the state estimation, the generalized plant input vector $\mathbf{i_q}$ and the generalized output vector $\mathbf{q}$ are required (Fig. 7.12). The generalized input vector $\mathbf{i_q}$ is obtained by current measurement and transformation into generalized coordinates. The generalized output vector $\mathbf{q}$ is obtained by the sensed air gaps $\mathbf{z} = [z_1, z_2, z_3, z_4]^T$ that are converted by the Moore-Penrose pseudo inverse Jacobian $\mathbf{J}_s^+$. It computes a "best fit" solution which minimizes the sum of squared differences between the measured gap values and their modeled geometrical relationship to the three DOFs. The differences of the generalized control forces and moments $\mathbf{Q_{set}} = [F_{z,set}, M_{\varphi,set}, M_{\vartheta,set}]^T$ and the prefiltered desired DOFs are then converted to the four magnetic control forces $\mathbf{F_{set}} = [F_{1,set}, F_{2,set}, F_{3,set}, F_{4,set}]^T$ based on the principle of least error squares. The sensor-actuator matrix $\mathbf{J}_{z\delta}$ represents the geometrical relationship between the measured gaps below the traveller and the gaps between the magnets and the traveller. The control currents $\mathbf{i_{set}}$ can be determined on the basis of the inverse non-linear individual models of the electromagnets. The control voltages $\mathbf{u_{set}}$ are then calculated using the inverse models of the power electronics that were obtained by a previous calibration of the power electronics.

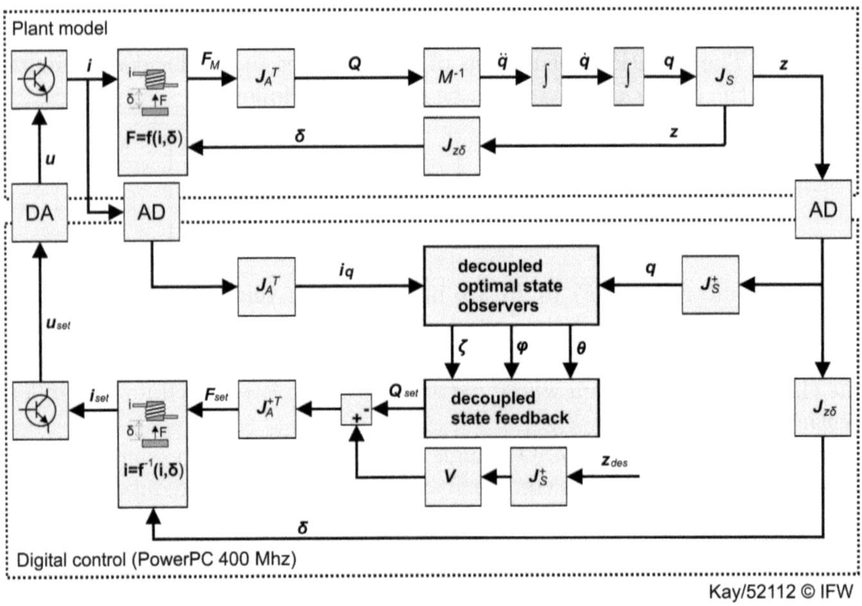

Fig. 7.12 Control plan for decoupled state feedback control

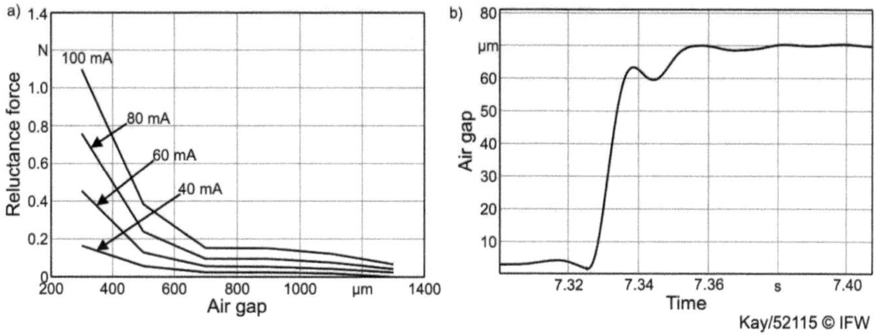

Fig. 7.13 (**a**) Characteristic diagram of a guiding magnet; (**b**) Step response to a given step of 70 µm

The behaviour of the magnetic guide can be seen in (Fig. 7.13.b). A step response to a given step of 70 µm is shown in the figure. The initial levitation height is 3 µm. The slide reaches the end position within 20 ms. The remaining oscillations are due to missing lateral guiding.

7.4 Capacitive Displacement Sensors for Active Guides

To control the air gap between the traveler and the stator, a miniaturized measurement system is mandatory. The air gap for the microactuator should be variable in the range between 0 to 15 µm. Due to a minimal sensor diameter of 1 mm given by the microactuator's geometry, commercial capacitive distance sensors cannot be integrated. According to the demands of the microsystems concerning the available space, the travel range, the gap measurement range, the resolution and the dynamics, a capacitive displacement sensor is developed. The main challenges that have to be met in the calibration process are the small air gaps and low measurement capacitances below one picofarad.

7.4.1 Design

Due to the higher possible resolution and lower complexity of capacitive sensors compared to e.g. inductive sensors, the capacitive principle was chosen to detect distances up to 15 µm with an adequate resolution. The easiest way to create such a sensor is to apply a parallel-plate capacitor using the traveler's surface as one electrode. This would assume an electrical connection to the traveler and therefore affect the actuator's motion.

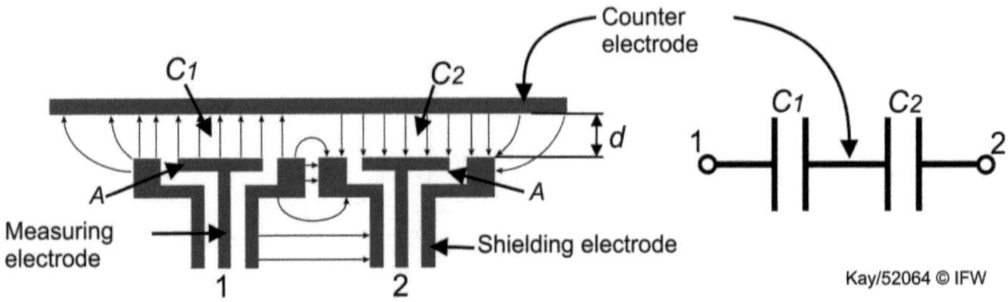

Fig. 7.14 Principle of the displacement sensor

To avoid any electric leads, the arrangement shown in Fig. 7.14 has been chosen. In this case, the traveler forms the middle electrode in a series connection of two capacitors. The equation for the gap d depending on the capacities C_1 and C_2 with the dielectric constant of vacuum ϵ_0 is given as:

$$d = \epsilon_0 \cdot \frac{A}{C_{total}}, \text{ with } C_{total} = \frac{C_1 \cdot C_2}{C_1 + C_2} \tag{7.3}$$

In order to suspend the inhomogeneous electrostatic field at the borders of the measuring electrode and therefore to obtain a higher sensitivity, shielding electrodes are surrounding the measuring electrodes. For this, the potential of the shielding electrode is driven to the measuring potential by a voltage follower. To assure a working shielding, both potentials have to be equal even at high frequencies. Another advantage of shielding is the reduction of influences of unknown environmental stray capacities. This effect can be shown in circuit simulations. But, due to the necessity of reducing complexity in the microfabrication of the sensors, the measuring electrodes are only laterally shielded. The results of finite element simulations show a 2% increase in sensitivity ($S = \delta C/\delta h$) at 10 µm air gap by applying shielding electrodes compared with unshielded measurements. Field homogenization is more ef-

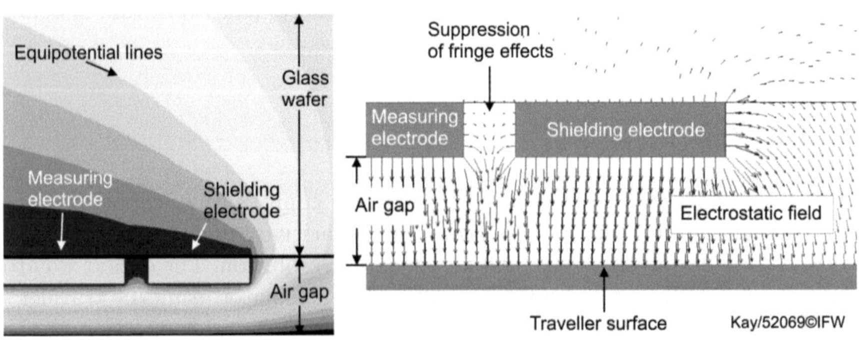

Fig. 7.15 Simulation results of field homogenization by shielding electrodes

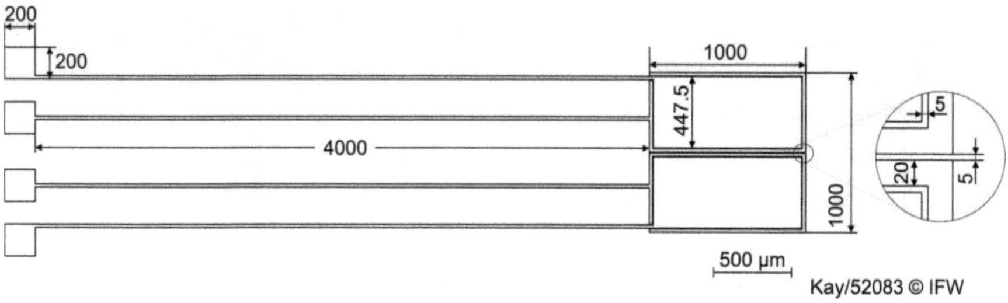

Fig. 7.16 Design of sensor electrodes used in the active magnetic guides

fective, if the shielding electrode is close to the measuring electrode, but it is independent of the thickness of the shielding electrode (Fig. 7.15).

The available space for the sensor is predetermined by the actuator design. According to the geometrical constraints, the outer dimensions of the system are $1 \times 1\,\mathrm{mm}^2$ (Fig. 7.16). Considering manufacturing limits, the critical spacing between the shielding and the measuring electrodes is defined as 5 µm. The same distance is defined for the spacing between the two surrounding shielding electrodes. The thickness of the shield electrodes and the feed lines is 20 µm. The active area for each capacitive electrode results in $447.5 \times 950\,\mu\mathrm{m}^2$. To provide an actuator travel range of 4 mm, the feed lines have the same length. The feed lines end at the soldering pads with a size of $0.2 \times 0.2\,\mathrm{mm}^2$.

7.4.2 Calibration

In order to prove the functional capability of the displacement sensor and to analyze its properties, an appropriate test rig was designed (see Fig. 7.17). By using a one-axis micro-translation stage it is possible to provide precisely defined gap spacings with a resolution of 50 nm. Another advantage of this stage is the controllability by a computer.

To assure an accurate functionality of the capacitive sensor, undesired tilting between the measuring electrodes and the target electrode has to be minimized. In the case of 0.5°degree tilting and an air gap of 7 µm, the error would be approximately 5%. By applying an adjustment procedure as shown schematically in Fig. 7.18, tilting can be avoided without time-consuming adjustment efforts. For any reasons tilting can occur between the sensor carrying wafer and the ground, e.g. by imperfect fixation (Fig. 7.18.a). Assuming clean surfaces, a counter electrode that is covered at the front side with insulating Si_3N_4 is applied on the sensor electrodes (Fig. 7.18.b). The back side of the counter electrode is wetted with a metal adhesive (Fig. 7.18.c). A ball joint is mounted below the translation stage and centered above the glue point manually by xy-stages (Fig. 7.18.d). The reference position is

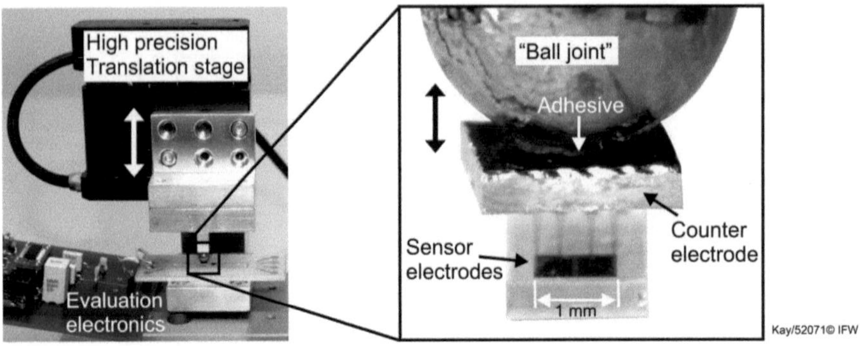

Fig. 7.17 Test rig for sensor calibration

defined by moving the translation stage down to the contact point between the ball and the counter electrode (Fig. 7.18.e). The calibration process starts by varying the air gap *d*. Thus, a high degree of parallelism between the counter electrode and the sensor electrodes is obtained.

The integrated circuit *PS021* from *Acam messelectronic* is used to evaluate the sensor's capacitance. The capacitance C_m is charged to an upper reference voltage V_{high} and then discharged through a fix resistor R_{dis} to a lower reference voltage V_{low}. By measuring the discharge time T_{dis}, the capacitance is given by:

$$C_m = \frac{T_{dis}}{R_{dis} \cdot \ln \frac{V_{high}}{V_{low}}} \tag{7.4}$$

Since the average measuring capacity is only about 1 pF, the integrated 24-bit analog-digital converter (ADC) is very useful to reduce stray capacitances that would occur on feed lines to external ADCs. The data can be obtained via the serial peripheral interface (SPI) at sample rates up to 50 kHz.

In order to get a first impression, a simple trajectory was chosen for the micro-translation stage. Starting from 10 µm, the gap should be decreased step-wise with a stepsize of 1 µm until contact and then turn back to the start position by increasing the gap. As can be seen in Fig. 7.19, the steps

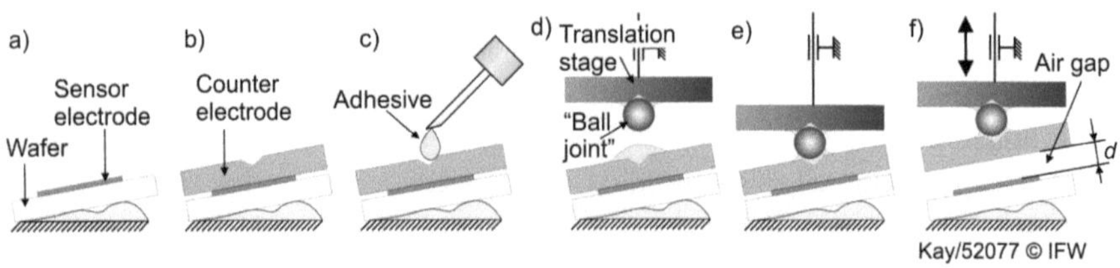

Fig. 7.18 Assembly concept of the sensor calibration rig

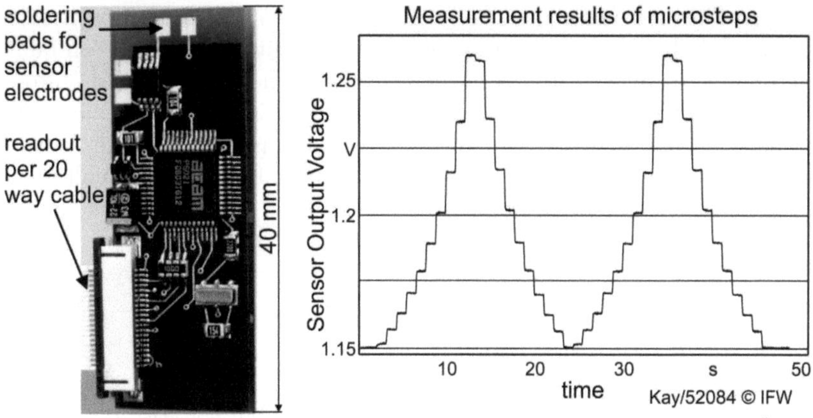

Fig. 7.19 Measurement result of the capacitive displacement sensor

are well-defined. As expected, there is a nonlinear relation between the sensor output and the distance due to the inverse proportionality of both factors (Eq. 7.3). Due to the slackness of the threaded spindle of the translation stage, hysteresis at the reverse points is found. Because of the high repeat accuracy, the slackness can be quantified as a constant value of approximately 1 µm. The standard deviation with respect to the air gap size depends on the actual gap itself and on the filter parameters. A deviation of 23 nm at a reference air gap of 5 µm and a sample frequency of 5 kHz could be achieved. With a prototype of an electromagnetic guide, the functionality of the system with a sample rate of 2 kHz and a sensor noise deviation of 150 nm can be shown (Sect. 7.3.2). Therefore, this sensor is applicable for levitation-based guides for microactuators.

7.5 Conclusion

In the sub-project C3, several active guides for the application in microsystems have been designed, built and tested. Both aerostatic and magnetic guides allow frictionless movement and are suitable for use in microsystems and for future applications in microfabrication concepts [6] as well. Aspects must be considered in material selection, design and integration that only occur due to miniaturization. For example, sintered air bearings that are well-proven in macro-bearings are not applicable in microscopic guides because local irregularities in the crystalline structure are particularly disturbing. The design of air bearings has been supported by numerical calculations that have been validated with measurements. In this way, adequate bearings with high stability have been obtained using discrete nozzles. Special alignment rigs with automatic stages have been constructed for guide-adjusting and sensor calibration. The presence of gaps below 10 μm requires specific approaches for measuring and aligning. A displacement measurement system has been built to determine gaps below 1 μm. A novel concept for the integration of sensors in micro-air bearings has been proved. Different solutions for magnetic guide design have been researched and proved with the use of adequate control concepts. The research work in this subproject has shown the difficulties in the design of active microguides and has contributed novel solutions for the design and manufacturing of active microsystems.

Acknowledgements The authors would like to thank D. Kopp and Dr.-Ing. J. Li for their research work and S. Brehmeier for his support. Last but not least, the authors gratefully acknowledge the funding of the reported work by the German Research Foundation within the Collaborative Research Center 516 "Design and Manufacturing of Active Micro Systems".

References

[1] Denkena B, Möhring HC, J L (2007) Entwicklung aktiver Linearführungen für Mikroaktoren. In: 5. Paderborner Workshop, pp 209–222
[2] Denkena B, Brehmeier S, Möhring HC (2008) Active linear guidances for micro actuators: alternative concepts and first prototypes. Microsystem Technologies 14(12):1961–1973
[3] Heinzl J, Zehentbauer H (1990) Neue Ergebnisse bei der Entwicklung aerostatischer Lager. Feinwerktechnik und Messtechnik 10
[4] Hesselbach J, Schöttler K, Raatz A (2008) Precision assembly of active microsystems with a size-adapted assembly system. In: IPAS, pp 199–206
[5] Ruffert C, Li J, Denkena B, H-H G (2007) Development and evaluation of an active magnetic guide for microsystems with an integrated air gap measurement system. IEEE Transactions on Magnetics 43(10):2716–2718
[6] Wulfsberg JP, Redlich T (2010) Square foot manufacturing: a new production concept for micro manufacturing. Production Engineering 4(1):75–83

Chapter 8
Design of Sensors for Position Control of Microactuators

A. Kornfeld, C. Kolleck, A. Ostendorf

Laser Zentrum Hannover
c.kolleck@lzh.de

Abstract For position control of a microlinear motor, an interferometrical principle was combined with an incremental sensing method, in order to achieve a resolution below 100 nm. A reflection grating placed on the moving slide is coherently illuminated by a laser beam, to allow the first two orders of the reflected beam to interfere with each other in an optical 3×3 coupler. The beat signal induced by the movement of the grating is detected as an intensity modulation in three phase-shifted output signals. The evaluation of these signals enables the measurement of both the distance and the direction of the slide's movement. The required technologies for the manufacturing, preparation, alignment, and assembly of the involved optical elements were applied. This includes refractive-index modification of glass, the adaptation of microgrippers, the microlens layout and selection, and the UV-curing of elements. The detection and the real time signal processing were realized, based on photodiode arrays and *LabVIEW* algorithms, respectively.

8.1 Introduction, Requirements and Measurement Principle

Microsystems with linear actuators such as the electromagnetic microactuators described in Chap. 3 require sensors for the accurate measurement of the moving element's position. Potential applications of the micromotor essentially depend on the spatial resolution of its position control. Since the motor is designed with a positioning accuracy of $< 1\,\mu m$, a measurement resolution < 100 nm is required. Additionally, the motor actuation involves electromagnetic fields and a steady thermal stress. Because of potential interference, inductive, electrostatic or magnetic principles were not taken into closer account. A holographic principle combines surface reflection and a transmission grating, and has been demonstrated in [3], but requires special polarization-sensitive signal evaluation, which is difficult to reproduce. An

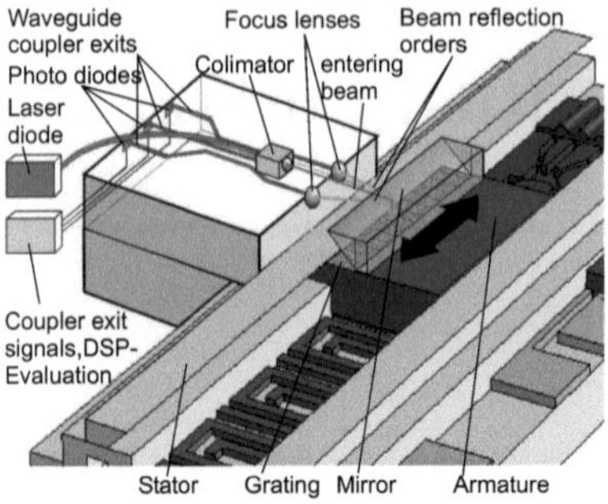

Fig. 8.1 Schematic position sensor, adapted to the microlinear motor

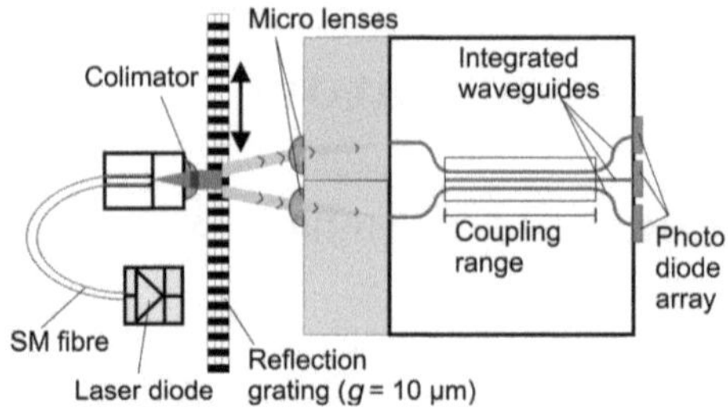

Fig. 8.2 Sketch of position sensor setup and beam path

integrated sensor based on interferometry is insensitive to magnetic fields, and thus is well suited, also with respect to increasing the accuracy which can be expected from general examinations in [1].

The sensor designed here is based on a miniaturized, optical interferometer, and it is connectable laterally to a micromotor, as shown schematically in Fig. 8.1. A laser beam illuminates a reflecting diffractive grating, which moves with the motor's slide, see left part of Fig. 8.2. This movement induces a frequency difference between the first diffraction orders. When interfering with each other in a coupler, these frequency-separated signals lead to a beat signal which can be detected as an intensity modulation at the coupler outputs. In contrast to an absolute position measurement, this kind of interferometer exhibits an incremental measurement principle. Thus, the measurement of displacement as well as direction has to be integrated to determine the motor's position. A 3×3 directional coupler is used for interference which also provides access to information about the direction of movement by measuring the relative optical phases of three modulated intensity signals.

This chapter is organized as follows. The most important step in order to minimize the sensor's extension is to design an integrated waveguide coupler. The effect of the coupler's geometry and optical parameters on beam propagation is simulated in Sect. 8.2, the subsequent production is described in Sect. 8.3. Considering the beam propagation from the source to the moving surface and onwards to the coupler, the position, layout, and production of optical guiding elements like grating, lenses, and prism are dealt with in Sect. 8.4. After interaction of the light in the coupler, three output intensity signals have to be detected, as described in Sect. 8.5. The resulting motor position has to be evaluated from the electrical output signals, using a real-time DSP algorithm, enabling the operation of the central position control (Sect. 8.6). Finally, all components have to be aligned, to be fixed towards each other and to be adapted to the micromotor by methods described in Sect. 8.7.

8.2 Simulation of Integrated Laser Beam 3×3 Couplers

For the design of the coupler geometry, simulations were performed. Parameters have been taken as input for the simulations which are achievable using the waveguide (WG) writing process with ultrashort laser pulses for a refraction modification in order to define a waveguide and, consisting of three neighbouring waveguides, forming a structure with three inputs and three outputs, called a 3×3 coupler.

8.2.1 Requirements

The coupler is necessary to couple in and guide two diffracted single modes, generated by a laser diode at a wavelength of 1550 nm, which limits the maximum mode field diameter to 15 μm according to formulas (1) and (2) in [4]. It should also enable mutual WG interference by cross-coupling into the third WG and back to the two initial WGs. The coupling length K of the coupler is defined as the propagation section parallel to the WG where the mode field maximum of one guide cross-couples into a neighboured guide, and back into the first one. This range has to be smaller than the complete central parallel range of all WGs. Hence, due to the limited total coupler length of about 7 mm, a coupling length of $K < 1$ mm should be realized by an appropriate combination of WG diameter B and center distance A along the middle propagation region where the cross-coupling takes place. Also, the optimum entrance distance E of each pair of waveguides has to be determined, which affects the curvature radii of the waveguides between the entrance and coupling region. For illustration of the WG parameters, see Fig. 8.3. An optimized coupler with high interference contrast produces three equal-intensity output signals when excited with just one input signal.

8.2.2 Simulations

First simulations were performed on glass materials such as soda-lime glass, with a constant refractive-index change of the order of $\Delta n \approx 4 \cdot 10^{-3}$. Later, for more detailed simulations, further calculations based on the material *Eagle2000* were accomplished for different magnitudes of index change. The minimum curvature radius decreases significantly with increasing refractive index shift. Hence, larger shifts of Δn enable larger entrance distances E.

For the simulation of two defined TEM_{00} beam modes propagating in refractive index modified glass media as in the mentioned coupler, 2D beam propagation method (BPM) simulations were accomplished with the software *OptiBPM* by *OptiWAVE* and *OptoDesigner* by *Phoenix BV*, successor of *OlympIOs* from *C2V*. Particularly *OlympIOs* also includes 3D simulation options as required for triangular WG coupler geometries. The results show the following qualitative dependencies.

Variation of Entrance Distance E and Curvature. Keeping the central WG distance A constant and increasing the entering distance E along the same WG propagation distance, the WG's curvature is increased. This increases also the beam energy loss by scattering into the surrounding substrate. For a maximum achievable index difference of $\Delta n \approx 6 \cdot 10^{-3}$, this effectively limits the maximum entering distance E to roughly $E \approx 100\,\mu\text{m}$. Larger values lead to extreme losses and a poor output signal contrast, which also reduces the maximum interference contrast. Thus for manufacturing, the parameter E was set to values within a range of $50\,\mu\text{m}$ to $100\,\mu\text{m}$. In Fig. 8.3, the coupling losses for $E = 50\,\mu\text{m}$ and $100\,\mu\text{m}$ can be compared qualitatively.

When Δn is increased to $\approx 1 \cdot 10^{-2}$, also smaller curvatures and hence, larger E are achievable, as could be seen from additional 2D simulations. However, the chosen material *Eagle 2000* for the final manufacturing series of $3{\times}3$ couplers does not allow $\Delta n > 6 \cdot 10^{-3}$ by the laser pulse processing.

Variation of Center Distance A, Border Distance d, and Coupling Length K. The coupling length K sensitively varies with the guide's border distance $d := A - B$ between the cylindrical surfaces which are identified by refractive index steps. Because their diameter B is determined by the process, d must be controlled by the center distance A. These relations were calculated for various Δn, e.g. from $4 \cdot 10^{-3}$ up to $6 \cdot 10^{-3}$, demonstrated in Fig. 7 of [5]. The output power of each waveguide is modulated quasi-periodically with the border distance d, with decreasing period length with increasing Δn. The respective center distance A should be adapted to the condition of equal powers $P1 = P2$. Thus the condition $P1 = P2$ is fullfilled only for discrete $d \approx 4\,\mu\text{m}$, which also depends on B and Δn, corresponding to a natural number of cross-couplings.

Relation Between 2D and 3D Couplers and Their Simulations. Most simulations were performed with a 2D model, assuming 3 equidistant planar WGs in the same central plane. For an optimum symmetry and equivalence to the triangular cross-section geometry of a fused fiber coupler as used in a reference sensor setup, the WG samples in *Eagle2000* were arranged in a triangle shape not only along the central part, but also along the curved entrances and outputs before and after the cross-coupling region. Thus, the simulation also has to consider this geometry, but only the *OlympIOs* software enables to model a 3D setting in this way. Some examples of results obtained are given by the 2D simulation in Fig. 4 of [5], and the 3D simulation in Fig. 8.4. For the first, a coupling length of $K = 1.33\,\text{mm}$ is obtained, and the latter one exhibits a coupling length of $K = 1.65\,\text{mm}$, as can be derived from parts 2 ($z = 3.80\,\text{mm}$) and 4 ($z = 4.625\,\text{mm}$) of Fig. 8.4, covering half of one K period. Here a small proportional factor between the parameter K in the planar and in the triangle model can be assumed as $c = K_{planar}/K_{triang} \approx 1.3$, all other parameters being kept constant. The qualitative behaviour of the beam propagation for both models shows similar cross-coupling in a symmetric manner between the two incoupled and the third WG. Hence, the 2D simulation results can easily be transferred to the 3D model.

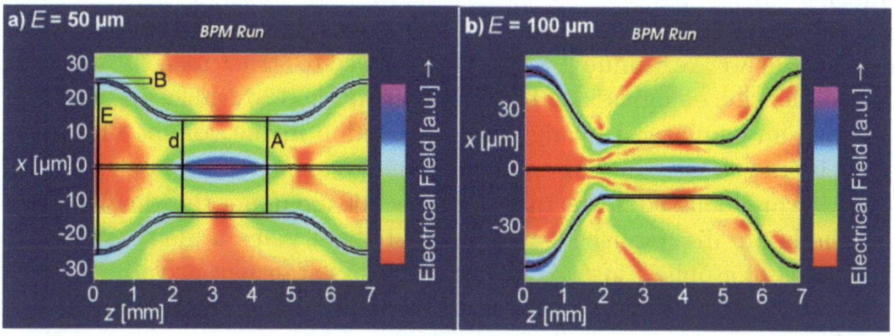

Fig. 8.3 3×3 coupler simulation by *OptoDesigner (Phoenix)* with WG center distance $A = 14\,\mu\text{m}$, WG diameter $B = 8\,\mu\text{m}$, mode field diameter $M = 11\,\mu\text{m}$, and entering distance: (**a**) $E = 50\,\mu\text{m}$; (**b**) $E = 100\,\mu\text{m}$

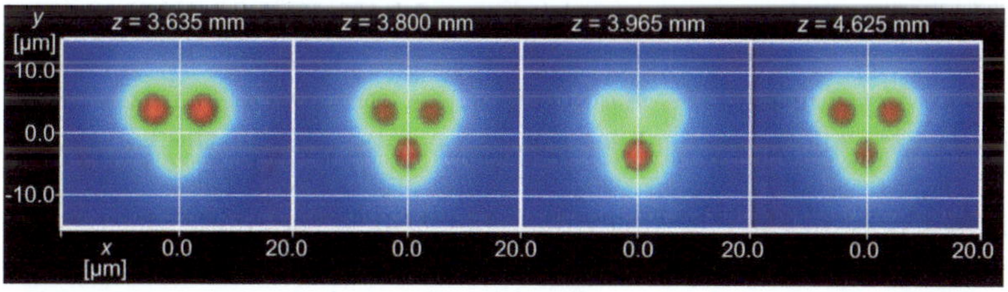

Fig. 8.4 3×3 triangular coupler simulation by *OlympIOs (BBV)*: 3D resulting coupling cross-section with mode energy distribution at positions 1–4 with $z = 3.635$ mm up to $z = 4.625$ mm

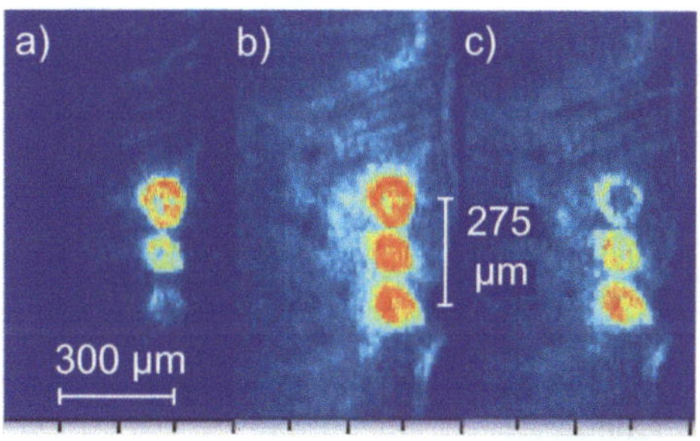

Fig. 8.5 Multi-mode coupler transmission, coupling into: (**a**) Top; (**b**) Both; (**c**) Bottom waveguide; taken from [5]

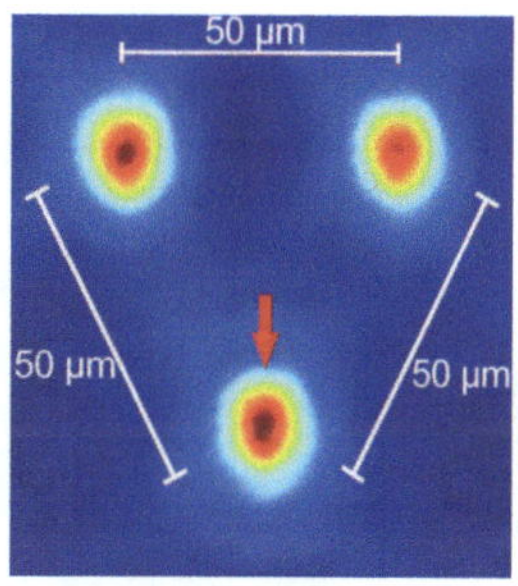

Fig. 8.6 Single-mode coupler transmission image

8.3 Production of Integrated 3×3 Couplers in Glass

8.3.1 Manufacturing Methods

First technological attempts to produce 3×3 waveguide couplers were surface-based methods, e.g. by using an excimer laser to ablate grooves from a sample and fill them with a polymer. As these attempts, described in [4], were not smooth enough and had the principal limitation of planar configurations, further methods were examined, which also allowed to structure a substrate in the third dimension of a glass sample. The only method to realize a waveguide coupler with an equilateral triangular cross-section in the volume of a glass substrate is its refractive index manipulation by means of ultra-short laser pulses, which are focussed in the range of 100−500 µm under the glass surface. Two methods are distinguished, and both of them place a glass sample so that the focus of a pulsed laser is placed inside its transparent volume.

In the first method, the sample is scanned parallelly to the axis of the entering laser beam path, enabling only short guiding structures of less than 1 mm length. In order to obtain coupling ranges of several millimetres, the second method was applied, which scans the sample in the plane perpendicular to the laser beam. First attempts were accomplished with an xy axis setup which enabled planar structures (system 2 in Table 8.1). With system 1 on an xyz stage, the central waveguide was set to a lower, but constant level. System 3 was used later, including an air cushion-born 3-axes stage (Fig. 8.7.a) which allows vertical scanning of components in the waveguide paths. This was demonstrated to produce 3×3 couplers with total triangular symmetry not only in the coupling range, but also in the input and output sections as in Fig. 8.7.c.

Some attempts were made not only with different laser systems, but also different glass materials, opening a high-dimensional parameter space of beam parameters such as wavelength, pulse energy E_p, duration t_p and repetition

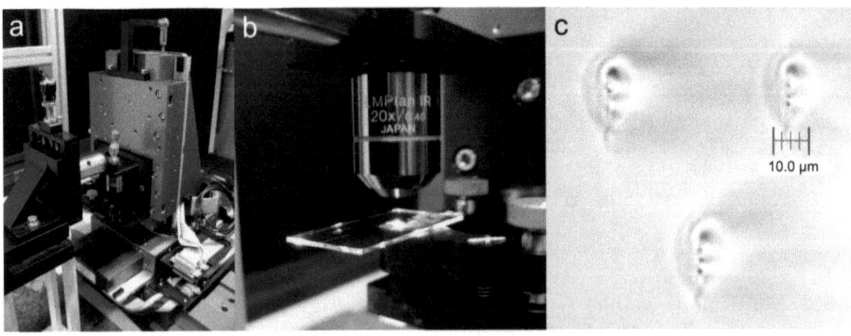

Fig. 8.7 3×3 triangular coupler production: (**a**) Setup of manufacturing system 3; (**b**) Substrate beneath the microscope objective; (**c**) Resulting waveguide structure

rate f_{rep}, as well as geometric parameters such as the scanning speed of the substrate v_s, and the energy deposition per length E_d. The lateral geometric properties of the written waveguide also depended on the shape of the beam focus, defined by the focal length f_{obj} and numerical aperture NA_{obj} of the focussing objective above the surface. A first orientation in this high-dimensional parameter space was enabled using the results of [7] and [9]. Later results were manufactured and examined also by [10], using very similar configurations.

Finally, the type of glass determined the exact refractive index profile of the intended waveguide. For the first attempts, conventional soda-lime glass and BK7 glass were used. Later, a special display glass, *Eagle2000*, replaced the first, due to better refractive index cross-sections.

Three different systems were available for the pulse laser source, showing parameters as listed in Table 8.1. They covered a large range of pulse repetition rates $f_{rep} = 1/t_r$ (5 kHz up to 10 MHz) and, respectively, pulse energies E_p (30 nJ up to 2 µJ per pulse). The pulse energy deposit per length E_s is calculated from these parameters, together with the scan speed v_s.

8.3.2 Produced Couplers

Multi-Mode Couplers. Channels with modified refractive index in glass (soda-lime glass) were produced using femtosecond pulse laser system 2, with the beam being focussed onto a 2-axis stage with parameters as in Table 8.1. Both the systems 1 and 2 have similar pulse durations, but provide very different pulse repetition rates f_{rep} and pulse energies E_p. For focussing, in system 2 a microscope objective with a focal length of 10 mm is applied. Its *NA* comes close to the *NA* set by [7]. The resulting waveguide cross-sections (Fig. 8.5) exhibit large diameters B of > 50 µm, and elliptical cores with an axis relation of 1:4 inside circular, but ring-formed WG structures. The appearance of irregular, ringshaped transmission as shown in Fig. 8.5

Table 8.1 Pulsed laser parameters for manufacturing of waveguide couplers

Laser parameter	System1	System2	System3
Wavelength λ	800 nm	1030 nm	1040 nm
Pulse rep.rate f_{rep}	5 kHz	10 MHz	1 MHz
Pulse energy E_P	2 µJ	60 nJ	300 nJ
Pulse duration t_P	150 fs	100 fs	70 fs
Energy deposit E_s	10 J m^{-1}	300 J m^{-1}	30 J m^{-1}
Scan speed v_s	1.0 mm s^{-1}	1.0 mm s^{-1}	10.0 mm s^{-1}
Numerical aperture *NA*	1.25	0.6	0.4
Substrate material	Ca-Na-glass	Ca-Na-glass	Eagle2000

indicates a multi-mode guiding and also cross-coupling from the lateral to the central WG.

Single-Mode Couplers. Further production used fs-laser system 1 ($f_{rep} = 5\,\text{kHz}$, $E_p = 2\,\mu\text{J}$) with changeable focus objectives. To obtain circular cross-sections, beam pre-shaping as applied by [9] with cylindrical lenses was an elaborate option. Alternatively, an oil immersion objective (*Zeiss Apochromat*, 100×) with a high $NA = 1.25$ to obtain a greatly reduced focus length was inserted. Thus, the diameter B and the ellipticity was reduced to about 10 µm and, respectively, a WG diameter ratio of 1:1.5 was achieved. However, the Δn profile obtained did not show a central maximum.

To obtain an intermediate parameter combination, and to manufacture the total length of the waveguides with a cross-section shape of an equilateral triangle, a pulse energy of 300 nJ was combined in system 3 with high scanning velocities $v_s = 10\,\text{mm}\,\text{s}^{-1}$ in three dimensions using an air-guided 4-axis system from *Aerotech*. A microscope objective from *Mitutoyo* (numerical aperture $NA = 0.4$, magnification $M = 20$) was used (see Fig. 8.7.b) to create a focus 150 µm beneath the surface of a glass substrate, a special material *Eagle2000* from *Corning*. In this case, oil immersion was not necessary, and the working distance above the glass surface was kept in the order of 1.0 mm. For lateral energy deposition, the linear beam polarization plane had to be kept parallel to the WG scanning direction, as confirmed later in [8]. This configuration revealed more homogeneous results than in any tilted case.

Qualification of Single-Mode Couplers. The structure obtained contained three waveguides which provided cross-coupling when feeding only one of the waveguide entrances with a single mode (SM) fiber beam, placed at a very small distance of roughly 10 µm in front of it. In most cases, both other waveguides yielded output signals with different power relations, but some of the 15 couplers generated nearly equal output powers, as for the sample in Fig. 8.6.

8.4 Beam Guiding by Refractive and Diffractive Elements: Grating, Prism, Lenses

In Fig. 8.8, all optical sensor elements are placed in a two-block-configuration in side view. In contrast to [6], this eases the adaption of the sensor towards the micromotor. The interaction between the measuring beam and the moving slide of the motor is realized by using a grating with metal stripes on glass substrates with a grating constant of $g = 10\,\mu\text{m}$, placed in the bottom part of Fig. 8.8. This grating with outer dimensions of $5 \times 6\,\text{mm}^2$, fitting the movement range of the micromotor ($\pm 2\,\text{mm}$), is attached to the slide, and reflects a collimated beam.

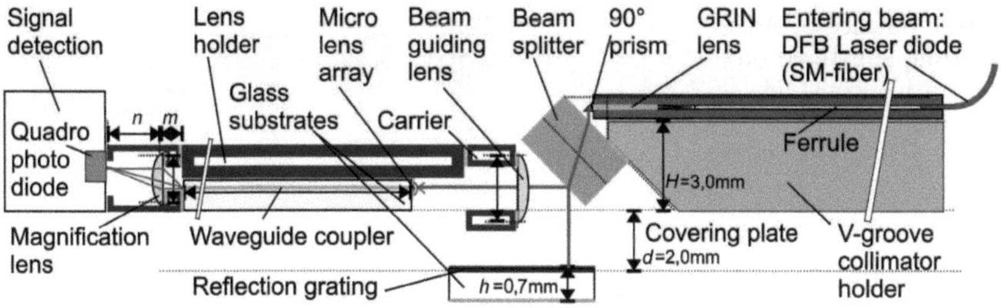

Fig. 8.8 Position sensor side view with complete design and optical path, 2-block-configuration

To guide the laser diode beam from the collimator to the grating, it is deflected rectangularly by a prism on the face of the collimator front surface. The size of the prism edges is $d_p = 1.0\,\mathrm{mm}$ on each edge or, for an alternatively prealigned collimator-prism combination, $d_p = 0.3\,\mathrm{mm}$.

Two beam orders diverge behind the grating, requiring a guiding lens which collects both orders to obtain converging beams, in order to couple them to the microlens array. This array focusses both orders into two different WG of the waveguide coupler. The guiding lens is a plano-convex lens for a designed distance of 8 mm between the grating and the lens array, being held by an aluminum frame. To focus both beam orders into the coupler, early attempts of lens manufacturing using 2-photon-polymerization as planned in [5] were not continued due to reflection problems in flat structures of a diameter of 50 µm. Hence, commercial microlens arrays with a pitch of 30 µm are applied. Due to their thin shape < 0.1 mm, they are combined with a gripping-border to be pneumatically aligned.

8.4.1 Grating as Beam Separation for Interference

The grating was manufactured by sputtering Au:Cr through a mask onto a sample surface. For a first sample, a silicon surface, no first order reflection could be obtained using a 1550 nm beam. Most likely, the reflection contrast between the silicon base and the Au-Cr layer in this IR region was not sufficient.

For this reason, a second attempt was done on glass as a basis carrier of the Au-Cr layer. In this case, a grating as shown in Fig. 8.9 which had a reflection per order of about $R = 6\%$ could be achieved. Compared to some metallic, commercial gratings with $R = 10\%$ per first order, but without a fitting grating constant and too thick substrate for micro-integration, this result was acceptable, though the beam shape showed a doubling effect by back reflection. This could be avoided later by using an AR coating on the back side of the glass sample.

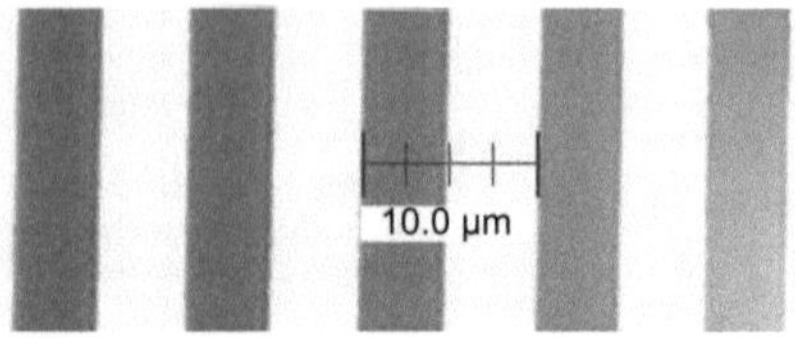

Fig. 8.9 Surface of Au-Cr grating, sputtered on glass, grating constant $g = 10\,\mu m$

8.4.2 Integration and Qualification of a 90° Prism

For the first version of the input collimator (length 20.0 mm, diameter 0.7 mm), manufactured by *Grintech*, an integrated metal tube including a front prism with equidistant edge lengths of 0.3 mm was qualified to have a beam waist diameter of 100 µm. However, a strong divergence leads to a beam diameter of roughly 200 µm at the end of the total setting's beam path range of 12 mm until entering the coupler. This diameter could not be adapted because the fiber-to-GRIN-lens distance had already been fixed by the manufacturer.

For a second collimator type, a GRIN lens and a single-mode fiber ferrule were placed in a glass tube. The GRIN lens (type: *GRIN2915* from *Thorlabs*) was fixed by UV curing (*NOA63* from *Norland*) first, leaving the fiber ferrule's axial position flexible. Hence, the beam waist position after leaving the collimator can be set individually in this configuration, i.e. to the grating position, which is advantageous to achieve a small focus on the waveguide.

In this case, a 90° prism of identical edge sizes of 1.0 mm was fixed on the front face of the GRIN lens using a drop of the UV-curing *NOA63*, being placed onto the lens center in Fig. 8.10.a, and then setting the prism onto this drop manually, (b) using a pair of tweezers, shifting the prism to the center position before curing it to the GRIN lens (c). After hardening, remaining adhesive was removed with a micro-knife (d). The mounted prism on the front face of the collimator, set into the V-groove which also holds the beam splitter, is shown in Fig. 8.11.

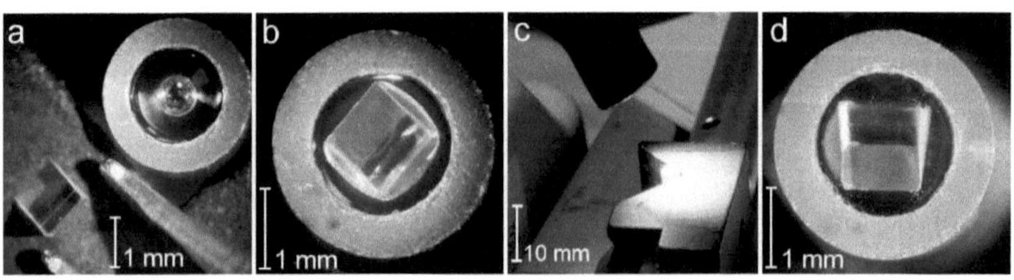

Fig. 8.10 (**a**) Collimator with UV adhesive and prism; (**b**) Prism placed on collimator; (**c**) UV curing process; (**d**) Remaining adhesive being removed

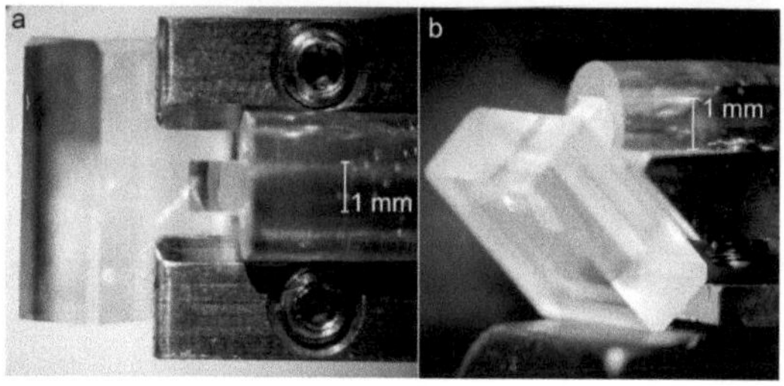

Fig. 8.11 Entering beam unit: V-groove, holding the collimator with prism and beam splitter: (**a**) Top view; (**b**) Side view

8.4.3 Beam Guiding Lens

To converge the divergent beam orders before coupling them into the microlens array and the waveguide coupler, a lens with a focal length of 2.0 mm is needed, with a diameter of less than 3 mm. These properties are fulfilled by the *Linos* lens of the type *G312002000* with a focal length of $f = 2.0\,\mathrm{mm}$ and a diameter $d = 2.5\,\mathrm{mm}$ which was fixed using UV adhesive (*NOA63*), with the flat surface onto an Al housing which measures $3.0 \times 3.0 \times 1.0\,\mathrm{mm}^3$. This housing was then adapted to a plate on the surface of the glass substrate of the coupler using a gripper, see Sect. 8.7.

8.5 Signal Detection

8.5.1 Photo Detectors

The photo detector applied receives a modulated optical intensity signal with required sampling frequencies of the order of 100 kHz; hence, it should have a bandwidth of > 1 MHz. For the interferometrical principle, single mode beams have to be detected, requiring a long coherence length $l_c > 100 \cdot r \approx 2\,\mathrm{m}$ with sensor beam range $r \approx 20.0\,\mathrm{mm}$. The small line band width of the laser diode $\Delta f = 10\,\mathrm{MHz}$ leads to $l_c = 6\,\mathrm{m}$, fulfilling the above condition. For the beam source, DFB laser diodes with a wavelength of $\lambda = 1550\,\mathrm{nm}$ are used, i.e., a 20 mW diode by *Anritsu*, or a 80 mW diode by *EM4*. Thus, the detector requires the photosensitive PD material InGaAs, due to its spectral response. The estimated output power of each of the three waveguide output signals ("raw signals") is expected to be about 1 μW, requiring an amplification after detection. The spatial requirements for the miniaturized sensor demand dimensions of less than 10 mm in each direction, and at least 3 sensitive areas in a triangular shape. Their optimum distance fits to the output distance $E_o = 50\,\mu\mathrm{m}$ of the waveguide coupler.

Since a photo diode array with the required detector pitch value, matching the 3 channel's distance of 50 μm, was not available, a quadrant photo diode (*Hamamatsu*, type *G6849-01*) with four quadrant sensitive areas of a radius $r = 0.5\,\text{mm}$ was selected for use as a detector. As its four sections are separated by 30 µm, the diagonal distance of 42 µm limits the image. Because the photo diode has a small spacing, including the electrical contacts, between the InGaAs layer and the covering glass anyway, this spacing is part of the distance between photo diode and waveguide which is used to achieve a magnification factor of 4 by using a small lens.

8.5.2 Positioning and Fixing

Due to the distance of the waveguide output channels and their magnification factor, the spots of the optical signal are expected to be separated by approx. 200 µm. Hence, the mounting tolerances perpendicular to the beam direction have to be kept within 50 µm, which is less critical than the input mounting tolerance of the guiding lens and focussing lens array. The position of the detector with a housing diameter of 8.0 mm and a length of about 4.0 mm was designed to be at the border edge of a micromotor, allowing a lateral detector shift on the side plane. This detector housing will be fixed by an UV adhesive after being intrinsically aligned to detect the three signals.

8.6 Realtime Signal Processing

In order to obtain the velocity and direction information about the motor, a processing of the photodetector signals is necessary. The development and verification of the algorithm was performed on a macroscopic interferometer system.

8.6.1 Reference Fibre Coupler Setup

A macroscopic demonstration setup was realized, using a 3×3 fiber coupler and one complete high speed amplified InGaAs PIN photo detector for each signal. The models *DET410, DET10C* and *PDA10CS* from *Thorlabs* were applied, with rise times of $t_r < 10\,\text{ns}$. With optical output signal powers of the order of 10 to 100 µW, a signal voltage U of about 0.2 to 2.0 V per channel could be examined using an oscilloscope, and evaluated using an algorithm based on a *LabVIEW FPGA* module.

In this reference setup which is depicted in Fig. 8.12, a fiber-based 3×3 coupler acts as signal interfering and recombination unit instead of the integrated glass chip coupler. The output signals showed sinusoidal voltage oscillations when shifting the grating by a piezo motor in different sinusoidal, triangular or different characteristic oscillations. These three signals had to be processed to derive the current position and velocity of the grating.

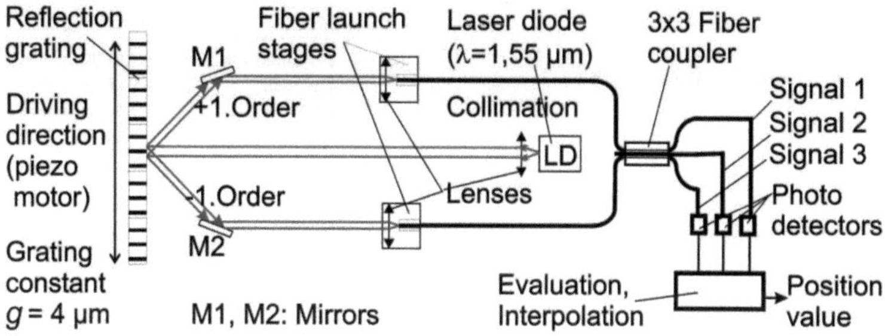

Fig. 8.12 Reference 3×3 fiber coupler setup for signal evaluation development

When evaluating these signals, the movement can be easily, but roughly derived from the number of signal maxima with a position accuracy of half of the grating constant (e.g., a grating constant $g = 4\,\mu\text{m}$ leads to a signal period length of $g/2 = 2\,\mu\text{m}$). Hence, a sub period phase evaluation is necessary to obtain a position resolution order of $0.1\,\mu\text{m}$. Following this, a phase step counter was realized. Any movement stop or directional change is indicated by the relative phase order of all three channels, which had to be evaluated.

8.6.2 *LabVIEW* Algorithm for FPGA

Digital signal processing (DSP) of all the steps described above on basis of *NI LabVIEW 7.0* is only possible offline. For a realtime system, a high processing speed of the operating system has to be combined with a realtime-capable module, processing all steps of the evaluating algorithm deterministically. A field programmable gate array (FPGA) is a deterministic calculating element using parallel processing in separate slices. The *NI PCI-7833R* model's calculating steps are performed with rates of 40 MHz. For a resolution of 100 nm at a speed of $v = 10\,\text{mm}\,\text{s}^{-1}$, the required position sampling rate is about $f_s = 100\,\text{kHz}$. Thus, for each position step, a number of 400 closed-loop clock cycles is available, being processed by the algorithm that runs this FPGA. Hence, the real time DSP condition is fulfilled, if the FPGA algorithm requires less than 400 clock cycles per sampling interval.

The algorithm developed operates with the following chronological steps, also shown in Fig. 8.13. The voltage signals are fed into the *NI PCI-7833R DAQ* module, which has an analog-digital transducer. After a first recursive digital noise filter, the maxima and minima in each signal are identified using a shift register mask. They are stored in two classes of extremes, typical ones with $< 12\%$ deviation towards the last extreme, and atypical ones with $> 12\%$ deviation. This distinction helps identifying phase jumps which can indicate directional movement changes. From their amplitudes (maximum value – minimum value), phase thresholds are defined at integer steps of 30°. Whenever one of the signals passes each threshold value, the phase increments

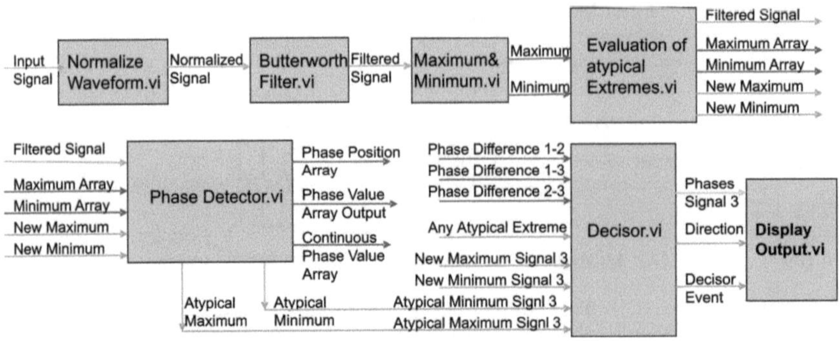

Fig. 8.13 Flow structure of FPGA algorithm for signal evaluation process

(e.g. with 30° or 1/12 of one period) are added, and each signal phase step is stored in a variable. In the next step, the differences between each of the three signal phase states are calculated.

A comparison of all three phase differences feeds the loop which decides about the direction. For a continuous movement, continuous phase states and differences are expected, whereas for a directional change, a mutual phase jump between the three phase differences e.g. from $(0°, 120°, 240°)$ to $(0°, 240°, 120°)$ can indicate it. This direction indicator, combined with counting the phase increments, calculates the current position of the grating. This position and also the velocity are displayed on the screen, or transduced to a port as input information for the central position control of the motor.

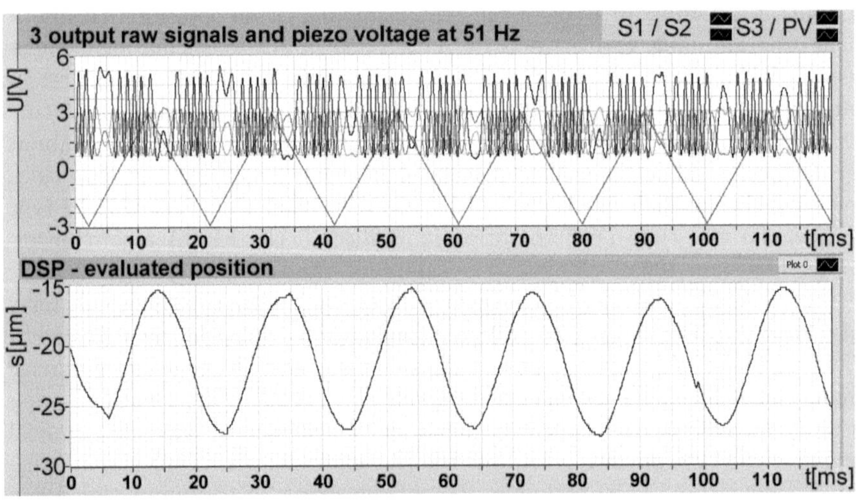

Fig. 8.14 DSP Signal Evaluation: basic output signals S1-S3 and triangular-shaped Piezo voltage PV (top), calculated position (bottom)

8.6.3 FPGA DSP Results

As a result, the algorithm described was applied on different movement types of the grating. The basic movement forms of sinusoidal and triangular piezo oscillation were combined with effects like different breaks with constant position, or different slopes on rising and falling parts of each signal. The results obtained were displayed on a PC interface, including the primary signals, the three phase states and differences, the derived amounts of velocity and position. Particularly, the maximum and minimum positions were sampled for longer times – several minutes – to judge about the measurement's stability.

For short time intervals as in Fig. 8.14, the resulting position is in coincidence with the expectations which arise from the displayed piezo voltage. For fast movements, a sinusoidal position modulation can also be detected, if the piezo voltage is triangular, due to the grating's and slide's mass inertia.

8.7 Alignment and Fixing of Optical Components

The alignment of test configurations was accomplished with use of one to three axis stages with an accuracy between 1 and 10 µm. The elements of the integrated sensor had to be aligned and fixed with very high accuracy in the order of ±5 µm. The easiest way to align the positions of optical elements was to induce passive marks and distance / tilt holders for well-defined element positions.

8.7.1 Optical Elements and Blocks to Align

For the alignment of the beam guiding and coupling elements, a first block 1 for fixing the collimator and the beam splitter plate was designed and manufactured as a V-groove, with the possibility of pitch adjustment using four screws, two of which are seen in Fig. 8.11. Using this pitch post-alignment, the beam is set exactly perpendicular to the grating, which is placed on the slide's surface, enabling the final alignment of entering and coupling block together on the top plate of the motor's stator. The beam splitter's 45°-tilt is realized passively using the V-groove's front end 45°-tilt. The splitter was fixed using the epoxy adhesive *TRA-Bond BA-F112* using a syringe and thin wires to insert adequate portions into small grooves on the tilted plane. The axis shift of the collimator defines the vertical position of the beam, whereas a collimator rotation in the V-groove sets the beam's lateral position at the second block 2. When the correct position was achieved by these motional degrees of freedom, the collimator was fixed to the V-groove by UV curing.

The second coupling block contains three optical peripherical elements with different mounting accuracy requirements. The beam guide lens for converging the two beam orders, the microlens array for focussing the orders into the waveguide coupler and the output lens for imaging three output signals onto a quadrant photodiode have to be aligned and fixed. Here, fixing the two input elements is most critical, because the waveguide entrances have to be coupled with an accuracy of $< 5\,\mu m$, estimated from the coupling of the single mode WG diameter of $10\,\mu m$.

8.7.2 Alignment and Fixation

For the coupler input and output elements alignment, the micro-grippers from [2] could not be used, because of the broad variation of element apertures. Thus, a three-axis vacuum gripper from *Klocke Nanotechnik* with an accuracy of 10 nm was deployed, picking the optical elements at smooth, plane surface positions using a canula, being set to a vacuum sleeving. This canula was carried by a head which can be moved by three linear stages (*Klocke*, type: *NMT 20B* and *NMT Z10*), each having an accuracy of 10 nm and controlled by a PC-based GUI.

The guiding lens, focussing lens array and the output magnification lens have to be aligned to a position which enables the detection of signals at the output photodiode. As this contains three elements, the first two are set when an IR camera replaced the detector. The guiding lens is pre-aligned passively to a position where its axis is identical with the waveguide coupler's central axis. The exact beam positions can then be set by the collimator block's alignment. As the last step, the microlens array's position must be set by the vacuum gripper intrinsically. This means optimizing the output power of all signals during fixation.

For this final position, all elements have to be fixed without any misalignment. Thus, any curing process of the two coupler block front lens elements has to be controlled by monitoring the transmitted power on the camera. Finally, the IR camera has to be replaced by the quadrant photodiode.

8.7.3 Adaptation to 1D and 2D Micromotor

The final application of the sensor requires its integration into the micromotor's housing. Here, the exact position and size of the grating are defined, and the sensor is aligned to the slide's geometry. In the basic 1D motor application, the spacing between the two sides was designed to be 8 mm, matching the spacing between the 90° prism of the collimator and the front face of the interfering waveguide coupler in the sensor's setup, shown in Sect. 8.4, Fig. 8.8.

For the 2D nano motor, the grating and sensor positions are designed to be at the edges of the application area, as shown in the top view in Fig. 8.15. A small border part of the application area is covered by the sensors and

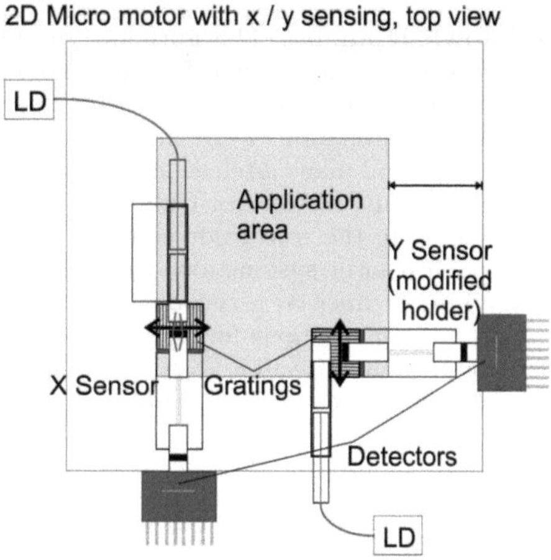

Fig. 8.15 Position sensor adaption to 2D micro motor, top view of x and y sensor

limits its range, because the gratings require illumination exactly from top. Alternative geometrical access, as e.g. sensing from the bottom, could not be realized because of the closed slide's carrier and its bearing structures.

8.8 Summary

A position sensor based on interferometry and microoptic elements was designed. The sensing principle could be demonstrated, first for a linear 1D movement with a resolution of 160 nm. This principle is transferable and adaptable to a 2D movement in x and y direction using two independent sensors with a tilt of 90° towards each other. A macroscopic demonstration setup, using a fiber coupler, showed the function of the principle and contributed to the development of the signal evaluation, first using a *LabVIEW* evaluation, later being supported by an FPGA module which enabled real time control. The coupler's function was implemented into a glass chip integrated 3×3 waveguide coupler. Its properties and manufacturing process were optimized and yielded couplers with only 7 mm length, hence a factor of 8 to 10 below the length of typical fiber couplers. By adding other microoptical elements, a complete sensor was designed and is being built, consisting of a beam guiding block and a coupling block. The most challenging part is the alignment of the beam orders in order to be focussed and coupled into single-mode waveguides with an optical path length of several milimeters. This sensor finally will be adapted to a linear micromotor.

Acknowledgements The authors gratefully acknowledge the funding of the reported work by the German Research Foundation within the Collaborative Research Center 516 "Design and Manufacturing of Active Micro Systems". We kindly thank J. Flügge for loaning an FPGA module, assistance e.g. during reference measurements at the PTB and many advices. Additionally, we thank G. Cerullo, S. de Nicola and R. Osellame from Polytecnico, University of Milano, for the friendly cooperation while manufacturing the waveguide couplers by system three. Technical assistance for applying the fs-laser systems one and two was provided from N. Bärsch and F. Will, whereas A. Neumeister assisted in alternative manufacturing methods for waveguides. Important contributions for the DSP by *LabVIEW / FPGA* evaluation were developed by H. Romann and F.G. Aliaga with their diploma theses. Several other students contributed to the optical components and alignment examinations.

References

[1] Flügge J (1996) Vergleichende Untersuchungen zur messtechnischen Leistungsfähigkeit von Laserinterferometern und inkrementellen Maßstabmesssystemen. PTB-Ber. F-23, Diss., Phys.-Tech. Bundesanst.

[2] Hoxhold B, Kirchhoff M, Bütefisch S, Büttgenbach S (2006) Sma driven micro grippers combining piezo-resistive gripping force sensors with epon su-8 mechanics. In: Proc. Eurosensors, vol 2, pp 190–191

[3] Kärcher B, Schreiber P, Fischer A, et al (2005) Mikrooptischer Positionssensor. In: Proc. of Mikrosystemtechnik-Kongress

[4] Kornfeld A, Stute U, Ostendorf A, et al (2004) Herstellung integriert-optischer Wellenleiter durch Excimer- und fs-Pulslaser für die interferometrische Positionsregelung. In: Jahrest. der DGaO, 105-B31

[5] Kornfeld A, Bärsch N, Kracht D, Ostendorf A (2008) Integrated optical micro structures for signal processing in the position metrology. J of Microsys Techn 14:1955–1960, dOI: 10.1007/s00542-008-0638-8

[6] Kornfeld A, Kolleck C, Ostendorf A (2009) Evaluation, Justage und Fixierung mikrooptischer Bauelemente zur Integration eines Interferometrischen Positionssensors. In: 4. Koll. Mikroproduktion, BIAS-Verlag, pp 171–176

[7] Kowalevicz M, Sharma V, Ippen E, et al (2005) Three-dimensional photonic devices fabricated in glass using a femtosecond laser oscillator. Opt Lett 21:1060–1062

[8] Morgner U (2008) Ultrakurz, Ultrastark, Ultrapräzise. In: 2. Workshop Optische Technologien, PZH-Verlag, pp 114–133

[9] Osellame R, Chiodo N, Maselli Vea (2005) Optical properties of waveguides written by a 26 MHz stretched cavity Ti:sapphire femtosecond oscillator. Opt Expr 13:612–620

[10] Szameit A, Dreisow F, Heinrich M, Pertsch T, Nolte S, Tünnermann A (2007) Coupling management of fs laser written waveguides. In: Proc. of SPIE, vol 6460, 0277-786X, 64600V-1 - V11

Chapter 9
Tactile Dimensional Micrometrology

S. Bütefisch, U. Brand, L. Doering, G. Dai, L. Koenders, G. Wilkening

Department 5.1 Surface Metrology
Physikalisch-Technische Bundesanstalt
uwe.brand@ptb.de; ludger.koenders@ptb.de

Abstract Tactile dimensional micrometrology faces different challenges in quality control of microsystems. Among all, the measurement in three dimensions and the great variety of different materials are great challenges. A microprobe has been optimized to solve both problems. It makes three dimensional measurements of coordinates and forces. Recently observed probing force dependent systematic deviations of tactile measurements are reported which reveal the necessity of exact probing force setting and calibration especially for microprobes. The probing force calibration by means of a high-resolution compensation balance and a new piezoresistive microforce sensor are described. For the complete calibration of micro measuring machines new dimensional standards for the 3D calibration of microprobes and positioning systems are presented.

9.1 Introduction

The excellent mechanical properties of silicon give silicon microprobes great metrological potential for quality control of microsystems. The development of an optimized piezoresistive silicon 3D microprobe, its metrological properties and its application for hardness and elasticity measurements are described in Sect. 9.2. The contribution is not restricted to the coordinate measuring properties but also concentrates on the force measuring properties of this probing system. The motivation for investigating the force properties are material dependent deformations during the tactile measurement process. Since modern microsystems are composed of different materials, different deformations occur and lead to systematic measurement deviations.

Problems occur if soft materials and materials of different mechanical properties have to be measured (Fig. 9.1). Microsystem structures which are composed of different materials such as Si, SiO_2, Cu, Al, *SU-8* or NiFe, deform during tactile measurement as a function of the Youngs modulus and the

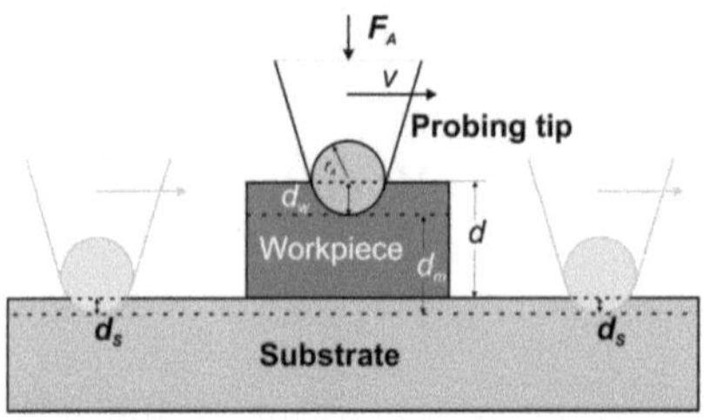

Fig. 9.1 Probing force dependent deformations at stylus instrument measurements (F_A: probing force, v: traverse speed, d_m: measured layer thickness, d: layer thickness, d_W: deformation of the layer, d_S: deformation of the substrate)

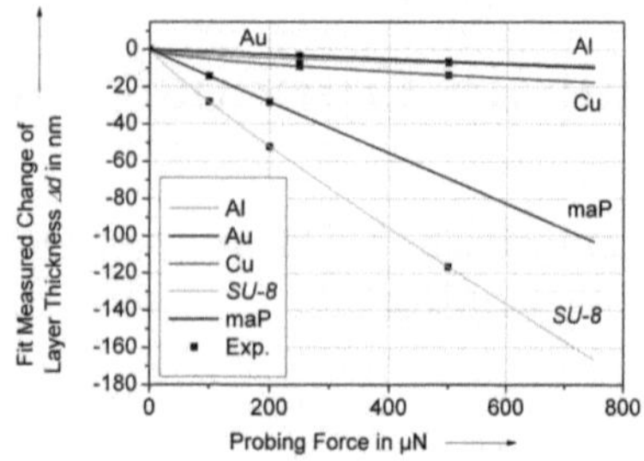

Fig. 9.2 Fit of the measured layer thickness change Δd in dependence of probing force for different layer materials on Silicon Substrates (stylus instrument with $r = 2\,\mu m$ and $v = 50\,\mu m\,s^{-1}$)

viscous modulus, the hardness and the scanning speed (Fig. 9.2) [3]. In the third section details are given of the calibration of probing forces using a compensation balance and a new piezoresistive micro force sensor, which is insensitive to the loading position. Application of 3D microprobes in a coordinate measuring machine leads to a demand for dimensional standards to measure both, the properties of the probing system and the properties of the positioning system. Dimensional standards which have been developed for that purpose are presented in Sect. 9.4.

9.2 3D Microprobes

Tactile metrology became more important for the manufacturing of hybrid microsystems in recent years. Apart from optical methods tactile metrology is an important tool for quality assurance. Optical measuring methods often reach their limits, especially for the dimensional measurement of three-dimensional objects, which can have undercut structures. In these cases tactile metrology is the only possibility to check the accuracy and to ensure the functionality of micro mechanical components. The conventional coordinate measurement technique offers 3D probing uncertainties down to 0.2 μm. Conventional probing-systems have probing spheres with a minimum diameter of 300 μm and probing forces of several mN. The measurement of hybrid microsystems requires smaller probing elements, smaller probing forces and smaller measuring uncertainties. Extensive research in development and optimization of tactile probing systems has been carried out in the last years [25].

Apart from a 1D cantilever probe which is capable of performing line scans on the inner walls of deep holes [19] a new patented so called "assembled cantilever probe" (ACP) has been developed and characterized here. This probe consists of a conventional AFM cantilever to which a second cantilever is attached [12]. Furthermore in collaboration with the Institute of Microtechnology (IMT) of Technische Universität Braunschweig (Chap. 6) a 3D microprobe, suitable for point and scanning probing in ultra high precision coordinate measuring machines was developed and optimized.

9.2.1 Working Principle of the Piezoresistive 3D Microprobe

The microprobe consists of a sensing element, a stylus with a probing sphere and a mounting. The sensing element is based on a silicon boss membrane which was fabricated by anisotropic KOH-etching. This wet chemical etch process is time controlled resulting in diaphragm thicknesses of 20 μm to 30 μm leaving a frame and a center boss of 400 μm thickness. The square opening is $3.4 \times 3.4\,\mathrm{mm}^2$, the top area of the center boss $1 \times 1\,\mathrm{mm}^2$ wide and the overall dimensions of the sense die are $6.5 \times 6.5\,\mathrm{mm}^2$. The tactile element of the sensor is a 7 mm long commercial stylus with a probing sphere with a diameter of 120 μm to 300 μm attached to the center boss by epoxy resin (Fig. 9.3). A deflection applied to the tactile element during the contact process is transmitted through the stylus to the center boss and results in a deformation of the diaphragm. The occurring deformations and the resulting states of mechanical stresses are shown in Fig. 9.4 for a vertically (z-direction) and a laterally (xy-direction) applied deflection.

Standard processing steps for insulation, diffusion and metallization have been carried out to fabricate piezoresistive elements on the backside of the membrane in the regions of high mechanical strains. The change in resistance is proportional to the mechanical strain. The design of the resistors is shown in Fig. 9.5. The working principle of this resistor configuration is described in detail in [5] and [7]. For the described sensor chip four of these structures in a Wheatstone configuration are placed on the regions between the membrane and the supporting frame (see Fig. 9.5, left). Neglecting crosstalk, the deflection in x-, y- and z-direction can be calculated with the signals of the four bridges:

$$a_z = a_1 + a_2 + a_3 + a_4 \tag{9.1}$$

$$a_x = a_2 - a_4 \tag{9.2}$$

$$a_y = a_1 - a_3 \tag{9.3}$$

$$\tag{9.4}$$

$a_{x,y,z}$ transformed signal
a_i signal of bridge i

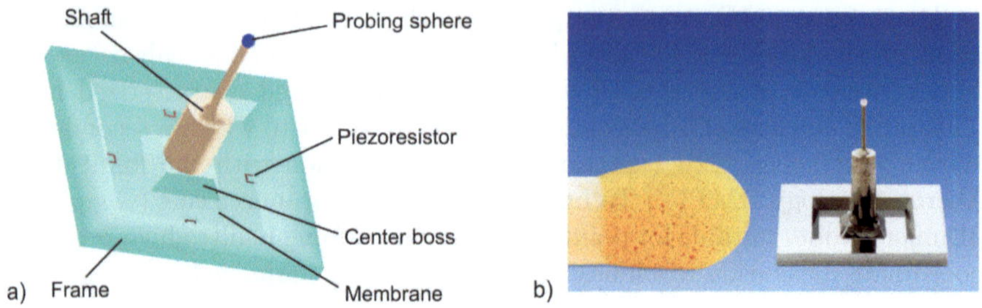

Fig. 9.3 Schematic (**a**) and photograph (**b**) of the initial microprobe design

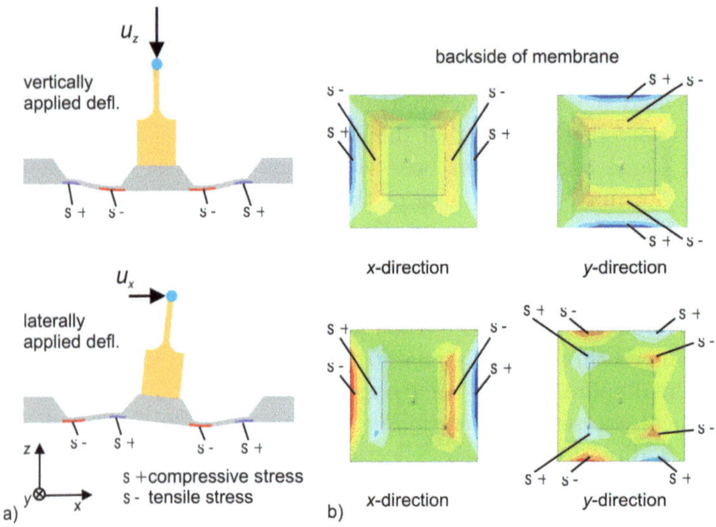

Fig. 9.4 (**a**) Motion of the center boss induced by a vertically and horizontally applied deflection of the probing sphere; (**b**) An FE analysis of the induced stress occuring in the membrane caused by this motion

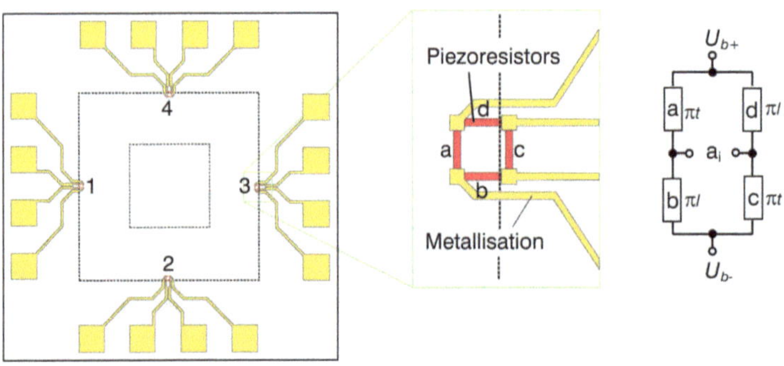

Fig. 9.5 Layout and circuitry of the piezoresitive elements

9.2.2 Optimized Fabrication Process

The piezoresistive elements were produced using a high temperature diffusion process to dope the silicon resulting in conductive paths [9]. To realize the electrical contact to the piezoresistors a gold layer was used with chromium as adhesive layer. As insulation layer silicon dioxide was applied. In a subsequent step the membrane was machined through a time controlled KOH wet etching. Fig. 9.6 shows a schematic cross-section of the layer system.

First versions of the microprobe showed large variations of the electrical performance from batch to batch. Apart from well working sensors there were batches with strongly detuned Wheatstone-bridges. In order to improve the reproducibility of the probe properties the fabrication process was optimized. The contacts between the piezoresistors and the metallic circuit path consisting of gold and chromium were identified as the error source [6].

Two actions were undertaken to improve the quality of the contacts:

- use of aluminum as metallization
- implementation of highly-doped contact areas

In the semiconductor industry aluminum is widely used as metallization due to its excellent contact qualities. It hasn't been used so far for the microprobe because it is non-resistant to the KOH solution, which is used for etching of the membrane. To be able to use aluminum as metallization a protection against the KOH etchant has to be provided. Therefore a special

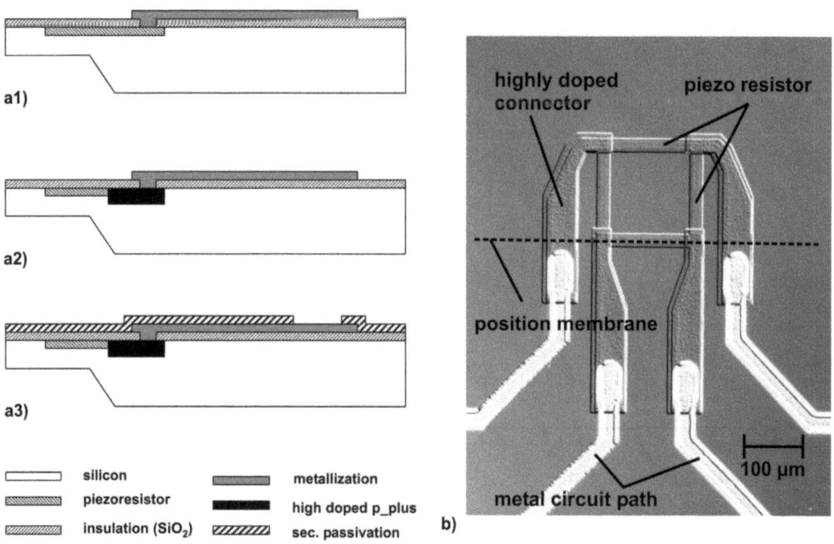

Fig. 9.6 Schematic of the layer-system: (**a1**) Initial process; (**a2**) High doped contact areas; (**a3**) Secondary passivation; (**b**) Micrograph of the sensing element

wafer-support was used [2], where the piezoresistors and the metallization are in a sealed cavity and only the backside of the wafer is exposed to the etchant.

The doping concentration in the contact area is another influencing factor for the quality of contact resistances. The contact resistance becomes smaller with increasing doping concentration. Unfortunately the sensitivity of the microprobe decreases with increasing doping concentration of the piezoresistors [9]. For the initial process a doping of about $1 \cdot 10^{19}\,\mathrm{cm}^{-3}$ for the contact area was chosen as a compromise between high sensitivity and high stability. A new approach consists in a two-stage diffusion: the contact areas are highly doped ($> 1 \cdot 10^{20}\,\mathrm{cm}^{-3}$) and for the actual piezoresistors a lower doping was chosen (see Fig. 9.6). This leads to higher bridge-resistances which is advantageous because the current and thus the heating of the sensor-chip decreases. Measurements showed that the contact resistance could be decreased by two orders of magnitude to $2.4\,\Omega$ [8].

Two more steps were taken to further increase the electrical stability of the sensor:

- replacement of the former diffusion process by a commercially available ion implantation process to improve reliability of the piezoresistors
- introduction of a secondary passivation layer to protect the piezoresistors against environmental influences like humidity or ion-contamination

Table 9.1 shows the electrical parameters "bridge resistance" and "offset" of sensors of wafers that have been fabricated using the diffusion process and of one wafer fabricated with the new implantation process. The stability of these parameters is a good indicator for the stability of the sensor itself and of the reproducibility of the fabrication process. Shown are the values for wafers with slightly different diffusion parameters, resulting in different bridge resistances. The standard deviation of the bridge resistance varies from $6 - 13\%$ full scale and the standard deviation of the offset from 1 to $4\,\mathrm{mV/V}$ for the diffusion process (wafer ID 5-8).

Table 9.1 Bridge resistance Rb and offset of sensors fabricated with the diffusion process (wafer ID 5-8) and implantation (wafer ID 9)

Wafer ID	Rb		Offset		Technology
	mean value [Ω]	standard deviation [%FS]	mean value [mV/V]	standard deviation [mV/V]	
5	954	13.2	-0.27	1.07	diffusion
6	698	5.7	4.05	3.95	diffusion
7	756	11.6	-1.97	2.56	diffusion
8	552	6.7	2.96	3.83	diffusion
9	1920	0.9	-1.25	2.79	implantation

The standard deviation of the offset of the implanted resistors (wafer ID 9) is 2.79 mV/V which is in the same range like the values for the diffusion process and a standard value for semiconductor sensors like e.g. pressure sensors. It is mainly caused by deviations occurring due to the lithography process. The mean value for the bridge resistance is 1920 Ω with a standard deviation of 0.9% FS which is an enormous improvement compared to the results of the diffusion process in terms of reproducibility of the fabrication process.

A further action has been undertaken to improve the electrical stability of the sensor by introducing a so called secondary passivation.

To insulate the piezoresistors and metallization a passivation layer is provided (see Fig. 9.6.a1). This layer is produced by simply thermally oxidizing the silicon resulting in an insulating silicon oxide layer. As a drawback the silicon oxide as insulation layer is permeable for small ions like e.g. potassium. These ions can diffuse through the oxide layer into the piezoresistors and change the electrical conductivity. This causes a drift of the output signal.

To minimize the environmental influence a second passivation layer was introduced. For this layer silicon nitride was chosen since it has excellent properties in terms of hermeticity and encapsulation providing protection against external environmental influences. The layer was applied on top of the layer system using a low temperature PECVD process. In a subsequent step the layer was opened over the bond pads enabling the wire bond (see Fig. 9.6.a3). This modification led to a further improvement of the electrical performance of the sensor [7].

9.2.3 Metrological Properties

Sensitivity Calibration

To perform the sensitivity calibration the microprobe was mounted on an ultra high precision micro/nano CMM based on the Nano Measuring Machine (NMM) [15]. The work piece can be moved in a volume of $25 \times 25 \times 5\,\text{mm}^3$ with nanometer accuracy. The precision movement is ensured using laser interferometers for all three axes.

Sensitivity calibration is a basic investigation task in the characterisation of the probe. It is described in detail in [13] resulting in a 4×3 matrix enabling the transformation of the four sensor signals into the deflection of the probing sphere in all three directions in space.

Probing Repeatability

The microprobe can be used in either point-probing or scanning mode. To investigate the repeatability of measurements, a point wise probing mode was

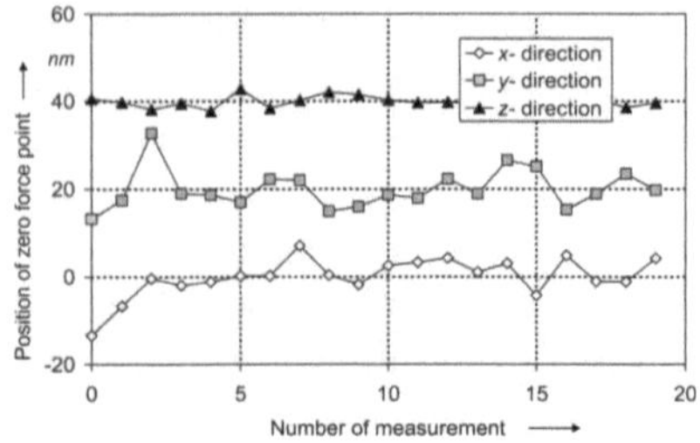

Fig. 9.7 Measured probing repeatability along the x-, y- and z-axis of the piezoresistive 3D microprobe

chosen. The measured probing point is calculated at zero probing force by linear fitting the probing curve. To examine the probing repeatability, several measurements at the same position of the sample were performed within a few minutes. The microprobe was retracted entirely from the surface between each measurement. The measured results are shown in Fig. 9.7. The standard deviations of the results are 1.3 , 4.4, and 4.4 nm along the z-, y- and x-axes, respectively. It can be seen that the probe has an improved probing performance along the z-axis compared to the x- and y-axes. This can be explained by the larger measurement sensitivity along the z-axis [13].

3D Measurement

In order to determine the 3D measuring capabilities of the 3D microprobe, a double ball artefact consisting of two ruby spheres with nominal diameters of 2 mm as described in [24] was fabricated. The two spheres are fixed on an *Invar* plate with a seperation D of 3 mm. The ruby spheres are measured using the microprobe at 30 probing points equally spaced on its upper hemisphere. The measurement on each sphere takes about four minutes.

The measured sphere radii R_1 and R_2 and the distance D between the spheres are illustrated in Fig. 9.8. Standard deviations of R_1, R_2 and D over 20 repeated measurements are 1.9 nm, 2.1 nm and 3.1 nm, respectively, indicating the high measurement performance of the microprobe. This per-

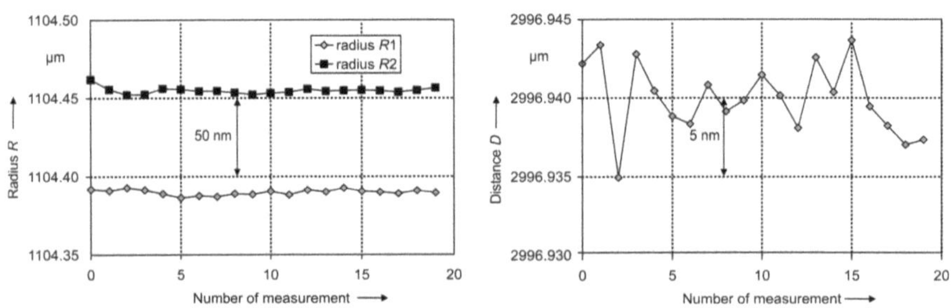

Fig. 9.8 (**a**) Measured radii R_1 and R_2 of the double ball artefact; (**b**) Measured distance D between the two spheres

formance is close to that stated for the ultra precision micro CMM developed at the National Metrology Institute of Switzerland – *METAS*, where a measurement repeatability of 1.5 nm for the determination of the inner diameter of a 3 mm micro ring was obtained.

9.2.4 Optimized Membrane Designs

The initial design of the microprobe showed a high stiffness compared to other microprobes. On the one hand this is an advantage since the resonance frequency is high and therefore the tendency to oscillations excited by the movement of e.g. the drives of the coordinate measuring machine (CMM) is low. Another advantage is that since even small deflections of the tactile element induce high mechanical stress in a stiff structure, and the electrical sensitivity is proportional to the mechanical stress, even very small deflections produce decent electrical signals resulting in a good signal to noise ratio. On the other hand the high stiffness of the microprobe is a disadvantage since high forces are necessary to deflect the stylus. In combination with small probing balls ($< 300\,\mu m$) these high contact forces lead to very high surface pressures resulting in elastic or even plastic deformation of the work piece [16]. Another drawback of such stiff structures is mechanical failure due to high stresses occurring in the structure. The maximum allowed deflection of the probing ball in z-direction is $30\,\mu m$ and $100\,\mu m$ for the x- and y-direction. The CMM has to be able to stop within this distance to avoid the breakage of the probe.

The isotropy of the stiffness is a parameter of great importance for microprobes. In the scanning mode a CMM controls its movements in reference to the deflection of the microprobe. If the stiffness of the microprobe is anisotropic a constant deflection results in anisotropic contact forces. Especially for soft materials this might lead to measuring errors due to the

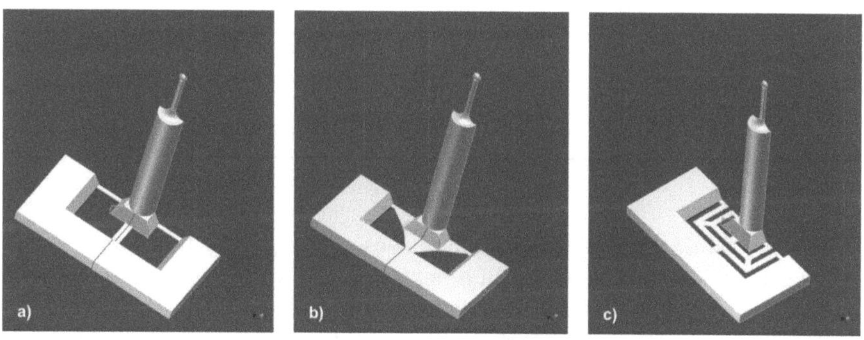

Fig. 9.9 New designs with structured membrane: (**a**) Beam design; (**b**) Triangle design; (**c**) Grid design

anisotropic depth of indentation of the probing ball. To overcome this problem the CMM has to be controlled using the contact force as parameter which makes the determination of the 3D stiffness of the microprobe and its implementation in the CMM controller necessary. Therefore it is desirable to make the stiffness of the probe as isotropic as possible to avoid the described problems. The ratio between the stiffness in z- and xy-direction for the initial design is 30–40 which makes the usage of this probe in a scanning machine with standard controllers impossible. Therefore the reduction of this ratio is a very important focus of all optimization efforts [20] and [23].

One possibility to modify the stiffness of the microprobe is to insert apertures in the membrane [5]. This leads to higher mechanical strains and a higher sensitivity.

Three approaches for a new probe design (see Fig. 9.9) will be presented and their mechanical behavior will be investigated using finite element analysis. As critical parameters the anisotropy of the stiffness and the sensitivity will be used as benchmark in comparison to the initial design.

To determine the sensitivity of the sensor the difference between the mechanical stress in $\langle 110\rangle$- and $\langle 1\bar{1}0\rangle$-direction of the longitudinal and transversal piezo coefficient in p-doped silicon at the position where the piezoresistors are located was analyzed [8]. Since this value is directly proportional to the output signal of the sensor it can be taken as equivalent for the sensitivity. The different designs are discussed in detail in [8] and the results of this analysis are given in Table 9.2. It can be observed that a dramatic reduction of the stiffness by a factor of 30 compared to the initial design could be achieved with the grid design whereas the ansisotropy could only be reduced by 18% to a value of 26.

Table 9.2 Mechanical stresses and stiffnesses of sensors fabricated by the diffusion process

Membrane design	$(\sigma_l - \sigma_t)_z$ [MPa]	$(\sigma_l - \sigma_t)_{xy}$ [MPa]	k_z [N m^{-1}]	k_{xy} [N m^{-1}]	k_z/k_{xy}
full membrane	54	10	6000	190	32
beam	62	12	700	20	34
triangle	80	16	1500	53	29
grid	8.9	1.1	175	6.8	26

For the fabrication of the described sensor designs the first steps of the standard process were used [4]. In a next step the wet chemically structured membrane was etched at the specific positions using a dry etch process [9]. As mask for this process photoresist was used. Fig. 9.10 shows the fabricated, fully functional sensor chips. For further reduction of the anisotropy of the stiffness of the microprobe a new appoach has been undertaken resulting in

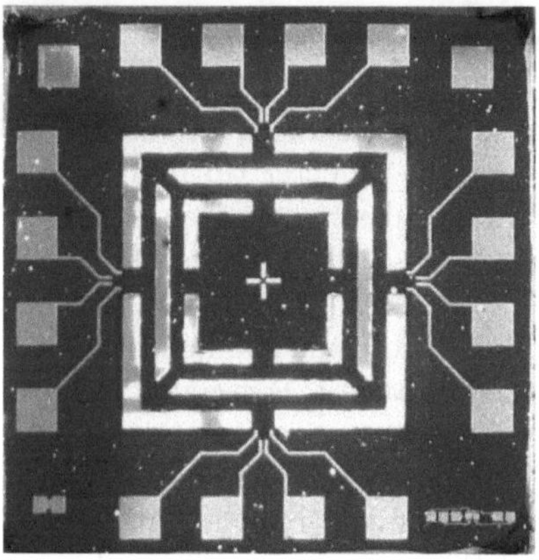

Fig. 9.10 Photograph of the top side of the "grid"-design

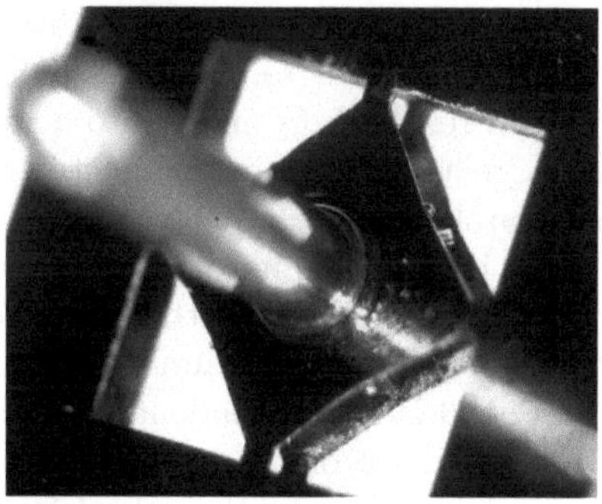

Fig. 9.11 Photograph of the isotropic "double triangle"-design

a new design. It consists of two structured boss membranes (triangle design), mounted face to face (Fig. 9.11). This double beam configurations leads to a dramatically increased stiffness in the x- and y-direction by blocking the tilting motion of the center boss through the eight supporting beams. For this "double triangle"-design a ratio between the xy- and the z-direction of two was achieved representing a nearly isotropic stiffness of the microprobe. In first tests this probe showed full functionality with approximately the same sensitivity like the initial design for the xy- and a four times higher sensitivity for the z-direction.

9.2.5 Application of the 3D Microprobe for Elasticity and Hardness Measurements

The measurements were made by a *LabView* guided measuring device consisting of a 3D microprobe with a pyramidal Berkovich-Indenter stylus, a compensation balance for calibrating the probing forces, a nanopositioner for creating forces, a temperature and humidity sensor, a home made bridge amplifier and a 100 kHz AD-converter board. The calibration of the 3D microprobe was carried out by simultaneously measuring the force with the compensation balance, the deflection of the nanopositioner (h_{Pifoc}) and the 3D microprobe signals. The actual indentation measurements were made without the balance. The deviation of the probing forces measured with the 3D microprobe were $< -1.3\%$.

The microprobes signal noise was 1.3 µN, increasing to 12 µN during probing a specimen at a probing force of 10 mN.

The tip rounding of the Berkovich indenter (angular pitch: 120°, interfacial angle: 65.3°) and the stiffness of the instrument k_{Instr} were determined by indentation measurements at two reference materials (Fused Silica and Sapphire).

The unloading curves were fitted using the Hertzian equation. An instrument stiffness $k_{Instr} = 20400\,\mathrm{N\,m^{-1}}$ and the indenter tip radius $R = 1.14\,\mu m$ resulted. Compared to a commercial hardness tester the setup used has a relatively low instrumental stiffness, because the stiffness of the 3D microprobe ($k_z = 19464\,\mathrm{N\,m^{-1}}$) is relatively low.

The indentation hardness H_{IT} was calculated from the maximum contact force F_{max} divided by the projected contact area A_k. For contact depth values greater than $0.7\,\mu\mathrm{m}$ the measured areal function corresponds to that of an ideal Berkovich indenter $23.96 \cdot h_c^2$. To calculate the indentation depth the compliance of the instrument h_{Instr} has to be subtracted from h_{Pifoc}. It is calculated from the actual force F and the instrument stiffness k_{Instr}. Starting from the model of Oliver and Pharr [18] the contact area A_k is calculated from the stiffness S of the material to be measured, the reduced Young's modulus E_r and a term β related to the geometry of the contact:

$$A_k = \frac{\pi \cdot S^2}{4 \cdot \beta^2 \cdot E_r^2} \tag{9.5}$$

The contact stiffness S is calculated from the measured unloading curve at the point of maximum testing force F_{max}. This is implemented by determining the slope of the upper 60–80% of the unloading curve. The contact depth h_c follows from the maximum indentation depth by subtracting the elastic deformation $\epsilon \cdot \frac{F}{S}$. The constant ϵ depends on the geometry of the indenter and has been determined by Oliver and Pharr [18] to 0.75 for Berkovich indenters.

Table 9.3 Experimental indentation modulus E_{IT} and indentation hardness H_{IT} in comparison to reference values

Material	Exp. indentation modulus (GPa)	Reference modulus (GPa)	Poisson's ratio	Exp. indentation hardness (GPa)	Reference hardness (GPa)
Fused silica	-	72 [10]	0.17	10.2 ± 6.8	9.3 [11]
Sapphire	-	410 [10]	0.234	48 ± 40	13 [17]
Aluminum, single crystal	59 ± 11	70.4 [22]	0.347	0.66 ± 0.27	0.59 [17]
Polycarbonat	3.5 ± 0.7	3 [10]	0.37	0.25 ± 0.07	0.17 [17]

The indentation modulus E_{IT} was calculated using Eq. (9.5) and using the contact area of an ideal Berkovich indenter. Fig. 9.12 shows the mea-

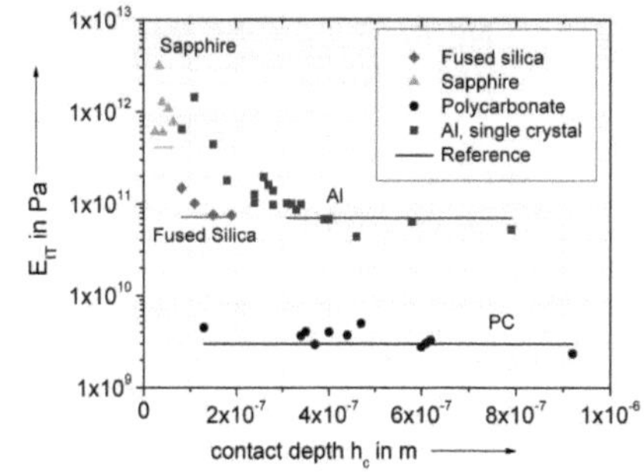

Fig. 9.12 Measured indentation modulus E_{IT} versus contact depth h_c

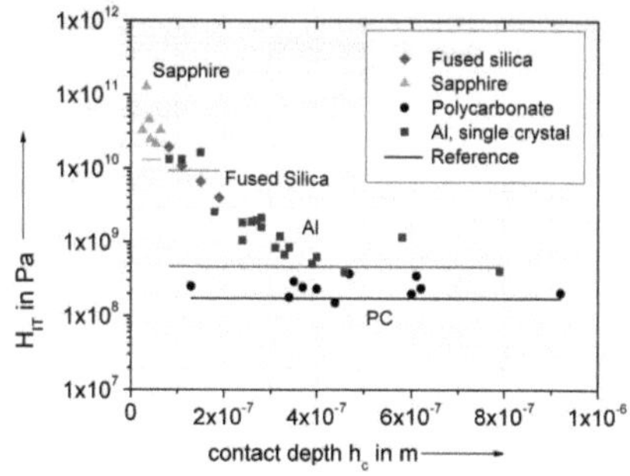

Fig. 9.13 Measured indentation hardness H_{IT} versus contact depth h_c

sured reduced indentation modulus E_{IT} as a function of contact depth h_c. For all investigated materials it can be seen that for small contact depths h_c the measured indentation modulus was too large. This corresponds to observations of Chudoba et al. [11]. When the measured indentation moduli are averaged for $h_c > 261$ nm then the values listed in Table 9.3 result. For the very compliant polycarbonate the measured modulus compares very well with the reference value. For the aluminum single crystal only from $h_c > 340$ nm a constant indentation modulus of 59 GPa is measured. The measured value deviates from the reference by only −16%. For Fused Silica a constant indentation modulus can be observed already from $h_c > 150$ nm, which deviates by only a few percent from the reference value. For Sapphire no indentation depths > 62 nm and thus no reliable measurement values could be realized due to the force limit of the 3D microprobe of 10 mN.

The indentation hardness H_{IT} was calculated from the maximum force divided by the projected contact area ($A = 0.207 \cdot h_c^{1.658}$). The measured values are shown in Fig. 9.13 in comparison with reference values. The measured indentation hardness corresponds quite well with the reference values (Table 9.3).

9.3 Calibration of Probing Forces

9.3.1 Compensation Balance Based Micro Force Measurement

For the calibration of bending stiffnesses with forces in the nN range a measuring device based on a compensation balance and a nanopositioner has been set up (see Fig. 9.14 (a)). The compensation balance with its measurement range from 2 g to 0.1 µg was used to measure probing forces with a resolution of 1 nN. The reproducibility of the balance was 2.5 nN, the linear-

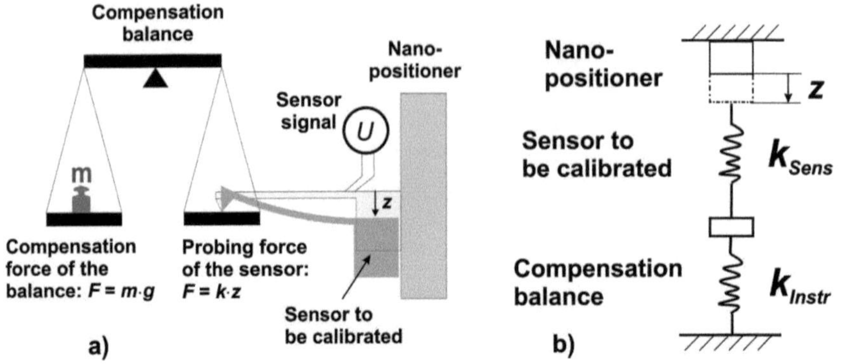

Fig. 9.14 (**a**) Sketch of the nanoforce measuring device; (**b**) Block diagram of the spring model of the nanoforce measuring device

ity deviations were 9 nN and its measuring uncertainty was estimated to be 0.1 μN.

In order to measure the bending stiffness k of sensors a coarse positioner and the nanopositioner have been mounted on the balance housing above the pan of the modified balance [14]. For tactile probing very flat and smooth probing areas were mounted on the balance pan. To determine the bending stiffness the probing force F necessary to achieve a certain deflection z of the sensor was measured. This is done by moving the sensor in fixed steps towards the balance pan until it touches the weighing platform. Then the balance measures the force corresponding mass. The measured bending stiffness k_{meas} results from the coupling of two mechanical springs, the one of the sensor to be measured k_{Sens} and the one of the nanoforce measuring device k_{Instr} (Fig. 9.14 (b)). From the measured stiffness ${k_{meas}}^{-1} = {k_{Sens}}^{-1} + {k_{Instr}}^{-1}$ the stiffness of the sensor can be determined if k_{Instr} (8900 N m^{-1}) is known.

The nanoforce measuring device is run in a thermally isolated box in order to reduce the thermal drift. In a typical measuring time of 20 minutes the temperature inside the box drifts by 1.3 mK. This leads with the thermal length change of the setup of 2.7 μm K^{-1} to a z-position drift of 4 nm. Besides the thermal length change of the setup the balance also shows a temperature induced drift of 0.38 μN K^{-1}. Subtracting this temperature effect from the measured force leads to the temperature compensated force resolution of the device of 1.5 nN in a measuring time of two hours.

9.3.2 Micro Force Transfer Standard for Probing Forces

Since for precise tactile dimensional metrology on soft surfaces the contact force is a parameter of crucial importance, there is a need of transfer standards to survey the contact forces exerted by tactile measuring equipment.

Therefore a new compact micro force transfer standard was developed and realized using bulk silicon fabrication techniques.

The sensor structure is based on the typical double beam configuration often used for conventional force sensors (Fig. 9.15). One side of the sensor is fixed and the force to be measured is applied to the opposite end of the structure. Due to the double beam configuration the point where the force is applied has ideally no influence on the result of the measurement which is a big advantage when using this sensor for the calibration of contact forces in tactile measuring equipment (stylus instruments/ CMMs).

The basic structure of the sensor consists of two bonded silicon chips with an overall dimension of $6.5 \times 3.5 \times 0.8\,\text{mm}^3$. At some positions the thickness of the material is reduced to 15–20 μm building the flexural hinges of the double beam structure. For the machining of the silicon the same fabrication processes like for the microprobe where used. To enable the free movement of the structure the thickness of the two center bosses have to be reduced in an additional etch step. To realize the sandwich structure two wafers of structured silicon are needed which are connected by silicon direct bonding. A standard dicing saw was used to separate the sensor dies leaving the compliant structure of the sensor.

By applying a force to the sensor structure mechanical stress occurs at the position of the flexural hinges. This mechanical stress can be converted into an electrical output signal using piezoresitve elements. The piezoresistors are fabricated by means of standard semiconductor fabrication techniques already described above. Fig. 9.15 shows a schematic and a photograph of the first prototypes. To demonstrate the functionality of the sensor principle, the first prototypes where assembled using modified sensor dies of the microprobe described above. Therefore there are two sets of piezoresistors on the chip, whereas only one set would be necessary leaving the other one without function.

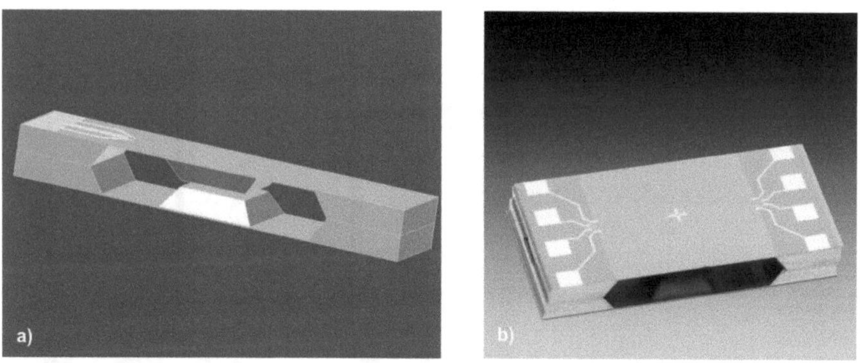

Fig. 9.15 Micro force sensor as transfer standard for the calibration of probing forces: (**a**) Schematic of the structure; (**b**) Photograph of the fabricated sensor

9.4 Dimensional Standards for Micro Metrology

For the validation of dimensional measuring equipment there is a need of dimensional standards to ensure the quality of the measurements on the one hand and to enable the possibility of comparison between different measuring equipment or even between different measuring principles (e.g. optical and tactile) on the other hand. The development of such standards is subject of intense research whereas a special focus lies on standards, which can be used for tactile as well as optical probing.

9.4.1 3D Silicon Artefact with Enhanced Groove Topography

A cooperation between PTB and the Institute of Microtechnology at the Technische Universität Braunschweig yielded in a concept for a lateral standard based on silicon, fabricated using wet chemical etching processes. It consists of an array of inverse truncated pyramids which are placed on a four inch silicon wafer.

The pyramidically shaped grooves are realized using a wet chemical anisotropic etching process with KOH. Using a ⟨100⟩ wafer the side walls of the rectangular grooves exhibit the typical angle of 54.7° to the wafer surface. The intersection of two adjacent side walls and either the bottom of the groove or the wafer surface results in a reference point. Eight reference points can be obtained by one pyramid. The array of pyramids leads to a set of reference points with the lateral size of the whole wafer.

This grid of reference points can be used to determine systematic deviations of CMMs like e.g. guidance errors or to perform comparisons between different measuring machines. A crucial parameter for the repeatability of the probing process is the flatness of the planes to be measured. The wafer surface exhibits a very good flatness of optical quality. The quality of the

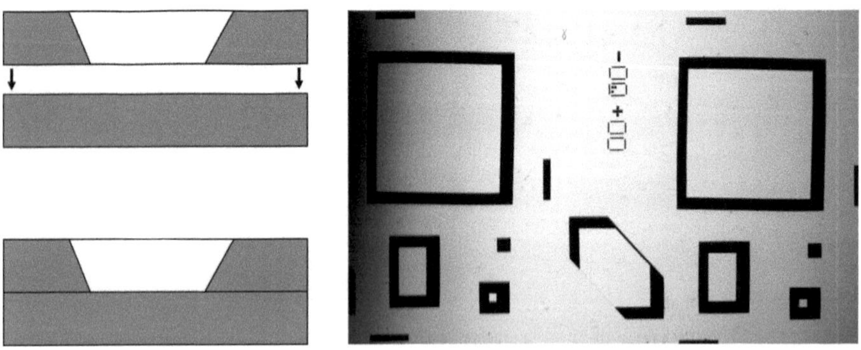

Fig. 9.16 Silicon artefact with enhanced groove topography using the silicon direct bond technology

side walls which are being build by the ⟨111⟩ planes in the silicon crystal depends mainly on the exact alignment of the etch mask to the crystallographic features [21]. The quality of the bottom of the groove depends on the uniformity of the etch process. The roughness of the surface is affected by the composition of the etchant and the etching parameters (temperature, concentration). It is possible to control these parameters yielding in a robust fabrication process capable to produce these structures on a commercial scale [1].

A new approach to decrease the surface roughness of the bottom of the groove was undertaken. The wafer was etched completely to build a rectangular hole with inclined side walls. Then a second – unstructured – wafer was bonded to the first one using the silicon direct bond method (see Fig. 9.16). This two wafer configuration also yields a pyramidal groove similar to the one described above. Since the (very smooth) surface of the lower wafer builds the bottom of the groove its topography is much smoother than the one of the etched version. Fig. 9.16 shows a photograph of the fabricated structure. The material is also a ⟨100⟩ silicon wafer which was etched using a KOH solution with a concentration of 40% and a temperature of 80°C. Besides the inverse truncated pyramids built by the ⟨111⟩ planes, structures with vertical walls where obtained by rotating the etch mask about an angle of 45° in respect to the ⟨110⟩-direction. The vertical walls of these structures are formed by ⟨100⟩ planes. Drawback of this structure is, that the ⟨100⟩ planes are not an etch stop like the ⟨111⟩ planes. Therefore the width of the structure is not only determined by the etch mask, but also by the etch time.

For the comparison of different probing principles like e.g. tactile probing and optical probing, standards which are suitable for use with both probing methods are of great interest. A first approach of a marker which can be probed by optical and tactile probes is shown in Fig. 9.17. The topology is also an inverse (complete) pyramid with a square footprint limited by ⟨111⟩ planes in silicon. The marker was fabricated in the same etching step like the rest of the structure. Since the whole structure is limited by ⟨111⟩ planes the dimensions are (relatively) independent of the etch time. For tactile probing the probing sphere rests on the flanks of the pyramid. For optical probes the edges of the etched groove provide a good contrast for the digital image processing.

9.4.2 3D Hybrid Artefact

Since spheres with extremely low form deviations are commercially available, they are widely spread as reference artefacts due to the easy way to determine their centre point with only three probing procedures. Optical probing tools often have problems performing dimensional measurements on spheres since they provide very poor contrasts for the edge detection in image based procedures and generally with the poor reflectivity of the smooth surface.

Fig. 9.17 Markers for tactile and optical detection

The hybrid artefact is a new approach to solve this problem by combining a high grade sphere with a silicon structure. The silicon structure was fabricated using the process described above whereas opto/ tactile markers on the one hand and a support for the sphere on the other hand were realized. Then the sphere was fixed in the support by an adhesive. Since the position of the markers and the support are very precisely defined by the lithographically structured etch mask and the sphere itself has a low form deviation a high precision reference system was obtained. The marker described above has the draw back that it is kinematically over determined, because the probing sphere potentially contacts four points at the side wall of the marker. In order to overcome this problem a rectangular marker was designed and realized in the hybrid artefact (see Fig. 9.18). To determine the reference point of this marker the probing sphere is lowered into the structure till it rests on the side walls of the V-groove. Then the probing sphere is moved towards one of the two end faces, always keeping in contact with the side walls. By touching one of the two end faces the probing sphere rests on three points resulting in a kinematically defined position. Since the other end faces together with the

Fig. 9.18 3D hybrid artefact, composed by a structured silicone base plate with rectangular markers and a steel ball

two side walls also builds a kinematic rest this marker provides two reference points.

Acknowledgements The authors gratefully acknowledge the funding of the reported work by the German Research Foundation within the Collaborative Research Center 516 "Design and Manufacturing of Active Micro Systems".

References

[1] (2010) http://www.simetrics.de/. Internet, accessed on May 12.
[2] (2010) www.silicet.de/. Internet, accessed on May 12.
[3] Brand U, Nesterov V, Doering L, Bütefisch S, Peiner E, Büttgenbach S, Frühauf J (2009) Neue taktile Sensoren für die Mikro- und Nanotechnik. TM pp 323–331
[4] Bütefisch S, Dauer S, Büttgenbach S (1999) Silicon three-axial tactile sensor for the investigation of micromechanical structures. Proc. Sensor 99, Nürnberg, vol 2, pp 321–326
[5] Bütefisch S, Solzbacher S, Ziermann R, Krause P, Büttgenbach S, Wilke R, Cao C, Pornnoppadol P, Brand U, Seitz K, Roth R (2003) New Micro Probe for Dimensional Metrology based on a Silicon Micro-structure. Proc. Sensor 2003, pp 75–80
[6] Bütefisch S, Brand U, Büttgenbach S, Wilkening G (2007) Taktile Messtechnik für aktive Mikrosysteme. In: Kolloquium Mikroproduktion – Fortschritte, Verfahren, Anwendungen, Karlsruhe, pp 201–206
[7] Bütefisch S, Dai G, Danzebrink HU, Brand U, Büttgenbach S (2009) Ultra high precision 3D microprobe for CMM applications. MikroSystemTechnik, Berlin
[8] Bütefisch S, Dai G, Feldmann M, Brand U, Koenders L, Büttgenbach S (2009) Neue Sensorstrukturen für die taktile Messtechnik an Mikrokomponenten. In: 4. Kolloquium Mikroproduktion, BIAS Verlag Bremen, pp 183–190
[9] Büttgenbach S (1994) Mikromechanik – Einführung in Technologie und Anwendung, 2nd edn. Teubner
[10] Chudoba T (2009) Referenzmaterialien für die instrumentierte Eindringprüfung, see: https://shop.strato.de/WebRoot/Store/Shops/61006128/Products/as_405001/Referenzmaterialien_0020_f_00FC_r_0020_die_0020_instrumentierte_0020_Eindringpr_00FC_fung.pdf
[11] Chudoba T, Herrmann K (2001) Verfahren zur Ermittlung der realen Spitzenform von Vickers- und Berkovich-Eindringkörpern. Härterei-Technische Mitteilungen (56):258–264

[12] Dai G, Wolff H, Weimann T, Xu M, Pohlenz F, Danzebrink HU (2007) Nanoscale surface measurements at sidewalls of nano micro structures. MST 18:334–341

[13] Dai G, Bütefisch S, Pohlenz F, Danzebrink HU (2009) A high precision micro/nano CMM using piezoresistive tactile probes. MST 20

[14] Doering L, Brand U, Peiner E, Behrens I (2003) Calibration of micro force setting standards using a new nano force calibration device. Proc. Mirror System Technologies, pp 492–494

[15] Manske E, Hausotte T, Mastylo R, Machleidt T, Franke KH, Jäger G (2007) New applications of the nanopositioning and nanomeasuring machine by using advanced tactile and non-tactile probes. MST 18:520–527

[16] Meli F, Kueng A (2007) AFM investigation on surface damage caused by mechanical probing with small ruby spheres. MST pp 496–502

[17] Menelao F (2009) Indentation modulus and hardness reference measurement. Priv. Comm.

[18] Oliver W, Pharr G (1992) An improved technique for determining hardness and elastic modulus using load and displacement sensing indentation experiments. J Mater Res (7):1564–1583

[19] Peiner E, Doering L (2009) MEMS cantilever sensor for non-destructive metrology within high-aspect-ratio micro holes. In: Collection of papers presented at the Symposium on Design, Test, Integration and Packaging of MEMS/MOEMS, Rome, pp 369–374

[20] Phataraloha A, Büttgenbach S (2006) A novel design and characterisation of monolithic three axial micro probe for dimensional metrology. In: APCOT, Singapur

[21] Shah IA, van der Wolf BMA, van Enckevort WJP, Vlieg E (2008) Wet chemical etching of silicon (111): Autocatalysis in pit formation. Journal of The Electrochemical Society 155:J79–J84

[22] Simmons G, Wang H (1971) Single Crystal Elastic Constants and Calculated Aggregate Properties: A Handbook, 2nd edn. The M.I.T. Press, Cambridge

[23] Tibrewala A, Hofmann N, Phataralaoha A, Jäger G, Büttgenbach S (2009) Development of 3D force sensors for nanopositioning and nanomeasuring machine. Sensors 2009 9:3228–3239

[24] Weckenmann A, Hoffmann J (2007) Long range 3D scanning tunnelling microscopy. In: Annals of the CIRP, vol 56

[25] Weckenmann A, Peggs G, Hoffmann J (2006) Probing systems for dimensional micro- and nano-metrology. MST pp 504–509

Part III
Manufacturing and Fabrication

Chapter 10
Fabrication of Magnetic Layers for Electromagnetic Microactuators

J. Chen, C. Ruffert[1], H. H. Gatzen[1], R. Bandorf, G. Bräuer[2]

[1] *Institute for Microtechnology (now: IMPT)*
Leibniz Universität Hannover
ruffert@impt.uni-hannover.de; gatzen@impt.uni-hannover.de

[2] *Fraunhofer Institute for Surface Engineering and Thin Films*
Fraunhofer-Gesellschaft
braeuer@ist.fhg.de

Abstract For designing and fabricating magnetic microactuators, both soft and hard magnetic materials may be required. For soft magnetic materials, a high saturation flux density B_s and a great relative permeability μ_r is desirable; for forming efficient permanent magnets, hard magnetic materials require a high maximal energy product $|BH|_{max}$. In the area of soft magnetic materials, investigations on NiFe81/19, NiFe45/55, and CoFe were carried out, while an example for a hard magnetic material providing a high energy product is SmCo. Since patterned thin-film magnets feature inferior magnetic properties compared to bulk magnets, a method of determining their magnetic properties was developed.

10.1 Introduction

A successful design and fabrication of active magnetic Micro Electro-mechanical Systems (MEMS) like electromagnetic microactuators strongly depends on the availability of appropriate technologies for magnetic materials. While soft magnetic materials are required for highly permeable magnetic circuits like flux guides, cores, and poles, magnetized hard magnetic materials serve as permanent magnets. Fig. 10.1 shows a comparison between the hysteresis loops of soft and hard magnetic materials. The technologies required for creating appropriate magnetic films include deposition techniques, patterning techniques to shape them into microstructures, and post processing techniques for annealing processes or hard magnet magnetization. A typical MEMS application requiring a soft magnetic material is the linear variable reluctance (VR) micro step motor [19], requiring hard magnetic materials is the linear synchronous micromotor, and requiring both is the linear hybrid micro step motor. All three are described in chapter 13.

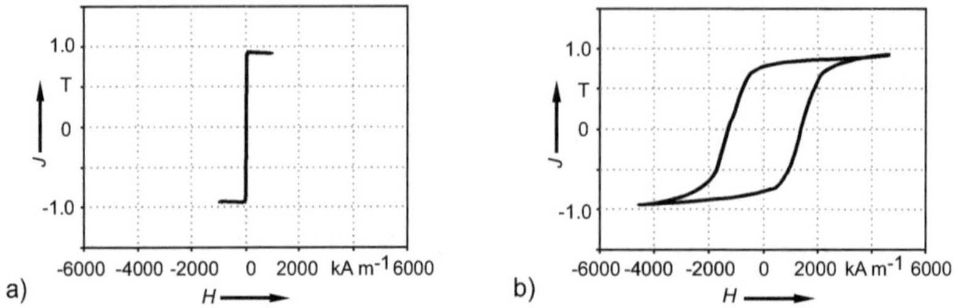

Fig. 10.1 Comparison of magnetic hysteresis loops: (**a**) Soft magnetic material (NiFe81/19); (**b**) Hard magnetic material (SmCo11/89)

10.1.1 Soft Magnetic Thin-films

To achieve a strong driving force in microactuators, a high magnetic flux is required. Furthermore, due to the small device dimensions, the risk of saturation effects in soft magnetic materials has to be considered. For these reasons, both a high relative permeability μ_r and great saturation flux density B_s are the key properties of soft magnetic materials aimed for the application in microactuators. Additionally, a low coerctivity H_c is desirable as well. For an application in active microdevices, the mechanical and chemical properties of these thin-films, such as low residual stress and a good corrosion resistance are also necessary. In contrast to applications in read write heads in the data storage technology, for microactuators the soft magnetic thin-films are usually required in rather high thickness, typically above 10 µm. For this reason, the residual stress of the thin-film must be controlled to avoid a delamination of the films. For soft magnetic applications in active microdevices, the ferromagnetic materials Fe, Co, Ni or their alloys are most commonly chosen. Widely used are NiFe alloys, also called permalloy. NiFe81/19 combines a high relative permeability μ_r, low coerctivity H_c, but a moderate saturation flux density B_s of only 1 T. On the other hand, NiFe45/55, with its substantially increased content of Fe has a B_s of 1.6 T. However, this material is also magnetostrictive, which typically is no drawback for actuator applications. Fig. 10.2 shows the dependence between the saturation flux density B_s and the atomic composition of Fe, Co, and Ni in bulk materials [9].

10.1.2 Hard Magnetic Thin films

For hard magnetic materials, the most important magnetic property is the maximal energy product $|BH|_{max}$. This value describes the energy, which can be stored in the material once it is totally magnetized. Due to their very high maximal energy product $|BH|_{max}$, rare-earth alloys became some of the most interesting hard magnetic materials since their discovery. Although NdFeB hard magnetic material features an even higher $|BH|_{max}$, its extreme

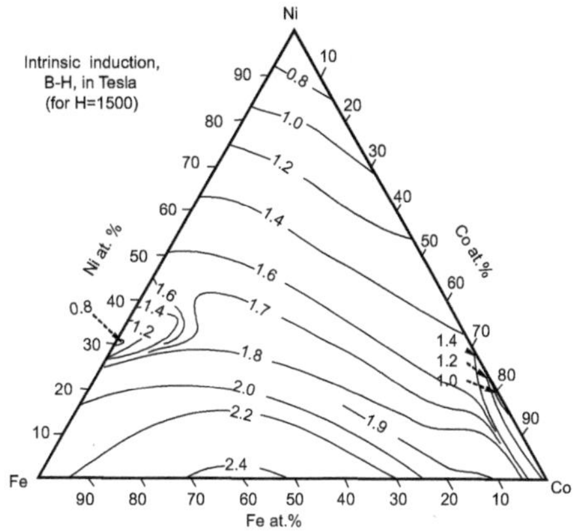

Fig. 10.2 Saturation flux density B_s of annealed bulk NiFeCo alloys for different compositions, by R.M. Bozorth [9], reproduced with permission in the IEEE copyright line © 1993

corrosion makes it difficult to fabricate thin-film magnets. SmCo magnets, widely used in the industry, are considered to be suitable for applications in microactuators because of their well-established deposition process, patterning technology, and acceptable corrosion resistance.

10.2 Fabrication Technologies

For the preparation and deposition of the magnetic films, various methods were used. Especially for soft magnetic thin-films, electroplating is applied most commonly. It is more favourable because of its high deposition rate and cost efficiency in comparison to any vacuum based deposition technique. For hard magnetic films, particularly rare-earth magnets, several challenges have to be solved to use electroplating techniques, since these materials are very susceptible to corrosion. Most electroplating processes in aqueous solution are not suitable for these applications. Vacuum based processes like physical and chemical vapor deposition (PVD, CVD) offer rather flexible alternatives. Due to their low deposition rates, a specific PVD process, gas-flow sputtering, was successfully investigated and applied.

Electroplating is an electrochemical process: the deposition of metals and alloys results from the reduction of metal ions from aqueous, organic, and fused-salt electrolytes [24]. The appropriate polarization between electrodes, which is determined by the electrode current density J, can be adjusted by the applied power. To improve the mass transportation and to achieve a homogenous deposition, using a pulsed current may be advantageous [25]. The agitation of the electrolyte is accomplished by using paddle cells during the process in order to improve the convection of electrolyte. The process temperature influences the diffusion of the ions in the electrolyte. All these process parameters determine the electrochemical kinetics of the plating process and

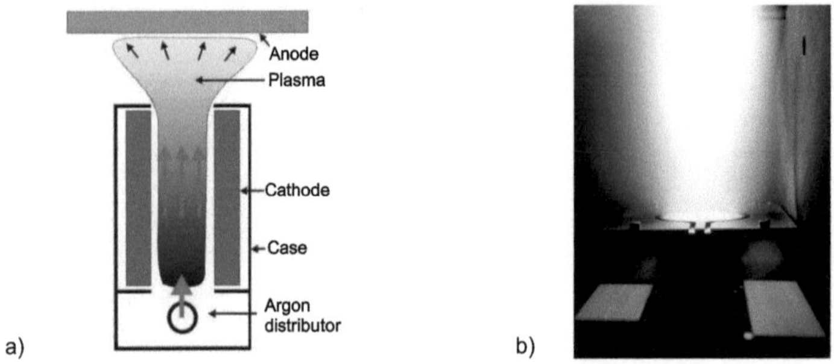

Fig. 10.3 Gas-flow sputtering (GFS): (**a**) Schematic representation of a planar GFS source; (**b**) GFS source in operation

thus the deposition results. For the deposition of alloys, the resulting chemical composition can be adjusted through those process parameters. An introduction into sputtering is already given in chapter 5. The sputtering process for magnetic films is often conducted with RF-power instead of DC-power due to the higher electron density.

Gas-flow sputtering (GFS) is a specific modification of the sputtering process. While most PVD processes require a high vacuum condition at least during pump-down, GFS operates in a coarse vacuum, reducing the costs of the deposition machines since less sophisticated vacuum pumps are required. The material source is a hollow cathode, e.g. a tube or two planar cathodes facing each other. The inlet of the inert gas is on the backside of the source. Due to the higher pressure in the process chamber, the sputtered particles in the hollow cathode are carried by a laminar inert gas flow towards the substrate surface. GFS creates high electron densities in the hollow cathode, since the anode is outside and the electrons need much more time to leave the source [8, 22, 23]. Fig. 10.3 depicts a schematic representation of a planar GFS source and an image of GFS source in operation.

For creating organic layers filled with magnetic particles, spin-coating techniques were applied. The magnetic particles are dispended in a photosensitive structural material, e.g. the negative photoresist *SU-8* and then spun onto the wafer by spin-coating. Depending on the desired thickness, *SU-8* resists with different viscosities are used. Furthermore, the thickness can be controlled accurately by the spin speed. The photoresist containing magnetic particles can be patterned by photolithography. This method is advantageous especially for fabricating hard magnetic thin-films, since there is no annealing process required, if the particles already feature the desired hard magnetic properties.

For most of the electroplating processes, a temporary photomask serves as a micromold, which is filled during the deposition procedure. Spin-coating the resist material onto the wafer and patterning it in a photolithography process creates the photomask. Another application of photomasks is their use as etch masks, i.e. protecting the structure intended to be created from etchant attack during etching. As processes, wet chemical etching or an ion bombardment and attack of radicals, respectively, in case of dry etching such as Ion Beam Etching (IBE) and Reactive Ion Etching (RIE) may be applied. After the etch process, a stripping process removes the photomask. Most of the wet chemical and reactive etching processes are isotropic, i.e. the etchant removes material in all directions at the same rate. This causes an underetching under the mask layer and limits the transfer accuracy of the patterns. The advantages of wet chemical etching are its selectivity and its low process temperature. Ion bombardment typically results in a directional etching, which leads to a higher etching resolution.

After the deposition and patterning of the magnetic thin-films, some further processes are usually required to achieve the desired magnetic properties of the thin-films. When sputtering hard magnetic thin-films, the material structure may be amorphous. The atoms or molecules in the material do not orient themselves in a crystalline structure. In such a case, the material may only have poor hard magnetic properties. A remedy may be an annealing process, which allows the atoms to re-arrange themselves in a crystalline structure. By crystallization, the coercivity is enhanced. Thus, the hard magnetic properties are improved. Annealing is also utilized for electrodeposited soft magnetic materials. During the annealing, the crystals relax. This results in a reduction of the stress in the thin-film. Since the coercivity has a strong dependence on the stress in the material, the soft magnetic properties can be improved this way. To achieve a permanent magnetic field, the hard magnetic material must be magnetized. A magnetic field may also be useful during the deposition. Especially in an electroplating process, a magnetic anisotropy can be generated by applying an external static magnetic field.

10.3 Test Systems

For testing and characterising the fabricated thin-films, several analysis techniques are available.

10.3.1 Magnetic Properties Analysis

The BH-looper technique is a method to determine the hysteresis loop of the material [27]. A homogenous magnetic field is generated by a pair of Helmholtz coils. This method is only suitable for a material, which can be saturated by a weak magnetic field, namely soft magnetic materials. Another limit of the BH-looper is its restriction to the specimen dimension dependent on the sensitivity of the pickup coils.

During a vibrating sample magnetometry (VSM) measurement, the sample to be characterized is placed in a homogenous magnetic field and exposed to a sinusoidal mechanical vibration. The electric field induced in the sample is detected by a pickup sensor. The magnetic moment of the sample can be derived. Thus, the dependence of the magnetic flux B on the applied magnetic field H in the material is plotted. Using this method, the sample's hysteresis loop is obtained. The magnetic characteristics like saturation flux density B_s, coercivity H_c, remanence flux density B_r as well as relative permeabilityμ_r can be derived this way.

Once the magnetic thin-film is integrated in a microsystem, the measurement with a VSM and BH-looper is no longer possible. To analyze the magnetic moment generated by the thin-films in the whole system, a magneto-optical indicator film (MOIF) technique is applied, which is based on the Farraday effect. During the measurement, a magneto-optical indicator is integrated in the microsystem. The polarization and intensity of the light will be changed by a magnetized thin-film. The magnetic domains can be observed on the indicator by a light microscope. The contrast is dependent on the strength of the magnetization [21].

The ferromagnetic resonance (FMR) technique is a widely used method for the characterization of magnetic materials. The characteristic value of damping parameters, the saturation magnetization, the anisotropy constants can be determined by using this measurement. The measurement is based on the principle of the absorption of microwaves in ferromagnetic materials [28]. With its high resolution, this technique is very suitable for characterizing ultra-thin films. A method for determining the magnetic properties of ferromagnetic thin-films with a thickness in the micrometer range, such as electroplated ferromagnetic films for MEMS actuators/sensors, was developed [31].

10.3.2 Composition and Structure Analysis

Since the magnetic properties of an alloy are strongly dependent on the composition of the material, a respective characterization of the magnetic thin-film is necessary. Energy dispersive X-ray spectroscopy (EDX) is a commonly used analysis technique to identify the chemical composition of a sample. An EDX analysis system is frequently integrated in a scanning electron microscope (SEM). The electron beam bombards the sample during the measurement and kicks out some of the electrons in the sample by electron collisions. The vacant position of the electron will be filled by a higher-energy electron. The energy difference released by this transfer is emitted as X-ray radiation. Since the energy is characteristic for every element, the chemical composition of a sample can be identified.

Wavelength dispersive X-ray (WDX) spectroscopy, as well as electron probe microanalysis (EPMA) follows the same working principle as EDX. The difference to the EDX method is that the detector classifies the X-ray energy by measuring the radiation's wavelengths [26]. Using an electron beam, the atoms are excited and characteristic X-ray emission is stimulated as before. For the WDX/EPMA analysis, a wavelength dispersive crystal spectrometer is used offering a 10 times higher energy resolution compared to energy dispersive measurements. The high sensitivity of the wavelength dispersive technique also allows the highly accurate quantification of lightweight elements like carbon, nitrogen, or oxygen. Furthermore, the chemical analysis and a thickness measurement of films between 5 and 500 nm is possible [30].

Secondary ion mass spectroscopy (SIMS) allows conducting a quantitative analysis of a film composition and also provides depth-profiling capabilities. Using Cs^+-ions for the bombardment enables the detection of a wide range of elements including very lightweight ones [18, 20]. The calibration and quantification of the measurement is executed using reference standards that were characterised by EPMA.

X-ray diffraction (XRD) analysis is a method to analyse the crystal structure of solid state materials. This technique is based on the scattering of X-rays in a crystal, with the crystal planes serving as a grating. For every crystal, a maximal intensity can be reached at a certain angle, which corresponds to the lattice spacing in the analysis direction. By comparing diffraction data with a standard database, the material as well as the crystal structure in the sample can be identified.

10.3.3 Methods for Residual Stress Determination

The residual stress is an important mechanical property of thin-film materials. A thin-film under high stress can cause the formation of cracks in the film and a delamination from the substrate. It is known, that a high stress also increases the coercivity H_c in soft magnetic materials [32]. For these reasons, the stress of the deposited thin-films should be analysed. To do so, the methods, e.g. XRD, curvature measurements, etc. were developed. The method applying curvature measurements is based on the Stoney equation, which is suitable for an analysis at the wafer level. The Stoney equation is usually applied for the measurement on an unpatterned substrate. It can be extended to patterned structures, if some preconditions are fulfilled [17].

10.4 Experimental Investigations on Soft Magnetic Materials

There are three soft magnetic alloys of interest for being used in active magnetic microdevices: NiFe81/19, NiFe45/55, and CoFe. Various processes to fabricate these thin-film materials were investigated.

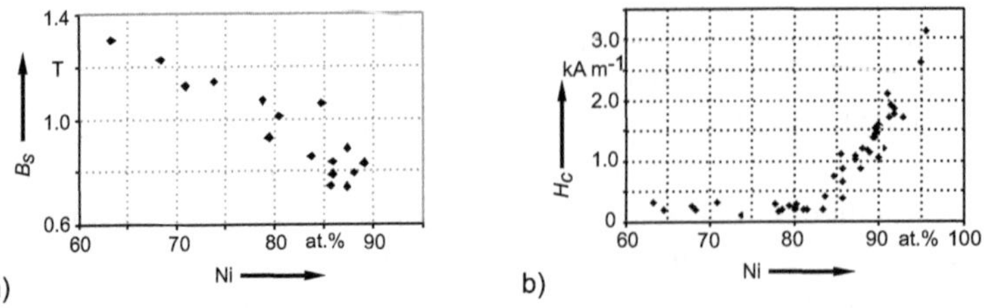

Fig. 10.4 Magnetic properties of electroplated NiFe thin-films as a function of the film composition: (**a**) Saturation flux density B_s; (**b**) Easy-axis coercivity H_c

10.4.1 Electroplated Soft Magnetic Materials

Electroplating processes were established to deposit soft magnetic thin-films. A horizontal paddle cell were used in all plating experiments. The first results presented are on NiFe81/19. The plating parameters, such as the Fe content in the solution, pulse ratio, current density, paddle frequency, and bath temperature were varied [16]. The influence of the Fe content in the films on B_s and H_c was investigated, which is depicted in Fig. 10.4

Wafers were coated with films of different thickness in consecutive electroplating steps, while keeping the plating parameters constant. It was found, that the maximal relative permeability $\mu_{r,max}$ decreases with increasing layer thickness d. Fig. 10.5 depicts this dependence. This decrease is caused by a growing film stress in thick NiFe81/19 layers.

A beneficial and an efficient parameter to control the film composition is the magnitude of the applied reverse current. Other parameters such as current density, bath agitation, and temperature can not be varied widely enough without a negative influence on the surface quality caused by hydrogen deposition. Table 10.1 shows the main plating parameters for the optimal process in both magnetic properties and surface quality. Fig. 10.6 depicts a hysteresis loop of an electroplated NiFe81/19 thin-film.

NiFe45/55 was the second soft magnetic material subjected to electroplating investigations. As already mentioned, it has a higher saturation flux density B_s than NiFe81/19, but is also magnetostrictive, which typically is no major disadvantage in active magnetic microdevices. The electrolyte recipe

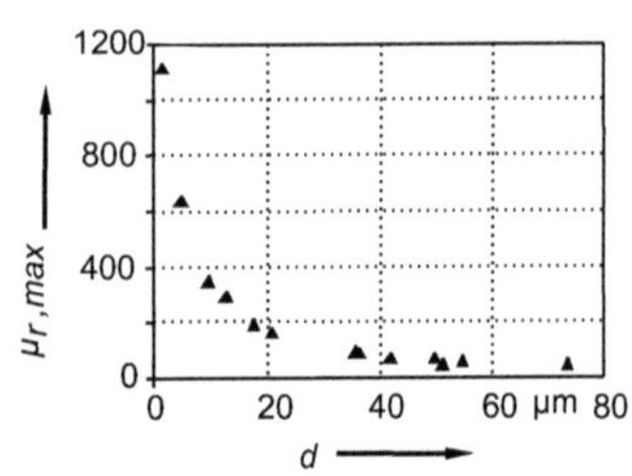

Fig. 10.5 Maximal relative permeability $\mu_{r,max}$ as a function of the layer thickness d

Table 10.1 Main parameters of NiFe electroplating for optimum results

Composition	Fe electrolyte content [$g\,l^{-1}$]	Pulse ratio T_{fwd}/T_{rev}	Current density [mA cm^{-2}]	Reverse current I_{rev} [mA]	Paddle frequency f_{paddle} [Hz]	Bath temperature [°C]
81/19	1.6	9/1	13	200	0.35	30
45/55	4.6	9/1	50	150	0.35	30

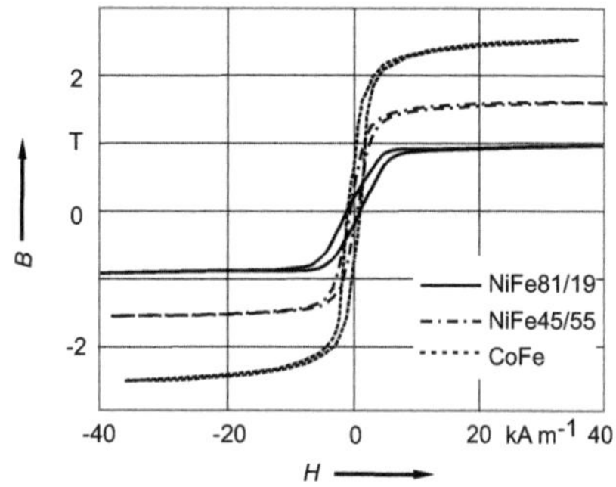

Fig. 10.6 Hysteresis loops of electroplated thin-films, a 10 µm thick NiFe81/19, a 12 µm thick NiFe45/55, and a 10 µm thick CoFe, respectively

was based on the NiFe81/19 electrolyte, but the Fe content was increased. Table 10.1 shows the optimized process parameter to electroplate NiFe45/55. The magnetic properties of plated thin-films were determined by VSM measurement. Fig. 10.6 shows a hysteresis loop of a 12 µm thick NiFe45/55 thin-film. A saturation flux density B_S of 1.5 T was achieved.

Further investigations focused on the reduction of Fe^{3+} ions in the electrolyte and their integration in the deposited films. They were based on injecting inert gas (Ar, N_2) in the electrolyte through a bubbler [14]. Deposition processes using various bubbling concepts were compared with a reference process. Derived from the magnetic properties, a short exposure to Ar or N_2 after the deposition process reduces the oxidation of the Fe^{2+} ions as well as the incorporation of $Fe(OH)_3$ and results in an improvement of the magnetic properties of the deposited NiFe45/55 thin-films.

CoFe promises yet another increase of the saturation flux density B_s – ideally up to 2.5 T. Initial investigations on electroplated CoFe thin-films resulted in a saturation flux density B_s greater than 2 T and a maximal thickness of 10 µm [6]. Further investigations focused on avoiding the oxidation of Fe^{2+} in the electrolyte and the integration of $Fe(OH)_3$ in the deposited layers [15]. For this purpose, a number of additives like sulfur bearing agents and reducing agents was applied. VSM measurements were used to characterize the magnetic properties. The composition of Fe and Co in the deposited films was measured by EDX. EPMA was used to determine the impurities (O, S, and B) in the deposits. First, the dependence of the saturation flux density B_s and coercivity H_c on the content of Co and Fe in the deposits was

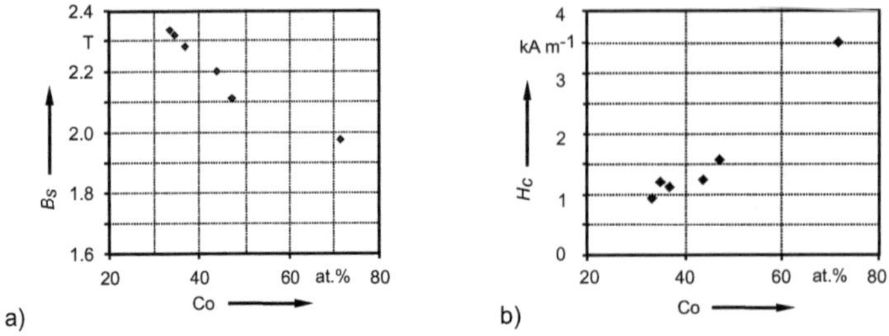

Fig. 10.7 The magnetic properties of electroplated CoFe thin-films as a function of film composition: (**a**) Saturation flux density B_s; (**b**) Easy-axis coercivity H_c

investigated. The result is shown in Fig. 10.7. For 35 at.% Co, the deposited CoFe films achieved the maximal value of B_s and the lowest H_c.

Table 10.2 Main parameters of CoFe electroplating for optimal results

Fe electrolyte content [$g\,l^{-1}$]	DMAB [$g\,l^{-1}$]	Saccharin [$g\,l^{-1}$]	Pulse ratio T_{on}/T_{off}	Current density [$mA\,cm^{-2}$]	Paddle frequency f_{paddle} [Hz]	Bath temperature [°C]
14	2	8–15	3/7	10	0.35	20

To further optimize this electroplating process, experiments with organic additives were conducted. Three electrolytes were mixed with sodium benzene sulfinate (SBS bath), ammonium citrate (citrate bath), and dimethylamine borane (DMAB bath), respectively. An additional bath was based on the DMAB bath by adding $0.5\,g\,l^{-1}$ saccharin for reducing the residual stress in thin-films. The highest B_s achieved was 2.45 T with additives either DMAB or SBS. The results also show that the addition of saccharin to the electrolyte is essential to minimize the stress in the film, allowing to obtain a thickness above 10 µm, which is desirable for MEMS applications. Table 10.2 depicts the main parameters for electroplating CoFe on account of optimal mechanical and magnetic properties. Fig. 10.6 shows a hysteresis loop of an electroplated CoFe thin-film.

10.4.2 PVD Deposited Soft Magnetic Thin-films

NiFe45/55 is not only interesting as an electroplated material, but also as a material deposited by PVD processes. Therefore, fabrication techniques for thick sputter deposited NiFe45/55 films were investigated as well [11]. The thickness range for the investigations was between 1 µm and 15 µm, representing typical thin-film thicknesses for the flux guides and poles. The films

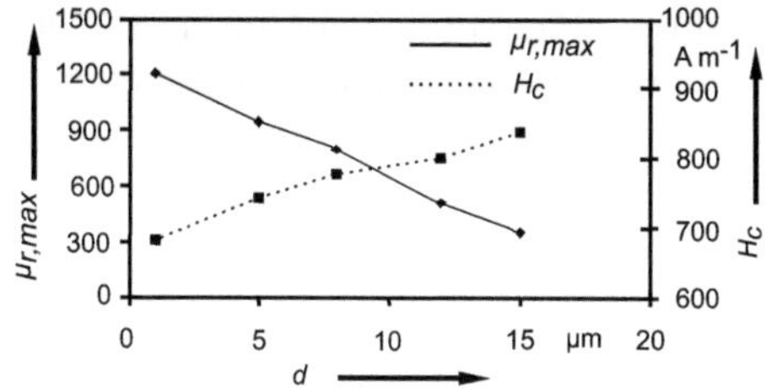

Fig. 10.8 Maximal relative permeability $\mu_{r,max}$ and coercivity H_c of NiFe45/55 as a function of the layer thickness (Al_2O_3 ceramic substrate).

were deposited on Al_2O_3 and B270 glass substrates, respectively. The structures were patterned by wet chemical etching. The photoresist AZ5214 as well as AZ4562 were applied as etching mask material and patterned by photolithography. The investigations included a variation of the Ar pressure in the range of 0.41 Pa to 1.04 Pa and a sputtering power density of 1.45 W cm^{-2} to 2.42 W cm^{-2}. At a power density of 1.93 W cm^{-2}, the film with best soft magnetic properties, a maximal permeability $\mu_{r,max}$ as well as a minimal coercivity H_c was achieved. Applying a lower Ar pressure of 0.41 Pa led to better soft magnetic properties as well. The dependence of the soft magnetic properties on the film thickness is shown in Fig. 10.8.

For the high-rate deposition of soft magnetic films, gas-flow sputtering was investigated [2, 3]. NiFe 81/19 targets were used for the GFS process. For a feasibility study of the process, a small source with two planar cathodes of 60 × 100 mm^2 size was used. For a maximal deposition rate, the gas flow and the applied power were varied. With increasing power, the deposition rate also increases. For the gas flow, an optimal dependance on the source design exists. With increasing power the static deposition rate increases from 28 μm h^{-1} for 1.2 kW to more than 50 μm h^{-1} for 2.5 kW applied power and 3 slm Ar flux.

Additionally, a bias voltage can be applied to the substrate to modify the properties of the growing film. Fig. 10.9 shows the resulting permeability of the NiFe films using 4 slm Ar flux and 2.0 kW applied power for varied substrate bias. The relative permeability μ_r varies from approximately 75 at 30 V bias to approximately 240 at 60 V bias.

Furthermore, the relative permeability was modified by the deposition conditions. The influence of the deposition temperature, applied bias, and an

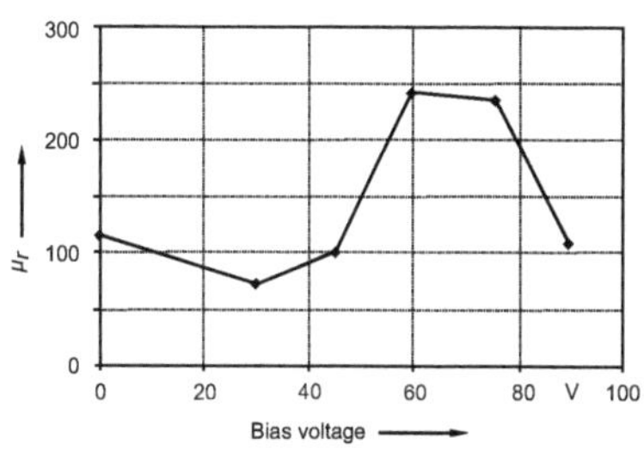

Fig. 10.9 Relative permeability of NiFe films deposited by GFS for a gas flow of 4 slm Ar and a target power of 2.0 kW as the function of the substrate bias

additional magnetic field during the deposition were investigated. Table 10.3 summarizes the optimum parameters for the static deposition of NiFe81/19 by GFS. Fig. 10.10 shows the hysteresis loop of the a 2.3 µm thick film.

Table 10.3 Optimized parameters for the GFS deposition of NiFe81/19 thin-films

Power [kW]	Ar gas flow [slm]	Bias voltage [V]	Pressure [Pa]	Dep. temperature [°C]
2	4	-60	44	< 100 (unheated)

10.4.3 Polymer Embedded Soft Magnetic Materials

Polymer embedded soft magnetic thin-films can be fabricated and patterned using spin-on techniques. NiFe nano and microparticles were mixed with the photoresist *SU-8*. To obtain a homogenous distribution of the particles, an ultra-sonic enhanced dispersion was used. The films were patterned photolithographically. The thickness of the film could be well adjusted by the viscosity of the *SU-8* and the spin-on process. Fig. 10.11 depicts an optical micrograph of patterned *SU-8* including magnetic particles and the respective hysteresis loops measured by VSM. The saturation flux density B_s corresponded with the percentage of particles in these thin-films very well. Only a very low relative permeability μ_r was measured, this system does not lend itself for a practical use in magnetic microactuators; the reason is a demagnetization effect between adjacent particles [1].

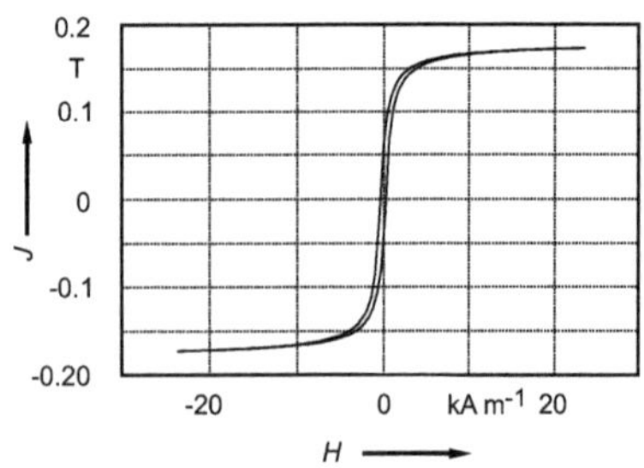

Fig. 10.10 Hysteresis loop of NiFe films deposited by GFS, measured by VSM

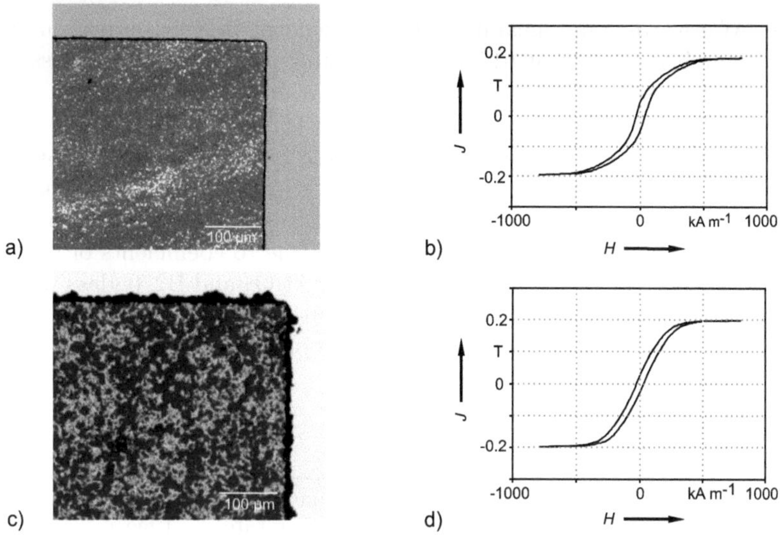

Fig. 10.11 *SU-8* film filled with 15 vol.% of NiFe particles: (**a**) Optical micrograph of a patterned *SU-8* film with nanoparticles; (**b**) Hysteresis loop of the pattern shown in (a); (**c**) Optical micrograph of a patterned *SU-8* film with microparticles; (**d**) Hysteresis loop of the pattern shown in (c)

10.5 Experimental Investigation on Hard Magnetic Materials

PVD processes (RF-sputtering and GFS) for fabricating hard magnetic Samarium Cobalt (SmCo) thin-films were investigated. A electroplating process to fabricating this materials were also established. Polymer embedded hard magnetic materials using Spin-on techniques offer another alternative.

10.5.1 SmCo Deposited by PVD

Samarium Cobalt (SmCo) is a hard magnetic material well suited for thin-film fabrication, since this material features a very high energy product $|BH|_{max}$ and an acceptable corrosion resistance. It is mainly used in the compositions SmCo17/83 ($SmCo_5$) or SmCo11/89 (Sm_2Co_{17}). For sputtered films, a typical value of $|BH|_{max}$ for SmCo17/83 is $83\,kJ\,m^{-3}$ and for SmCo11/89 is $125\,kJ\,m^{-3}$ [13]. Therefore, SmCo11/89 is the more desirable alloy for microactuators requiring a permanent magnet. The technology to deposit this material as well as the patterning using wet chemical etching and ion beam etching were investigated. Both SmCo compositions depict acceptable hard magnetic properties only in their crystalline phase. When using sputtering as a deposition technique, an amorphous SmCo thin-film is created on a cooled substrate. Afterwards, an annealing step was executed in a vacuum oven at 560°C. Although using a SmCo17/83 target, after the annealing process a SmCo11/89 film exists. This is due to the fact that during the process an appropriate percentage of the Sm is oxidized at the chosen process conditions. As a result, the process creates a film containing SmCo11/89 and Sm_2O_3. The presence of the latter has no major negative impact on the material's magnetic properties except on the SmCo composition. The wafer materials investigated included Si, B270 glass, and Al_2O_3. Due to coefficients of thermal expansion (CTE) close to the ones of SmCo, Al_2O_3 and B270 glass were best suited as wafer materials. The optimal process parameters, such as the sputtering pressure and the power density, are depicted in Table 10.4.

Table 10.4 Main parameters for RF-sputtering SmCo11/89 thin-films for optimal results

Power density [W/cm²]	Sputtering pressure [Pa]	Deposition rate [µm/h]	Annealing temperature [°C]	Annealing time [hour]
3.8	0.867	10	560	6

Films with a thickness of up to 50 µm could be achieved without tending to delaminate. The magnetic properties of these thin-films were measured in plane as well as perpendicular to the substrate using a Superconducting Quantum Interference Device (SQUID). Fig. 10.12 depicts the hysteresis loop of a hard magnetic SmCo11/89 layer with a thickness of 30 µm which was sputter deposited on an Al_2O_3 substrate. A $|BH|_{max}$ of 90 kJ m^{-3} was measured. Obviously, this thin-film shows a magnetically isotropic behavior.

To analyze the crystal phase, samples were prepared on Si substrates with the tri-layer system Cr/SmCo/Cr. The SmCo11/89 phase in the thin-films

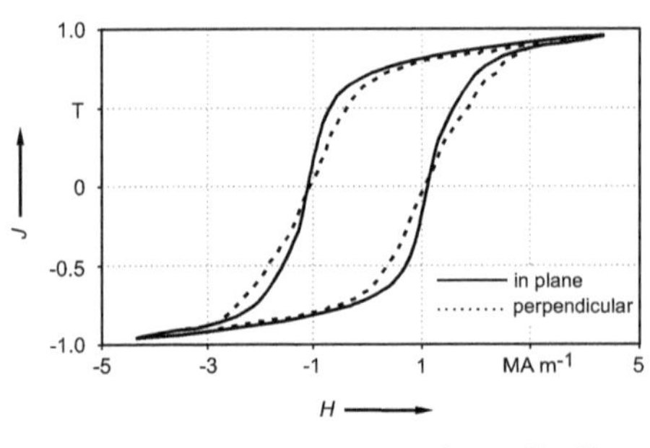

Fig. 10.12 Hysteresis loop for a SmCo11/89(Sm_2Co_{17}) layer with a thickness of 30 µm measured by a SQUID. Measurements were taken in plane and perpendicular to the layer.

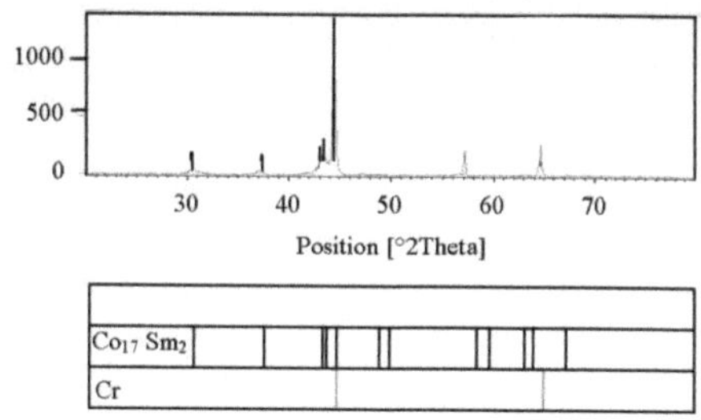

Fig. 10.13 XRD analysis result of a Cr/SmCo/Cr tri-layer after annealing

was determined by XRD analysis. Figure 10.13 shows the XRD-diagram of a SmCo film with the tri-layer system.

10.5.2 Gas-flow Sputter Deposited SmCo

For the deposition of hard magnetic films, RF-sputtering is commonly used. Besides RF-sputtering, GFS offers a new method for high rate deposition. Since GFS is carried out under coarse vacuum conditions, and since Sm highly reacts with oxygen, the question arises whether a stable hard magnetic phase can be synthesized by this method. The deposition rates using a SmCo17/83 target for RF-sputtering were improved up to approximately $10\,\mu m\,h^{-1}$ [10]. Using GFS, the static deposition rates were enhanced up to approximately $85\,\mu m\,h^{-1}$ [3]. Investigations of the chemical composition of the SmCo films by EPMA showed that depending on the preparation conditions, up to 17.3 at.% oxygen was incorporated in the films. The optimized preparation parameters are summarized in Table 10.5. Nevertheless the films with the highest oxygen content also show the highest coercitivity. The intrinsic coercitivity H_{ci} of the GFS films was up to $1.2\,MA\,m^{-1}$. Fig. 10.14 depicts the hysteresis loop of a SmCo film, deposited by GFS.

Table 10.5 Optimised preparation parameters for GFS deposition of SmCo films

Power [kW]	Ar gas flow [slm]	Bias voltage [V]	Pressure [Pa]	Dep. temperature [°C]
3	3	-	37	< 450

10.5.3 Electroplated Hard Magnetic Materials

Although a process to sputter deposit SmCo was developed successfully, research on an electroplating process was also conducted. As a rare earth element, Sm theoretically can't be plated from an aqueous solution because of

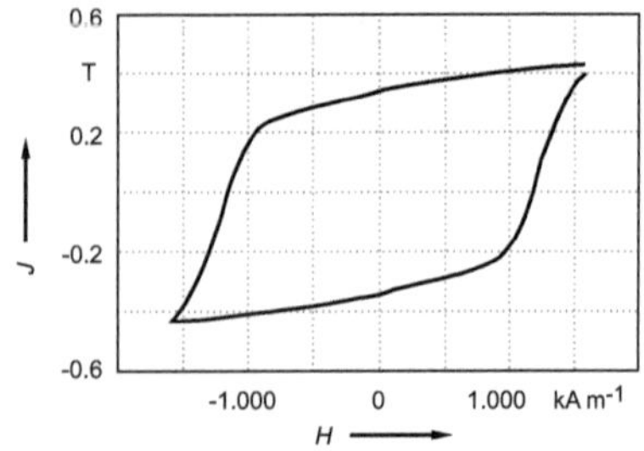

Fig. 10.14 Hysteresis loop of a 13.5 μm SmCo film, deposited by GFS with the parameters in Table 10.5

its more negative electrochemical potential compared to H_2. Recently, it was reported, that SmCo films were successfully deposited by electroplating by using an organic additive, which results in a formation of a metal Sm complex [29]. The electroplating of SmCo thin-films from an aqueous solution was investigated on a beaker level. The organic additive glycine served as a complexing agent. Al_2O_3 was used as substrate material due to its CTE almost matching the one of SmCo, which is important for the annealing process. 200 nm Cr layer and 50/200 nm Cr/Au bilayers were sputtered as seed layer, respectively. A cathode and a Pt anode with the same dimension of $2 \times 2\text{mm}^2$ were arranged in parallel. The cathodic current density was varied between $5\,\text{mA cm}^{-2}$ and $800\,\text{mA cm}^{-2}$ while the total electric charge was held at 24 C. A weak adhesion of deposited SmCo thin-films on the Au seed layer could be observed. On the other samples, a SmCo layer could be detected after the plating process. The Sm at.% in the films was determined by EDX. A Sm content of up to 12 at.% in the films was measured by EDX. Table 10.6 shows the parameters to electroplate SmCo thin-films.

Table 10.6 Main parameters of SmCo electroplating for optimum results

Sm electrolyte content [mol l^{-1}]	Co electrolyte content [mol l^{-1}]	Glycine electrolyte content [mol l^{-1}]	Current density [mA cm^{-2}]	Bath temperature [°C]
0.6	0.12	0.36	100–200	60

The deposited films were annealed at 560°C. The magnetic properties were measured by a VSM. For an electroplated thin film, deposited on a chip at a current density of $200\,\text{mA cm}^{-2}$, the intrinsic coercivity H_{ci} was $35\,\text{kA m}^{-1}$. Fig. 10.15 shows the hysteresis loop of this sample. Compared to SmCo thin-films using RF sputtering, the electroplated SmCo thin-films showed only a moderate H_c and B_r. The incorporation of O or $(OH)^-$ resulting in oxides and hydroxides could be the reason, since the contamination of O could not be avoided by electroplating from an aqueous solution. An EPMA measurement showed, that up to 40 at.% of O may be integrated in the SmCo films.

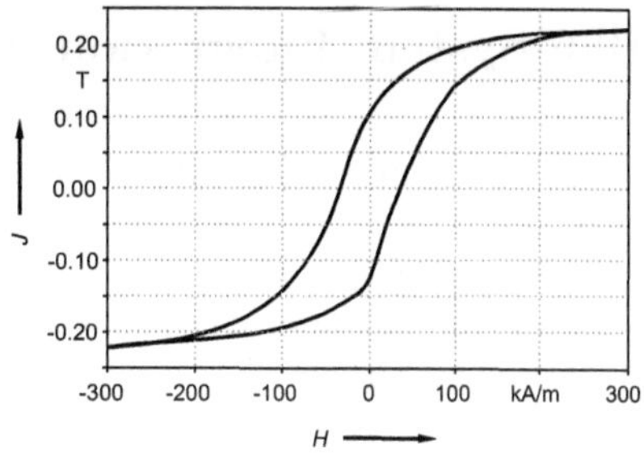

Fig. 10.15 Hysteresis loop of an electroplated SmCo film after annealing

10.5.4 Polymer Embedded Hard Magnetic Materials

Investigations with neodymium iron boron (NdFeB), barium ferrite (BaFe), and strontium ferrite (SrFe) hard magnetic particles were carried out [4]. The particles were embedded in *SU-8* photoresist. The maximal content of NdFeB, BaFe, and SrFe particles was 24 vol.%, 24 vol.%, and 13 vol.% respectively. The filler content was limited by the coating and patterning processes. Up to this filler respective content, *SU-8* with embedded particles can still be patterned photolithographically. An overview of their properties is shown in Table 10.7.

Table 10.7 Properties of the applied hard magnetic particles

Material	Content with particle size >32 µm	B_r [mT]	H_{ci} [A m^{-1}]
NdFeB	>50%	750	670–750
BaFe	>10%	160–190	200–220
SrFe	>0.3%	180–240	205–220

The properties of the patterned *SU-8* films containing hard magnetic particles are shown in Table 10.8.

Table 10.8 Properties of *SU-8* films containing hard magnetic particles

Material	Filler percent [vol.%]	B_r [mT]	H_{ci} [A m^{-1}]
NdFeB	24	55	500
BaFe	24	30	185
SrFe	13	30	220

10.6 System Integration Aspects

For design and fabrication of electromagnetic microactuators, some system integratons aspects must be considered. The following investigations involve the process compatibilities as well as the magnetic measurements for modeling and simulation of whole system.

To ensure the process compatibility by integrating hard magnetic SmCo and electroplated NiFe thin-films, investigations were carried out [12]. Two process sequences were used: The NiFe thin-films were first electroplated on the substrate before the SmCo was sputtered and vice versa. Since the hard magnetic SmCo thin-films must be annealed at 560°C to achieve their desired crystalline state, an investigation of the influence of this annealing process on electroplated NiFe thin-films was conducted. Hard magnetic tri-layers of Cr/SmCo/Cr were deposited onto NiFe thin-films by RF-sputtering. Then, the Cr/NiFe/Cr/SmCo/Cr multi-layer was annealed. Cr was commonly used as an interlayer material because of its diffusion inhibiting properties. A SIMS analysis was utilized to study the boundary layer. A diffusion zone of around 300 nm thickness was determined. Compared to the NiFe and SmCo thin-films with a thickness of a couple of 10 µm, such a small diffusion zone does not result in a great impact on the NiFe/SmCo thin-film system. To determine the influence of the annealing process on the soft magnetic properties of the NiFe thin-films, Cr/NiFe/Cr layers were fabricated and annealed. The magnetic properties were measured by a VSM before and after the annealing process. The results show that the annealing had a positive influence on electroplated NiFe thin-films, which results in an increase of the permeability μ_r by 10% due to a decreasing stress in the NiFe layer. In a second set of investigations, the hard magnetic properties of the tri-layer system Cr/SmCo/Cr, deposited either directly on the substrate or on a NiFe layer, were compared. After annealing, good hard magnetic properties could be achieved for both systems. Therefore, the process compatibility of electroplated NiFe thin-films and RF sputtered SmCo layers is assured.

Two etching techniques were investigated for patterning sputtered SmCo thin-films: wet chemical etching and IBE [13]. By means of wet chemical etching, ammonium-cernitrate (ACN) was utilized as an etchant. Since this etching process is isotropic, an underetching occurred. This effect limited the transfer accuracy of the patterns and influenced the sidewall profiles. Therefore, this underetching effect must be taken into account when designing the etching mask. The advantage of wet chemical etching is its high etching rate. The etching rate could be increased from $2\,\mu m\,min^{-1}$ to $12.5\,\mu m\,min^{-1}$ by increasing the agitation of the etchant, as well as rising the etchant temperature from 20 to 50°C. Fig. 10.16 shows an SEM micrograph, depicting the sidewall of a patterned SmCo magnet. By utilizing IBE, a variation of the etching angle had a substantial influence on the etching rate for SmCo. The change of etching angle from 7° to 45° resulted in an increase of the etching rate from 14 to $20\,nm\,min^{-1}$.

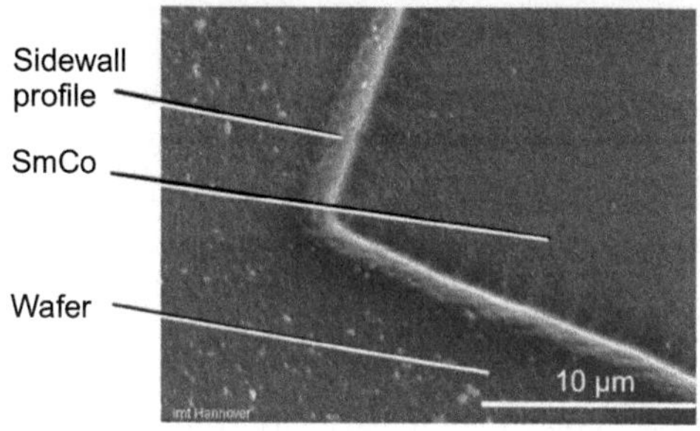

Fig. 10.16 SEM micrograph of the sidewall formation for an isotropic wet chemical etching of a SmCo thin-film

For modeling and simulating magnetic microactuators, the exact knowledge of magnetic properties of the magnetic material is required. Generally, a microactuator core consists of a lower flux guide, poles, and in some cases teeth, fabricated using soft magnetic material. In most microactuators, the magnetic flux flows in lateral direction to the lower flux guide and perpendicular direction the poles as well as the teeth. Thus, the relative permeability μ_r in these preference directions is of substantial interest. Furthermore, the lower flux guides often show dimensions of some millimeters, while the poles and teeth are only a few micrometers. The measurement methods should not be limited by the dimensions and shapes of the samples. Initial investigations showed, that the relative permeability μ_r of permalloy structures show a strong directional dependence, when measured by a VSM. This is caused by the measurement method, which is open loop. Since there are air gaps between the sample and the excitation coils, the thin-film patterns are subjected to demagnetizing fields, which influence the measured hysteresis loop as well as the relative permeability μ_r. To determine the correct μ_r, corrections are required. A method for compensating this effect was proposed [7]. It takes the demagnetization effects into account, described by the following equation:

$$H_{int} = H_{ext} - N_d M \tag{10.1}$$

H_{int}, H_{ext}, N_d and M are the internal magnetic field in sample, the applied magnetic field, demagnetizing factor and the magnetization in the sample, respectively. The influence of demagnetization effects can be corrected by using appropriate demagnetization factors. This way, approximated solutions for the magnetic behavior of the flux guides, as well as the poles are achieved [5].

10.7 Conclusion

The deposition and patterning of magnetic thin-films in electromagnetic microactuator applications were investigated. In the area of soft magnetic thin-films, the NiFe alloy with an atomic composition of 45/55 was found to be most suitable as flux guiding material for magnetic microactuators. Electroplating is the favourable process for fabricating this material, because of its high deposition rate and the opportunity of using resist patterns as micromolds. A promising future soft magnetic material is CoFe, which exceeds NiFe45/55 in saturation flux density B_r. An appropriate electroplating process could be developed. For hard magnetic applications, research on depositing and annealing SmCo was conducted. It could be demonstrated that both RF-sputtering as well as gas flow sputtering are suitable technologies to fabricate it. An annealing of the material after its deposition at a rather high temperature of 560°C proved to be necessary for the material to achieve its full potential. To accurately determine the magnetic values and in particular the permeability of patterned soft magnetic films, a method based on VSM measurements was developed. It helped to resolve design issues in magnetic microactuators originally occurring and caused by using data of unpatterned magnetic films.

Acknowledgements The authors acknowledge the contributions of several colleagues involved in the Collaborative Research Center 516 "Design and Manufacturing of Active Micro Systems": Thomas Budde, Martin Föhse, Matthias Bedenbecker, Bruno Majer, Michael Balke, and Holger Lüthje. Our appreciation also goes to Zbigniew Celinski et al. for the cooperation works on FMR measurements. Additional the authors gratefully acknowledge the funding of the reported work by the German Research Foundation within the Collaborative Research Center 516.

References

[1] Anhalt M, Weidenfeller B, Mattei JL (2008) Inner demagnetization factor in polymer-bonded soft magnetic composites. Journal of Magnetism and Magnetic Materials 320(20):844–848

[2] Bandorf R, Bloche A, Ortner K, Jung T (2007) High rate deposition of soft magnetic thick films by gas flow sputtering. In: Proceedings of the Annual Technical Conference of the Society of Vaccum Coaters SVC, vol 50, p 473

[3] Bandorf R, Bloche A, Ortner K, Lüthje H, Jung T (2007) High rate deposition of magnetic material by gas flow sputtering. Plasma Processes and Polymers 4(S1):129–133

[4] Bedenbecker M, Gatzen HH (2005) Herstellung polymergebundener Hartmagnete durch in SU-8 eingebrachtes Magnetpulver. In: Mikrosystemtechnik, Begleitband zum Kongress 2005, Freiburg, VDE Verlag, pp 581–584
[5] Bedenbecker M, Gatzen HH (2009) Investigations on the permeability of soft magnetic micro strucutres. SMM
[6] Bedenbecker M, Bandorf R, Lüthje H, Bräuer G, Gatzen HH (2006) Development and fabrication of magnetic thin films. Microsystem Technologies 12(7):655–658
[7] Bedenbecker M, Celinski Z, Gatzen HH (2006) Directional permeability dependence in electroplated permalloy layers. In: Symposium on Magnetic Materials, Processes and Devices, 210th Meet. of The Electrochemical Society, Cancun, Mexico, pp 123–135
[8] Birkholz M, Albers U, Jung T (2004) Nanocomposite layers of ceramic oxides and metals prepared by reactive gas-flow sputtering. Surface and Coatings Technology 179(2-3):279–285
[9] Bozorth RM (1993) Ferromagnetism
[10] Budde T, Gatzen HH (2002) Patterned sputter deposited SmCo-films for MEMS applications. Journal of Magnetism and Magnetic Materials 242:1146–1148
[11] Budde T, Gatzen HH (2003) Thick Sputter Deposited NiFe45/55 Films for Flux Guides in Magnetic MEMS. Soft Magnetic Materials Conference
[12] Budde T, Gatzen HH (2004) Magnetic properties of an SmCo/NiFe system for magnetic microactuators. Journal of Magnetism and Magnetic Materials 272:2027–2028
[13] Budde T, Gatzen HH (2006) Thin film SmCo magnets for use in electromagnetic microactuators. Journal of Applied Physics 99:08N304
[14] Chen J, Hansen S, Gatzen HH (2009) Long term investigations on NiFe electroplating processes. In: ECS Transactions /Volume 16/Issue 45/ Magnetic Materials, Processes, and Devices 10, The Electrochemical Society, pp 155–165
[15] Chen J, Flick E, Gatzen HH (2010) Minimizing oxygen inclusion when electroplating high saturation density cofe for microelectromechanical system. Journal of Applied Physics 107:09A311
[16] Föhse M, Gatzen HH (2002) Optimizing the magnetic properties of electroplated permalloy for flux guides in micromotors. In: Proc. 7th International Symposium on Magnetic Materials, Processes and Devices, 202nd Meeting of The Electrochem. Soc., Salt Lake City, UT, USA, pp 125–136
[17] Freund LB, Suresh S (2003) Thin film materials: stress, defect formation, and surface evolution. Cambridge Univ Press
[18] Gao Y (1988) Influence of experimental conditions on matrix effect in sims. Applied Surface Science 32(4):420–430
[19] Gatzen HH, Stölting HD, Büttgenbach S, Dimigen H (2000) A novel variable reluctance micromotor for linear actuation. In: Proc. Actuator, pp 363–366

[20] Gnaser H, Oechsner H (1991) Sims depth profile analysis using mcs$^+$ molecular ions. Fresenius' Journal of Analytical Chemistry 341(1):54–56
[21] Hubert A, Schäfer R (1998) Magnetic domains: the analysis of magnetic microstructures. Springer
[22] Jung T, Westphal A (1993) High rate deposition of alumina films by reactive gas flow sputtering. Surface and Coatings Technology 59(1-3):171–176
[23] Ortner K, Birkholz M, Jung T (2003) Neue Entwicklungen beim Hohlkatoden-Gasflusssputtern. Vakuum in Forschung und Praxis 15(5):236–239
[24] Paunovic M, Schlesinger M (2000) Fundamentals of electrochemical deposition, john wiley & sons
[25] Puippe JC, Leaman F, Electroplaters A, et al (1986) Theory and practice of pulse plating. American Electroplater's Society
[26] Scott VD, Love G (1983) Quantitative electron-probe microanalysis
[27] Traisigkhachol O, Rissing L, Gatzen HH (2009) A concept for the characterisation and analysis of the permeability of soft magnetic thin-films. The Electrochemical Society
[28] Vonsovskii SV (1966) Ferromagnetic resonance. Pergamon Press
[29] Wei JC, Schwartz M, Nobe K (2008) Electrodeposition of Sm-Co permanent magnets from aqueous media. Electrochem Soc 10:D660
[30] Willich P, Bethke R (1995) Performance and limitations of electron probe microanalysis applied to the characterization of coatings and layered structures. Fresenius' Journal of Analytical Chemistry 353(3):389–392
[31] Zagorodnii VV, Hutchison AJ, Hansen S, Chen J, Gatzen HH, Celinski Z (2010) Broad-band ferromagnetic resonance characterization of lossy ferromagnetic metallic elements. Journal of Applied Physics 107:113,906
[32] Zou P, Yu W, Bain JA (2002) Influence of stress and texture on soft magnetic properties of thinfilms. IEEE Transactions on Magnetics 38(5 Part 2):3501–3520

Chapter 11
Fabrication of Excitation Coils for Electromagnetic Microactuators

F. Pape, A. Waldschik, C. Ruffert[1], O. Traisigkhachol, M. Feldmann, T. Kohlmeier, V. Seidemann, S. Büttgenbach[2], H. H. Gatzen[1]

[1] *Institute for Microtechnology (now: IMPT)*
Leibniz Universität Hannover
ruffert@impt.uni-hannover.de; gatzen@impt.uni-hannover.de

[2] *Institute for Microtechnology*
Technische Universität Braunschweig
s.buettgenbach@tu-braunschweig.de

A key component for electromagnetic micro devices is the coil. Exposing it to an electric current results in exciting a magnetic field. The three coils most commonly used in microactuators are the meander, the planar spiral, and the helical type. For fabricating these coils in thin-film technology a combination of photolithography (for creating a micromold) and electroplating (for creating a Cu coil pattern) is applied. As insulation materials, either a photosensitive epoxy (*SU-8*) or stress compensated Si_3N_4 lends itself for usage. For helical coils, both horizontal and vertical patterns are required. For such an application, an electrophorctic photoresist is used.

11.1 Introduction

Excitation coils are a basic component of any electromagnetic or electrodynamic microsystem. Their purpose is to convert an electric current into a magnetic field H. By doing so, they are creating a magnetic flux Φ. Coils may be applied being surrounded by air or in conjunction with a highly permeable magnetic circuit. Fig. 11.1 depicts the most common types of excitation coils fabricated in thin-film technology. Fig. 11.1.a shows a meander coil. Due to its planar structure, it is the most simple thin-film coil: its whole conductor is created in a single fabrication step. A meander coil induces a magnetic flux Φ perpendicular to the wafer it is fabricated on. If used in conjunction with a magnetic circuit, its poles are located inside the meanders [7]. Fig. 11.1.b depicts a planar spiral coil. Like the meander coil, it uses planar technology. However, it cannot be fabricated in a single plane, because the inner end of the spiral coil has to be tabbed. Since a second plane is required anyhow, in most cases a second coil layer is used rather than a lead. A planar spiral coil also

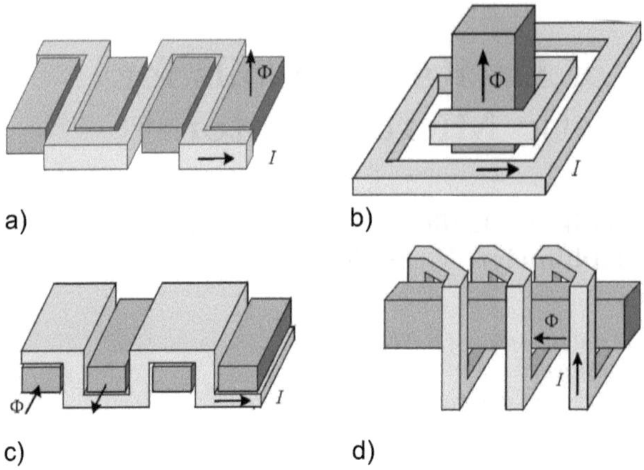

Fig. 11.1 Schematic view of the most common types of thin-film excitation coils and the direction of magnetic flux Φ they are creating: (**a**) Linear meander coil; (**b**) Spiral coil; (**c**) Vertical meander coil (flux direction ambiguous); (**d**) Helical coil [9]

induces a magnetic flux perpendicular to the wafer it is fabricated on. If used in a magnetic circuit, there is a magnetic core at its center [10]. Fig. 11.1.c presents a vertical meander coil. It has conductors at the bottom and on top, connected by vias. A vertical meander coil induces a magnetic flux Φ parallel to the wafer it is fabricated on. Due to its 3D design, a magnetic circuit used in conjunction with it may be chosen as a planar structure [10]. Fig. 11.1.d illustrates a helical coil, which is a 3D spiral coil. Similarly to the vertical meander coil, there are conductors in two planes, also connected by vias. A helical coil induces a magnetic flux Φ parallel to the wafer it is fabricated on. If used in a magnetic circuit, it has a magnetic core at its center [11, 10].

The coils were fabricated using a combination of photolithography, deposition, and etching techniques. Most commonly, coils were electroplated. In this case, photolithography was used for creating a temporary micromold for the deposition process. Since this deposition process requires a conductive substrate, a seed layer is necessary, which typically was sputter deposited before the micromold was created. After the deposition, the micromold was removed by solvents and the seed layer by dry etching.

The typical coil material is Cu, due to its excellent electric conductivity and its low tendency to electromigration. As favourite material for lateral and vertical insulation as well as general embedding processes, photosensitive epoxy (*SU-8*) was applied. Rather than being used as a temporary photomask, it serves as a structural material. This is also due to the fact that there is no solvent available for stripping it.

For 3D meander and helical coils, both horizontal and vertical patterning techniques are required. For such an application, an electro-depositable (ED) photoresist may be advantageous since it allows for an even layer deposition on the whole surface of the device. In many cases, rather great building heights are required for the magnetic micro devices. For this reason, quite commonly coil fabrication processes take advantage of a High Aspect Ratio Micro Structure Technology (HARMST).

11.2 Photoresist Pattern Creation

Photolithography is the standard technology for transferring patterns photo-optically. By creating a temporary photomask, an additive or subtractive patterning process may be conducted. Additive processes are electroplating in micromolds as described previously as well as the lift-off techniques. For the latter, a photomask is created first, followed by a coating of the whole wafer with the material to be patterned. When stripping the resist after the deposition, the thin-film on top is removed with the resist, while the one in the resist windows remains. A subtractive process is used after a thin-film deposition: the photomask protects film areas intended to remain from an etchant attack in case of wet-chemical etching or from an ion bombardement in case of dry etching. After the etching process, the photomask is stripped. In this case, the emphasis was on creating micromolds through photolithography, typically with rather high resist film thicknesses. A typical photolithography process sequence consists of the following steps: substrate cleaning, spin-coating of the photoresist, soft baking (to drive out the solvent), aligning the photolithography mask and exposure, development, and post exposure bake (not mandatory, depending on the respective process). For the exposure, UV light depth lithography was applied using a Hg vapor discharge lamp featuring a spectrum in the near UV area.

In general, there are two basic types of photoresist: positive tone and negative tone resist. Using a positive tone one, the exposed areas are removed during the development, while a negative resist is polymerized in the areas exposed to light. In this case, the unexposed areas are removed during the development. For the applications described here, *DNQ/Novolak*-based positive tone resists of the AZ series of *Clariant Inc.* were subjected to investigations and ultimately applied, due to the excellent compatibility to acid electrolytes and an easy removability.

Initially, the positive tone resist AZ4562 was used for micromolds. This resist was optimized for enabling a fabrication of high aspect ratio microforms not only on smooth substrate surfaces, but also on a topography of several tens of micrometers height [15]. It was found that 24 µm to 27 µm thick micromolds created in a single coating process of ness were optimal for creating 15 µm high electroplated structures. Deeper molds were realized by multiple coatings. In this case, drying on a hot plate was performed after each coating step, slowly rising the temperature to 100°C, followed by a slow cool down. For a good exposure result, several hours of dwell time are necessary to

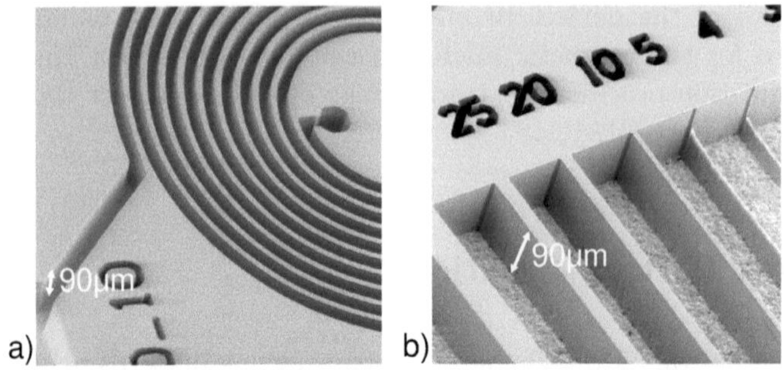

Fig. 11.2 SEM-photographs of a coil micromold: (**a**) 90 µm thick AZ9260-resist; (**b**) Test structures for determining the flank angle

ensure a constant and sufficient water content in the resist before exposure. An aspect ratio of 9:1 was achieved with optimized process parameters for thicknesses of about 90 µm using multiple coatings.

Advanced developments were carried out with the positive tone photoresist AZ9260. This resist features a high optical transparency, while layer thicknesses of more than 90 µm could be patterned with an enhanced aspect ratio of 12:1 [4]. Fig. 11.2 presents micromold tests patterns. Fig. 11.2.a shows a 90 µm high coil micromold pattern with 10 µm both for the conductor width, as well as a lateral distance between the turns, featuring a good resolution. The limit of the resolution and the flank angle could be determined by means of test structures ranging between 5 µm and 10 µm for a 90 µm thick layer (Fig. 11.1.b). A flank angle of 88° could be achieved. These results constitute a considerable widening of the application spectrum as well as the performance of *DNQ/Novolak*-based positive tone resists.

For accomplishing structural features in the sub-micrometer range, AZ1505 was applied. It was used at a film thickness of 0.5 µm and allowed a smallest pattern size of 0.7 µm on wafer level patterned by direct writing laser lithography. Using the patterned resist as etching mask, submicrometer coils were created in the sputtered Au layer underneath.

The application of AZ resist on patterned wafers is non-trivial, especially during spin-coating and exposure. On top and between existing structures, the height of the AZ layer may vary, which results in differences in the exposure time for one wafer. However, when exposing with the maxima time required for the total resist height, the structures already generated will be over exposed minimizing their resolution. Moreover, the development time of small and deep structures increases due to a continuous exchange of the developer, which is mainly controlled by diffusion rather than convection. These phenomena result in an over development of the structures located furthest above.

As a resist allowing negative tone imaging, the image reversal resist AZ5214 was used. The process for this resist is as follows: first the exposed areas are cross-linked in a bake cycle, then the unexposed areas are converted soluble in a flood exposure.

For applications requiring a vertical wall coverage, electro-depositable (ED) photoresists like *Shipley's PEPR2400* (now called *InterVia3D-P*) and *EAGLE2100ED* (*InterVia3D-N*), taking advantage of electrophoresis effects, were investigated [4]. These resists were developed for the printed circuit board (PCB) industry and gained importance in MEMS [2]. Their main advantage is that arbitrary geometries can be coated with uniform layers of photoresist; particularly on the vertical sidewalls of high structures, resist layers may be created and patterned. These photoresists consist of an aqueous emulsion of a solid compound. The solid particles are made of micelles of an acrylic co-polymer, which are stabilized by positive or negative surface charges according to the type of the resist.

For applying the ED photoresist, a coating system was developed [5]. Fig. 11.3 depicts both the system as well as coating results. The setup is composed of two stationary electrodes integrated in a small basin containing the photoresist diluted in deionized (DI) water. The basin can be easily placed inside an ultrasonic bath, which serves as a heater and allows for a temperature control. The wafer coated with an electrically conductive material used as a seed layer was placed in a holder and served as complementary electrode. This holder is retained by pins within the ED-basin (Fig. 11.3.a). When applying an electrical field, the micelles migrate to the wafer (electrode). When reaching the wafer electrode, the emerging ions are neutralizing the micelles, become destabilized and lead to the deposition. This process is maintained until the deposited film is self-insulating, which has the advantage that the layer thickness is highly reproducible. Furthermore, the resulting resist thickness depends on the conductivity of the solution, the applied voltage (or current, respectively,) and the bath temperature.

Initial coating experiments were performed on unstructured substrates with various seed layers, e.g. Cu, Au, and Cr, to determine the influence of their conductivity. Results showed that the current has to be limited depending on the conductivity to avoid pinholes and an uneven or rough resist surface. In further experiments, a good linear relation between bath temperature and film thickness was found (Fig. 11.3.b). In a temperature range from 21°C to 33°C, the resulting layer thickness could be varied between 7 µm and 28 µm with good reproducibility. Additionally, at each temperature, the film thickness could be influenced by increasing or decreasing the applied voltage. After the deposition, the solvents and remaining water were evaporated from the resist by a prebake step, which had a direct influence on the coverage problems at sharp edges. During the baking process, the photoresist started to reflow. As a result, at high temperatures, the resist was pulled down from the edges. Furthermore, the thickness of the resist was altered, becoming thinner close to the edges and increasing further away from the edges due to the surface tension of the resist. A substantial reflow of the photoresist is not desirable when using the structures for an electroplating mold. Vari-

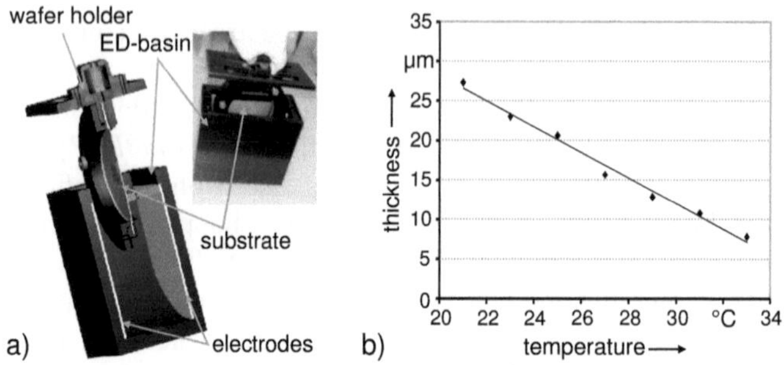

Fig. 11.3 Electrodeposition of photoresist: (**a**) Coating equipment; (**b**) Relation between bath temperature and resulting layer thickness

ous baking processes were tested using a hotplate and a convection oven at different temperatures. The best results to minimize the reflow of the resist were accomplished by replacing the baking step by a vacuum process. This process did not impede the photosensitivity. It was found that a deposition should not be conducted at temperatures higher than 34°C to get good results and avoid a reflow. The applied exposure dose depends on the thickness of the resist layer and was varied in a range from 500 to 3,000 mJ cm^{-2}. The optimal exposure parameters with respect to the development conditions were determined. They are comparable to the specification found in the Shipley data sheet. Both resists are easily removable by using acetone for *PEPR2400* and an appropriate remover for *EAGLE2100ED*. Aspect ratios of 5:1 (*PEPR2400*) and 8:1 (*EAGLE2100ED*) were achieved by using the optimized process parameters (Fig. 11.4). The deposition tests were conducted on *SU-8*.

SU-8 is a high contrast transparent photosensitive epoxy which allows to processing thick and planar layers. Due to a change of its refractive index during exposure, *SU-8* allows to creating patterns with high aspect ratio and nearly vertical sidewalls. However, *SU-8* is a structural material rather than

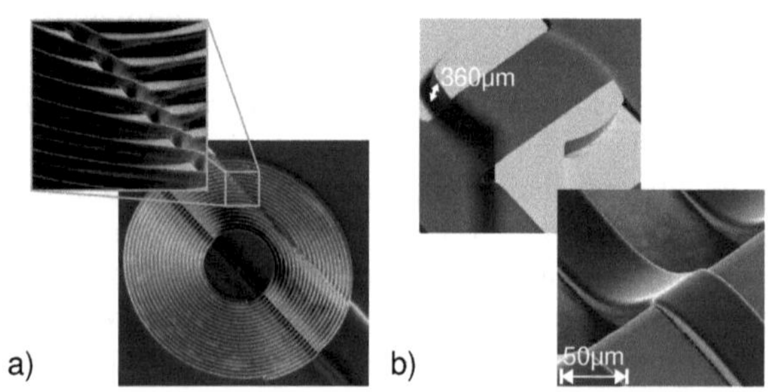

Fig. 11.4 *SU-8* structures covered by ED resists: (**a**) *PEPR2400*; (**b**) *EAGLE2100ED*

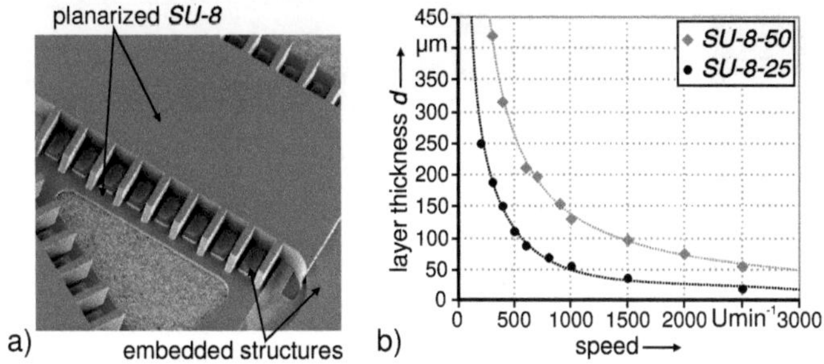

Fig. 11.5 Coating and planarization properties of *SU-8*: (**a**) Structures embedded and planarized by *SU-8*; (**b**) Layer thickness of *SU-8* depending on viscosity and rotational velocity

a photoresist, since no agents to strip the epoxy material are available. A removal is only possible by plasma ashing [18].

SU-8 is available in a variety of viscosities, expressed by the respective formula name (e.g. *SU-8-5*, *SU-8-25*, and *SU-8-100*). Fig. 11.5 demonstrates the coating and planarization properties of *SU-8*. Fig. 11.5.a shows an example of a structure embedded in *SU-8*. The thickness of the created layer depends on the coating speed and on the viscosity of the *SU-8* (Fig. 11.5.b). There is another drawback in using *SU-8*: a substantional batch-to-batch variation in the photo-optical behavior. For each new lot, the key process parameters have to be analyzed and documented, allowing to adjust the parameters in the further fabrication. The importance of this quality management is due to the irreversible processing of *SU-8*. Continuous optimizations were carried out regarding process parameters like softbake, exposure, and post exposure bake (PEB). Doing so, the aspect ratio of *SU-8* could be improved to 60:1 compared to earlier results delivering 36:1. Furthermore, the lateral resolution was enhanced to 8–9 µm for a 600 µm thick layer. Fig. 11.6 shows high aspect ratio test structures. Fig. 11.6.a shows a test, up to which height *SU-8* remains still standing. For a 9 µm wide test structure, the tallest height without toppling was 600 µm [3]. A flank angle measurement on a height aspect ratio pattern resulted in an angle of 90.2° (Fig. 11.6.b) [17].

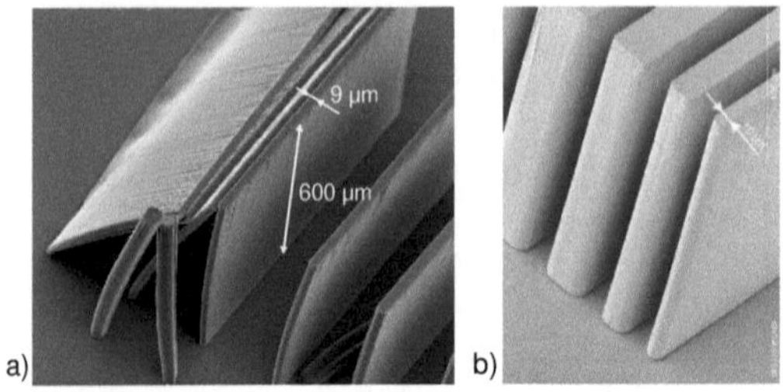

Fig. 11.6 *SU-8* test structures: (**a**) Investigation of toppling; (**b**) High aspect ratio patterns [17]

11.3 Electroplating of Cu Microcoils

The main advantage of a plating process is the achievement of height cross sections, as required for most microcoils. The coil material has to be highly conductive and has to show a low tendency to electromigration. For these two reasons, Cu was used as conductor. Furthermore, Cu lends itself to electroplating. As an electrolyte for this process, *CUBATH SC* was chosen in a room temperature process. As described before, photoresist micromolds were used for patterning the coils. Before preparing the micromold, an adhesion layer and a conducting seed layer were sputter deposited. As adhesion layer, Cr and as seed layer, Au or Cu were used. Next, a micromold was prepared and wetted with DI-water before the electroplating process started. For achieving a sufficient bath excitation, the process was conducted in a plating cell with a moving paddle. When the intended coil height was reached, the wafer was removed from the cell, cleaned by DI water, and spin-dried. Fig. 11.7 presents the plating process sequence [8]. To reduce the surface roughness, an addi-

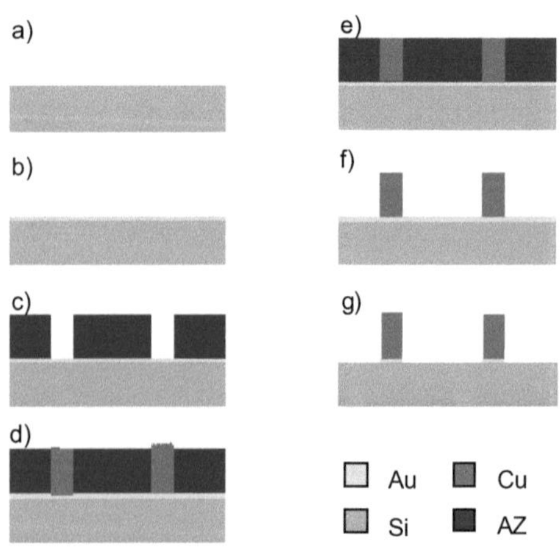

Fig. 11.7 Cu electroplating process: (**a**) Wafer cleaning; (**b**) Seed layer deposition; (**c**) Photolithography; (**d**) Electroplating; (**e**) Chemical-mechanical polishing (CMP); (**f**) Photoresist stripping; (**g**) IBE of the seed layer [8]

tional Chemical-mechanical Polishing (CMP) step was performed, whenever required. Afterwards, the micromold was removed by a solvent. Finally, the seed layer was removed by a wet-chemical etching process or Ion Beam Etching (IBE).

11.4 Insulation

For insulating the coils, either an organic or inorganic material was used. The organic material was *SU-8* [9, 11], as inorganic materials Al_2O_3 or Si_3N_4 were employed [14]. The use of *SU-8* as an insulator requires that the photoresist micromold as well as the seed layer under a coil layer already have been removed. The task of the *SU-8* is to fill the gaps between coil conductors or coil conductors and magnetic cores, respectively, to accomplish insulation. This is achieved by spin-coating *SU-8* [13]. Depending on its thickness, the *SU-8* either only fills the gaps (if the height of the *SU-8* matches the height of the conductors) or exceeds the conductor material in height. In the second case, the *SU-8* also creates a top insulation layer, which may be the basis for a second coil layer. In this case, openings for vias are photolithographically patterned in the *SU-8*. If Al_2O_3 is used as an insulation material, the material is typically sputter deposited [7]. Compared to other insulating processes, the sputtering process of alumina is comparatively slow. Si_3N_4 is an inorganic insulator wich may be deposited by Plasma Enhanced Chemical Vapor Deposition (PECVD). PECVD layers show an excellent edge coverage. Si_3N_4 films with very low stress can be accomplished using a PECVD two-frequency mixing process. In this case, a high frequency (HF) and a low frequency (LF) are applied in an alternating mode. A film deposited in the HF mode exhibits tensile stress. In the LF mode, the ion bombardment during the LF phase causes a compacting of the deposited layer, resulting in a compressive stress in the deposited Si_3N_4 film. By choosing a multilayer design consisting of two alternating layers (i.e. applying a HF or a LF for plasma generation), a minimal stress level may be achieved. The optimal ratio of HF (at 13.56 MHz) and LF (between 100 kHz and 300 kHz) was found at 13 s to 7 s. With such a frequency mixing process, the tensile stress could be reduced down to some 10 MPa.

Since the costs for the precursor gases for this process are rather high, the use of the inorganic layers was limited to create thin intermediate insulations between the single coil layers and not applied for embedding the whole microstructures. Beside the low mechanical stress, Si_3N_4 has another advantage when used as an intermediate insulation film: a coefficient of thermal conduction (CTC) much greater than for organic insulators (29 $Wm^{-1}K^{-1}$ for Si_3N_4, while the CTC for *SU-8* is 0.2 $Wm^{-1}K^{-1}$ [12]). This provides a better transfer of the heat which is generated by the microcoils whenever excited. However, the main advantage of Si_3N_4 is a smaller film thickness required compared to *SU-8* when used as insulation material between two coil layers. After the deposition of an inorganic insulation layer, it has to be patterned (unless it is the final coverage, where only the pads for the electric contacts have to be free). While *SU-8* is patterned by photolithography directly, PECVD layers are patterned by wet chemical or dry etching requiring an adequate mask. Wet chemical etching is an isotropic process resulting in an underetching of the film. Since in general, an underetching is undesirable, a dry etching process based on an ion beam was chosen, which features a directional etching profile. The photomasks used for the process were made of AZ series resist [11].

11.5 Integration of Coils into Microactuators

The fabrication technologies developed were applied for four types of excitation coils: linear meander coils, planar spiral coils, 3D meander coils, and helical coils. The examples chosen are: (1) a linear meander coil in a linear variable reluctance (VR) micro step motor; (2) planar spiral coils used in a linear synchronous micromotor (double layer spiral coil), a magnetic levitation system (four-layer spiral coil), and a technology study for a submicrometer coil; (3) a 3D meander coil, also for a linear VR micro step motor; (4) a helical coil used in a microinductor. The following paragraphs present the respective fabrication technologies.

11.5.1 Planar Meander Coil

The planar meander coil is the simplest coil design, because the coil conductor can be fabricated in a single process step. For the chosen example, the planar meander coil is intended for the use in a linear micro step motor. Therefore, the coil has to be integrated into a soft magnetic circuit consisting of bottom flux guides, poles between the meanders, and teeth on top of the poles (Fig. 11.8). Since the soft magnetic material is a conductor (NiFe81/19), an insulation between coil and magnetic components is required.

The first step of creating the coil was depositing a seed layer consisting of an adhesion layer and the conductive material, because the coil was fabricated on non-conductive wafer material. Next, a photolithography process created the micromold for the coil itself; an electroplating process filled the micromold with Cu, thus creating the coil conductor. Afterwards, the micromold was stripped and the seed layer is removed by IBE. Then, the coil was embedded in *SU-8*, which also served as a reliable insulation material between coil and poles. This step concluded the linear meander coil fabrication process. Fig. 11.9 shows the coil after its deposition.

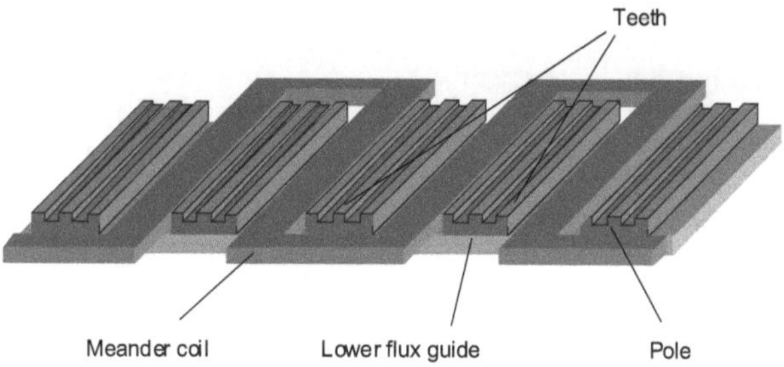

Fig. 11.8 Schematic view of a linear meander coil, integrated into the stator of a VR micro step motor.

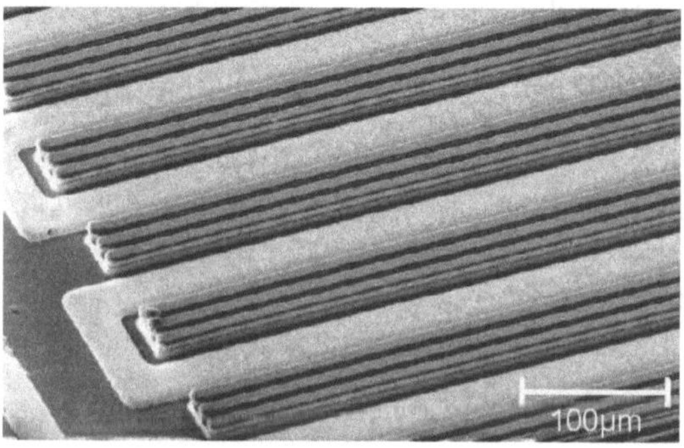

Fig. 11.9 REM micrograph of a meander coil, integrated into the stator of a VR micro step motor [7]

11.5.2 Planar Spiral Coils

Planar spiral coils require a much lower driving current than meander coils to achieve the same magnitude of ampere turns. They are, however, more complex, requiring at least two conductor layers. Due to their higher impedance, they require a higher voltage. For magnetic microactuators and levitation systems, planar spiral coils are used without as well as with soft magnetic cores. To emphasize the coil technology requirements, the following examples feature coils without magnetic cores.

Double-layer Spiral Coil

For demonstrating the double-layer spiral coil technology, its application in a linear synchronous micromotor was selected as an example [9]. Fig. 11.10 presents a schematic representation of the device. The spiral coil fabrication started with a sputter deposited seed layer made of Au. On the seed layer, micromolds were prepared by photolithography. Then, the first Cu coil layer, as well as leads and bond pads were electroplated into the micromolds. To

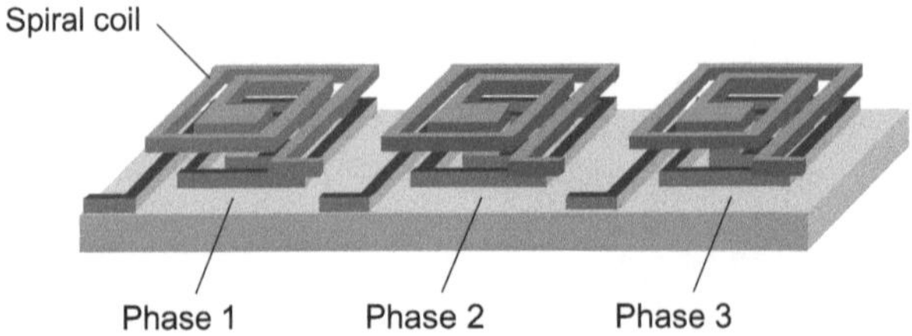

Fig. 11.10 Schematic view of a two layer spiral coil system for a linear synchronous micromotor

achieve a high surface flatness for the following process steps, a CMP process was conducted. Next, the micromold was stripped and a new one was created. Its purpose was creating a via connection between bottom and upper layer. The Cu via was deposited by electroplating. Afterwards, the micromold was stripped and the seed layer was removed by IBE. For insulating the first coil layer as well as embedding the whole coil structure, *SU-8* was used. This step concluded the fabrication of the first coil layer. The process steps for the second coil layer were equal to those of the first one. The completed coil system was embedded in *SU-8*, with the bond pad left uncovered by photolithography. In a final series of tests, the bond pads were reinforced by electrodepositing a Ni diffusion barrier and an Au bond surface. This fabrication sequence for bond pads was typically applied to all coil systems for magnetic micromotors.

Fig. 11.11 presents coil systems for two types of synchronous motors [6, 1]. The coils not only vary in size, but also in shape. While one features rectangular coils (Fig. 11.11.a), the other has circular ones (Fig. 11.11.b). The change in shape was introduced based on the experience of the data storage industry, which found out that corners in coil structures are prone to initiating defects.

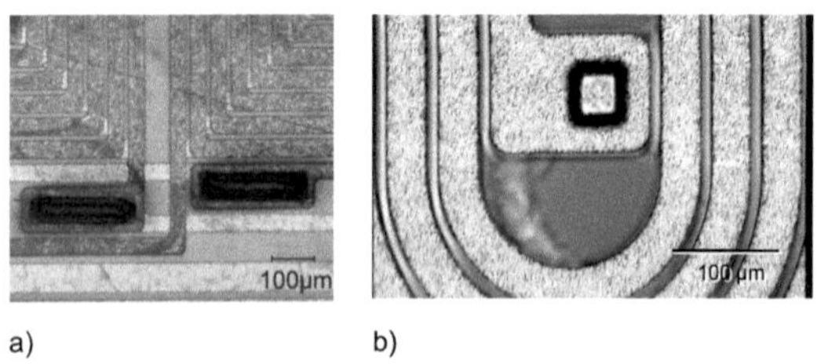

Fig. 11.11 Coil systems for linear synchronous micromotors: (**a**) Rectangular coil (large version); (**b**) Circular coil (small version) [1]

Four-layer Spiral Coil

As an example of a four-layer spiral coil, a component for a magnetic levitation system was chosen. During the fabrication of coil systems with four layers, the resulting height of the complete system has to be kept at a minimum. By using thin inorganic vertical insulation layers, a four-layer system with a height of 82 µm could be fabricated. The fabrication of the four-layer spiral coils started with the sputter deposition of a seed layer with a 50 nm Cr adhesion layer and a 200 nm Au conduction layer. Next, an AZ photoresist layer was patterned by UV depth lithography to create a micromold. The first coil layer was fabricated by Cu electroplating. After stripping the AZ resist, the seed layer was removed by IBE. Then, the coil layer was embedded in *SU-8*. This step also created the lateral coil insulation. To achieve a high flatness, the embedded first coil layer was planarized by CMP, down to the top surface of the coil conductor. For a vertical insulation between the coil layers, a 500 nm thick low stress Si_3N_4 layer was deposited by a PECVD process. Openings for the vias were created by photolithographically patterning an AZ etch mask, followed by an IBE process. This step concluded the fabrication of the first of the four coil layers. For fabricating the second, third, and fourth layer, the same process sequence was repeated three times [14]. The developed four-layer spiral coils were used for an electromagnetic guide. Fig. 11.12 depicts the four-layer spiral coil [14]. Fig. 11.12.a provides a top view of the third coil layer after CMP and before the deposition of the insulation layer, while Fig. 11.12.b shows a cross section of the system. The thin Si_3N_4 insulation layers are clearly visible.

Technology Study for a Spiral Coil with a Submicrometer Conductor Width

A technology study for a coil with submicrometer size conductors was carried out. For the fabrication, first an Au layer serving as the future conductor was sputtered. Next, the positive tone AZ1505 with a thickness of 0.5 µm was spin-coated. The exposure was done on wafer level using the direct writing laser lithography equipment *Heidelberg Instruments DWL66*. The submicrometer coils were created by IBE. The test patterns featured 0.8 nm wide conductors with a spacing of 0.7 µm (Fig. 11.13).

11.5.3 Vertical Meander Coil

While planar coils are generating a magnetic flux perpendicular to the wafer plane, a vertical meander coil creates a horizontal one, parallel to it. The vertical meander coil consists of lower conductors, vias, and upper conductors. These coils are necessary for energizing single-plane cores. Fig. 11.14

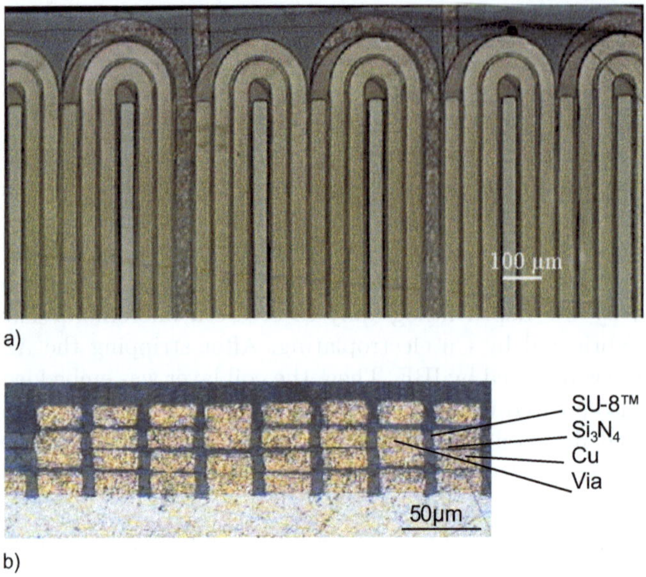

Fig. 11.12 Four-layer spiral coil: (**a**) Top view of the third coil layer after chemical-mechanical polishing and before the deposition of the insulation layer [13]; (**b**) Cross section of the system

depicts one of these vertical meander coils used for the creation of a variable reluctance (VR) step motor with a horizontal flux (see Chap. 12).

Technologies for the fabrication of such coils were investigated [16]. Figure 11.15 depicts the fabrication steps. The fabrication of the vertical meander coil started with the sputter deposition of a seed layer. Afterwards, an AZ photoresist layer was patterned by UV depth lithography to create a micromold. The lower conductor was created by Cu electroplating. Next the coil was embedded in an *SU-8* insulation layer. (Fig. 11.15.a). Subsequently, the core structures were fabricated and embedded in a second *SU-8* layer (Fig. 11.15.b). In both insulation layers, orifices were provided for the via contacting. In the last step, both the vias and the upper conductors were re-

Fig. 11.13 Coil with submicrometer conductors

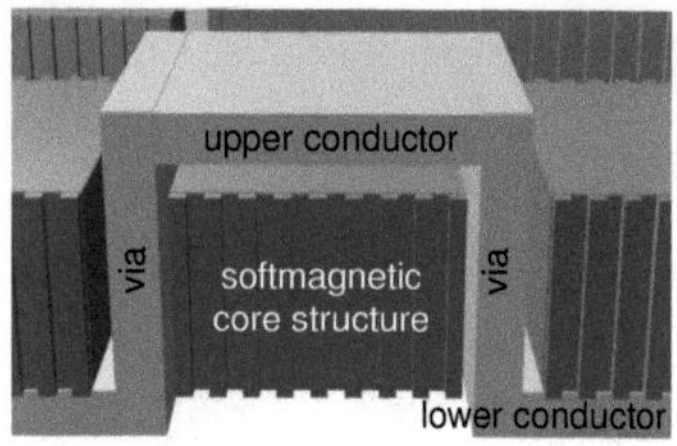

Fig. 11.14 Schematic view of a vertical meander coil, integrated in a VR step motor with a horizontal pole plane

alized simultaneously by Cu electroplating in an AZ micromold (Fig. 11.15.c). Therefore, the Cu seed layer thickness was increased to achieve a low resistance all over the substrate to avoid varying deposition rates. In Fig. 11.16, the multilayer composition, especially the hollow-shaped vias are illustrated.

Alternatively to the use of *SU-8*, the vertical meander coils were fabricated mainly by the application of the electro-depositable photoresist *InterVia3D-N*, whereby the fabrication of the vertical meander coils (and also the helical coil presented in the next paragraph) could be greatly simplified. Thereby, vertical and upper conductors could be exposed in one step. For the integration of this technology in existing process sequences for the coating of structured substrates, the deposition parameters already identified were adjusted to reflect the respective requirement. The exposure and prebake parameters were optimized to allow a precise pattern transfer with high aspect ratios. Furthermore, the thickness of the seed layer was increased to obtain an optimal conductance all over the substrate and mainly the vertical sidewalls to provide a sufficient resist deposition.

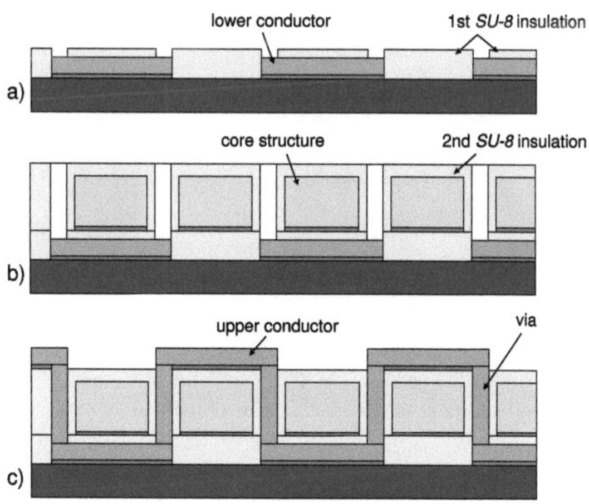

Fig. 11.15 Process sequence for the fabrication of a vertical meander coil

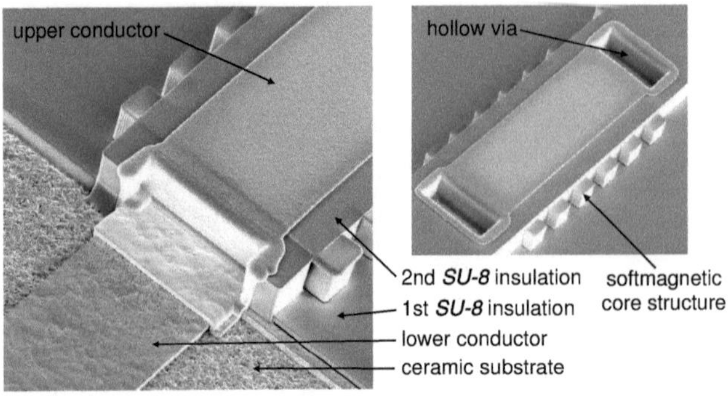

Fig. 11.16 SEM-photograph of hollow conductor vias as a part of the vertical meander coil

11.5.4 Helical Coil

Generally, the process steps for fabricating a helical coil (Fig. 11.17) were: creating the lower conductors, the first *SU-8* insulation, the core, and the second *SU-8* insulation [19]. Subsequently, the sample was covered by a Cu seed layer, onto which the ED-resist was deposited homogeneously, followed by a prebake. For a simultaneous exposure of the mold for vertical and upper conductors, the exposure dose was optimized to obtain a sufficient exposure of the sidewall structures, but without a resolution loss of the upper conductor structures (Fig. 11.17.a). Both vertical and upper conductors were deposited in the same step by Cu electroplating. Afterwards, the resist was removed using a solution at 60°C supported by ultra sonic. Consecutively, resist residuals were ashed in an oxygen plasma and the seed layer was removed by etching (Fig. 11.17.b).

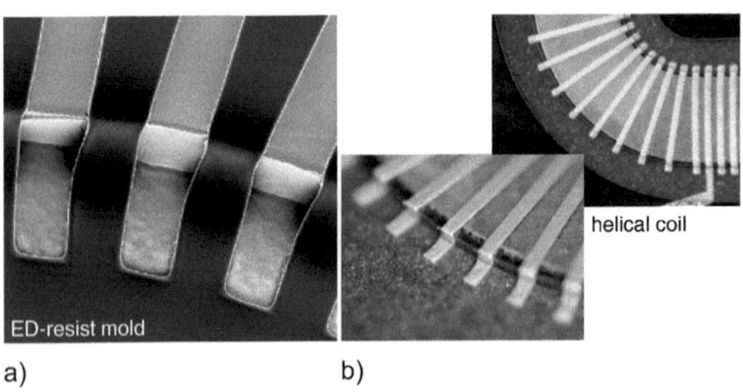

Fig. 11.17 Helical coils fabrication: (**a**) ED-resist mold; (**b**) Completed helical coil around a core

11.6 Conclusion

The microcoils developed enable the realization of various magnetic Microelectro Mechanical Systems (MEMS). High Aspect Ratio Micro Structure Technology (HARMST) for the coil fabrication could be established. Fabrication processes could be suggested for four types of thin-film coils: a linear meander coil, spiral coils, a vertical meander coil as well as a helical coil. Each one is predestined for specific applications. *SU-8* was found to be a very useful structural material, which lends itself both as insulator, as well as a high aspect ratio embedding. It could equally be applied in any type of coils, in the fabrication of all types of coils, no matter if planar or 3D. Electro-depositable resists proved to be very valuable when fabricating 3D coils. A planarization through a Chemical-mechanical Polishing (CMP) process was found to be a precondition for fabricating most high aspect ratio coil components in any planar technology. It further could be demonstrated, that the vertical building height of multilayer coils can be substantially reduced by applying low-stress Si_3N_4 as insulation material. Compared to the technology based on thick organic insulation layers, this is an important improvement.

Acknowledgements The authors gratefully acknowledge the funding of the reported work by the German Research Foundation within the Collaborative Research Center 516 "Design and Manufacturing of Active Micro Systems".

References

[1] Bedenbecker M, Hahn M, Ruffert C, Gatzen HH (2007) Linear synchronous micro motor with further miniaturized dimensions. In: Proc. ASME 2007, Las Vegas, NV, USA, pp DECT2007–34,677
[2] Boellaard E, Pham NP, van den Brekel LDM, Burghartz JN (2001) RF-devices realized in MEMS by using electrodepositable photoresist. vol 28-29, pp 1–7
[3] Feldmann M (2007) Technologien und Applikationen der UV-Tiefenlithographie: Mikroaktorik, Mikrosensorik und Mikrofluidik. Dissertation, TU Braunschweig, Shaker Verlag
[4] Feldmann M, Büttgenbach S (2006) The potential of novel photo resists in MEMS applications: Technology and characteristics of electrodepositable photo resists and CAR44. In: APCOT 2006, Singapore, p 148
[5] Feldmann M, Waldschik A, Büttgenbach S (2006) Technology and application of electro-depositable photo resists to create uniform coatings

needed for complex 3D micro actuators and sensors. Microsystem Technologies 13(5-6):557–562

[6] Föhse M, Stölting HD, Gatzen HH (2003) A batch fabricated linear synchronous motor. In: Proc. Int. Mech. Eng. Congr. and Expo IMECE 2003, Washington, D.C., USA, pp 1–8

[7] Gatzen HH, Stölting HD, Büttgenbach S, Dimigen H (2000) A novel variable reluctance micromotor for linear actuation. Proc Actuator 2000 pp 363–366, Bremen, Germany

[8] Gatzen HH, Klocke F, Kamenzky S, Traisigkhachol O (2008) Electroplated Cu micro electrode for application in micro electrostatic discharge machining (EDM). ECS Transactions 16(45):255–268

[9] Kohlmeier T, Gatzen HH (2002) Challenges in using photosensitive embedding material to planarize multi-layer coils for actuator systems. Journal of Magnetism and Magn Mat 242-245:1149–1152

[10] Kohlmeier T, Seidemann V, Büttgenbach S, Gatzen HH (2004) An investigation on technologies to fabricate microcoils for miniaturized actuator systems. Microsystem Technologies 10:175–181

[11] Pape F, Creutzburg T, Traisigkhachol O, Gatzen HH, Büttgenbach S (2009) Thin-film fabrication of a rotating synchronous motor. Proc MST-Kongress 2009, Berlin, Germany pp 828–831, VDE-Verlag

[12] Petersen KE (1982) Silicon as a mechanical material. Proceedings IEEE 70(5):420–457

[13] Ruffert C, Gatzen HH (2008) Fabrication and test of multi layer micro coils with a high packaging density. Microsystem Technologies 14(9-11):1589–1592

[14] Ruffert C, Feldmann M, Büttgenbach S, Gatzen HH (2006) Fabrication of functional components for electromagnetic microactuators. Microsystem Technologies 12:670–675

[15] Seidemann V, Büttgenbach S (2002) Fabrication technology for closely coupled micro coils with integrated flux guidance and their application to proximity and magnetoelastic force sensors. In: Proc. IEEE Sensors 2002 Conference, Orlando, USA, pp 580–584

[16] Seidemann V, J E, Büttgenbach S, Stölting HD (2003) Linear variable reluctance (VR) micro motor with horizontal flux guidance: Concept, simulation, fabrication and test. In: Proc. 12th Int. Conf. on Solid-State Sensors and Actuators, Boston, USA, pp 1415–1418

[17] Seidemann V, Kohlmeier T, Föese M, Gatzen HH, Büttgenbach S (2003) Fabrication of exciting coils. In: Mikroproduktion, Vulkan-Verlag, Essen, pp 1415–1418

[18] Traisigkhachol O, Schmid H, Wurz MC, Gatzen HH (2010) Applying SU-8 to the fabrication of micro electro discharge machining electrodes. Microsystem Technologies DOI: 10.1007/s00542-009-1011-2

[19] Waldschik A, Feldmann M, Büttgenbach S (2007) A novel reluctance actuator employing an embedded ferromagnetic foil. Microsystem Technologies 13(5-6):551–555

Chapter 12
Development and Fabrication of Electromagnetic Microactuators

A. Waldschik, M. Feldmann, V. Seidemann, S. Büttgenbach

Institute for Microtechnology
Technische Universität Braunschweig
s.buettgenbach@tu-braunschweig.de

Abstract In recent years several photoresists were introduced, tested, and optimized for the fabrication of components for electromagnetic microactuators. They serve as insulation layers as well as mold for electroplating of different materials like copper for conductors and nickel-iron alloys for soft magnetic functional structures. Several electromagnetic microactuators were developed and fabricated using these techniques. Over the past few years the actuators have become increasingly complex through the integration of additional components. The range of electromagnetic actuators could be completed by special polymer magnets which enable creation of any shape of hard magnetic structure. Thus, all types of electromagnetic actuators based on reluctance as well as the electro-dynamic principle can be realized.

12.1 Introduction

Through the development and optimization of new technologies, a variety of complex MEMS applications are feasible. In special case, electromagnetic microactuators have been developed and realized. Their general construction consists of electrical conductors or coil systems (Chap. 11), soft or hard magnetic materials (Chap. 10) as well as insulation and embedding dielectrics (Chap. 11). During the past few years, several technologies were introduced and consecutively optimized for effective application in the fabrication of components for microactuators. Electro-magneto-mechanical and electro-dynamic working principles in the form of microactuators, micromotors, and microrobots have successfully been realized and tested. This chapter describes the combination of the developed technologies which has lead to complete systems in the form of three demonstrators. Specifically depicted is the development and fabrication of linear and rotating variable reluctance (VR) actuators providing horizontal magnetic flux as well as synchronous motors providing axial magnetic flux (see Fig. 12.1).

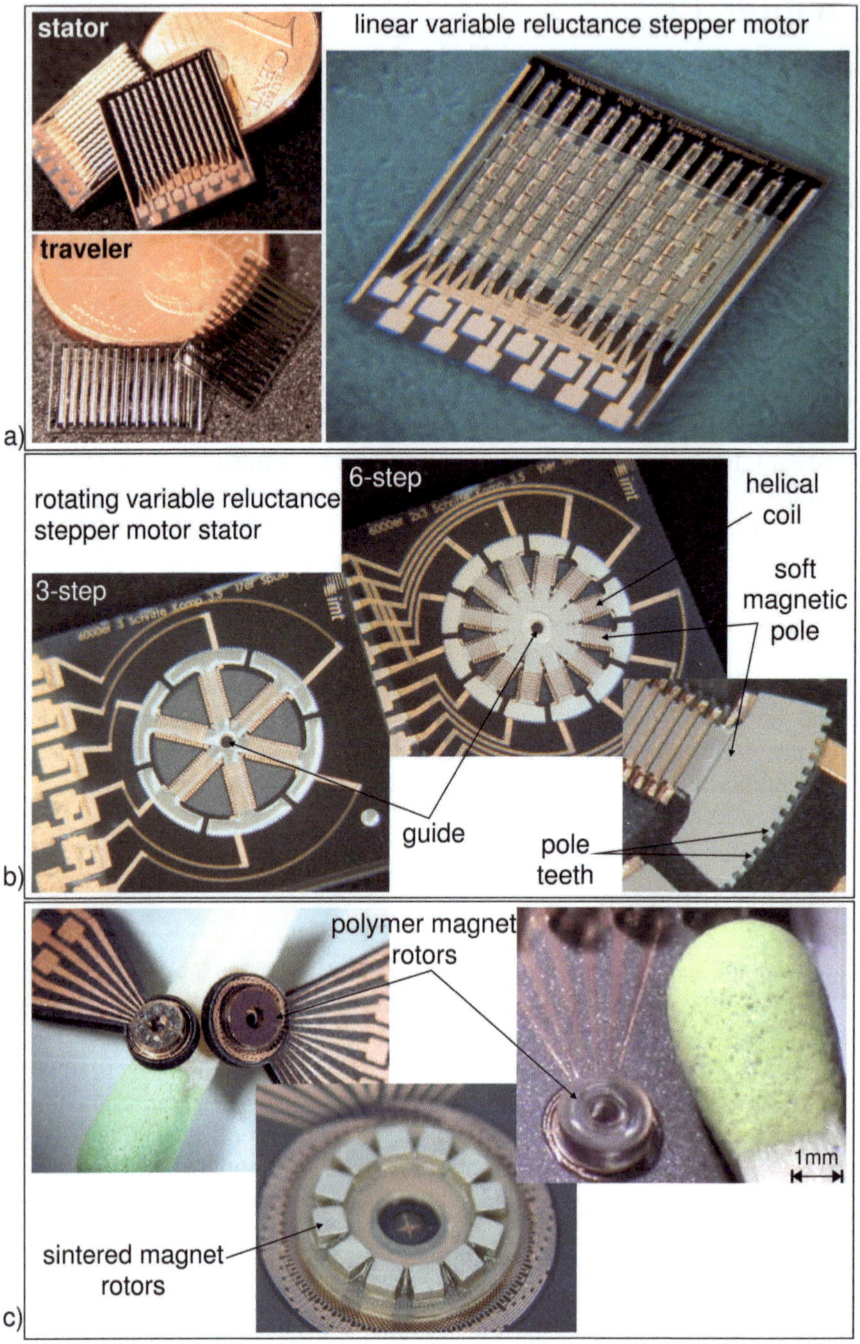

Fig. 12.1 Overview of various fabricated microactuators: (**a**) Components of linear VR stepper motor [2, 6]; (**b**) Rotating VR stepper motor stators; (**c**) Synchronous micromotor [11]

12.2 Linear VR Stepper Motor

The general mode of operation of these miniaturized stepper motors is equal to that of macroscopic motors. The forces generated by the variable reluctance principle are used for realizing a continuous stepping movement [8, 9]. This is achieved by a complementary arrangement of air gaps between the active parts of the stator and the traveler (Fig. 12.2). The stator consists of vertical meander coils wound around soft magnetic stator poles. The traveler is composed of a traverse with soft magnetic teeth which is arranged on both sides of each stator pole. In this configuration, the coils generate a horizontal magnetic flux which is guided in plane and closes over the air gaps between the stator and the traveler.

Both stator and traveler poles are toothed with a pitch of 100 µm. The tooth width is half of the pitch. Depending on the step size, the motor has three or six phases located in parallel. Each is shifted by 1/3 or 1/6 tooth pitch. The consequent application of current to a suitable phase results in an alignment of the traveler to that excited phase and produces a stepwise linear motion. The important advantage of having a horizontal magnetic flux is that the complementary arrangement of the air gaps between stator and traveler provides a compensation of the typically normal forces perpendicular to the drive direction. The stator is composed of six coil systems which are arranged symmetrically to the centerline. This is done in order to achieve a uniform force acting on the traveler and to avoid an undesirable torque. Additionally, the stator and traveler feature mechanical guide rail structures in order to avoid shifting perpendicular to the intended direction of motion.

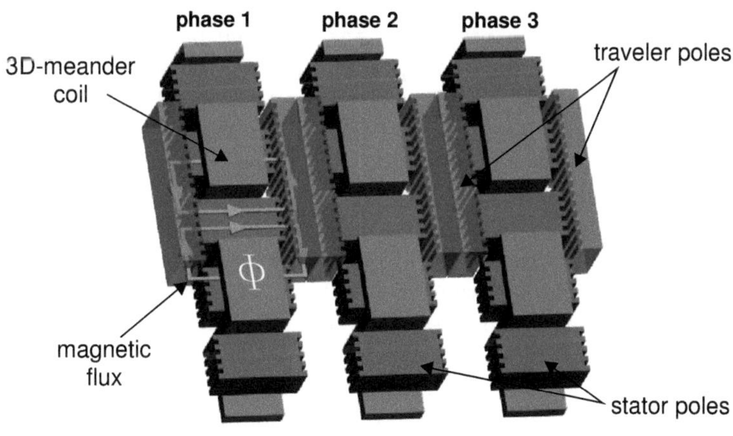

Fig. 12.2 Schematic view of the VR micromotor: first generation [2]

12.2.1 The First Generation

The concept of the first generation of this motor is shown in Fig. 12.2. In this design, the stator comprises three (or six) separately excitable phases. A phase consists of 40 (or 20) poles. Each pole is wrapped by a three quarter winding of the vertical meander coil which is wound alternately beneath or above the pole. The coil is run along a single pole row in one direction and is subsequently led to the symmetrically arranged part of the phase where it meanders back. The distance between the tooth surfaces of stator and traveler poles is about 5 μm. In this design the stator poles have a thickness of 51 μm and consequently the traveler poles must be 75–80 μm thick in order to overlap the upper conductor and insulation of the stator. Substantial research into the electroplating of copper for the coil structures and nickel iron for the soft magnetic poles was conducted for the initial fabrication process. Additionally, photo resists and their process parameters were evaluated and optimized for their use as molds, in planarization steps and for electrical insulation. The optimization was aimed at high aspect ratios structures which are necessary to achieve the adjustment of the small air gap. As the thickness of electroplating mold is increased to the maximum, the lateral precision must not degrade in order to ensure exact air gap geometry. Much effort was applied to the optimization of AZ4562 (*Clariant Inc.*) processing by adapting the softbake process times, exposure dose and development. The latter was optimized to be used for precise mold creation of approximately 100 μm thickness at an aspect ratio of 8. Additional process development was pursued for the realization of via interconnects through 80–100 μm thick dielectric. Due to its high aspect ratio patterning capabilities, *SU-8* proved to be best suited for this, since it allows the subsequent fabrication of via holes in each insulation layer [8].

The stator fabrication process begins with the pattering of AZ4562 as mold for the electroplating of copper for the lower conductor structures. A subsequent *SU-8* layer is used for embedding and insulation. This layer provides openings for the later via interconnects. A 100 μm thick AZ4562 layer is deposited by multiple coating and is structured to form a mold. The mold is filled with 50 μm nickel iron alloy by electroplating. In the same step, the guide rails are realized, guaranteeing the high adjustment precision necessary for the tight air gap dimensions of only a few micrometers over the stator length of 8 mm. In the following step, a thick *SU-8* layer is applied for the planarization and insulation providing via holes (Fig. 12.3.a). At the end of the process, a microform consisting of AZ-resist on top of the second insulation layer and the via holes in the *SU-8* serves for electroplating of the via and upper conductor in a single step. Fig. 12.3.b shows the completed stator consisting of toothed stator poles extending sideward out of the second insulation layer and the surrounding 3D-meander coil.

The first generation traveler simply consists of a 700 μm glass substrate upon which the comb shaped traveler teeth are electroplated in a 92 μm high

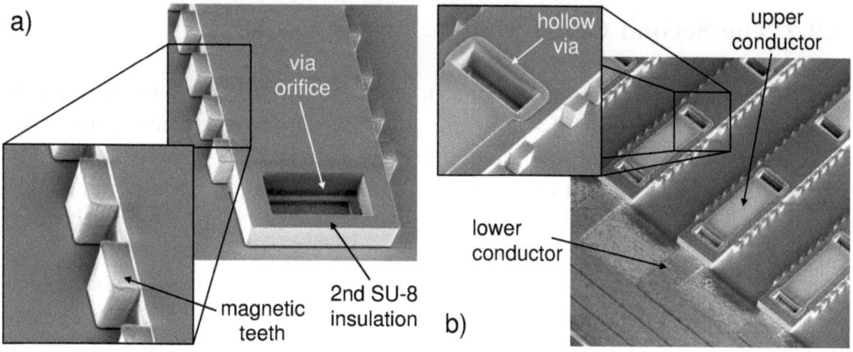

Fig. 12.3 (**a**) *SU-8* planarization; (**b**) Finished stator [9]

AZ4562 resist mold. To reduce magnetic property degrading tensile stress in the nickel iron structures, the mold is filled partly with copper before NiFe-electroplating. In this way only the overlapping top 45 µm of structure height that magnetically face the stator poles after assembly are yielded. The realized device was successfully mounted with the traveler. The rail shaped guides, which were designed at a higher precision than the air gap, proved to securely guide the traveler poles along the stator. Functional tests were carried out incorporating a ruby ball bearing for friction reduction (Fig. 12.4). The bearing was realized by a silicon base plate with three anisotropically etched v-shaped grooves [8]. The grooves for the 800 µm diameter ruby balls to the left and right of the stator die provide a fixed guide geometry due to the fact that the balls roll along the etch stop (111)-planes. The stator is placed in a large groove that can be subsequently adjusted in depth until the traveler exactly fits on top. In this way it is possible to compensate the weight of the traveler and greatly reduce the friction in the traveling direction. In this configuration, the traveler could initially be moved in a stepwise fashion.

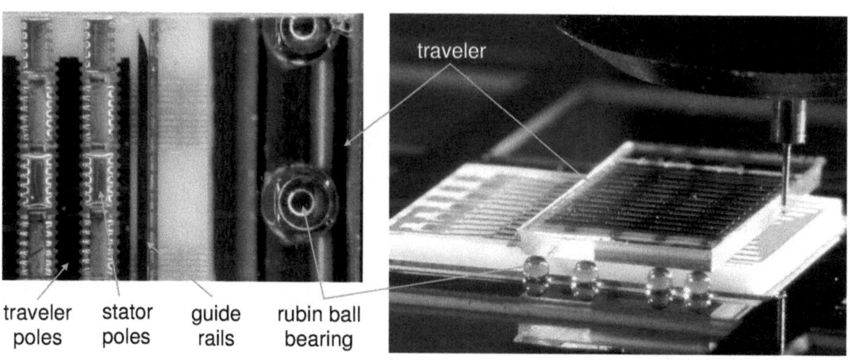

Fig. 12.4 Photographs of mounted linear micro motor with ruby ball bearing

12.2.2 The Second Generation

Upon evaluation of the general functional capability as well as the process suitability of fabrication of the stator and traveler components, further investigations were carried out concerning the fabrication process and design optimization with the aim of increasing motor performance. In doing so, the second generation of the linear reluctance motor has been realized with several improvements.

In comparison to the single turn meander coil of the first generation, a novel double winding 3D-meander coil was integrated (Fig. 12.5.a), which reaches the same flux intensity despite a reduced current (Chap. 2) [2]. In the first design, each pole is wrapped by the meander coil alternately beneath or above the pole. In this new design however, the coil is led back and forth in the same pole row (Fig. 12.5.b). The coils of two symmetrically arranged pole rows are then connected by additional system connections. Each pole is thereby encased by a lower and upper conductor as well as four via connections. It was found that the optimal design, e.g. the optimal number of teeth per stator pole, strongly depends on the permeability of the soft magnetic material. Furthermore, the structure height itself turned out to have an optimum value, since it strongly influences the permeability of the flux guidance because of restrictions in the micro technological process flow. The best results for driving forces in relation to diminishing permeability and gap surface were determined at a structure height for the stator poles of 30–35 μm and 8 teeth per pole [1]. The traveler pole height was correspondingly adapted.

Considering the results, the complex process sequence for stator fabrication (Fig. 12.6.a–f) must be enhanced by additional electroplating and insulation

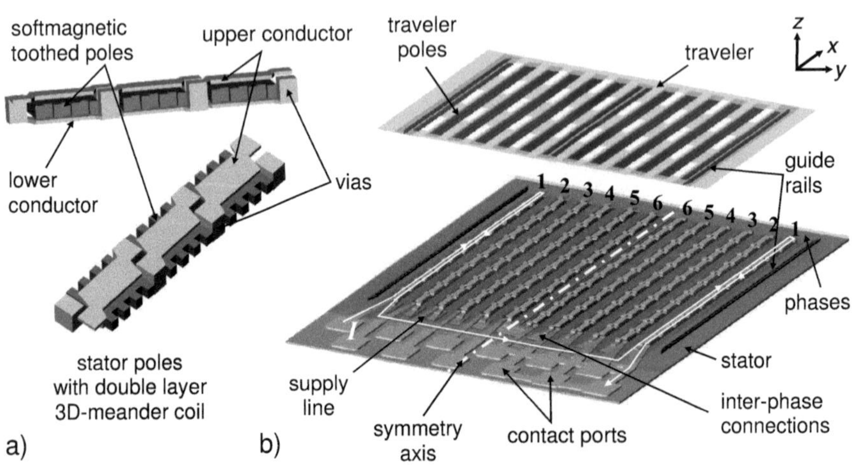

Fig. 12.5 (**a**) Schematic view of the double layer meander coil; (**b**) The second generation design [6]

steps to realize the system connection layer [2]. These system connections are initially created by an electroplated copper layer. They are next covered with a structured *SU-8* insulation layer (a). Subsequently, the lower conductors are deposited on top of the insulation layer, thus creating an electrical contact to the system connections (b). A second *SU-8* insulation layer with openings for the vias, connecting the lower with the upper conductors, serves as a base for electroplating nickel-iron for the stator poles and the guide rail structures in a patterned micro resist mold (c). For this mold, AZ9260 was introduced and optimized enabling structures with aspect ratios of 12. After electroplating and removing the resist mold, a final *SU-8* layer is spun on. This layer insulates the stator poles and provides bars for the following deposition of ED-photo resist (e) which was especially optimized for the coil fabrication (Chap. 11). These bars separate the contact pads of the lower conductors so that, by simultaneous electroplating of vias and upper conductors (f), a short circuit can be avoided.

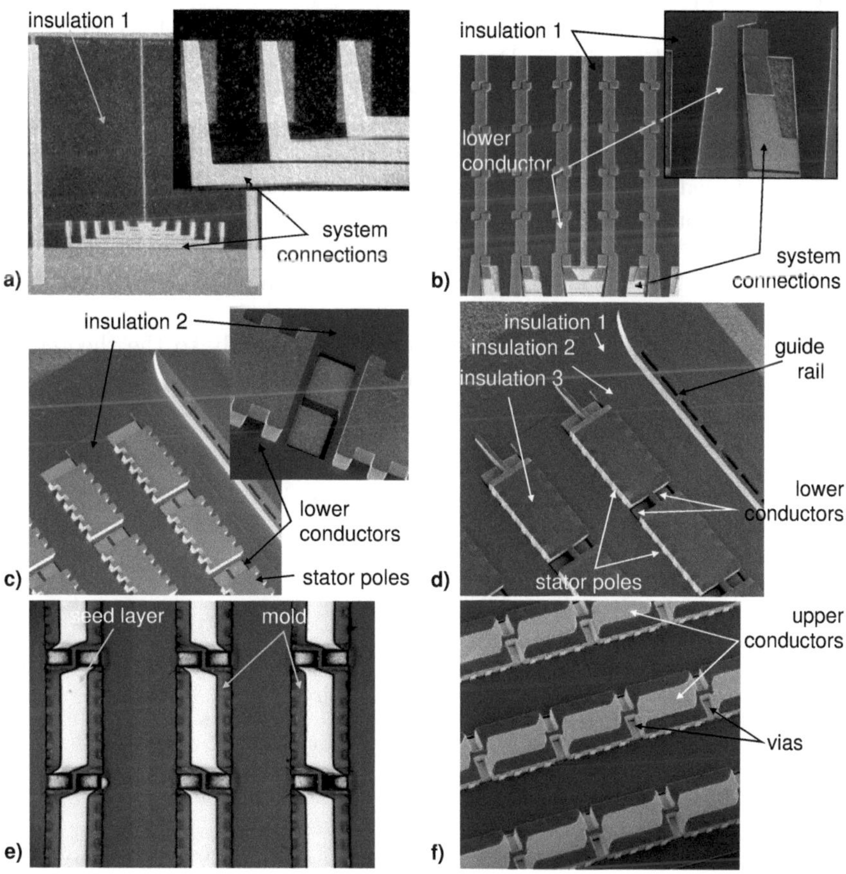

Fig. 12.6 Photographs of stator fabrication steps [6]

The traveler fabrication process was also enhanced. Instead of the original 700 µm thick glass substrate, a 210 µm glass substrate was used to achieve a large reduction in weight [6]. *SU-8* structures are then deposited onto this substrate which serve as plateaus. The structure height depends on the thickness of the upper conductors and the third insulation of the stator. The traveler pole height could thereby be adapted to the stator pole height, ensuring an optimal pole overlap. Furthermore, the tensile stress of the nickel iron structures is reduced. These structures are fabricated by AZ molding and electroplating. The new traveler has a weight of 42 mg in comparison to 124 mg of the former one. The lighter traveler enables a substitution of the ball bearing with a tribological bearing. Additionally, both stator and traveler are covered by a nearly frictionless a-C:H-layer (Chap. 6).

12.2.3 Characterization

This linear VR micro motor has been characterized extensively. Measurements of thermal behavior, driving forces, and friction as well as functional tests were carried out [6, 7]. During these investigations, measurements of the driving force showed a nearly sinusoidal characteristic with a period of 100 µm, which conforms to the pitch of the soft magnetic poles. Furthermore, the new motor design results in a force which is nearly four times higher when compared to the former design, despite a larger air gap, and requires only half the previous drive current (Fig. 12.7.a). The motors were successfully tested, and continuous bidirectional movements over a path of 3.5 mm were achieved. Fig. 12.7.b shows the stepwise movement of the traveler with the typical transient effect.

The measured step widths correspond almost exactly to the theoretical values of 16.7 µm and 33.3 µm. Measurement of the friction forces were carried out for ball bearings as well as for a tribological bearing with and without

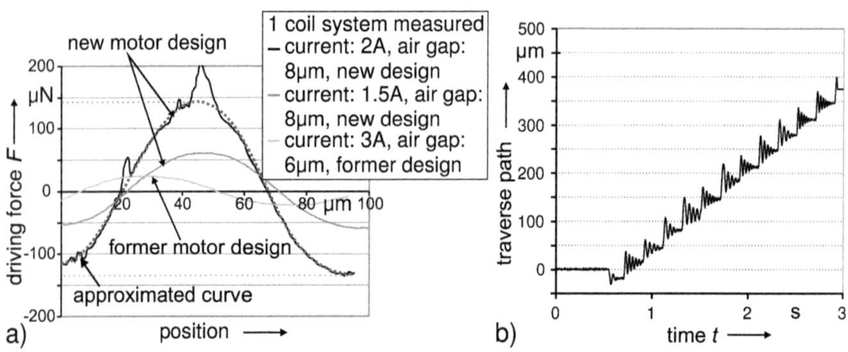

Fig. 12.7 (**a**) Comparison of driving forces of the new and old VR motor; (**b**) Characteristic of stepwise movement [6]

low friction layers. The traveler was moved over a path of 1 mm using a force sensor. Friction forces were measured at 0.6–0.8 mN without the low friction layers and show a high variability caused by the stick-slip effect. Reduced friction forces in the range of 0.1–0.3 mN were observed using a-c:H layers. Lower values of 0.02–0.05 mN were achieved with the use of ball bearings, but required considerable assembly effort.

12.3 Rotating VR Stepper Motor

The general concept of the linear VR stepper motors has also been adapted to a rotating version (Fig. 12.8.a). This rotating stepper motor is configured as an external rotor motor and consists of 6 or 12 soft magnetic pole shoes which exhibit a fine toothed structure on the outer diameter. Each pole shoe is wound with a three-dimensional helical coil of 7 to 11 turns. Two opposite coils are each connected to one electrical phase, whereby a horizontal magnetic flux is induced into the pole shoes. The magnetic flux closes over the soft magnetic rotor yoke, which is also toothed, and the air gaps between the stator and rotor teeth (Fig. 12.8.b). A compensation of the normal reluctance forces is attained due to the complementary arrangement of the air gaps and the centered guide structures. By sequentially energizing each phase, a continuous movement with discrete steps is obtained as a result of the tangentially acting reluctance forces. The motor features small step widths of 33.3 µm and 16.7 µm corresponding to incremental rotation angles of 0.64° and 0.32° for a pole circle of 6 mm in diameter [12].

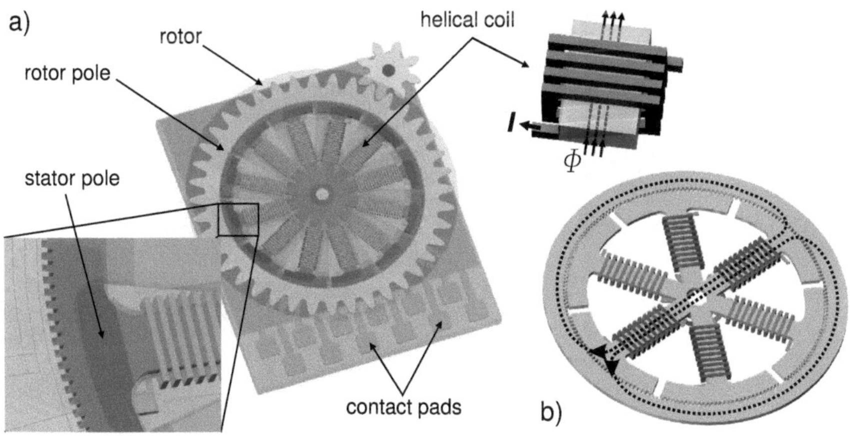

Fig. 12.8 (**a**) Concept of rotating VR stepper motor; (**b**) Schematic path of the magnetic flux

12.3.1 Fabrication

The fabrication of both components – the stator and rotor – is completed separately. The full process sequence for the fabrication of the stator is shown in Fig. 12.9.

Initially, the system connections are provided by structuring an AZ9260 layer and electroplating copper. These conductors connect the two coils of one phase and the coils with the contact pads. These structures are first insulated by a patterned *SU-8* layer. On this layer, the lower conductors of the coils and the contact pads were made by means of AZ9260 molding and copper electroplating (a) followed by a second patterned *SU-8* insulation layer. For the soft magnetic pole structures, a mold of AZ9260 is prepared by multiple coatings. This 90 μm thick mold is filled with 35–40 μm nickel iron alloy in the composition of 80% nickel and 20% iron (b). A following third *SU-8* layer serves as insulation between stator poles and the electrical conductors. This layer contains comb structures in the areas of the lower conductors (c). Finally the upper and sidewall conductors are made. The electro-depositable photo resist *InterVia 3D-N* is applied. This negative resist allows a conforming coating of an arbitrary shape (d). The sidewall and upper conductors can be patterned and electroplated simultaneously in this way. In this process, the *SU-8* comb structures prevent a short circuit between the separate sidewall conductors (e). The finished stators are shown in Fig. 12.1.b.

For the first rotor design, *SU-8* was used as a carrier material. A copper sacrificial layer is first electroplated onto a ceramic substrate. In the following step, two *SU-8* layers are created. The first one serves as a carrier structure

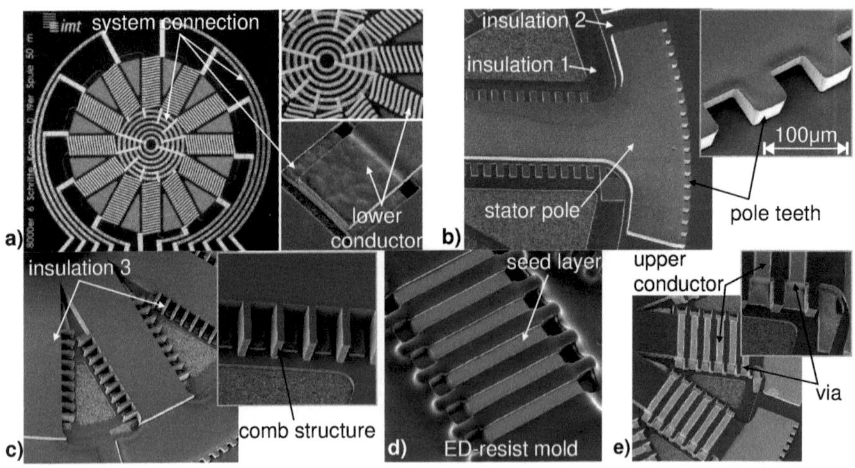

Fig. 12.9 Photographs of the process steps for the rotating stepper motor stator fabrication

and the second one as a plateau. The plateau layer has a height of 25–30 µm, corresponding to the sum of the thickness of the third insulation and the upper conductors of the stator. It is thereby ensured that the rotor and stator cores have the same height. Soft magnetic pole structures are deposited onto the plateau by nickel-iron electroplating in a 90 µm thick AZ9260 mold. The final separation of each rotor is achieved by etching of the sacrificial copper layer, while the nickel iron structures are protected by an AZ9260 mask.

First tests were carried out with these *SU-8* rotors, whereby several rotational steps were observed. A continuous movement was not obtained. Extensive analysis has shown that these rotors exhibited a non-negligible bending. Two stable states are obtained when the rotor is assembled on the stator (Fig. 12.10.a). In the first case, there is an overlap between stator and rotor cores, but there is no guiding in the center. In the second case, guiding is achieved, but no overlap. Both cases affected the functionality negatively and are caused by intrinsic stress in the nickel iron structures. This stress occurs while electroplating and increases with the layer thickness. Thus, the stiffness of the carrier had to be enhanced without an essential increase of rotor weight to counteract the deformation.

In this way, a 210 µm thin glass substrate is used as the carrier. The same process sequence is used, but without the copper sacrificial layer. In the last process step, the glass substrate is etched from both sides by masking the backside with gold. The rotors were also separated in this process. The glass carrier rotor offers a slightly higher weight, but the bending is strongly minimized. Total deflection amounts to 11 µm over a rotor diameter of 7 mm. Consequently, both the functionality of the guide structure and the pole overlap are ensured. The assembled system is shown in Fig. 12.10.b.

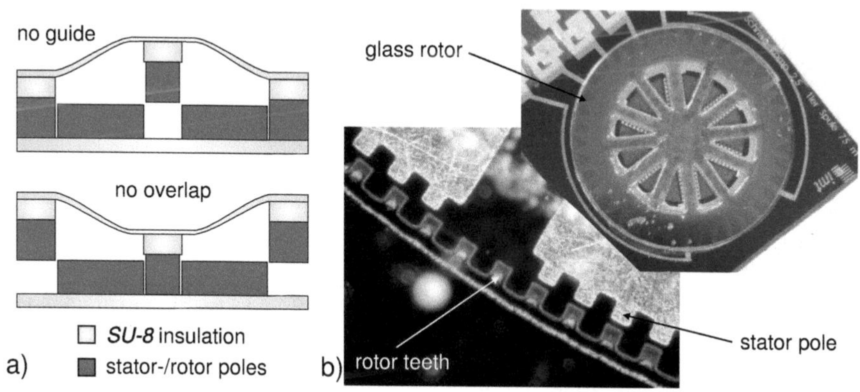

Fig. 12.10 (**a**) Bending states of *SU-8* rotor; (**b**) Assembled system with glass rotor

12.3.2 Characterization

By using the described processes, stators and rotors are successfully fabricated. Depending on the number of turns, the helical coils possess resistances between 0.8 Ω and 1.8 Ω. They are able to be loaded temporarily with currents up to 2 A without destruction. The guidance structures allowed the components to be assembled and tested for temperature behavior, friction, and resulting torque. The frictional torque varied strongly depending on the quantity of contact points between the rotor and stator structures. In the case of contact between only the rotor pin and stator guide, the friction torque reached a minimum and ranged from 0.03 to 0.04 µNm. In this case, a movement with small discrete steps is achieved. The rotor could be rotated continuously and bidirectionally with switching frequencies up to 200 Hz, corresponding to a rotational speed of 21 rpm.

The driving torque was determined only for the static case by force measurements in a defined distance to the center of rotation. Therefore, the first phase was energized resulting in tooth to tooth positioning. The rotor is then touched by a 3D-force sensor, and the other phases are energized sequentially with increasing currents. In doing so, the forces are measured for a three-phase motor for 1/3 of the tooth pitch. In Fig. 12.11.a the results are shown for a motor with 42 µm pole overlap depicted for different currents. The average values of the measured torques with respect to the current are shown in Fig. 12.11.b. The maximum is achieved with a current of 1 A and amounts to 0.30 µNm. The following decrease of torque is based on an increase of friction due to heating.

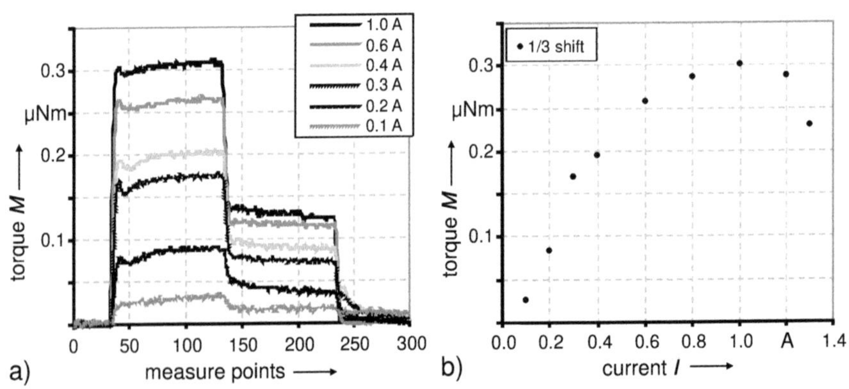

Fig. 12.11 (**a**) Driving torques for different currents; (**b**) Dependence on the applied current determined for a three-phase external rotor motor

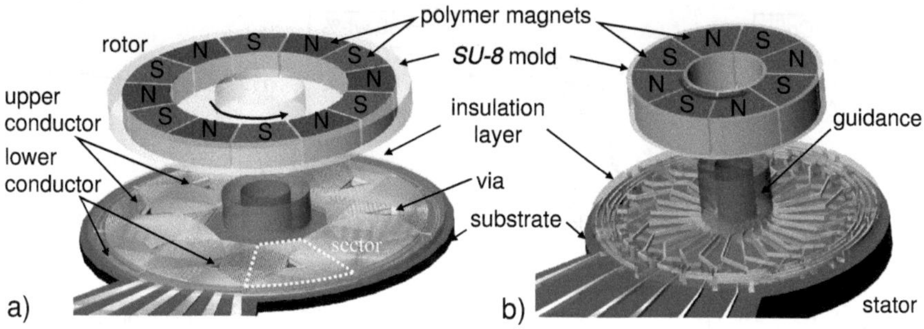

Fig. 12.12 Concept of rotating synchronous motors with (**a**) concentrated coils (**b**) overlapping coils [11]

12.4 Rotating Synchronous Micro Motors

Besides the reluctance based motors, research into motors based on Lorentz force has been performed. In this way, permanent magnet synchronous motors were designed and fabricated in disc rotor design, including planar spiral coils and special hard magnetic structures [5, 11]. The basic setups of these brushless motors are shown in Fig. 12.12. The stator consists of double layer coils, which are arranged as concentrated coils or overlapping coils. The coils have 6–30 turns per phase depending on the motor size and number of poles. The arrangement of the coils, arranged in three or six phases, and magnets allows rotation in continuous and stepping modes. The rotor is made of an *SU-8* form, which contains alternate magnets. These magnets are realized by polymer magnets or commercial magnets. Both magnet types feature an alternate axial magnetization which, in the case of the polymer magnets, is impressed using special magnetization equipment. The rotors have diameters between 1 mm and 5.5 mm. For the adjustment of the rotor and the stator, a centrically arranged circular guide is integrated.

12.4.1 Polymer Magnets

Polymer magnets are micro composite materials which are fabricated by embedding powdery, magnetically hard materials in a polymer matrix. During the past years, three different techniques for micro scale structuring of these composites have been developed. These techniques include direct structuring (a), lift-off process (b) and soft lithographic molding (c) [3]. Different qualities and properties can be obtained depending on the used technique.

For the fabrication of the composites, various magnet powders were used and characterized for the applications as polymer magnets, like rare earth materials (neodymium iron boron, samarium cobalt), ferrites and MQP^{TM}. Depending on the material, the grit sizes of the powders are between 0.8 µm

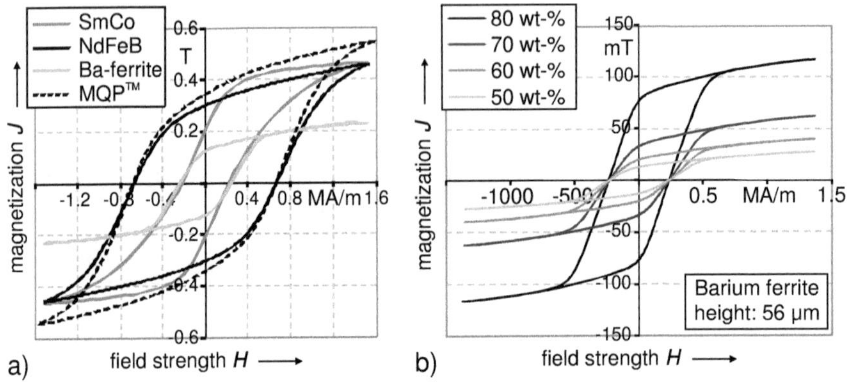

Fig. 12.13 (**a**) Magnetization characteristics of different polymer magnets; (**b**) Magnetization depending on powder ratio [3]

and 9 µm whereas ferrites have the finest grits. The fabricated polymer magnet structures are characterized regarding their magnetic properties with a vibrating sample magnetometer. In Fig. 12.13.a the magnetization curves are compared for various used polymer magnets with 80 wt-% powder ratio. Furthermore, the results show that the achievable residual magnetism augments with increasing powder ratio (Fig. 12.13.b) or structure thickness, whereas the coercivity stays constant. In comparison to electroplated or sputtered layers, the lift-off process provides magnetically hard structures with high thicknesses of some 100 µm combined with high edge quality and high aspect ratios. Consequently, the magnetic rotor structures are made using this technique.

12.4.2 Fabrication

The fabrication process of stator and rotor includes UV-depth lithography using AZ9260 for electroforming and *Epon SU-8* for insulation, planarization and embedding. The process sequence for the fabrication of the stator starts with the lower conductors of the double layer coil. Therefore, a chromium-copper layer is deposited followed by an AZ9260 layer. The AZ9260 is patterned as mold and filled with 12 µm thick copper using electroplating. After stripping the AZ9260 mold and etching of the seed layer (Fig. 12.14.a), a 20 µm layer of *SU-8* is spun onto these structures as an insulation. This layer provides openings for through connections to the upper coil layer (b). In the next step, via connections as well as upper conductors are likewise structured in the same step by copper electroplating (c). A following second *SU-8* layer serves as insulation between upper conductors and rotor magnets and as a bearing layer (d). In the last step, the circular guidance structures are made by pattering a 200 µm thick *SU-8* layer (e). Two rotor types are fabricated

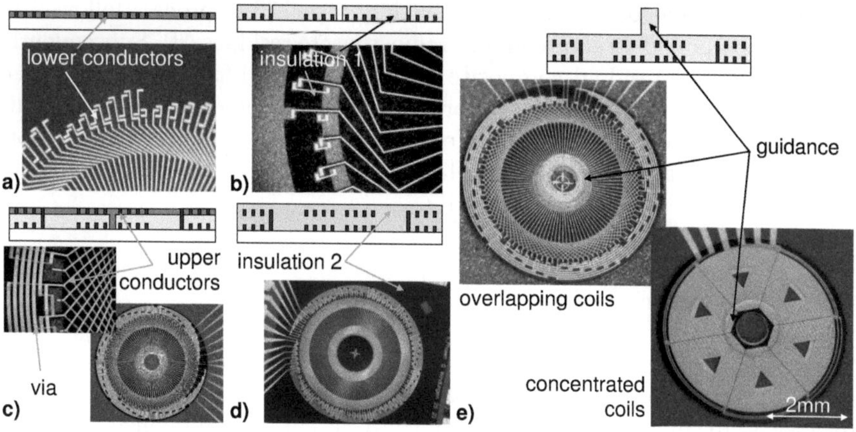

Fig. 12.14 Process steps for stator fabrication

including rotors with polymer magnets and with commercial magnets. In comparison, polymer magnets possess lower magnetic properties but offer a flexible shaping, whereby high area filling is possible. Furthermore, polymer magnets could be prepared in dimensions and forms which are unavailable commercially. The rotors are realized by *SU-8* creating a form for the magnets with high precision adjustment.

For the fabrication of the rotor (Fig. 12.15.a–e) a sacrificial copper layer is electroplated onto the substrate followed by a thin patterned *SU-8* layer. It serves as base plate and as protection in a later etching step (a). After that, a 400 μm high *SU-8* layer is patterned to provide the filling form (b). The commercial magnets are directly mounted inside the form and held by a fit. In contrast, for rotors with polymer magnets, the above mentioned lift-process is applied. Therefore, the not required areas around the *SU-8* mold are filled with a solvable resist like AZ9260 (c). The pasty polymer magnet is inserted

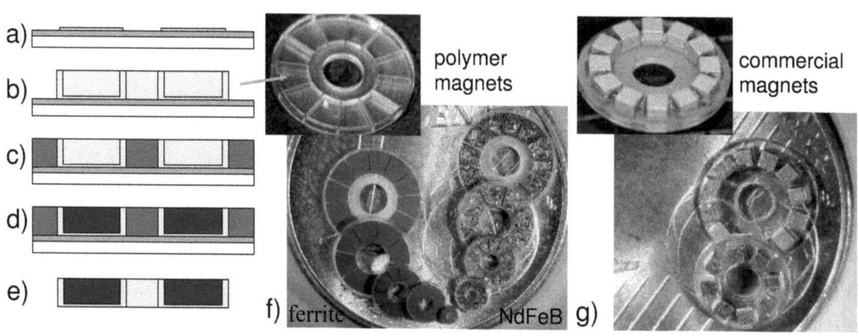

Fig. 12.15 (**a**–**e**) Schematic process steps for rotor fabrication; Rotors with (**f**) integrated polymer magnets and (**g**) with commercial magnets [11]

into the mold and baked out. After baking, a polishing process followed to level the compound structure and to remove waste residues (d). Subsequently the AZ9260 and residual composite are removed by lift-off while the used magnet structures remain in the form. By etching the sacrificial layer, the rotors are detached from substrate (e). The weight of the so fabricated rotors (Fig. 12.15.f–g) ranges from 1 mg to 45 mg depending on diameter and magnet type.

12.4.3 Characterization

These systems have been characterized regarding the thermal behavior, the resulting torques and the functionality. For thermal measurements, the stator coils were supplied for 180 s with a constant current. In defined intervals, the system temperature was measured using an infrared camera. After 180 s the temperature reached a constant value. At this point, the current was switched off whereby the stator was passively cooled down. The initial temperature was already attained after 180 s. The working temperatures were constituted between 27°C and 62°C for current amplitudes between 50 mA and 200 mA. Torque measurements were performed with specific test rotors for each stator type and magnet type fabricated in the same process sequence. They had notches in a fix distance to the center of rotation. The stylus of the force sensor is adjusted into these notches to detect friction and driving forces. The friction torque varies depending primarily on the rotor weight and additionally on the fit between rotor hole and stator guidance. The values amount 0.12–0.16 μNm for rotors composed of commercial magnets (36.3 mg) and 0.04–0.05 μNm for polymer magnet rotors (14–17 mg), respectively. The torque characteristics are indicated by the typical slip-stick-effects.

For the determination of the driving torque, one phase of the motor is energized whereby the rotor is turned and centered to this phase. Then, the stylus of the force sensor is adjusted in the notch. After that, current amplitudes of phases 2 and 3 were changed with the same amount but in opposite direction. The current was increased and the forces in x- and y-direction were measured. The resulting torque was calculated. In Fig. 12.16, the torques for different motor configurations are depicted including the friction torque. The torque depends on the applied current and the coil type. The measurements on rotors composed of commercial magnets have shown that torques up to 10 μNm were achieved by applying currents up to 70 mA (Fig. 12.16.a). The achievable torque for polymer magnet rotors is one order of magnitude lower because of the lower residual magnetization of the polymer magnets (Fig. 12.16.b). The values range from 0.8–1.35 μNm at 70 mA. Thus, all torque characteristics present a linear dependency to the applied current corresponding to the linear correlation of the Lorentz force.

The functional tests show impressively that only 3–10 mA are necessary for driving the rotors. Further increase of the current leads to homogeneous and

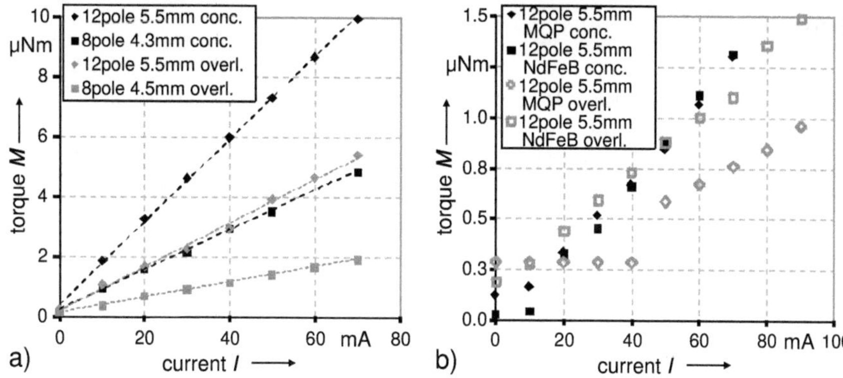

Fig. 12.16 Resulting torques of the synchronous motors using rotors composed of (**a**) commercial magnets and (**b**) polymer magnet rotors

smooth running whereby rotating speeds over 7000 rpm have been achieved. The motors could be driven over a long period with only minor abrasion of the *SU-8* layer bearing layer.

12.5 Conclusion and Outlook

Different electromagnetic actuators have been fabricated on the basis of several developed fabrication technologies. Important technologies were shown using the example of VR actuators. Due to further developments in the technologies, the fabrication of electrodynamic actuators was enabled mainly by the introduction of polymer magnets. On this base, Lorentz-force based actuators like synchronous motors and complex microrobots have been successfully realized [4]. First applications are realized on the basis of the synchronous micromotors. For example, micropumps and microstirrer are fabricated for complex microfluidic systems [10]. Furthermore, new pumps are under investigation using the developed technologies.

Acknowledgements The authors gratefully acknowledge the funding of the reported work by the German Research Foundation within the Collaborative Research Center 516 "Design and Manufacturing of Active Micro Systems".

References

[1] Demmig S, Gehrking R, Feldmann M, Büttgenbach S, Ponick B (2006) Specific design rules for micro linear motors. In: Symposium on Power Electronics, Electrical Drives, Automation and Motion, pp 352–357

[2] Feldmann M, Büttgenbach S (2007) Linear variable reluctance (VR) micro motors with compensated attraction force: concept, simulation, fabrication and test. IEEE Transactions on Magnetics 43(6):2567–2569

[3] Feldmann M, Büttgenbach S (2007) Novel microrobots and micromotors using Lorentz force driven linear microactuators based on polymer magnets. IEEE Transactions on Magnetics 43(10):3891–3895

[4] Feldmann M, Waldschik A, Büttgenbach S (2007) Electromagnetic micro actuators, micro motors, and micro robots. In: Proc. of SPIE, vol 6798, pp 11/1–11/10

[5] Feldmann M, Waldschik A, Büttgenbach S (2008) Synchronous micro motors: concept, fabrication and test. In: Proc. of Actuator 2008, pp 669–672

[6] Feldmann M, Waldschik A, Ruffert C, Gatzen H, Büttgenbach S (2008) Fabrication of drive and guide components for micro motors. Microsystem Technologies 14(12):1941–1947

[7] Gehrking R, Demmig S, Ponick B, Feldmann M, Büttgenbach S (2006) A micro linear motor with integrated passive magnetic guidance. In: 32nd Annual Conference on IEEE Industrial Electronics, pp 1245–1250

[8] Seidemann V, Edler J, Büttgenbach S, Stölting H (2003) Linear variable reluctance (VR) micro motor with horizontal flux guidance: concept, simulation, fabrication and test. In: TRANSDUCERS, Solid-State Sensors, Actuators and Microsystems, vol 2, pp 1415–1418

[9] Seidemann V, Kohlmeier T, Föhse M, Gatzen H, Büttgenbach S (2004) High aspect ratio spiral and vertical meander micro coils for actuator applications. Microsystem Technologies 10(6):564–570

[10] Waldschik A, Büttgenbach S (2010) Micro gear pump with internal electromagnetic drive. Microsystem Technologies 16(8):1581–1587

[11] Waldschik A, Feldmann M, Büttgenbach S (2008) Novel synchronous linear and rotatory micro motors based on polymer magnets with organic and inorganic insulation layers. Sensors & Transducers Journal, Special Issue on Microsystems: Technology and Applications 3:3–13

[12] Waldschik A, Feldmann M, Büttgenbach S (2008) Rotary variable reluctance (VR) micro motors: concept, simulation, fabrication and test. In: Proc. of Actuator 2008, pp 453–456

Chapter 13
Development and Fabrication of Linear and Multi-Axis Microactuators

S. Hansen, H. H. Gatzen

Institute for Microtechnology (now: IMPT)
Leibniz Universität Hannover
gatzen@impt.uni-hannover.de

Abstract To create an xy-micro- or nanopositioner, three types of micromotors were evaluated: a linear variable-reluctance (VR) microstep motor, a linear hybrid microstep motor and a linear synchronous motor. The most apropriate drive scheme found for an xy-positioner was the linear VR microstep motor. By combining four of these systems with magnetic levitation devices, an xy-positioner with frictionless motion could be created. In two alternative versions, a pole-based xy-microactuator with a resolution of 500 nm and a tooth-based xy-nanoactuator with a resolution of 100 nm were designed. Both versions use a microstepping drive mode.

13.1 Introduction

A particularly promising approach for electromagnetic microactuators is using linear devices [6]. They lend themselves to hybrid system integration, with separate parts forming the stator, the traveler and the guide. This way, the challenges of micromachined rotating motors, which are handicapped by bearing issues, can be avoided. If designed appropriately, such a linear device might even lend itself to nanoactuation, typically associated with a minimal displacement of 100 nm. For drive schemes, ideas may be borrowed from the macro-world. Three drive schemes particularly suited are the variable reluctance (VR) step motor, the hybrid step motor and the synchronous motor. The advantage of a step motor is its capability of a rather small incremental motion; the advantage of a synchronous motor its simple design.

In each case, a proper match between the necessities of the drive scheme and the fabrication challenges of micromachining technologies has to be found. Since the power an active system is able to convert depends on the device's volume, rather great building heights may be expected. Another advantage of linear motors is the fact that they may be used to create an xy-actuator. By arranging four linear actuators in a square and combining them with an appropriate guide, this becomes possible.

13.2 Linear VR Microstep Motor

A simple way of creating incremental microsteps in a linear system is applying the variable reluctance (VR) step motor drive scheme, using a motor design, which features teeth on top of the poles and on the traveler [4]. This way, the tooth pitch dictates the step width rather than the distance between the poles. Fig. 13.1 presents a schematic representation of a stator for a linear VR microstep motor. On the bottom is a stator with poles energized by a coil. On top of the poles is a row of teeth. The traveler on the top also consists of poles with teeth; they are located below a flux guide. Energizing a motor phase (i.e. a group of poles energized jointly) induces, regardless of the current direction, opposite magnetic polarities in the stator and the traveler teeth. A north pole in the stator teeth effects a south pole in the traveler teeth and vice versa. The opposite polarities result in a magnetic force, which pulls the traveler teeth on top of the stator teeth.

Typically, a VR step motor has three phases with the teeth offset by plusminus one third of the tooth pitch. Switching the current between the phases results in a traveler step of one-third of the tooth pitch in either forward or backward direction [6].

To fabricate the linear VR microstep motor, High Aspect Ratio Micro Structure Technology (HARMST) was applied. It combines the use of rather deep photoresist patterns serving as micromolds and electroplating. This way, both the soft magnetic as well as the conductive micromotor components were created. For both stator and traveler, Al_2O_3 was used as a wafer material. For the stator, the fabrication process started with creating the bottom yoke. It is made of the soft magnetic permalloy (NiFe81/19). To allow a deposition on the non-conductive wafer, a conductive seed layer is required. It was created by sputter deposition and ultimately removed by Ion Beam Etching (IBE).

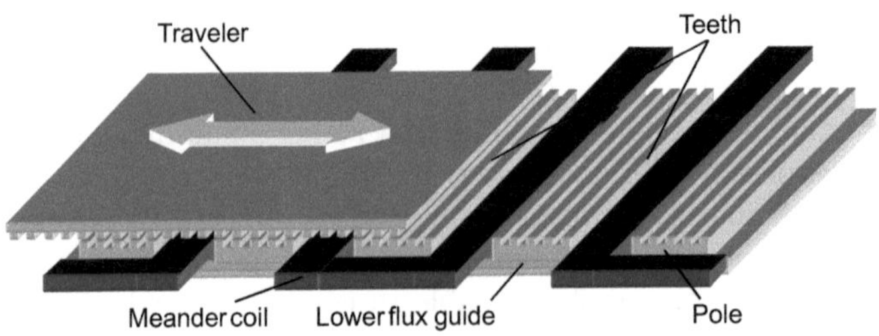

Fig. 13.1 Schematic representation of a VR microstep motor's stator (only one phase is shown)

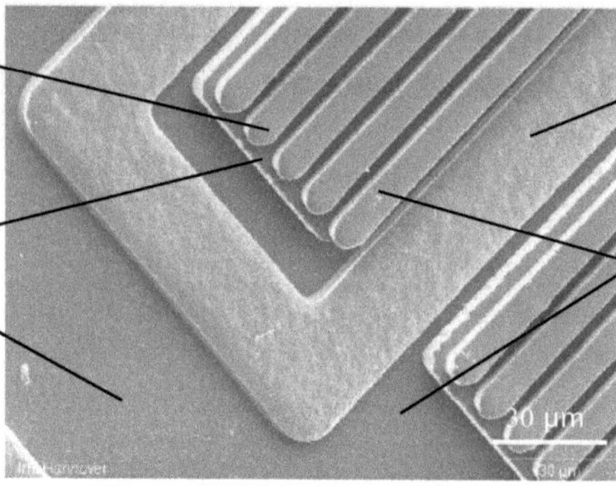

Fig. 13.2 Stator section of a VR microstep motor (REM micrograph) [4]

In the next step, the poles were created. To conclude the fabrication of the magnetic path, the teeth were created on top of the poles. The next task was fabricating the single-layer meander type coil. First, it was necessary to create an insulation layer, which was created by sputter depositing Al_2O_3. It serves as non-conductive coat between the conductive permalloy and the coil. Next, electroplating Cu created the meander coil itself. The finished coil was embedded in the photosensitive epoxy *SU-8*. It served both as a lateral insulation between the Cu conductor and the permalloy poles, as well as a plane surface serving as a pole face. Its thickness defines the air gap between the stator teeth and the teeth of the traveler ultimately sliding on it. To achieve good tribological behavior, the surface was coated with wear-resistant diamond-like carbon (DLC) (see Chap. 5). Fig. 13.2 shows an SEM micrograph of a completed stator system.

For fabricating the traveler, the same process sequence was applied as for creating the magnetic path of the stator. The fabrication was completed by an *SU-8* embedding process and a DLC coating as described at the end of the stator fabrication (see Chap. 5). When executing functional tests, it was found that there was a tremendous discrepancy between the forces predicted by the FEM simulation and the actual measurement results. As will be discussed later, the permeability values for small patterns, particularly tooth structures with high shape anisotropy, are substantially lower than those for depositing films (see Chap. 10).

13.3 Linear Hybrid Microstep Motor

One way to increase the motor forces in macroscopic step motors is to extend the VR step motor drive scheme to a hybrid one [3]. This is done by additionally integrating permanent magnets into the traveler. In VR motors, a north pole on a stator tooth induces a south pole in its traveler counterpart and vice versa. By appropriately integrating permanent magnets into the traveler, a constellation of opposing north versus north as well as south versus south poles becomes possible. By appropriately combining opposite poles (e.g. north versus south) and even poles (e.g. north versus north), the forces induced are substantially increased.

Fig. 13.3 shows a schematic representation of a hybrid microstep motor. On the bottom is a stator with poles. As in the case of the linear VR microstep motor, there is a row of teeth on top of the poles. Typically, a stator has two phases (as shown), each one energized by a coil. The traveler on the top consists of two flux guides with teeth; sandwiched in between is a permanent magnet magnetized in cross direction. As a result, all teeth in one row are north poles and in the other row south poles. The offset between the two rows of teeth is half a tooth pitch. The stators are driven in a bipolar mode. For instance, exciting one phase with a negative current induces a south pole while a positive current creates a north pole in the stator teeth. As a result, traveler and stator teeth with opposite polarization are trying to line up, while in the case of even polarization, the traveler teeth are drawn into the gap between the stator teeth. Reversing the current results in a change of the polarization direction in one phase. In the simplest drive scheme, only one of the two phases is energized. Each change results in steps of a quarter of a tooth pitch. After four steps, the traveler has moved one full tooth pitch.

Without exciting the coils the traveler is held in a detent position. This is due to the flux created by the permanent magnet located in the traveler.

To fabricate a hybrid microstep motor, a design with six phases, which were all always energized, was chosen. Its travel sequence consists of 12 steps. With a chosen tooth pitch of 68 µm, the hybrid microstep motor features a step width of approx. 5.7 µm.

The process sequence for the stator is very similar to that of the linear VR microstep motor, with the exception that NiFe45/55 was used instead of NiFe81/19. The reason is a higher saturation flux density B_s (1.6 T instead of 1 T) and a greater relative permeability μ_r. Fig. 13.4 presents sections of both the stator and the traveler. The stator (Fig. 13.4.a) has three teeth per pole, except for poles between phases. To minimize lateral flux leakage, there are pole overhangs with a pair of teeth at the rim.

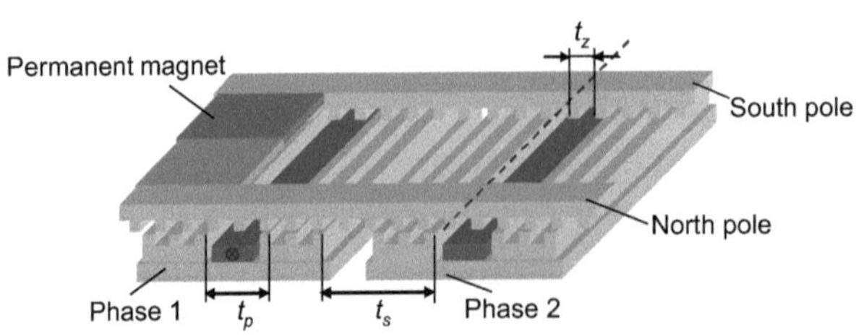

Fig. 13.3 Schematic representation of a hybrid microstep motor's stator with two phases

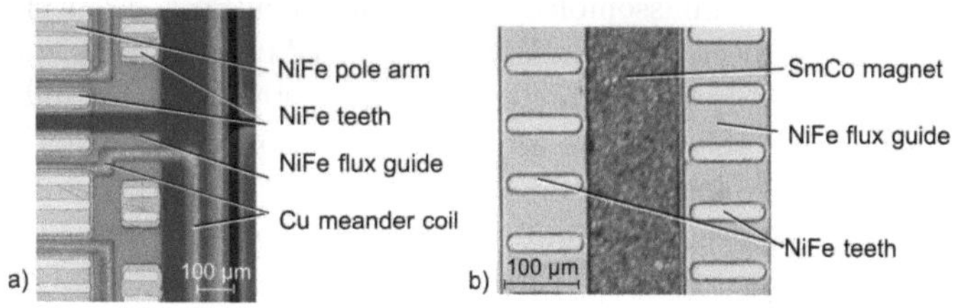

Fig. 13.4 Components of a linear hybrid microstep motor: (**a**) Stator section; (**b**) Section of a traveler strip [10]

A traveler consists of three strips, each one composed of a pair of soft magnetic flux guides with teeth straddling a Sm_2Co_{17} permanent magnet magnetized in cross direction. This way, it was possible to magnetize the permanent magnets of batches of travelers.

The first traveler fabrication step was sputter-depositing Sm_2Co_{17}. In its deposited state; the material is amorphous. To achieve the desired hard magnetic behavior, an annealing step under vacuum conditions at a temperature of 560°C is required. To pattern the Sm_2Co_{17} layer, a wet chemical etching process was applied. The soft magnetic flux guides and teeth were fabricated using electroplating of NiFe45/55 in micromolds. Fig. 13.4.b depicts a section of a basic traveler element. After its fabrication, the traveler was subjected to a strong magnetic field to magnetize the SmCo magnets. The degree of magnetization has to match the field created by the stator coils.

Due to the similarity of fabrication processes for the stators of a linear VR microstep motor and a hybrid microstep motor, a fabrication sequence may yield either type of motor. Contrary to the linear microstep motors discussed so far, the motors shown in Fig. 13.5 feature double-layer spiral coils and pole arms. The only difference in the process steps is that a different tooth geometry is required for the two versions.

An evaluation of the performance of linear hybrid microstep motor prototypes was conducted. Among other, force measurements [11] and magneto-

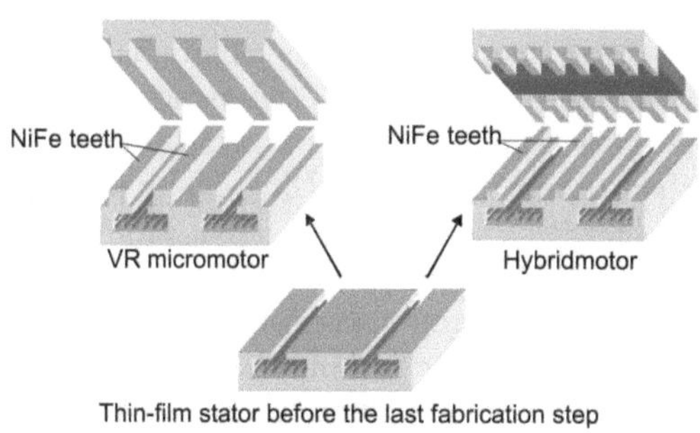

Fig. 13.5 Micromotor Family [9]

optical field measurements on unassembled stators and travelers [12] were executed. There were major discrepancies between the FEM simulations and the experimental data, similarly to the linear VR micro step motor results mentioned above. Therefore, the reasons for these deviations were investigated. As it turned out, the relative permeability μ_r of the patterned soft-magnetic structures (flux guides, poles and in particular teeth) were substantially lower than the values measured for films of the same material at a comparable thickness. The main reason was the shape anisotropy of the structures. By coming up with a method to determine the actual magnetic behavior of flux guide and pole structures based on VSM measurements, an approximate solution for this issue could be found.

Another contribution to the deviations were challenges in maintaining the desired air gap length between stator and traveler. This issue was ultimately resolved by developing an appropriate, self-adjusting assembly technique for achieving the desired air gap length.

Another issue was caused by the inherent hybrid step motor design. A precondition for a proper functionality of the hybrid step motor is that the magnetization of the stator teeth (accomplished by energizing a stator phase) perfectly matches the magnetization of the traveler teeth (induced by the permanent magnets residing in the traveler). It turned out to be very difficult to appropriately magnetize the SmCo hard magnets in the traveler. Since this issue was considered hard to control in any future linear hybrid microstep motor, a VR step motor approach was selected for the nanoactuator ultimately to be developed.

13.4 Linear Synchronous Micromotor

The third drive scheme investigated is a linear synchronous micromotor. Typically, these motors use permanent magnets to create a strong magnetic field in the air gap, which interacts with a much weaker field created by the energized coils [5]. Usually, synchronous motors are rotational devices, driven by a three-phase current. Similarly, the linear synchronous micromotor was also designed with a three-phase coil system. Since a synchronous motor is identical in its construction with a permanent magnet step motor, it may also be driven in a stepping mode. This seems a much more appropriate drive scheme for a linear micromotor than connecting it to a three-phase power supply. In this mode, the motor is driven with a rectangular current profile rather than a sinusoidal three-phase current. Fig. 13.6 presents a schematic representation of the linear synchronous micromotor and also explains the motor's function in a stepping mode for a step-by-step excitation of the phases.

In the first step, a current is applied to phase 1. The current direction in the coil results in a south pole above the coil. As a result, a north pole of the permanent magnets aligns to this phase. Next, phase 2 is excited; the

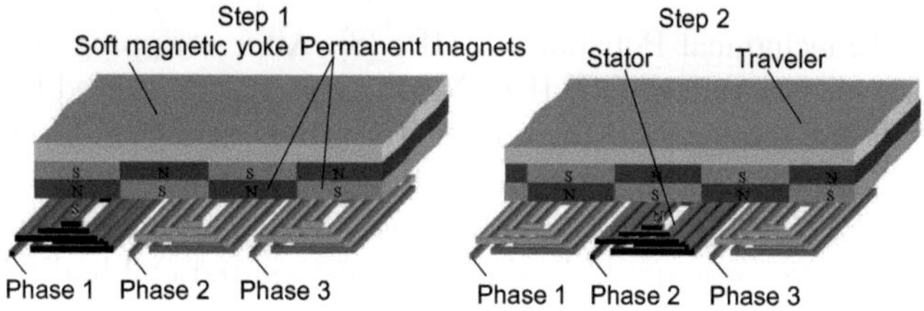

Fig. 13.6 Schematic representation of the linear synchronous micromotor [5]

current direction is opposite to the one of the previous step, inducing a north pole above the coil. This attracts the south pole of the permanent magnets. Not shown is the third step, when phase 3 is excited. The current direction is appropriate for a south pole above the coil, which attracts a north pole of the permanent magnets. There are three more steps to complete one sequence; they are the same as steps 1 through 3, except for a reversed polarity of the excitation current in the coils. After six steps, the sequence starts anew.

Compared to the step width of the VR and hybrid microstep motors discussed earlier, the step width is rather coarse – after all, the micromotor pitch is not dictated by a tooth structure, but rather by the pole arrangement of the permanent magnets. There is, however, a way of reducing the step width, which is regularly applied to macroscopic rotational step motors: microstepping. Instead of exciting the coils with a constant current, the current values of the phases are varied. This allows dwelling on any location between the full step positions.

The stator of the linear synchronous micromotor features nine double-layer coil systems placed in a row. The coils are arranged in groups of three, with each coil in a group representing one of the three phase (i.e. the first coil represents phase 1, the second phase 2, the third phase 3 and the fourth phase 1 again and so on.) This way, each phase is represented by three double-layer coil systems. Each coil has 50 turns.

The linear synchronous micromotor was fabricated in two generations. Additionally, there were variations in the wafer material and in material and type of the permanent magnets. The wafer material for both stator and traveler was either Si or MnZn-ferrite. In the ferrite version, the magnetic wafer provided an additional flux guide function, minimizing the system's reluctance. However, there is a drawback: due to the permanent magnets in the traveler, a magnetic stator substrate causes a permanent vertical force.

The wafer fabrication process for the first generation stator of the linear synchronous micromotor started with a thermal oxidation of the Si or an Al_2O_3 sputter deposition on the MnZn-ferrite, respectively, to create an insulation layer. Next, the wafers were sputter-coated with a Cu seed layer. In the following step, a photomask was created for the bottom coil layer, as well as the via connection. A Cu electroplating process created the first coil layer. Due to non-uniform Cu growth over the wafer surface, the coil structure was subjected to Chemical-mechanical Polishing (CMP) [8]. After stripping the photomask, the seed layer was removed by IBE. Next, *SU-8* was deposited by spin coating and patterned by photolithography. It served as insulation layer both between the coil leads as well as between the first and second coil layer. For fabricating the second coil layer, the same process sequence as previously applied for the first coil layer was used. Again, the electroplated coil had to be planarized by CMP. After the seed layer removal, the whole coil system was embedded in *SU-8*. For completing the stator wafer process, the bond pads were fabricated. Fig. 13.7 shows stators of the two linear synchronous micromotor generations.

MnZn-ferrite was also used as wafer material for the traveler, respectively. Two types of travelers were built: with sputter deposited and annealed SmCo11/89 films and with plastic magnets filled with NdFeB. The films were magnetized with a customized magnetization system, the NdFeB-filled magnets were already magnetized, cut to shape and attached on the traveler substrate with an adhesive.

One disadvantage of electromagnetic and electrodynamic linear motors is the great vertical force between stator and traveler. In case of a tribological guide between stator and traveler, this results in rather high frictional forces. A remedy for this challenge was the implementation of a microball guide with V-shaped grooves serving as races ground in both stator and traveler and ruby microballs used as rolling elements. Fig. 13.8 presents a schematic representation of a ball guide (Fig. 13.8.a) as well as a ball guide's ground V-groove and 200 μm diameter ruby ball (Fig. 13.8.b). An application of this ball guide system was the linear synchronous motor described above. Fig. 13.8.b clearly shows the V-groove, which is interrupted at the center for allowing a connection between the bond pads and the coils.

After assembly, the linear synchronous micromotors were subjected to tests. Table 13.1 shows the results of the first generation micromotor in its two versions with discrete, 200 μm thick NdFeB filled *Vacophan* permanent

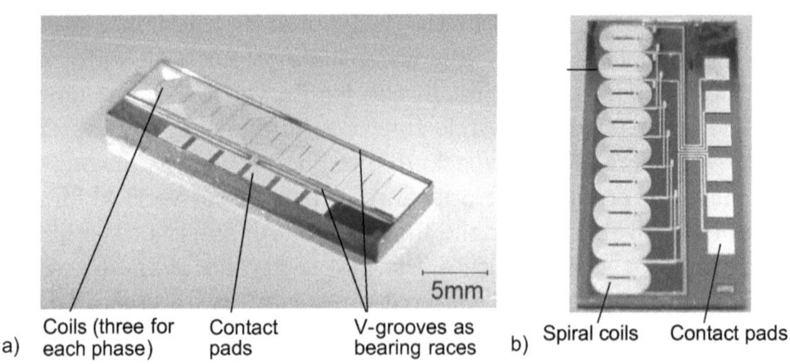

Fig. 13.7 Stator of a linear synchronous micromotor: (**a**) First generation; (**b**) Second generation [1]

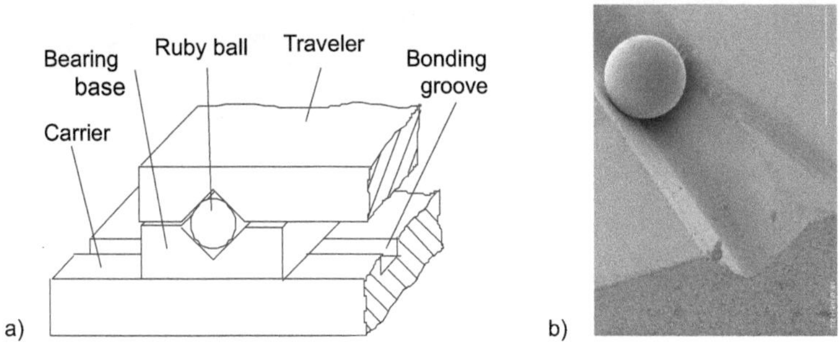

Fig. 13.8 Ball guide: (**a**) Schematic representation; (**b**) SEM micrograph of the ball guide's ground V-groove and 200 μm diameter ruby ball [7]

magnets (Type A) and with sputtered SmCo17/83 permanent magnets on a 30 μm-thick NiFe underlayer (Type B).

Table 13.1 Comparison of simulation and experimental results for a linear synchronous micromotor

First generation linear synchronous micromotor	Type A	Type B
Footprint	$7.7 \times 36\,\mathrm{mm}^2$	
Travel	11 mm	
Permanent magnet	*Vacophan*	SmCo17/83 thin-film
Permanent magnet, BHmax	$28\,\mathrm{kJ\,m^{-3}}$	$20\,\mathrm{kJ\,m^{-3}}$
Driving force, calculated (max)	8 mN	1.6 mN
Driving force, measured (max)	7.4 mN	1.5 mN

While there were substantial differences between the FEM simulation results and the actual experimental data for the VR and hybrid step motors, no such deviations were observed for the linear synchronous micromotor in any of its versions. This is yet another indication that the reason for deviations was erroneous data for the soft magnetic micromotor components. With no patterned highly permeable components within a linear synchronous micromotor, this source of error was not present. Except for the soft magnetic parts, there is very good agreement between FEM simulations and the micromotor performance.

a)
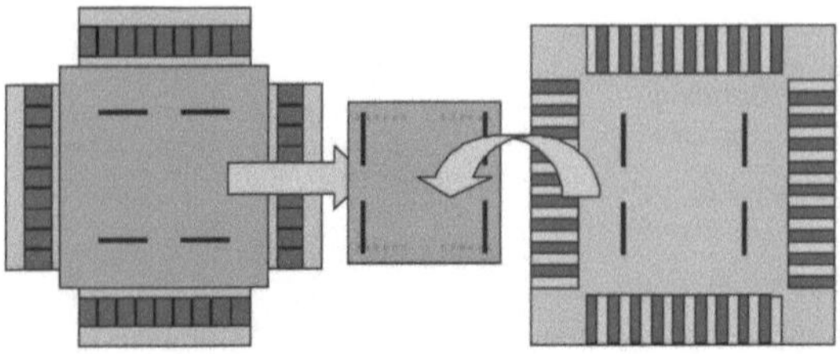
b)
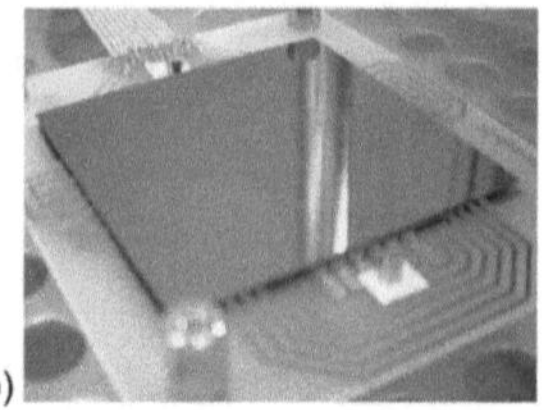

Fig. 13.9 Linear synchronous micromotor based xy-actuator (first generation Type B): (**a**) Schematic representation; (**b**) Execution

13.5 xy-Actuator Investigations

13.5.1 xy-Actuator Concept and Prototype

Systems for assembling microparts typically are macroscopic, thus looking disproportional compared to the parts handled. Particularly the actuation components are rather large in size [14]. The micromotors developed so far open up an opportunity for achieving a miniaturized positioning system: arranging four of the micromotors in a square creates an xy-actuator serving as a positioner for a microassembly unit (Fig. 13.9). To evaluate this concept, preliminary investigations taking advantage of the synchronous motors developed were conducted. Figure 13.19 presents a schematic representation of the xy-microactuator. A base plate holds the four stators of synchronous motors arranged in a square. It features two pairs of inline V-grooves. An interruption at the groove's center allows for the tracing of leads between stator and contactpads. An x slide positioned on top has four matching V-grooves at its bottom and another four V-grooves oriented at an angle of 90 degrees at its top, defining the y-direction. A traveler plate on top contains the four motor travelers appropriately oriented on the top of the stators facing down. It also contains the V-grooves for the y-direction travel. Fig. 13.9.b presents the completed system using first generation linear micro synchronous motors Type B.

13.5.2 Investigations on Magnetic Levitation

While the linear synchronous micromotor based xy-actuator prototype verified the basic feasibility of such a positioner, such a system would not lend itself to nanoactuation for two reasons: first, a synchronous motor does not allow a sufficient step resolution, not even when applying micro stepping; second, the microball guide causes too much friction for a submicrometer positioning accuracy. To remedy the second item, a combination of a magnetic levitation during actuation and a lowering to a tribological contact plane was

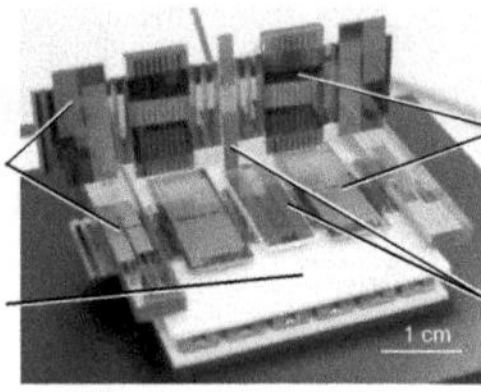

Fig. 13.10 Linear test setup for magnetic levitation [2]

chosen. This way, the actuation itself occurs frictionless, while in the lowered state, a stable rest position is maintained.

Initial investigations on magnetic levitation were conducted in one axis only and (due to the simplicity of the buildup) with a repelling system. Fig. 13.10 represents the test setup. The height control is accomplished by a capacitive system (Chap. 7).

13.6 VR Micro- and Nanopositionier for xy-Actuators

13.6.1 Approach

With the initial investigations on xy-actuators completed, it is obvious which of the linear micromotors is best suited for this application: a linear VR microstep motor. It excels in simplicity and allows, when equipped with teeth, rather small step widths. Applying microstepping, either open loop or with the feedback of a length measurement system, should allow actual positioning steps below 100 nm. As a backup version to a system with teeth, a pole based device was also developed. Rather than using a planar meander coil, a double-layer spiral coil was applied.

13.6.2 Actuator Geometry

Based on simulation results as described in Chap. 2, the actuator geometry was defined. Fig. 13.11 depicts the cross section through the two linear VR micro- and nanomotor versions. The pole-based version (Fig. 13.11.a) with a step width of 66 µm represents a linear VR micropositioner. The toothbased version has a step width of 13.2 µm (Fig. 13.11.b). In case a mode with 132 microsteps between incremental steps is chosen and proves to be feasible, a nanoactuator with a minimal positioning resolution of 100 nm is created. Both geometries where chosen so that the stator fabrication is identical except for the teeth. A basic motor element contains four poles with each pole surrounded by a double-layer coil with a total of six turns. There is a total of six motor elements. By grouping them in pairs, three phases are created. The

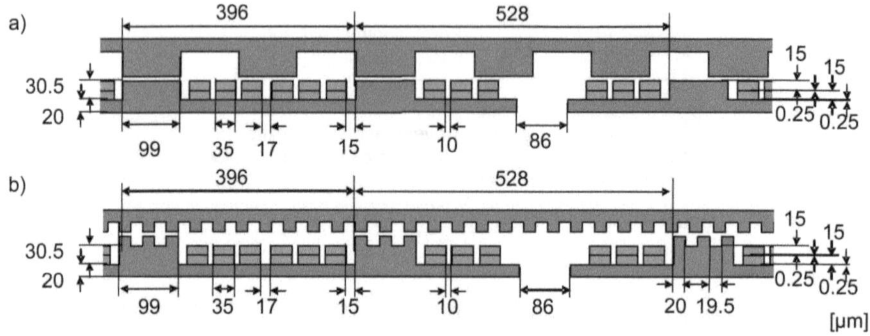

Fig. 13.11 VR actuator cross section: (**a**) Pole based microactuator; (**b**) Tooth based nanoactuator [13]

soft magnetic components of the flux guide and poles are made of NiFe45/55. The currentcarrying coil is made of Cu and the insulation material is *SU-8* (laterally) and Si_3N_4 (vertically), respectively. The traveler consists of a flux guide and poles, both fabricated of NiFe45/55.

Fig. 13.12 shows the overlay of the lithography masks for the stator. The rather long leads between the bond pads and the active part are required to ultimately allow x and y traveler motion. Each stator system is surrounded by four alignment marks required for automated assembly (Chap. 15). A stator system's footprint is $15 \times 8\,\text{mm}^2$.

Fig. 13.13 shows the mask overlay of the micromotor's traveler for the polebased system. As in the case of the stator, it includes alignment marks for automated assembly. A traveler's footprint is $14.6 \times 6.7\,\text{mm}^2$. The smaller width compared to the stator dimensions is necessary for avoiding a traveler protrusion during a lateral displacement.

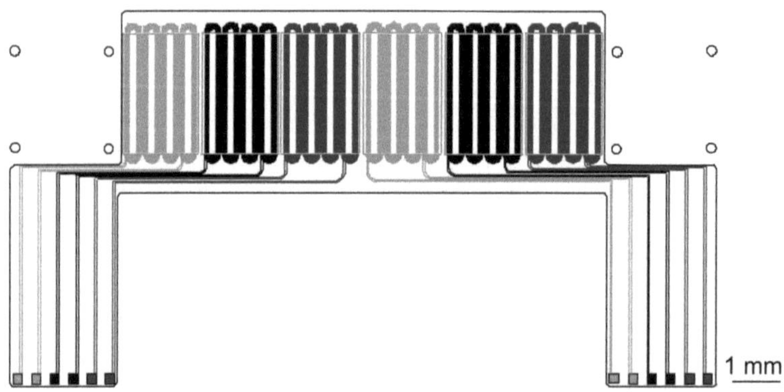

Fig. 13.12 Stator mask overlay for the pole-based system

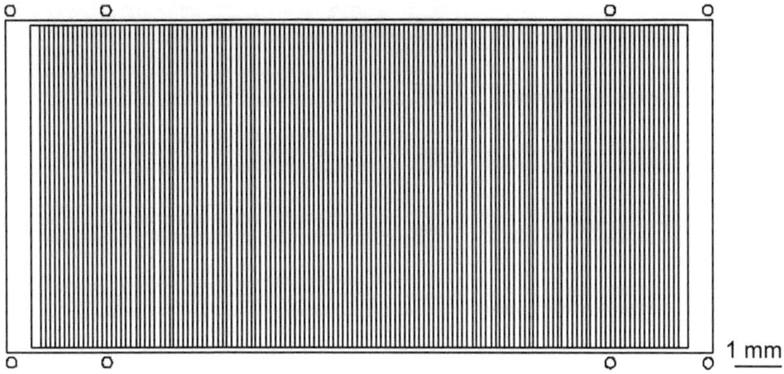

Fig. 13.13 Mask overlay of the pole-based traveler

13.6.3 Stator Fabrication

The fabrication was conducted on a ceramic Al_2O_3 wafer (Rubilat). To fabricate the microstep motor, High Aspect Ratio Micro Systems Technology (HARMST) was applied. Fig. 13.14 shows the sequence of the fabrication process. First, a Cr (50 nm) / NiFe (400 nm) seed layer for the electroplating process was sputter deposited. Then, AZ9260, a positive tone resist was spin-coated with the thickness of 27 µm, exposed and then developed. The rim of the substrate was stripped with acetone to create the contact area required for electroplating. In the next step, 15 µm of Ni was deposited by electroplating. A CMP step was applied to polish the alignment marks, increasing their reflectivity. Finally, the resist was stripped.

In the next step, a 27 µm-thick AZ9260 resist pattern was created serving as a micromold for the lower flux guide. The flux guide was fabricated by electroplating 20 µm of NiFe45/55. Afterwards, its surface was planarized by CMP. After stripping the photomask, the seed layer was removed by IBE (Fig. 13.14.a). Then, the lower flux guide was embedded in *SU-8* followed by a CMP planarization (Fig. 13.14.b). To create a vertical insulation layer, 250 nm of Si_3N_4 was deposited by Plasma Enhanced Chemical Vapor Deposition (PECVD) step.

The first step in creating the lower coil layer was sputter depositing a Cr (50 nm) / Au (200 nm) seed layer combination. Next, a 27 µm-thick AZ9260 resist pattern was created as a micromold for the first coil layer. The 17 µm-thick coil layer was fabricated by Cu electroplating. After stripping the micromold, the seed layer was removed by IBE (Fig. 13.14.c). As a first coil insulation layer, Si_3N_4 was deposited at a thickness of 250 nm. To create windows for the poles at the center of the coil, photolithography was applied in conjunction with IBE. Next, a seed layer of Cr (50 nm) / NiFe81/19 (400 nm) for the poles was sputter deposited and the AZ9260 micromold for the poles was created. A NiFe45/55 plating process created the first pole level. To achieve a plane surface, it was planarized by CMP. After stripping the micromold,

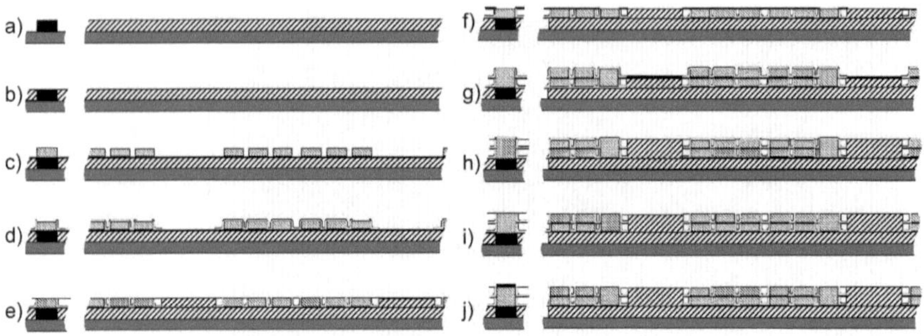

Fig. 13.14 Stator Fabrication Sequence

the seed layer was removed by IBE. Afterwards, the coil layer was embedded in *SU-8* which also formed the insulation between the coil conductors as well as the conductors and the poles. Planarizing the *SU-8* with CMP lowered the *SU-8* surface to the upper conductor level. Coating the wafer with Si_3N_4 concluded the fabrication of the first coil level (Fig. 13.14.e).

The fabrication of the second coil layer started with opening windows for the poles and the vias in the Si_3N_4 (Fig. 13.14.f). This was accomplished by a combination of photolithography (AZ9260) and IBE. Otherwise, the fabrication steps for the second coil layer were the same as for the first (Fig. 13.14.h). In case the stator version with teeth was fabricated, a seed layer of Cr (50 nm) / NiFe81/19 (400 nm) for the tooth fabrication was deposited. Then, the resist pattern (AZ9260) for the teeth was created, followed by electroplating NiFe45/55 (Fig. 13.14.i). Conducting a CMP process planarized the teeth. Stripping the micromold and removing the seed layer by IBE concluded the tooth fabrication. For creating a pole type stator, this process was omitted.

In the final sequence, the bond pads were fabricated. To do so, the windows for the bondpads were fabricated in the same way as the windows for the poles and vias. Then, an Au seed layer was sputter deposited. Next, an Ni diffusion layer and then the Au bond layer were deposited by electroplating. A removal of the seed layer by IBE concluded the wafer process for the stator (Fig. 13.14.j).

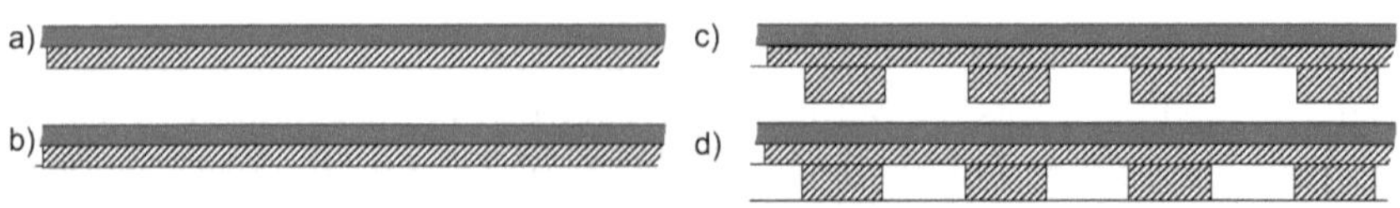

Fig. 13.15 Traveler Fabrication Sequence

13.6.4 Traveler Fabrication

The traveler fabrication follows along the same line as the stator fabrication described previously, with the exception that all coil fabrication steps were omitted. This way, the process sequence is both much simpler and substantially shorter. Fig. 13.15 shows a schematic representation of the process steps.

13.6.5 System Integration

To allow a maximal flexibility during the experimental investigations, both a number of test setups and the xy-positionier itself where fabricated. The first group was force measurement test systems, allowing one-dimensional travel. It could be used either with ball guides, keeping the air gap at a length of approximately 8 µm (i.e. the nominal value) or in conjunction with an xy-stage allowing to adjust the gap length freely. A second group, also with a one-dimensional travel was a levitation measurements test system (Fig. 13.16). Each system contained a single stator and traveler of the drive system and parallel to it, a pair of stators and flux closures of the levitation system. It also contained a capacitive gap length measurement system with four pairs of capacitor plates.

Fig. 13.17 shows a photograph of the xy-actuator components. The base plate (bottom right), its four stators and its four capacitor plates represent the xy-actuator's bottom. The top plate (top, for reason of clarity shown upside down) houses the four levitation systems. the traveler plate (bottom left) contains the four actuator travelers and the capacitor plate counterparts at their bottom (neither one visible) and the flux closure for levitation at their top. After assembly, the traveler plate will be sandwiched between base and top plate. The version shown is a polebased system. Assuming 132 microsteps between incremental steps, a polebased system represents a micropositioner with a resolution of 500 nm while a toothbased version with the same microstepping drive scheme results in a nanopositioner with a resolution of 100 nm.

Actuation activity is conducted as follows: at the beginning of a step sequence, the actuated platform rests on the stators. To initiate the actuation, the magnetic levitation system raises the actuation platform, a process during which the capacitive height measurement system controls the air gaps between the active part and the flux guide. With the resulting spacing between all active parts of the magnetic levitation system and its flux guides, the desired air gap is reached. For actuation, the nanopositioning system executes appropriated positioning steps. Once the actuation platform has reached its desired location, the magnetic levitation system lowers the actuation platform, allowing it to rest on the stators.

Fig. 13.16 Components of one-dimensional test system

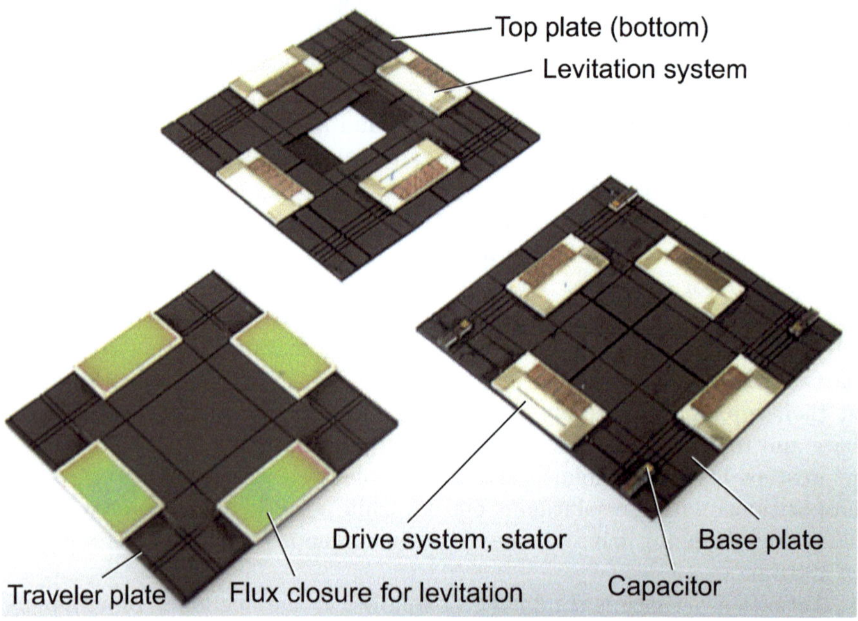

Fig. 13.17 Photograph of the xy-actuator components

To assemble the xy-nanopositioner, self-aligning stacking techniques in conjunction with an adhesive bonding of the components are applied. For a controlled creation of air gaps, micrometer sized mica foils are temporarily inserted. For an industrial scale assembly, a robotic process is envisioned (Chap. 21).

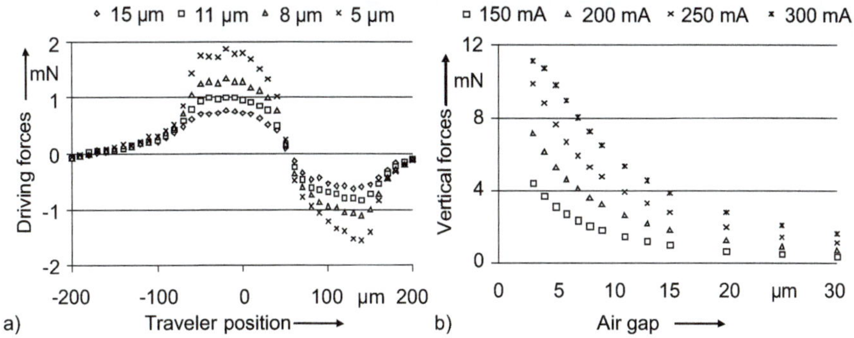

Fig. 13.18 Force Measurements: (**a**) Driving force; (**b**) Vertical force

13.7 Experimental

13.7.1 Driving force measurement

Using the force measurement test system, both the driving force F_x, as well as the normal force F_z of a polebased system were determined (Fig. 13.18).For the tests, a standard stator and an experimental traveler with a reduced number of poles, which allowed particularly accurate measurements in the vicinity of the stator poles, were used. During the test, only one of the six stator coils was energized. Fig. 13.18.a shows a family of curves of the driving force F_x over a travel range of 400 µm for a number of air gaps. For an air gap of 8 µm, the maximal driving force was 1.2 mN. The setup also allowed measurement of the vertical force F_z. For conducting the test, the traveler was substituted by a flux closure mounted on a vertical force gauge. During the measurement, the flux closure was lowered towards the stator while measuring the vertical force F_z. Fig. 13.18b shows the family of curves for various air gap lengths as a function of the excitation current I. As previously, only one of the six coils was energized. These are the forces that ultimately have to be overcome to achieve a levitation of the unloaded system. Due to the fact that the test setup coincides with the magnetic levitation arrangement, the stator also represents the expected levitation forces.

13.7.2 Levitation test

Using the levitation measurements test system, the minimal current required for achieving levitation of an unloaded system was determined. The stator of the levitation system is the same one as the stator of the polebased drive system. For the tests, all stator coils of one levitation stator system were excited by a constant current I. During the test, the capacitive gap measurement

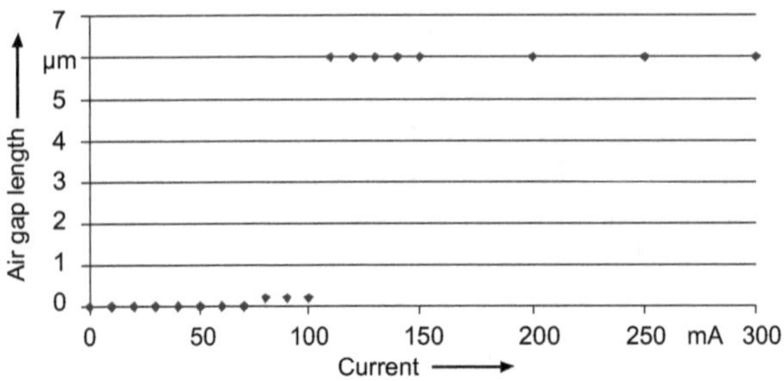

Fig. 13.19 Measurement of levitation current

system was inactive (Fig. 13.19). As soon as the current reached 100 mA, the traveler was attracted and snapped on to the levitation stator system.

13.7.3 xy-nanopositioning system

As this publication was nearing completion, the assembly of the xy-nanopositioning system approached completion. Ultimately, the tests conducted on the linear systems will be duplicated on the xy-nanopositioning system. Furthermore, the system will be evaluated on the microstepping conditions and the minimal achievable positioning resolution will be determined.

13.8 Conclusion

By developing three types of linear micromotors, alternative microactuation principles could be evaluated. The three types were a linear VR microstep motor, a linear hybrid micromotor and a linear synchronous micromotor. Initially, there was a substantial mismatch between the FEM simulation results of Leibniz Universität Hannover's Institute for Drive Systems and Power Electronics (Antriebssysteme und Leistungselektronik, IAL) and micromotor measurement results. It was found that the reason were much lower permeability values in soft magnetic microstructures than expected based on data of continuous films. By applying a methodology taking demagnetizing fields of small structures into account (Chap. 10), the issue could be resolved. It was further found that a linear VR microstep motor was the most promising motor design for a xy-nanopositioner. Such a positioner could be designed and fabricated by combining four linear positioners, four magnetic levitation systems and an appropriate capacitive measurement system to adjust the air gap between stators and travelers.

Acknowledgements This work was sponsored in part by the DFG (German Research Foundation) within the Collaborative Research Center 516 "Design and Manufacturing of Active Micro Systems". The authors would like to thank S. Demmig, J. Edler, R. Gerking, G. Janssen and B. Ponick of IAL for simulating the micromotor and S. Bütefisch of the Physikalisch Technische Bundesanstalt (PTB) Braunschweig for conducting force measurements. The authors would also thank S. Cvetkovic (IMT-H) for supporting the system assembly. Additional the authors gratefully acknowledge the contributions of several former colleagues involved in the Collaborative Research Center 516 "Design and Manufacturing of Active Micro System": Matthias Hahn and Martin Föhse.

References

[1] Bedenbecker M, Ruffert C, Hahn M, Gatzen HH (2007) Linear synchronous micro motor with further miniaturized dimensions. In: Proc. of the ASME 2007, pp 1–6

[2] Cvetkovic S, Ruffert C, Gatzen HH (2007) Assembly concept of a magnetic levitation system for a linear micro actuator. In: Proc. ASME 2007, Las Vegas, NV, USA 2007, DETC2007-35266,

[3] Deaconu D, Chirila A, Navrapescu V, Albu M, Ghita C, Popescu C (2008) Two hybrid stepper motor models. In: 9th WSEAS International Conference on Automation and Information (ICAI 08), pp 129–134

[4] Föhse M, Edler J, Stölting HD, Gatzen HH (2002) Investigations on the pole geometry optimization of a variable reluctance microactuator. In: 47. Internationales Wissenschaftliches Kolloqium 2002, Ilmenau, Germany, pp 215–216

[5] Föhse M, Edler J, Stölting HD, Gatzen HH (2003) A batch fabricated synchronous motor. In: Proc. Int. Mech. Eng. Congr. and Expo IMECE 2003, Washington D.C., USA, pp 1–8

[6] Gatzen HH, Stölting HD, Büttgenbach S, Dimigen H (2000) A novel variable reluctance micromotor for linear actuation. In: Proc. Actuator 2000, Bremen, Germany,, pp 363–366

[7] Gatzen HH, Morsbach C, Karyazin A (2002) High precision machining of a longitudinal bearing for a linear microactuator. In: Proc. EUSPEN 3rd Int. Conf. 2002, Eindhoven, The Netherlands, pp 325–328

[8] Gatzen HH, Morsbach C, Kourouklis C (2003) Chemical-mechanical planarization of a *SU-8*/copper combination for mems. In: Proc. ASPE 18th Ann. Meet. 2003, Portland, OR, USA,, pp 575–578

[9] Gatzen HH, Stölting HD, Ponick B (2004) Alternatives for micromachined linear actuators. In: Proc. 9th Int. Conf. on New Actuators 2004, Bremen, Germany, pp 317–320

[10] Gatzen HH, Hahn M, Bedenbecker M, Ponick B, Gehrking R, Demmig S (2006) Advances in the development of a linear hybrid micro actuator.

In: Proc. 10th Int. Conf. on New Actuators 2006, Bremen, Germany, pp 207–210

[11] Hahn M, Bedenbecker M, Brand U, Gatzen HH (2004) Evaluation of a linear hybrid micro step motor. In: Proc. 8th Int. Symposium on Magnetic Materials, Processes and Devices, 206th Meet. of The Electrochemical Society 2004, Honolulu, HI, USA, pp 493–505

[12] Hahn M, Bedenbecker M, Gatzen HH (2007) Evaluation of a linear hybrid microstep motor by means of magnetic flux measurements. IEEE Trans on Magn 43(6):2588–2590

[13] Hansen S, Norpoth J, Joss C, Rissing L, Gatzen HH (2010) Magneto-optical characterization of the stator of a variable reluctance (vr) micro step motor. In: Proc. Actuator 2010, Bremen, Germany, pp 706–709

[14] Probst M, Flückiger M, Pan S, Ergeneman O, Nagy Z, Nelson BJ (2007) Manufacturing of a hybrid acoustic transmitter using an advanced microassembly system. IEEE Transactions On Industrial Electronics 56(7):2657–2666

Chapter 14
Micromachining of Parts for Microsystems

D. Hahmann, M. Rüggeberg, R. Wittmer, M. Reichstein, M. Hlavac, B. Denkena[1], H.-W. Hoffmeister[2]

[1] *Institute of Production Engineering and Machine Tools*
Leibniz Universität Hannover
denkena@ifw.uni-hannover.de

[2] *Institute of Machine Tools and Production Technology*
Technische Universität Braunschweig
h.hoffmeister@tu-braunschweig.de

Abstract This article describes microgrinding processes for the machining of materials used in microsystems. An example of such microsystems is an air guide, which is used as an example for the presented machining processes. High workpiece surface qualities and shape accuracies are the aims in micromachining. To accomplish this, chemical vapor deposition (CVD) coated grinding wheels and pins, as well as multi-layered metal bonded grinding wheels, were developed and investigated, including novel dressing technology. In addition to conventional grinding processes, ultrasonic-assisted grinding is analyzed regarding the resulting surface quality. Microgrinding tools with CVD-coating were developed for structuring materials like glass, ceramics and cemented carbides. The influence of the process parameters and tool design on the workpiece quality is investigated. Finally, the micro-air guide was produced with different micromachining processes within the required tolerances. The investigated microgrinding processes with grinding pins enable the manufacturing of microholes with low chipping. Several microstructures can be machined in parallel when grinding with multiple microprofiled grinding wheels. Furthermore, dressed undercut geometries enable the grinding of perpendicular microstructure flanks.

14.1 Introduction

A wide spectrum of machining processes is available for the manufacturing of microparts or microsystems. Photolithographic processes and laser machining increase accuracy, but three-dimensional structures are hard to produce economically by these processes [1, 26]. Electro discharge machining (EDM) enables the manufacturing of three-dimensional structures with a high accuracy. However, electroconductive workpiece materials are needed [24, 34, 35].

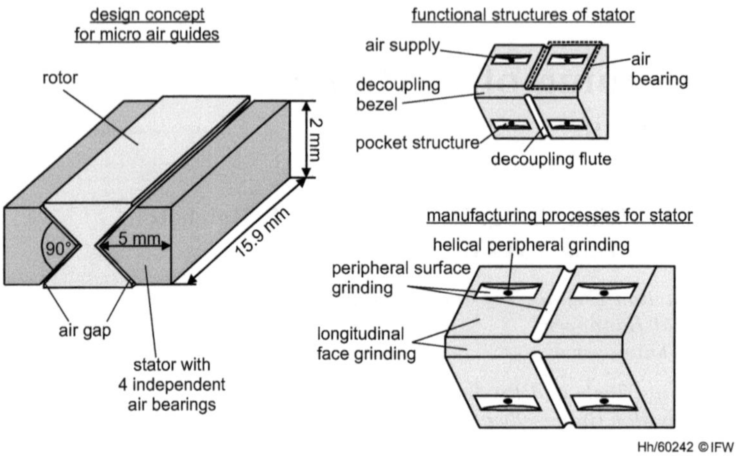

Fig. 14.1 Design concept of a micro air guide and the used manufacturing processes

Machining processes like microgrinding guarantee the machining of a wide spectrum of materials with a good accuracy [33, 36]. Furthermore, these processes enable high productivity. A micro-air guide is machined to show the potential of these processes. The micro-air guide consists of two stators and one rotor. Each stator has four independent air bearings (Fig. 14.1). The air bearings are decoupled by a bezel at the top of the stator and a flute on both stator flanks. Small holes connect the air bearings to the compressed air supply on the backside of the stators. A pocket structure disperses the air into the air bearing.

During the machining process, face grinding will be used to machine the surface of the air bearings and the decoupling bezel to gain a planar and smooth surface. The pocket structures and decoupling flutes will be machined in peripheral surface grinding due to the small structure dimensions. The holes for the air supply will be manufactured by peripheral grinding. However, these processes could not be used for such workpieces prior to the fundamental work presented in the following chapters.

14.2 Microgrinding

For microgrinding processes the influence of the grinding process on the part quality is investigated to accurately manufacture planar surfaces. Two different strategies are traced to machine the functional microstructures of the air bearing. On the one hand, CVD-coated grinding wheels with beneficial wear properties in grinding are developed. On the other hand, multi-layered metal bonded diamond grinding wheels are proved to machine multiple microstructures. For this reason, an electro contact discharge dressing (ECDD) process is developed. Afterwards, both grinding wheel types are used for the machining of microstructures in hard and brittle materials.

14.2.1 Workpiece Surface and Geometry Quality in Microgrinding

The micromachining of hard and brittle functional components provides high surface quality and geometrical accuracy. These requirements can only be met by grinding processes. There are several grinding processes, like peripheral surface grinding, peripheral surface grinding with a tilted tool, ultrasonic assisted peripheral surface grinding and face grinding [2, 27, 28, 30, 32]. All of them provide different workpiece surface topographies, even if the same grinding wheel, or rather the same grain size and grain concentration, are applied. The surface quality of the workpiece is limited by the tool geometry and topography in peripheral surface grinding. The influence of the tool on the surface of the workpiece can be reduced by tilting the grinding wheel. A tilt angle of 10° reduces the surface roughness by a factor of three to $Ra = 30\,\mathrm{nm}$. A further increase of the tilt angle only leads to a small decrease of the surface roughness. It is also possible to reduce the influence of the tool on the workpiece surface using ultrasonic assistance parallel to the workpiece surface. Ultrasonic assistance can reduce the surface roughness by a factor of three to four to $Ra = 25\,\mathrm{nm}$ [2, 8]. The lowest surface roughness and the lowest influence of the tool topography and geometry on the workpiece are gained in face grinding with a surface roughness of $Ra = 10\,\mathrm{nm}$ [27, 28, 30, 31, 32]. Thus, face grinding was chosen for the machining of the air-bearing surface and the decoupling bezel.

14.2.2 CVD-Coated Grinding Wheels

Microcutting operations are distinguished from other processes like lithography by their high flexibility, short manufacturing time as well as the machinability of a broad spectrum of materials. Thus, they are very appropriate for the manufacturing of high-precision structures and components of microsystems technology [15]. This is shown by the example of coated single layer CVD-diamond grinding wheels.

This section provides the results of grinding tests with these novel tools. The process enables fast and high-quality grinding of microstructures in various hard and brittle materials [15, 37, 38].

CVD-coated tools for microgrinding operations were formerly only used for shaft tools [14]. The advantages of CVD-coatings, which have been known for many years, were transferred to miniaturized grinding wheels [21, 18]. They have the same advantages as CVD-microgrinding pins, but they can even be used with higher cutting velocities and higher material removal rates. Because of their small grain size, hard and brittle materials can be machined

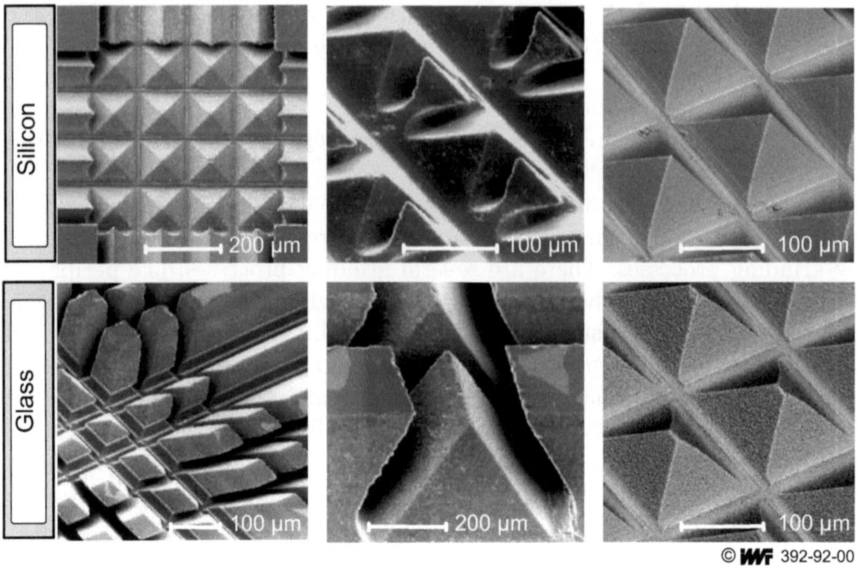

Fig. 14.2 Examples of microstructured surfaces in silicon and glass [38]

with low edge breakouts when optimized machine parameters are used. This means that micro-cracks and chipping can be minimized, and better surface qualities can be achieved.

First, the tool manufacturing process consists of machining the cemented carbide tool body. Characteristics of the CVD-layer and the results of the grinding tests are presented below. The development of these tools offers new possibilities of microstructuring hard and brittle materials (Fig. 14.2) [22, 16].

Tool Development and Specifications of CVD-Coated Grinding Wheels

The grinding wheels consist of a cemented carbide tool body and a rough CVD-diamond coating. After designing an appropriate geometry for these novel microgrinding tools (Fig. 14.3), the inner diameter with transition fit size was ground. Subsequently, the cemented carbide tool bodies were produced with cup wheels while the machine table was rotating. They have a diameter of 48 mm and a width of 0.16 mm. The design, which has already been tested, is a grinding wheel with an even standard profile.

The tool bodies were machined at the Institute of Machine Tools and Production Technology (IWF) and afterwards coated with a CVD process at the Fraunhofer Institute for Surface Engineering and Thin Films (IST). The examination of the CVD-diamond microgrinding wheels with a scanning

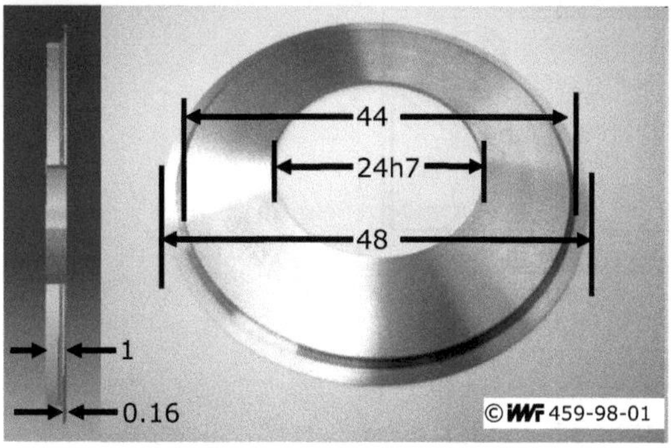

Fig. 14.3 Geometry of the cemented carbide tool body [20, 19, 23]

Fig. 14.4 CVD-layer of the microgrinding wheel [20, 19]

electron microscope (SEM) showed that the diamond crystals have a size of 5 µm (Fig. 14.4).

Grinding with CVD-Coated Grinding Wheels

The materials aluminum nitride, aluminum oxide, glass, silicon and cemented carbide were ground with CVD-diamond grinding wheels, while the infeed as well as the cutting speed and the feed rate were varied. In addition to the in-process measurement of normal and tangential forces with a dynamometer during grinding, the edge breakout, the surface roughness and the tool wear of the microgrinding wheels were evaluated and documented.

The analysis of the cutting forces showed that they were lower than 1 N for all investigated materials and process parameters. The development of the tangential and normal forces as a function of the cutting forces by grinding silicon showed that higher feed rates led to lower cutting forces when the same specific material removal rate is ground (Fig. 14.5).

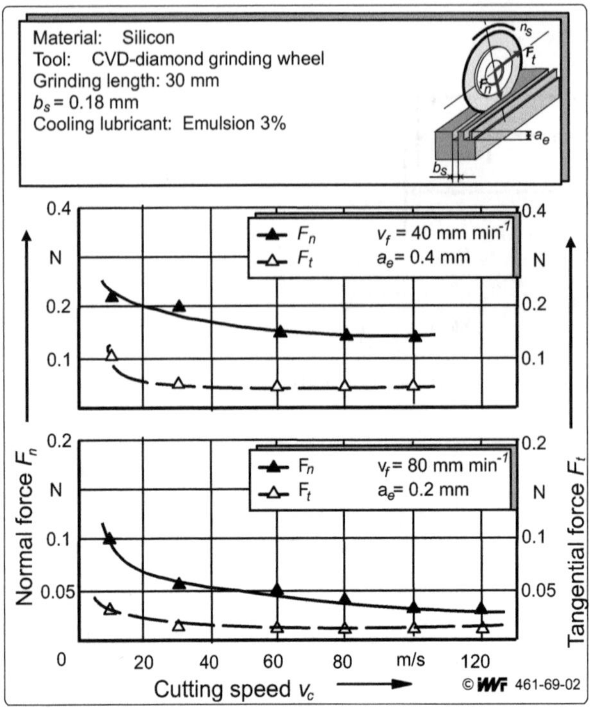

Fig. 14.5 Cutting forces [19, 23]

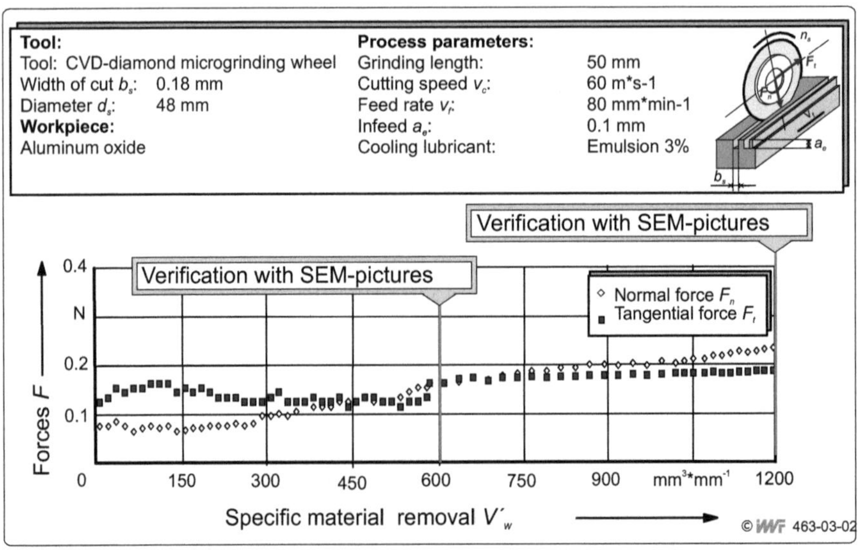

Fig. 14.6 Cutting forces by microgrinding aluminum oxide [19, 23]

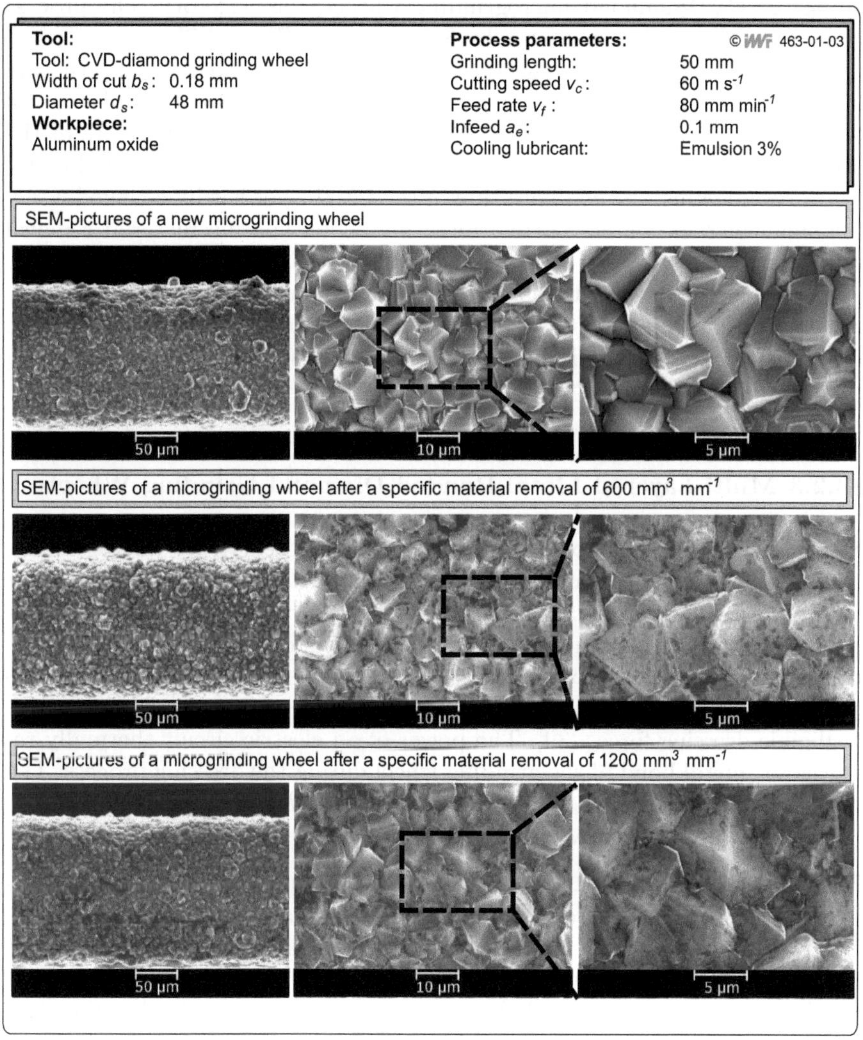

Fig. 14.7 Wear behavior by microgrinding aluminum oxide [19, 23]

The normal forces during the tests were approximately twice as high as the tangential forces. Moreover, it can be stated that the forces decrease when the cutting speed is increased. This is due to the fact that the equivalent chipping thickness decreases for higher cutting speeds [38].

Fig. 14.6 highlights the development of the cutting forces as a function the specific material removal by microgrinding aluminum oxide. The normal and tangential forces are relatively stable for the area of a specific material removal of 600 $mm^3\,mm^{-1}$ and afterwards there is a slight increase of the cutting force. As there is only a small increase, the process and wear behavior of the CVD-diamond microgrinding wheels is satisfactory. The process behavior of the tool is evaluated by SEM-pictures (Fig. 14.7).

The new microgrinding wheel shows a crystalline, sharp-edged structure of the CVD-diamond coating. The results after a specific material removal of 600 $mm^3 mm^{-1}$ and of 1200 $mm^3 mm^{-1}$ are similar. The structure of the crystallites stays well preserved, and except for a few adhesions of cooling lubricants and chips, the configuration of the CVD-layer hardly changed. Only the small increase of the normal force for higher specific material removal rates by cutting suggests little wear.

It can be summarized that there was a maximum chipping of 5 µm of the workpiece. The chipping was 1 µm for grinding cemented carbide and 5 µm for grinding glass. The chipping of the workpiece was higher for up-grinding than for down-grinding [20].

14.2.3 Multi-Layered Metal Bonded Diamond Grinding Wheels

Metal bonded grinding tools offer a remarkable potential for microgrinding due to their favorable wear behavior. The grinding tool has to be profiled for the grinding of microstructures. However, this is not possible using mechanical dressing methods because of a high dresser wear and high dressing forces. ECDD is a numerically controlled dressing process offering the possibility to create the geometry and topography of the grinding wheel simultaneously at negligible dressing forces [39]. The power of an electric circuit thermally removes the metal bond of the grinding wheel during the dressing process. The topography is generated at lower dressing voltages U_{d0} whereas the geometry is generated at higher dressing voltages U_{d0} [28].

Dressing of Single Microprofiles with Graphite Electrodes. The influence of the dressing parameters on the specific material removal rate of the grinding wheel and the G ratio was investigated. Therefore, the infeed of the electrode f_{rd}, the limitation of the dressing voltage U_{d0} and the limitation of the dressing current I_{d0} varied. The G ratio in dressing is defined as the ratio between the removed grinding tool volume and the used electrode volume. A higher G ratio means a more efficient use of the electrode.

The specific material removal rate Q'_{sd} and the G ratio are significantly influenced by the dressing parameters. The specific material removal rate Q'_{sd} increases by rising the dressing voltage U_{d0}, the limitation of the dressing current I_{d0} or the infeed of the electrode f_{rd} up to a certain value. A further increase of those dressing parameters does not increase in the specific material removal rate Q'_{sd} further. The G ratio is proportional to the infeed of the electrode f_{rd} if the dressing voltage U_{d0} and limitation of the dressing current I_{d0} are constant. It is proportional to the specific material removal rate in dressing Q'_{sd} for the dressing voltage U_{d0} and the limitation of the dressing current I_{d0} at constant infeed of the electrode [6, 7, 28].

The shape of the grinding wheel microprofile depends on the movement path of the electrode. Here, a linear relation was found [3, 28]. The size of the electrode limits the minimum size of the profiles due to the numerically controlled profile generation in ECDD. Thus, only single microprofiles can be dressed by ECDD if a graphite electrode is used. Nevertheless, single filigree roof-shaped profiles can be reproducibly dressed on the grinding wheel. The gained profile angle of 60° is dressed precisely and reproducibly. Furthermore, a minimum profile tip radius of about 3 µm is achievable [13].

Dressing of Multiple Microprofiles with Wire Electrodes. The use of a thin copper wire electrode in a new dressing unit significantly reduces the minimum profile size and enables the dressing of multiple microprofiles [10]. In the ECDD process, the feed direction of the wire electrode is perpendicular to the peripheral surface of the grinding wheel, in contrast to the tangential feed direction in wire electrical discharge machining (EDM). At the new dressing unit the wire electrode is guided close to the grinding wheel surface by a conical duct to give adequate mechanical support. The influence of the electrode feed rate f_{wire}, the limitation of the dressing voltage U_{d0} and limitation of the dressing current I_{d0} on the specific material removal rate was investigated. It was found that the process is most efficient and stable at a dressing voltage of 20 V, a limitation of the dressing current of 0.1 A and an electrode feed rate of 1 µm s^{-1}, if a grinding wheel with a grain size of 16 µm and an electrode diameter of 224 µm is used [10].

The dressing of concave rectangular and convex roof profiles was investigated. The relation between the electrode path and the flute width is analyzed at the rectangular profiles (Fig. 14.8).

The electrode oscillates parallel to the grinding wheel surface within the constant oscillation distance L_d and with the feed rate v_{fd} between position "1" and "2" to generate rectangular profiles. The feed rate is set to 1 mm min^{-1} as a result of preinvestigations. The oscillation distance is varied between 0 µm and 750 µm.

The SEM-picture clarifies the thermal character of the bond removal and can be clearly distinguished from the undressed grinding layer beside the flute. The detailed micrograph shows highly protruded grains. Thus, the grinding layer is profiled and sharpened with only one process step.

Furthermore, a linear relation between electrode path and flute width was found. The theoretical flute width is the sum of the oscillation distance and the electrode diameter. However, the measured flute width differs from the theoretical. The flute is slightly wider than the electrode diameter without oscillation. It is slightly narrower as the theoretical flute width for an oscillatory movement. It is assumed that this results from the low mechanical stiffness of the wire electrode and a small guide clearance within the duct. The electrode is pushed aside by a force acting radially to the electrode. Thus the electrode is pushed towards the center of the flute if an oscillation

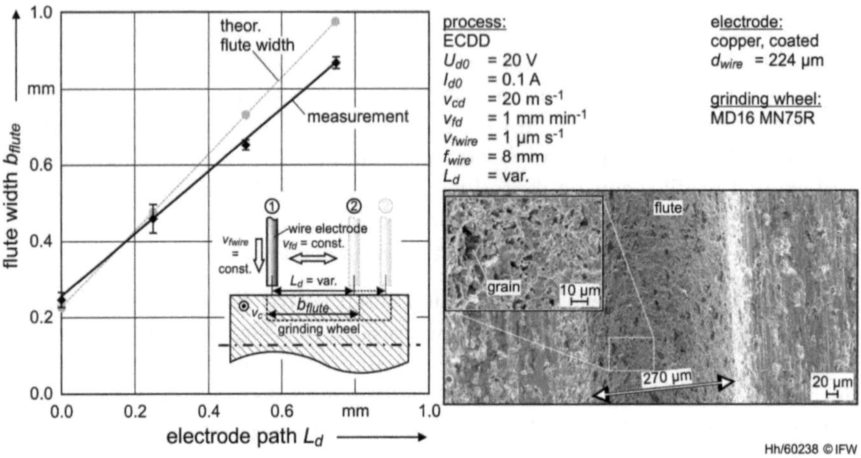

Fig. 14.8 Relation between electrode path and flute width [11]

movement occurs. Then the generated flute is narrower than the theoretical one.

The influence of the dressing strategy on the profile tip radius was investigated at roof shaped profiles (Fig. 14.9). The oscillation distance is stepwise reduced for dressing of the roof profiles. The cycle time is defined by the actual oscillation distance and the actual feed rate. The required number of cycles is defined by the quotient of total dressing time and cycle time. The total dressing time is kept constant for all investigations. The influence of two different strategies and the feed rate on the profile tip radius was investigated as well (Fig. 14.9, left side). The cycle time is kept constant in the first strategy. Therefore, the feed rate is stepwise decreased by the same factor as the oscillation distance. The feed rate is kept constant in the second strategy. Thus, an increased number of cycles is required to achieve the same total process time. In both strategies the outer reversal points, labeled number "1" and "3", are fixed. The inner reversal points "2" and "4" stepwise approximate the outer reversal points due to the incremental reduction of the oscillation distance.

The profile tip radius decreases with increasing feed rate for both strategies. However, the strategy with a constant feed rate generates smaller profile tip radii (Fig. 14.9, right side). This can be explained by the stepsize of the incremental reduction of the oscillation distance, which depends on the number of cycles. A high number of cycles results in a smaller stepsize and thus in a smaller profile tip radius. On the one hand, the number of cycles increases with increasing feed rate. On the other hand, the number of cycles is higher for the strategy with a constant feed rate. For example, the number of cycles is seven for the first and 26 for the second strategy at a feed rate v_{fd0} of 0.1 mm min^{-1}. This results in a stepsize between two dressing cycles

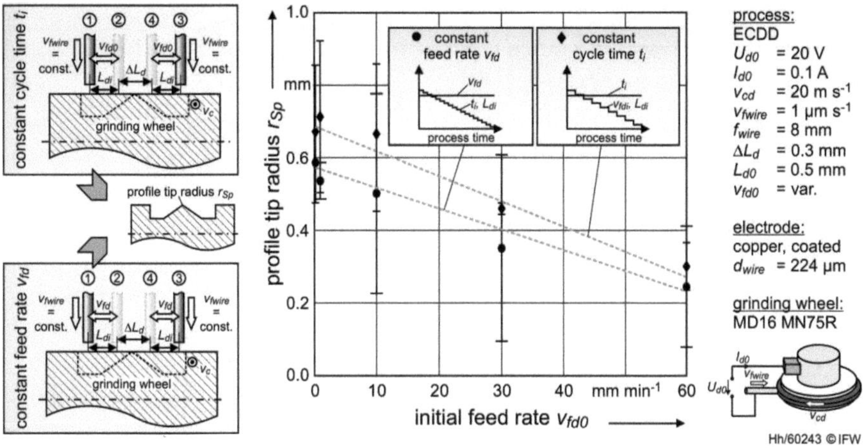

Fig. 14.9 Influence of the feed rate and dressing strategy on the profile tip radius

of 71 µm for the first and 19 µm for the second strategy. Thus, 71 µm of the profile tip are not dressed in the second circle for the first and only 19 µm are not dressed in the second strategy.

Grinding with Multiple Microprofiled Grinding Wheels

The ECDD microdressing process is used to generate multiple profiled grinding wheels. The grinding of microstructures is investigated with these tools. The dressing and grinding operations were carried out on one grinding machine.

The contact conditions in grinding are influenced by the tool properties like grain size and concentration and the process parameters. They are described by the chip thickness h_{cu} and the geometrical contact length l_g [25] among other things. Aluminum oxide samples are chosen, due to the favorable machinability by microgrinding and the use of this material for the manufacturing of the demonstrator. Crossed microstructures are machined by two orthogonal profile grinding processes (Fig. 14.10). Several microstructures are machined parallel in grinding with multiple micro-profiled grinding wheels in contrast to micro-grinding with single dicing blades. Thus, process economy is enhanced in grinding with multiple microprofiled grinding wheels.

Both structures shown in Fig. 14.10 are machined with the same material removal rate. One structure is machined with a low infeed and a high feed rate and the other structure is ground with a high infeed and a low feed rate. The SEM-pictures show that grinding with a high infeed and a low feed rate produces larger edge break-outs than grinding with a low infeed and a high feed rate. This is contrary to the results known for surface grinding [25]. The geometrical contact length is in the range of the structure dimensions

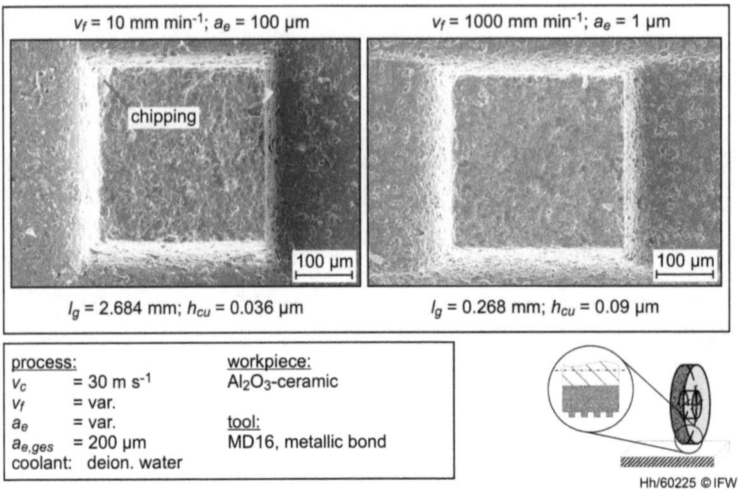

Fig. 14.10 Influence of the grinding parameters on the surface edge damage [12, 11]

for an infeed of 1 µm and a feed rate of 1000 mm min^{-1}. The geometrical contact length is about ten times longer than the structure dimensions for an infeed of 100 µm and a feed rate of 10 mm min^{-1}. The flutes machined in the first profile grinding process result in an interrupted cut in the second profile grinding process. Thus, the maximum contact length is shorter than the geometrical contact length if the geometrical contact length exceeds the microstructure dimensions. Considering this, the normal and tangential single grain forces, calculated from force measurements, are significantly higher in grinding with an infeed of 100 µm and a feed rate of 10 mm min^{-1}. This is the main reason for the increased edge break-out.

The minimum workpiece microstructure dimensions are limited by the flute width of the grinding wheel profile if the microstructures are machined with only one tool path per grinding process. The structure size can be reduced if two tool paths with the offset Δz are used (Fig. 14.11). Two cases are distinguished. The bar width of the grinding wheel profile is larger than the flute width in the first case. Then, the microstructure dimensions are reduced proportionally to the offset between the two tool paths. The bar width of the grinding wheel profile is smaller than the flute width in the second case. In that case, the microstructure dimensions are reduced until the offset between the tool paths corresponds to the bar width of the grinding wheel profile. The microstructures machined in the first tool path are segmented in the second tool path if the offset exceeds the bar width of the grinding wheel profile. The detailed SEM-picture in Fig. 14.11 shows a group of four microstructures that were machined by the segmentation of one microstructure produced in the first tool path. However, the flat tops of the 200 µm-high microstructures with an edge length of 20 to 40 µm show brittle material

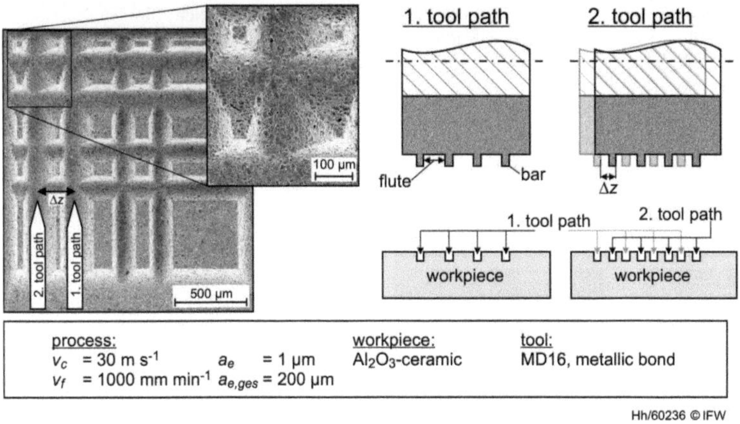

Fig. 14.11 Reduction of structure size by two grinding paths with the offset Δz [11]

breakouts in the range of the ceramic material grain size. This leads to the assumption that the workpiece material properties, e.g. the grain size and the fracture toughness, have a significant influence on the quality of the machined microstructures. Thus, the edge quality of microstructures machined in aluminum oxide, zirconium oxide and cemented carbide is analyzed below.

All samples are machined with the same process parameters. The samples differ in the initial state of the material before and after sintering beside the material itself. The zirconium oxide and cemented carbide are fine grained and have a low porosity. The aluminum oxide has coarse grains in the range of 10 to 20 µm and a higher porosity. The fracture toughness of zirconium oxide is twice as high and the fracture toughness of cemented carbide is up to five times higher than for aluminum oxide. The SEM-pictures show that the edge quality is higher for zirconium oxide and cemented carbide (Fig. 14.12). The chipping for aluminum oxide reaches more than 10 µm. The edge breakout is about 3 µm for zirconium oxide and less than 2 µm for cemented carbide. The edge quality increases with increasing fracture toughness and decreasing material grain size. Thus, the further reduction of the microstructure dimensions by using different offsets Δz or higher infeeds and the increase of the height to width ratio is investigated at cemented carbide.

Fig. 14.13 shows the minimum microstructure dimensions machined in cemented carbide. The microstructure top is reduced to a sharp tip with a radius of about 3 µm at a structure height of about 400 µm. However, the flanks of the structure are not perpendicular due to the grinding wheel microprofile. Thus, further investigations were carried out to gain perpendicular profile flanks.

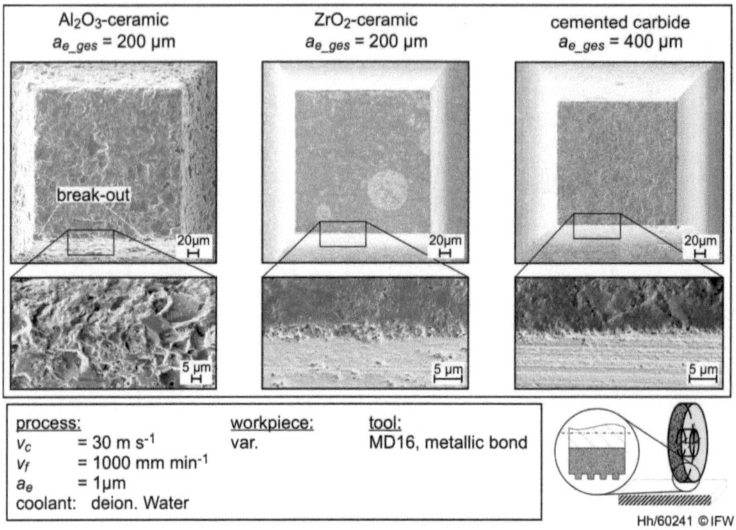

Fig. 14.12 Influence of the workpiece material on the surface damage

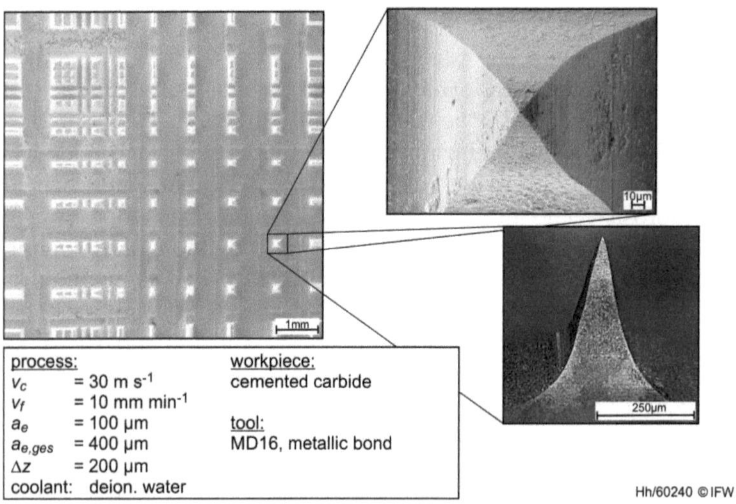

Fig. 14.13 Microstructure

Grinding with Undercut Profiles

A grinding wheel with undercut profiles is used for grinding microstructures with perpendicular flanks. Only one side of the grinding wheel microprofile is dressed with an undercut to clarify the influence of the undercut profile on the machined microprofile shape. The undercut profile is dressed in two steps. The feed direction of the microelectrode is perpendicular to the grinding wheel surface in the first dressing step. The electrode is fed under the tilt angle γ in the second dressing step to generate the undercut (Fig. 14.14). The tilt angle γ is set in the range of 15° to 30°. The dimensions of the machined microstructures are not the imprint of the grinding wheel microprofile. They are generated by the projected grinding wheel microprofile resulting from the grinding wheel rotation. Thus, two flanks of the machined microstructures are expected to be perpendicular when grinding with the undercut microprofiled grinding wheel described above.

Fig. 14.14 shows a machined microstructure. The flanks machined with an undercut profile are perpendicular except for the lower third of the microstructure. The flanks machined with a standard grinding wheel profile have a flank angle below 90°. The detailed SEM-pictures of the corners machined without and with undercut profile underline the significant difference in the shape of the edges. The corner machined without undercut profile shows a smooth crossover from the top area to the flanks. The corner machined with undercut profile shows a sharp crossover from the top area to the perpendicular flanks. Thus, grinding with an undercut profile enables the machining of perpendicular microstructure flanks and sharp edges.

The achieved microstructure dimensions are smaller than required for the air guide, but there are further applications for such microstructures. For example, an array with multiple equidistant and equal pins can be used as storage for microparts in microfactories to assure defined micropart positions. Furthermore, multiple small flutes can be used as a component of microreactors or micro-heat exchangers [29].

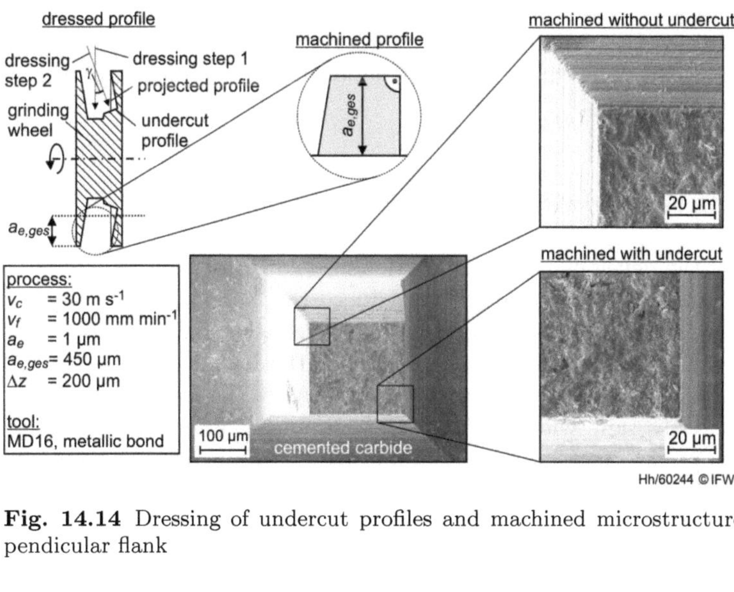

Fig. 14.14 Dressing of undercut profiles and machined microstructures with perpendicular flank

14.3 Microgrinding of Boreholes

In contrast to microgrinding wheels, microgrinding pins offer the possibility to machine boreholes and various closed structures. At present, electro-plated diamond hollow-grinding pins are used to machine small drill holes into hard and brittle materials. These tools have a diameter of 0.9 mm. One major disadvantage of hollow tools when grinding a blind hole is that the tool leaves a workpiece-pin in the middle of the borehole [4].

Compared to CVD-diamond coated grinding pins, electro-plated diamond grinding pins are known to obtain a relatively low grinding quality concerning the surface roughness and chipping level [17]. The difference in the machining-quality can be explained with the characterization of the CVD-diamond coating, which is known to consist of small, sharp crystallites with a size of a few micrometers [37, 38]. In order to improve the machining quality of boreholes in hard and brittle materials new grinding tools were developed.

14.3.1 Development of Microgrinding Pins

The geometry was optimized based on the successful development of CVD-diamond coated microgrinding pins. The variation of the tool-geometry was determined to be an essential magnitude of influence for the quality of the ground borehole. The starting tool geometry was a pin which was dressed in a first step to a ball end head. Further developed tools were shaped as cone-head geometries with varying angles of 15°, 30° and 60°. After fabrication, the tool bodies were coated with a CVD-process (Fig. 14.15).

Examination of the CVD-diamond microgrinding pins with a scanning electron microscope showed that the diamond crystallites have a width of approximately 2 – 6 µm.

14.3.2 Microgrinding with the Developed Tools

The results of the developed tool geometries showed that a slow widening of the borehole to the maximum diameter reduced the chipping and increased the quality at the edges of the ground holes. A steep, conical geometry with an angle of 15° generated the best quality with the lowest chipping at the edges. This can be explained with a good centering of the grinding-tool and the simultaneously slow widening of the borehole [9]. Figure 14.15 shows SEM-pictures of the tested geometries in silicon workpieces and compares the results with a conventional electro-plated diamond hollow grinding pin.

The application of electro-plated diamond hollow grinding pins cause a chipping level of about 150 µm and more. The surface roughness Sa of the borehole walls is more than 0.6 µm. CVD-diamond coated pins with the 15° cone were able to reduce the chipping to 10 µm with a surface roughness at

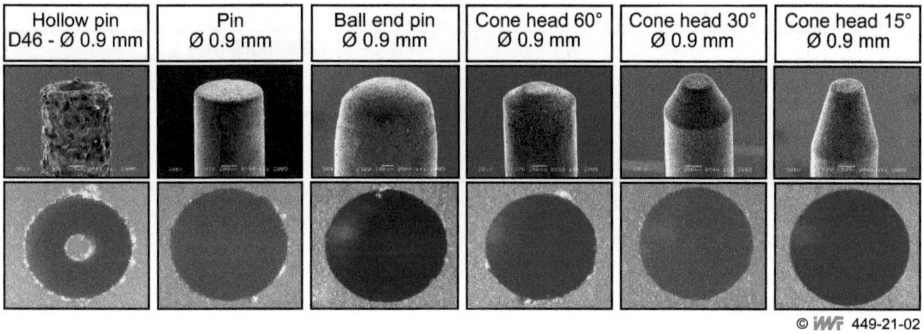

Fig. 14.15 Increasing borehole quality in silicon due to geometry variation [4, 5]

the borehole walls of $Sa = 0.2\,\mu\text{m}$. Additionally, the tool lifetime could be increased.

With this concept, grinding tools with a minimal diameter of 250 µm and an aspect ratio of 1:40 enable the successful grinding of boreholes [9].

14.3.3 Ultrasonic-Assisted Holegrinding with Profiled Tools

Process forces are reduced and coolant supply is enhanced by the use of ultrasonic vibration in drilling of brittle materials. However, the edge quality of the machined holes with the yet undressed tools is insufficient for microparts. Conical shaped grinding pins enhance the edge quality of the workpieces in conventional drilling of brittle materials as shown in the previous section. Thus, prior to ultrasonic assisted grinding, the standard profiled grinding tools are profiled by ECDD to transfer those advantages to this process. Tools with a diameter of 10 mm and 1 mm were used for the investigations. Conical shaped profiles and a reduced grinding layer thickness were dressed at the grinding tools with 10 mm diameter. Only conical shaped profiles were dressed due to the thin grinding layer thickness at the 1 mm diameter tools. First, ultrasonic-assisted grinding of aluminum oxide with a standard tool geometry was carried out in wide range of cutting speeds, feed rates and amplitudes, as shown in Fig. 14.16. The specific edge breakout volume and the specific normal grinding force were evaluated for all experiments. The specific edge breakout volume is defined as the edge breakout volume per millimeter circumferential distance of the outer diameter of the machined holes. The parameter settings resulting in the smallest, a medium and the largest specific edge breakout volume were used for ultrasonic-assisted grinding with the profiled tools afterwards.

The specific edge breakout volume is reduced for both profiles compared to the standard tool geometry at 10 mm tool diameter. However, the specific normal force, the measured force related to contact area, is only reduced for the conical shaped profile. The unprofiled tool machines better edge qualities

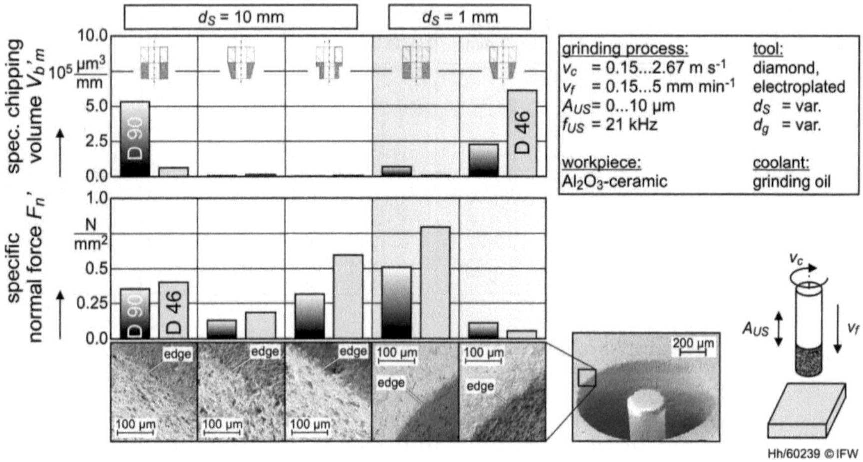

Fig. 14.16 Edge breakout in ultrasonic assisted grinding with profiled and standard tools

at 1 mm tool diameter than the 10 mm tools, as shown in the SEM-picture in the lower part of Fig. 14.16. The edge breakout is significantly increased for the conical shaped 1 mm tools even though the specific normal force is considerably decreased as for the 10 mm diameter tools. It is assumed that the increased edge breakout volume is the result of insufficient centering of the 1 mm grinding tools on the dressing machine. The eccentric profile leads to a tool deflection in grinding, which results in a hole diameter about 20% larger than the outer tool diameter and an instable grinding process. However, the increased edge quality for the profiled 10 mm tools and the same tendency in the normal grinding force for both tool diameters shows the potential of conical shaped tools to enhance the edge quality in ultrasonic-assisted grinding.

14.4 Conclusions

The workpiece surface quality is improved by peripheral surface grinding if the grinding wheel is tilted or ultrasonic assistance is applied parallel to the workpiece surface. The best surface quality is obtained by face grinding. ECDD was advanced for the microprofiling of multi-layered metal bonded grinding wheels. This process enables the dressing of complex multiple micro-profiles and undercut geometries on the grinding wheel. Several microstructures are machined in parallel by the use of multiple microprofiled grinding wheels. The dimensions of the microstructures are minimized by different grinding strategies. Grinding with undercut profiles enables the generation

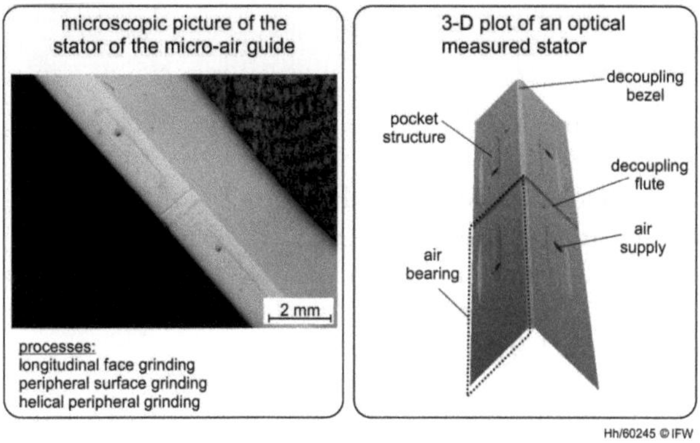

Fig. 14.17 Machined micro-air guide

of perpendicular microstructure flanks. Furthermore, a significant influence of the material properties and the microstructure quality was found.

Microgrinding with CVD-diamond microgrinding wheels is an appropriate method for microstructuring hard and brittle materials. In order to improve the machining quality when grinding these materials, new CVD-diamond microgrinding wheels were developed and the basic technological parameters were studied. The normal and tangential forces during grinding are lower than 1 N. Moreover, even by grinding the diffcult-to machine material aluminum oxide, they did not cause visible wear. Further advantages of these microgrinding wheels are low surface roughnesses at the groove base and marginal edge breakouts.

Boreholes in hard and brittle materials can be successfully machined with grinding pins. CVD-diamond grinding pins are able to increase the tool lifetime and the edge quality of the holes compared to electro-plated diamond grinding pins. Moreover, the tool geometry is an influencing parameter concerning the amount of chipping at the edges. The performed experiments show that conical dressed geometries reach the best grinding results. The edge quality of the holes can be enhanced with ultrasonic-assisted grinding if conical dressed tools are used.

The air-guide was manufactured with longitudinal face grinding, peripheral surface grinding and peripheral grinding (Fig. 14.17). All requirements are met by the advanced machining processes.

Acknowledgements The authors of this work wish to acknowledge the financial support of the German Research Foundation (DFG) within the Collaborative Research Center 516 "Design and Manufacturing of Active Micro Systems".

References

[1] Büttgenbach S (1994) Mikromechanik – Einführung in Technologie und Anwendung. B.G. Teubner, Stuttgart

[2] Denkena B, Friemuth T, Reichstein M (2003) Potentials of different process kinematics in micro grinding. Annals of the CIRP 52(1):463–466

[3] Denkena B, Reichstein M, Hahmann D (2006) Electro contact discharge dressing for micro grinding. In: Proceedings of the 6th International conference of the euspen, pp 92–95

[4] Denkena B, Hoffmeister HW, Hahmann D, Hlavac M (2007) Spanende Fertigungstechnik zur Herstellung von Mikrosystemen. Kolloquium Mikroproduktion, Forschungszentrum Karlsruhe

[5] Denkena B, Hoffmeister HW, Li J, Bütefisch S, Reichstein M, Hlavac M, Hahmann D (2007) Entwicklung und Fertigung eines aerostatischen Mikrolagers mit mehreren Düsen. wt Werkstatttechnik online 3:120–125

[6] Denkena B, Reichstein M, Hahmann D (2007) Dressing of filigree fine-grained metal bonded grinding wheels. Production Engineering – Research and Development 1(3):239–243

[7] Denkena B, Hoffmeister HW, Hahmann D, Hlavac M (2008) Machining of micro-systems. Microsystem Technologies 14(12):1909–1916

[8] Denkena B, de Leon L, Hahmann D (2008) Ultrasonic assisted grinding of brittle materials – surface topography and process forces. In: Proceedings of the 23rd Annual Meeting of the ASPE and the Twelfth ICPE, ASPE/ICPE, pp 384–387

[9] Denkena B, Hoffmeister HW, Hahmann D, Wittmer R (2009) Optimierte Werkzeuge treiben die Mikrosystemtechnik voran. Maschinenmarkt 39:56–60

[10] Denkena B, de Leon L, Hahmann D (2009) Dressing of micro profiles by wire electro discharge dressing. In: Proceedings of the 9th International conference of the euspen, pp 146–149

[11] Denkena B, de Leon L, Hahmann D (2010) Grossflächiges Schleifen von Mikrostrukturen. wt werkstattstechnik online 100(H. 6):487–493

[12] Denkena B, de Leon L, Hahmann D, Wang B (2010) Einsatz hochharter Schneidstoffe beim Schleifen. dihw – Diamant Hochleistungswerkzeuge 1/2010:36–43

[13] Denkena B, de Leon L, Wang B, Krawczyk T (2010) Manufacturing of riblet-structures by profile grinding with metal bonded wheels. In: Proceedings of the 10th International conference of the euspen, vol 2, pp 295–298

[14] Gäbler J, Schäfer L, Hoffmeister HW, Wenda A (1999) Development and application of cvd diamond micro tools for milling and grinding. In: Proceedings of the 1st International conference of the euspen, pp 434–437

[15] Hesselbach J, ed (2002) mikroPRO – Untersuchung zum internationalen Stand der Mikroproduktionstechnik. Vulkan-Verlag, Essen

[16] Hesselbach J, Hoffmeister HW, Hlavac M (2005) Micro-grinding – efficient technique for microstructuring hardened steels. Production Engineering – Research and Development 12(1):1–4
[17] Hoffmeister HW, Hlavac M (2006) New drilling tools for the manufacturing of micro-holes in hard and brittle materials. In: Proceedings of the 6th International conference of the euspen, pp 256–259
[18] Hoffmeister HW, Menze B (2004) Grinding and honing micro bores in hardened steels. In: Proceedings of the 4th International conference of the euspen, pp 205–206
[19] Hoffmeister HW, Wittmer R (2009) Grinding hard and brittle materials with cvd-diamond microgrinding wheels. American Society for Precision Engineering
[20] Hoffmeister HW, Wittmer R (2009) Verification of the process and wear behavior of cvd-diamond-microgrinding wheels by grinding hard and brittle materials. In: Proceedings of the 9th International conference of the euspen, pp 94–97
[21] Hoffmeister HW, Gäbler J, Menze B, Sachers M, Schäfer L (2003) Micro abrasive pencils with cvd diamond coating. Diamond and Related Materials 12(Issues 3–7):707–710
[22] Hoffmeister HW, Hlavac M, Gatzen HH, Morsbach C (2003) Mikrostrukturierung durch Schleifen. Abschlusskolloquium DFG-Schwerpunktprogramm 1012 pp 39–54
[23] Hoffmeister HW, Schäfer L, Gäbler J, Wittmer R (2010) Mikroschleifen mit CVD-Diamant-Schleifscheiben. Jahrbuch Schleifen, Honen, Läppen und Polieren pp 206–212
[24] Klocke F, Lung D, Thomaidis D, Antonoglou G (2004) Using ultra thin electrodes to produce micro-parts with wire-edm. Journal of Materials Processing Technology 149(1):579–584
[25] Lierse T (1998) Mechanische und thermische Wirkungen beim Schleifen keramischer Werkstoffe. VDI-Verlag, Hannover
[26] Meijer J, Du K, Gillner A, Hoffmann D, Kovalenko V, Masuzawa T, Ostendorf A, Poprawe R, Schulz W (2002) Laser machining by short and ultrashort pulses – state of the art and new opportunities in the age of the photons. Annals of the CIRP 51(2):531–550
[27] Okana K, Waida T, Suto T, Mizuno J, Kobayashi T (1993) Microgrinding of micromachine parts. Proceedings of the 1st International ABTEC Conference pp 100–105
[28] Reichstein M (2008) Schleifen miniaturisierter Bauteile. Dissertation Universität Hannover, Hannover
[29] Steinke ME, Kandlikar SG (2004) An experimental investigation of flow boiling characteristics of water in parallel microchannels. Journal of Heat Transfer 126(4):518–526
[30] Tönshoff H, Friemuth T, Reichstein M (2001) Ultra precision grinding of porous and non-porous super hard and brittle materials for compo-

nents of micro systems. Annals of the German Academic Society for Production Engineering 8(1):21–24

[31] Tönshoff H, Denkena B, Friemuth T, Reichstein M (2002) Precision grinding of components for active micro systems. Proceedings of the 3rd International Conference and 4th General Meeting of the euspen pp 803–806

[32] Tönshoff H, Denkena B, Friemuth T, Reichstein M (2003) Precision grinding of components for aerostatic micro guidance. Precision Engineering 27(2):185–188

[33] Uhlmann E (1998) Surface formation in creep feed grinding of advanced ceramics with and without ultrasonic assistance. Annals of the CIRP 47(1):249–252

[34] Uhlmann E, Piltz S, Doll U (2001) Electrical discharge grinding (edg) using microstructured disk electrodes. Production Engineering – Research and Development 8(1):25–30

[35] Uhlmann E, Piltz S, Oberschmidt D (2008) Machining of micro rotational parts by wire electrical discharge grinding. Production Engineering – Research and Development 2(3):227–233

[36] Weck M (2001) Ultraprecision machining of microcomponents. Proc International seminar on precision engineering and micro technology pp 113–122

[37] Weck M (2003) Mikromechanische Produktionstechnik. Abschlusskolloquium DFG-Schwerpunktprogramm 1012, Shaker-Verlag, Aachen

[38] Wenda A (2002) Schleifen von Mikrostrukturen in sprödharten Werkstoffen. Dissertation Technische Universität Braunschweig, Vulkan-Verlag, Essen

[39] Xie J, Tamaki J, Kubo A, Iyama T (2001) Application of electro-contact discharge dressing to a fine-grained diamond grinding wheel. Journal of the Japan Society for Precision Engineering 67(11):1844–1549

Part IV
Microassembly

Chapter 15
Size-Adapted Manipulation Robots for Microassembly

J. Ellwood, A. Burisch, K. Schöttler, G. Pokar, A. Raatz, J. Hesselbach

Institute of Machine Tools and Production Technology
Technische Universität Braunschweig
a.raatz@tu-braunschweig.de; j.hesselbach@tu-braunschweig.de

Abstract In order to assembly the microactuators which have been presented, unique microassembly techniques, robots, and peripheral sensors are needed. It has also been observed that an imbalance between the ever-decreasing product sizes to the size of currently implemented manipulators exists. These things as well as the large initial investment costs of such manipulators, have been the inspiration for new microassembly technologies. Here special emphasis is placed on two examples, one showing a size-adapted robot and the other extending this concept to a miniaturized design. Different considerations to obtain a repeatability on the order of 1 µm or better are presented. In the presentation of the size-adapted design, a special emphasis is placed on the integration of a three-dimensional vision sensor and resulting sensor-guided assembly control concept. The second robot discussed is a highly miniaturized robot designed for desktop factories. Here the high level of performance is obtained using miniaturized, backlash-free microgears. Further insight into the highly accurate microgears and their dynamic effects on the robot are covered.

15.1 Introduction

The goal of assembling parts which have dimensions of, or less than 1 µm with a comparable accuracy places unique challenges (also discussed in Chap. 22) on the assembly robots. As with almost all assembly tasks, the considered robots should at least be able to move freely in three-dimensional space (x, y, z) and have at least one rotational degree of freedom about the end effector (Ψ), referred to as *Schönflies* motion. Although conventional robots are able to achieve the required precision and *Schönflies* motion, there are several disadvantages which need to be addressed. Within the scope of this chapter, the size of these robots is seen as the main disadvantage, as it is not only wasteful with regards to energy consumption and have large initial overhead, but are also over dimensioned for the presented task.

This chapter thus presents a methodical approach to help describe a design process to create both a size-adapted robot and a robot which is further reduced in size, referred to as a miniaturized robot. After a brief presentation of the desired characteristics, a structural analysis is used to narrow the field of available robot structures. With the goal of increased assembly precision, the favorable stiffness offered by parallel structures, results in these being the only structures considered. In addition to saving space, these smaller robots have several inherent advantages which result in increased precision. For example, these smaller lengths are less sensitive to deflections which can result from temperature changes. During this phase, it is possible to take the required assembly area into consideration, and thus "adapt" the robots to these smaller tasks.

The next step is a sensitivity analysis, which is used to find and overcome limitations in the potential structures. One of the main optimization criterion is the resulting repeatability. Such a sensitivity analysis also allows the effects of different kinematic parameters to be explored. After the size-adapted structure has been selected and the lengths of the structure have been calculated, design considerations to allow the integration of additional sensors for assembly are presented.

Following the design phase, both the size-adapted robot, referred to as the $Micabo^{f2}$, and the miniaturized design, the *Parvus*, were created. A review of the important aspects of each of these robots is then presented, with repeatability measurements conducted according to the standard *ISO 9283*. For the $Micabo^{f2}$, a further discussion of the resulting assembly uncertainty is presented, with particular emphasis on how this error is improved through the incorporation of additional sensors. Here the integration of the vision sensor was a result of the team work between the Institute for Machine Tools and Production Technology (IWF) and the Institute of Production Measurement (IPROM). Due to the intricacies of the microgears, a more extensive look into how the sensitivity analysis is used to design the parallel planar structure ($\underline{R}RRR\underline{R}$) is shown. The precision of this robot is again found in accordance with the standard *ISO 9283*, with a further discussion of the unique transmission behavior of these gears.

There is a large conceptual basis which extends across both robot concepts. This includes the justification and considerations for a parallel robot, what a sensitivity analysis is and how it can be applied to the chosen structure, and ultimately how the precision of the resulting robots can be determined. A brief introduction to these topics and their applications is presented below.

Kinematic Considerations of Robot Structures

With precision being one of the main goals, some generalizations as to the range of acceptable robot structures can be made. One of the primary ways to differentiate robot structures is between those with open kinematic chains, serial robots, and those with closed kinematic chains, parallel robots [14].

Some of the classic differences are that parallel robots have redundant kinematic chains, which increase the structural stiffness in comparison with serial robots [7]. When properly designed, it is possible to greatly reduce the moved masses of parallel robots by placing most of the drives directly on the base frame. In theory, allowing for an increased maximum obtainable speed.

The advantages of parallel robots come at a cost, primarily in that parallel structures have a comparitively small workspace to footprint ratio. This limitation is compounded in that parallel robots have limited flexibility, and thus often designed for a given task. In order to get the best of both worlds, it is possible to take the favorable characteristics from both concepts to create hybrid structures. Such a combination could be achieved in that the main parallel structure gains a rotational degree of freedom (serial) about the end effector. Such structures can thus offer the precision gains from parallel robots and benefit from some of the dexterity of serial robots.

Kinematic Synthesis and Sensitivity Analysis

During the first stages of the kinematic synthesis, a broad range of acceptable hybrid structures can be considered. The complete selection process of the kinematic structure is outside the scope of this chapter, but is covered in more detail in [9]. From this, it can be seen that planar five bar structures are well suited for the desired application.

With the general structure, a further design criterion needs to be established which will enable the length parameters of the robot to be determined. For both the robots, this criteria includes the desired workspace, the theoretical maximal repeatability and the maximal obtainable speed. Although the workspace can be directly calculated and the obtainable speed can be found as a function of motor input, the resulting repeatability can only be estimated. To gain insight into the repeatability, a sensitivity analysis can be used to determine the effects that structural parameter changes will have on the resulting output.

In its simplest form, the sensitivity E is found through variations of the input Δx_I and through variations of the output Δx_O. In its simplest form, the sensitivity can be said to be

$$E = \frac{\Delta x_O}{\Delta x_I}.$$

Extending this to the design process for a size-adapted robot for microassembly, it is of interest to see how fluctuations of the lengths influence the end effector. Here, the forward kinematic problem (FKP), a set of transformation equations to calculate the end effector locations $\boldsymbol{\rho}$ as a function of the actuator locations $\mathbf{q}$, allow the sensitivity to be analyzed.

This leads to the structural sensitivity $\mathbf{J}_S$, which looks at the effect a deviation of the actuator input has on the end effector. Another important criterion is the dimensional tolerance $\mathbf{J}_{DT}$ of the end effector, which looks at how deviations of the length and angle parameters of the different structural members $\mathbf{l}$ effect the end effector. This insight allow a quantitative input into the resulting structure.

$$\mathbf{J}_S = \frac{\partial \rho}{\partial \mathbf{q}} \tag{15.1}$$

$$\mathbf{J}_{DT} = \frac{\partial \rho}{\partial \mathbf{l}} \tag{15.2}$$

Performance Calculations and Terminology

After each of the presented robots has been designed and constructed, a way of quantifying the robot performance needs to be established. As suggested by the international standard *ISO 9283*, the absolute accuracy of the pose (AP) and the repeatability or precision of the pose (RP) can be calculated by finding the difference between the desired and obtained pose for different points throughout the entire workspace [5]. "Multiple direction accuracy" and "path accuracy" are also terms used to describe the magnitude between the desired and obtained pose or path respectively.

15.2 Size-Adapted Robot for Microassembly

The main task of the larger of the two size-adaption concepts is the assembly of an active hybrid microsystem. As already stated, a multi stage design process is used to facilitate the designing of the structure, which is then followed by the creation of the robot. During the design phase the main design goal is to optimization the predicted obtainable repeatability of a specific task, here microassembly. A complete kinematic synthesis and sensitivity analysis of this robot has already been published in [9] and is presented in Sect. 15.3 for the second size-adapted robot.

With a defined structure and a general knowledge of which actuators and robot-dependent sensors are to be integrated, there are several constructional factors which need to be considered. Special emphasis is made on additional sensors, which are incorporated to gain more information about the parts being assembled, as well as the assembly scene. This is a key aspect of microassembly because a functional robot with a repeatability well below the desired assembly tolerances does not directly imply a lower assembly uncertainty. During the assembly task, the smallest of deviations can drastically effect the resulting assembled system. In order to overcome these inconsistencies, they first need to be identified and quantified during the assembly process. This is where the multitude of integrated sensors comes into play, as each is designed to detect a unique property. These sensors include the vision sensor, a laser sensor, as well as a force sensor between the gripper and the main robot structure, with these sensors being discussed in Sect. 15.2.2.

15.2.1 Robot Creation and Performance

Once the kinematic structure has been selected, special design considerations such as which sensors to incorporate as well as experience gained from similar structures need to be taken into account. Keeping the main task, microassembly, in mind, it is possible to make special design considerations which can help improve the resulting assembly precision. This is primarily with regard to the vision sensor, and its location. If the camera can be placed directly over the assembly scene, it is then possible to use a relative positioning strategy. This places some design constraints on both the vision sensor and robot structure. For the camera, this means that it needs to maintain a small design, which is discussed in Sect. 15.2.2. This effects the robot structure as it needs to be able to accommodate this unique sensor, and ideally allow an additional degree of freedom to overcome depth of field shortcomings. Additional insight was gained from the first model, the $Micabo^{f}$, in which carbon fiber arms were used. Testing showed that these thin arms could result in additional vibrations, and thus a stiffer arm design has been chosen in the redesign [12].

The parallel-hybrid robot $Micabo^{f2}$ has been constructed with the design considerations mentioned above and can be seen in Fig. 15.1. A unique modular head unit was designed, which enables both a serial degree of freedom to move the gripper, as well as an independent degree of freedom to allow the camera to focus. In addition to this, a force sensor has been integrated between the robot's gripper and the robot structure. A more detailed picture of this head unit can be seen in Fig. 17.1. This compact design with maximum flexibility is achieved through the redundant use of the linear guides which enable motion of both the camera axis and the gripper. To avoid collisions, additional limit sensors have been integrated, as discussed in [12]. As shown in Table 15.1, the overall design considerations have enabled a worst-case repeatability of 0.6 µm in the xy-plane, according to the standard *ISO 9283*. Although it is possible to increase the obtainable maximum speed of the

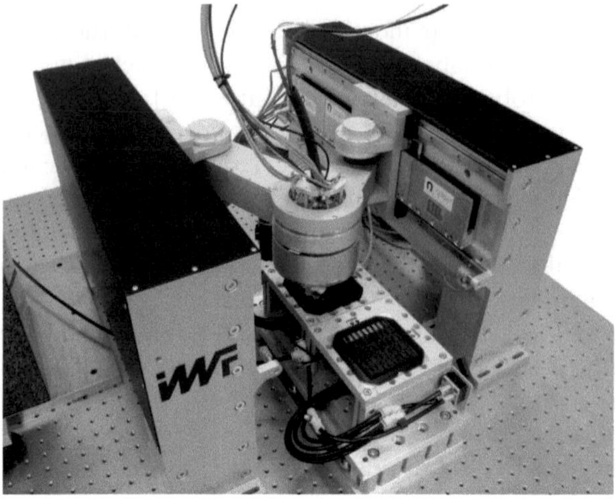

Fig. 15.1 The precision assembly robot $Micabo^{f2}$

robot through very powerful amplifiers, the small distances between parts will result in a minimal improvement in the assembly cycle time.

Table 15.1 $Micabo^{f2}$ technical specifications

Criterion	Value	Unit
Workspace (x,y,z, absolute)	$160 \times 400 \times 15$	mm^3
Footprint	500×600	mm^2
Resolution (max., xyz)	0.1	μm
Resolution (max., *focus*)	0.4	μm
Repeatability (best, worst)	0.5, 0.6	μm
Linear speed (max.,xy)	100	$mm\,s^{-1}$
Linear speed (max.,z)	4.28	$mm\,s^{-1}$
Rotational speed (Ψ-axis)	50	rpm
Angular resolution (Ψ-axis)	0.02	°
Payload	50	g

15.2.2 Additional Sensors and their Integration

Although the robot has impressive repeatability, this does not directly imply a comparable assembly uncertainty. The primary cause of the increased error of the assembly can be such things as small deviations in the function structure used by the vision sensor, part differences which can arise during manufacturing, as well as the placement tolerances of the parts within the assembly area. To overcome these differences, it is necessary to incorporate sensors into the robot which allow their identification and quantification. These sensors are referred to as "additional sensors", as they are not needed for the robot to function, but allow it to gain additional information about the assembly scene.

In order for enable the robot to react to all the additional sensor information, the control strategy needs to handle this information in a meaningful way. The chosen control strategy can be broken down into many different sub-controllers within a large cascade control loop, which has been achieved using a modular *dSPACE* system. Further reading on the details of the controller system can be found in [6]. The complexity of the low-level actuator controllers extends beyond the scope of this chapter, but is extensively covered in such texts as [11]. Of primary interest here is how the different sensors are integrated and used to modify and correct the robot's position.

Vision Sensor The most important of the additional sensors is the vision system, which is used to generate the relative positioning vector used to join the parts. A relative positioning strategy is favored over an absolute strategy, as it allows both parts to be seen at the same time. This not only reduces and simplifies the calibration requirements but will inherently improve the generated positioning vector.

The robot now needs to react to the quantifiable error reported by the vision sensor. Here a look-&-move strategy is favored over a visual servoing technique. The primary advantage of the chosen method is it's ability to decouple the vision system from the control loop. With visual servoing, the camera system is integrated directly into the control loop, which can complicate the stability of the resulting robot controller. As the vision system is not directly incorporated in the robot, an additional communication protocol is required. The chosen protocol can influence the total assembly cycle time.

In Process Laser Scanning Although the camera system is able to provide 3D information about the assembly scenario, its precision and accuracy are highly dependent on how well and how many reference marks can be seen. As with 2D cameras, the integration of a laser system can be used to gain information about the third dimension. In order to explore these ideas, in process laser scanning (IPLS) has been explored. To accomplish this, a laser sensor has been integrated into the head unit of a robot. This was then used to detect the outline and plane of a part placed randomly within a predefined area of the robot. A visualization of the resulting data can be seen in Fig. 15.2. The knowledge of the part plane can then be used to correct the angle of the part, using a tip-tilt table. Additional information about the parts location can be gained by looking at where the edges of the part are, ultimately allowing the parts relationship to the robot to be found. A further discussion of the incorporation of the laser system and its advantages during the assembly process is given in [10].

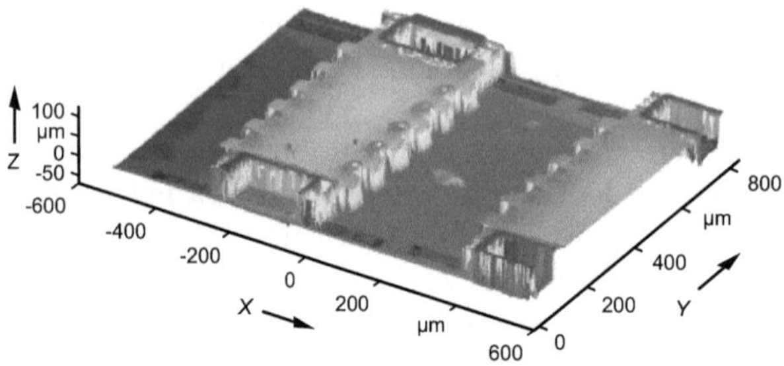

Fig. 15.2 An IPLS scan of a micropart

Force Sensor In addition to the vision and laser sensors, there is a force sensor which has been integrated within the wrist of the robot. This enables the robot to place a part with a predefined force, record the force during an assembly process, as well as act as preventing a maximum force from being reached. The wrist, or between the gripper and the structure, is advantageous as it is close to the end effector of the robot. With the physical integration of the force sensor within the robot, it is important to consider what information this sensor is giving. Here the force sensor is not located directly above where the contact point where the force is occurring. This results in an additional moment which needs to be considered. Knowing the dimensions of the gripper as well as its relation to the force sensor, allows a force and moment balance routine to find the forces acting on the gripper.

The integration of the force sensor should allow a part to be placed with a desired force. Again it is important to consider how this information should be handled by the control loop. As the force signal of the controller is an analog signal with a high frequency, it is integrated directly into the vertical control loop. The implemented cascade controller can be seen in Fig. 15.3. Within this controller, the desired position P_D can be modified by the force correction P_C, which is then compared against the actual position value P_A.

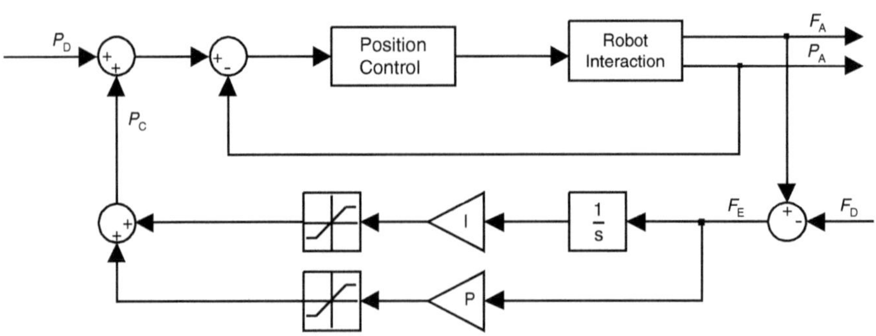

Fig. 15.3 The implemented position-force controller

The resulting value is then sent to the robot's position control system. The force correction term F_E is found from the difference between the actual force F_A and the desired or maximum force F_D. Special considerations are made so that the controller will only alter the vertical movement of the gripper if the interaction force is above a desired or maximum level, acting in a compliant manner. This compliance is achieved through a non-linearity, which can be seen after the proportional and integral gain terms.

As each of the three sensors give information about a different aspect of the assembly, it is possible to use them in a serial manner. For example, it is possible to use laser sensor to identify where parts are within a predefined area and to correct the xy-plane of the substrate. With the corrected part plane, it is then possible to pick a part and move it above the substrate. As both the handled part and substrate can be seen at the same time, a relative look-&-move strategy can then be used to correct the part errors. With information from the 3D vision system and the IPLS, it is then possible to place the part. If desired, the force between the substrate and part can be set to a fixed value.

15.2.3 Assembly Uncertainty

After the harmonization of the sensors within the $Micabo^{f2}$, parts made especially for demonstrative purposes were created and then assembled. These two demonstration parts are based on original active system components, and only have the reference marks. The handled guides with the dimensions 8.4 mm by 1 mm are placed on the rectangular substrate with dimensions of $3.92 \times 10.66\,\text{mm}^2$. Optimization of the assembly scenario and further optimizations, such as improving the visibility of the parts in the camera, have allowed the obtainable 3-sigma assembly uncertainty of less then 10 µm [4]. Although this is still more than a factor of ten greater then the repeatability of the robot, it represents a great improvement from the original assembly results. It was additionally found that the orientation of the camera to the parts being assembled can drastically effect the resulting assembly uncertainty. This was shown in [4], in which the gripper-to-camera orientation was changed to allow both sides of the handled part to be seen.

15.3 Miniaturized Robot for Desktop Factories

Size-adapted as well as miniaturized robot structures are well suitable for the use in desktop factory applications that will be later described in detail in Chap. 22. A first functional model of such a miniaturized robot prepared for use in a desktop factory, its structural design and associated technical challenges are all described below.

15.3.1 Concept of the Robot Structure

In the last few years, the development of miniaturized handling systems has been made possible due to the availability of new miniaturized drives, gears and other equipment which are described in Chap. 22. Based on the new equipment, it is possible to realize completely new specifications for miniaturized robots. To find a suitable robot structure, the basic specifications have to be determined and quantitative specifications have to be fixed (see Table 15.2). The aim is to find a simple modular structure with a small envelope and easy access to the working area of the robot based on the initially determined kinematic synthesis. The robot has to have the previously mentioned four degrees of freedom of *Schönflies* motion [1].

Table 15.2 Requirements for the robot structure

Criterion	Correlation	Value
Workspace	=	credit-card-sized ($80 \times 50\,\text{mm}^2$)
Footprint (area of robot base)	<	DIN-A4 ($210 \times 297\,\text{mm}^2$)
Theoretic Resolution	<	$1\,\mu\text{m}$
Linear speed	>	$100\,\text{mm}\,\text{s}^{-1}$

A pre-selection for a possible kinematic structure in the xy-plane of the robot was identified by taking a closer look at conventional industrial robots. In this principle design phase, two different plane kinematic structures were compared: the serial SCARA structure and the parallel five-bar linkage structure. A detailed positioning sensitivity analysis at all points in the workspace of the conceptual designs helps to determine whether the serial or the parallel structure should be used. The structural analysis described in Sect. 15.1 was done for the possible robot structures and gives information about their sensitivity and expected repeatability. In contrast to the sensitivity, the repeatability of a robot structure cannot be calculated, only estimated. Based on several estimated interference factors, the results of the sensitivity of the structure and the sensitivity of dimension tolerance were simulated and served in combination as qualitative information about the precision of the robot structure. The iterative analysis of different geometrical parameters is aimed at optimizing arm length by keeping a credit-card-sized workspace, high accuracy and a theoretical resolution of up to $1\,\mu\text{m}$. The sensitivity plots of the optimized geometrical structures for serial and parallel kinematics that meet those requirements respectively, are shown in Fig. 15.4. The sensitivity plot of the upper half of the workspace of a miniaturized serial SCARA robot is shown in Fig. 15.4.a. The sensitivity plot of a miniaturized parallel five-bar linkage with similar arm length as the serial SCARA is illustrated in Fig. 15.4.b. The sensitivity map of the serial structure only shows good values in very limited areas of the available workspace. The parallel structure,

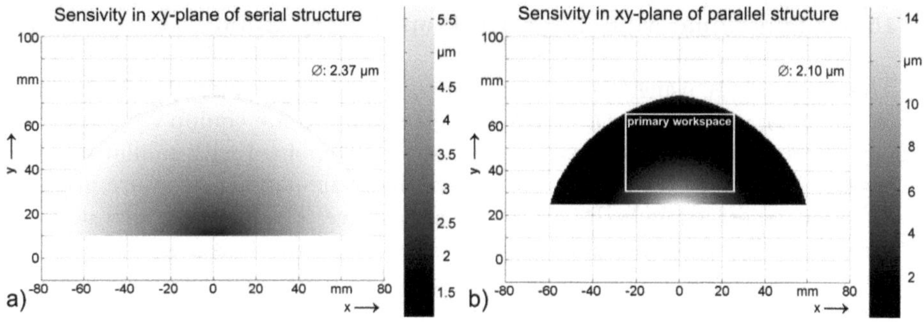

Fig. 15.4 Sensitivity plot: (**a**) Serial structure; (**b**) Parallel structure

in comparison, achieves good values over almost the complete workspace and also has a symmetrical sensitivity map. The theoretical repeatability of this parallel structure, based on accuracy data of the drive components (angular repeatability [13]) and ball bearings (radial deflection), prospects a Cartesian repeatability of 2–6 µm within the primary used workspace (see Fig. 15.4.b). Furthermore, the parallel structure offers a significantly better dynamic performance, since only the gear motor for the rotational Ψ-axis is carried by the moving arm. Additionally, the passive joints of the parallel structure are easier to miniaturize than active joints. For these reasons, parallel structures such as the five-bar linkage are well-suited for miniaturized robot designs [1].

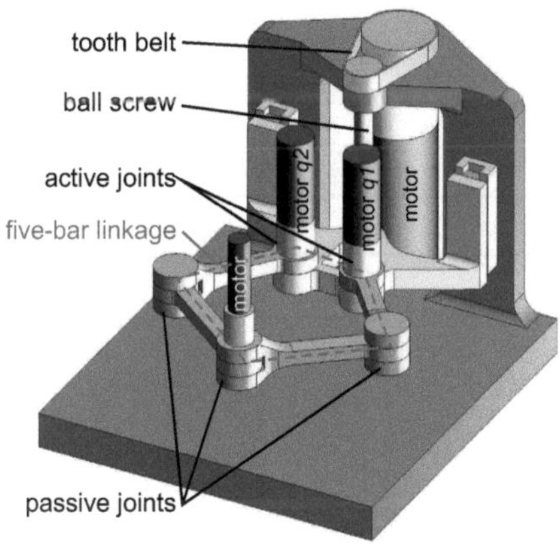

Fig. 15.5 Design study of the *Parvus*

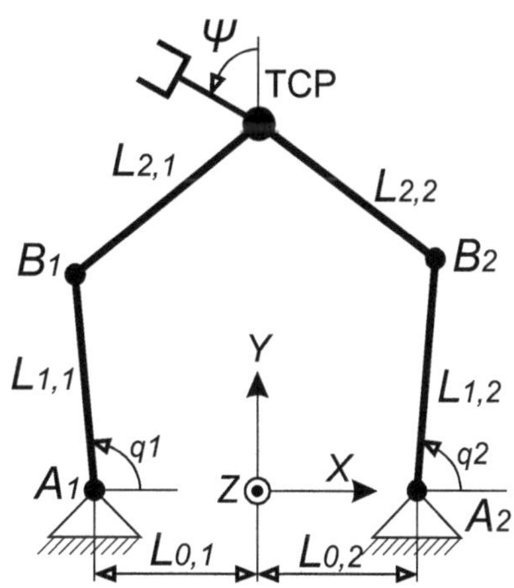

Fig. 15.6 Parallel structure

15.3.2 Functional Model of a Desktop Factory Robot

The design study using the parallel structure as the basic structure is shown in Fig. 15.5 and 15.6. Both active joints *A1* and *A2* are equipped with *Micro Harmonic Drive* gearboxes (see Chap. 22) combined with miniaturized electrical motors (*q1* and *q2*). They drive the plane-parallel structure in xy-direction, whereas joints *B1*, *B2* and C are passive. The plane-parallel structure offers two translational DOF in the xy-plane. The z-axis is integrated as a serial axis in the base frame of the robot. The easy handling of the whole plane-parallel structure driven in the z-direction is possible due to its minimized drive components and light aluminum alloy structure. The z-axis is driven by a *Harmonic Drive* servo drive combined with a precision ball screw. Additionally, the rotational hand axis Ψ was designed as a hollow rotational shaft integrated in the passive joint C as the Tool Center Point (TCP) of the parallel structure. This allows media such as vacuum or pressure to be passed along the hand axis. A *Micro Harmonic Drive* gearbox combined with a miniaturized electrical motor turns the Ψ-axis. This axis with a diameter of 2.5 mm can be equipped with several handling tools such as vacuum grippers. All joints are nearly free of backlash and have low friction due to preloaded angular contact ball bearings. All servomotors support a high resolution encoder feedback signal. The drives (*q1*, *q2*) of the parallel structure are equipped with an encoder (resolution of 256 increments) and a microgear (ratio of 500:1). That results in an angular resolution of the robot arm of 0.0028°. Additionally, the *Parvus* is equipped with two magneto-resistive position sensors for the initialization of the robot as well as an emergency stop function. The first prototype of *Parvus*, seen in Fig. 15.7, fulfills the following specifications shown in Table 15.3.

Table 15.3 Technical specifications of the *Parvus*

Criterion	Value	Unit
Workspace (xy, absolute)	4658	mm^2
Workspace (max., cubical)	$60 \times 45 \times 20$	mm^3
Footprint	100×53	mm^2
Robot cell	130×170	mm^2
Resolution (max., xy-plane)	< 1	µm
Repeatability (best, worst)	3.6, 5.7	µm
Linear speed	> 100	$mm\,s^{-1}$
Rotational speed (Ψ-axis)	187^a / 60^b	rpm
Angular resolution (Ψ-axis)	0.022^a / 0.007^b	°
Payload	50	g

[a]ratio 160:1 [b]ratio 500:1

Fig. 15.7 The functional model of the miniaturized robot *Parvus*

15.3.3 Analyses of the Robot Structure

To get information about the robot specifications, the precision characteristics mentioned in Sect. 15.1 have to be measured at the robot's end effector. Fig. 15.8 shows the primary and secondary workspace of the *Parvus*. The primary workspace is the used workspace for assembly tasks and therefore investigated concerning its characteristics meeting the *ISO 9283* standard [5].

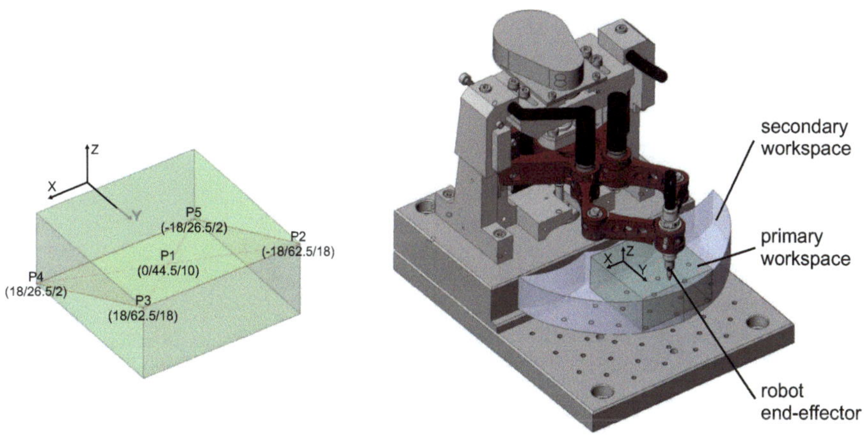

Fig. 15.8 Workspace of the *Parvus* and measuring points

Repeatability. The repeatability of the robot's end effector was measured within the primary workspace, as shown in Fig. 15.8. The first measurements of the *Parvus* equipped with the first version of the microgears showed a repeatability of 14 µm (worst value within the workspace at $3\,\sigma$). This relatively high value could have been improved by optimizing the stiffness of the microgears (see Chap. 22.2), that makes the gears less sensitive to disturbing frictional moments of the robot kinematics. Therefore a repeatability of better than 5.7 µm (worst value, at measuring point P3, at $3\,\sigma$) can be guaranteed. This measured value is in the range of the expected repeatability of 2–6 µm, as presented in Sect. 15.3.2. As stiffness of the microgears is important for the repeatability of the robot, it is also very important for the multiple direction accuracy.

Multiple Direction Accuracy. The multiple direction accuracy is greatly influenced by backlash and hysteresis in the robot drives and other stiffness-related factors. The accuracy has been measured within the primary workspace (Fig. 15.8) at points *P1*, *P2* and *P5*. These values are 10–30 times higher than the values of the repeatability. Investigations have been performed to determine why there is such a difference between the multiple direction accuracy of the robot and the repeatability, even though the microgears are backlash-free. In this case, the discrepancy is caused by the relatively low stiffness of 6.13 Nm rad^{-1} [8] of the microgears compared to the dimensions of the robot structure. The microgears have no backlash when they are unloaded; however, the *Parvus* structure contains preloaded ball bearings and is connected directly to the microgears. These ball bearings cause frictional torque in both the robot structure and the microgears, which stops the gear system before it can reach its neutral position. This phenomenon occurs when approaching from either the left or the right side, and it results in an elastic hysteresis when the robot structure is combined with the microgears. If this assumption is correct, placing an additional preload in the robot structure would reduce this effect and improve the multiple direction accuracy. Fig. 15.9 shows the results from an experiment that was carried out to prove this assumption. In the experiment, there is a spring force F_{spring} applied between the pas-

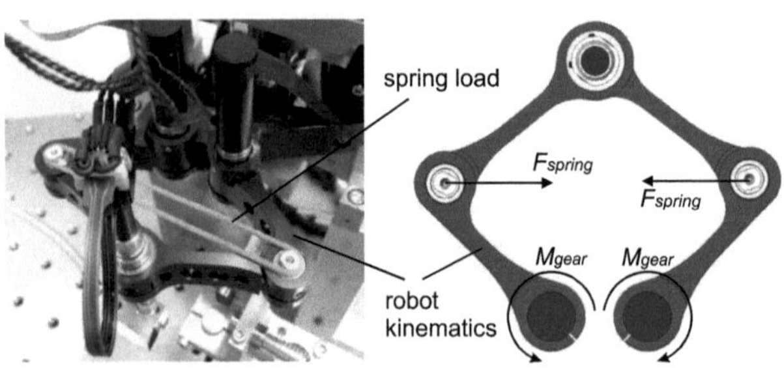

Fig. 15.9 Robot kinematics with spring load and reaction torque

sive joints of the robot structure, which induces a reactive torque M_{gear} in the microgears. This torque is higher than the frictional torques of the robot structure, and it is therefore able to deflect the gear system beyond its neutral position into a more stable state in both directions. The results, illustrated in detail in [3], verify the aforementioned assumption and show that the applied force can improve the multiple direction accuracy by a factor of up to 3, in this case from 55 µm to 18 µm. The experiment with the spring is actually possible to perform in only a very small area of the workspace. In order to implement a stable preload on the robot drives without limiting the amount of workspace, the torque has to be applied directly to the drives. In this case, a torsion spring or an active torque generator must be connected to the output of the microgears.

Path Accuracy. Aside from stiffness-related influences, there are influences on the microgears having to do with the characteristics of the robot's kinematics. In general, *Harmonic Drive* gears show special behavior in kinematic error and transmission compliance, as shown in Fig. 15.10.a [15]. Due to the high grade of miniaturization of the *Micro Harmonic Drive* gears and the robot, these effects greatly influence the path accuracy of the entire robot, as shown in Fig. 15.10.b. In most cases, the path accuracy is only important for machine tools, but not for pick-and-place robots; however, in some cases a deviation from the set path must be considered, as illustrated by the assembly task in Chap. 22.

To characterize the path accuracy of the robot, a sample path has been measured with a 3D laser tracker. Fig. 15.11 shows the results of the measured path accuracy (a) and the experimental setup (b). The actual path of the *Parvus* varies periodically with a maximum deviation of 100 µm from the expected linear path. This behavior is obviously caused by the transmission error of the microgears. The transmission error inside the gear system has kinematic and dynamic components. The kinematic effect can be illustrated by a measurement of the path accuracy at very low speed, showing cyclic deflections. At higher speed, however, this kinematic effect initiates higher

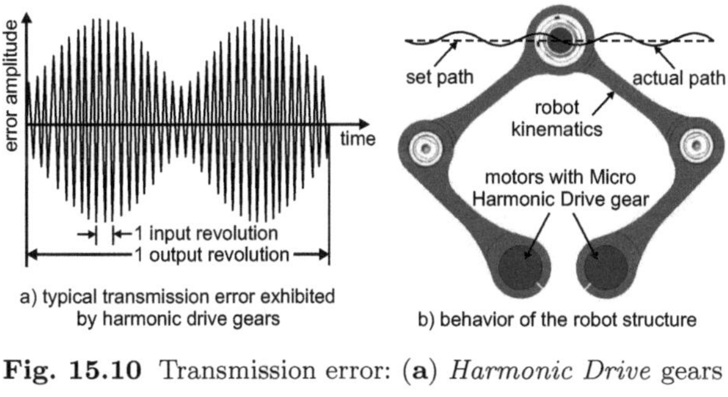

Fig. 15.10 Transmission error: (**a**) *Harmonic Drive* gears [15]; (**b**) Robot structure

vibrational deflections that rely on the dynamic behavior of the microgears and the robot structure.

A typical approach for reducing dynamic effects in robot control is the Computed Torque Feed Forward (CTFF) control strategy. To illustrate the potential of this strategy, an inverse dynamic model of the *Parvus* structure was calculated [2]. The simulation of the necessary driving torques for the robot structure showed that the necessary driving torque of the motors (precalculated by CTFF) was much lower than the frictional torque of the microgears and would therefore not be effective. Due to the miniaturization of the structure, the drives and the high transmission ratio of the microgears, CTFF is not applicable in this case.

A much more promising approach is to reduce or directly compensate the kinematic transmission error of the microgears that initiates the vibrations of the robot structure. A model of the transmission errors and elasticity characteristics of the microgears can thus help to optimize the kinematic and dynamic behavior of the *Parvus*. A possible approach is to measure the transmission error at the gear output using a small angular sensor, presented in Chap. 22. This allows information to be obtained about the driving angle during operation and generating a correction function. With this data the control can reduce (compensate) the transmission error of the gears by more than 50% and thus improve the path accuracy of the entire robot.

Absolute Accuracy. The results of the absolute accuracy measurement of the robot within the primary workspace are presented in Fig. 15.12.a. The regular grid demonstrates that the structure behaves as theoretically predicted. Before mounting the robot, the effective arm lengths were measured and considered within the control algorithm. Due to the reduced size of the robot, the dimensions of the structure can be measured with a microscope or coordinate measuring machine, and thus a calibration of the assembled structure is not necessary. The deviations from the set points to the measured points are the same magnitude (100–300μm) as the deflections of the path accuracy. The combination of both measurements illustrates that the path accuracy curve matches the measured absolute accuracy (see Fig. 15.12.b).

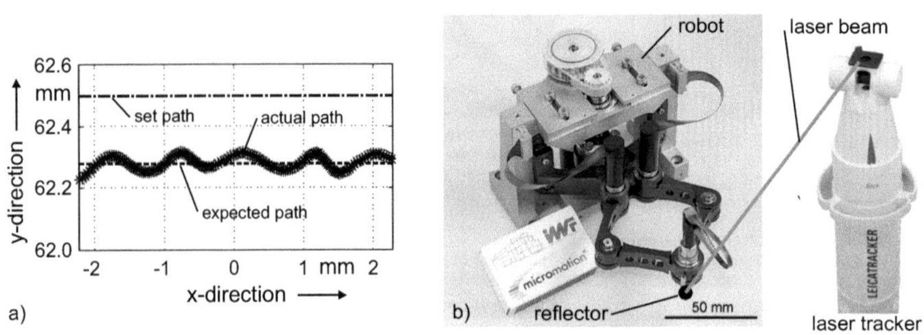

Fig. 15.11 Path accuracy: (**a**) Measurement results of *Parvus*; (**b**) Measuring setup

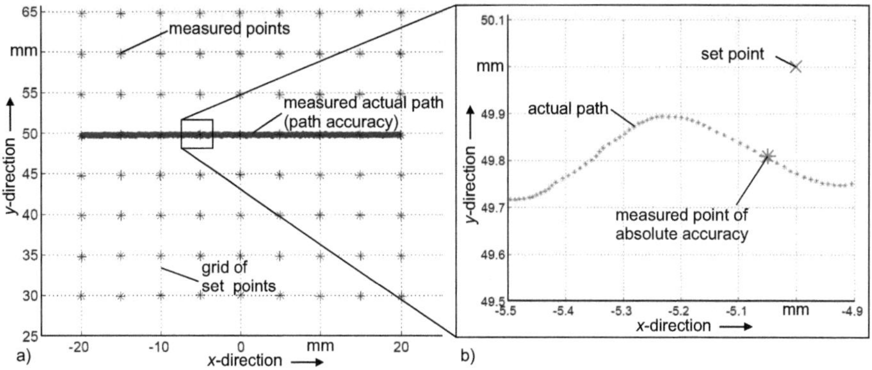

Fig. 15.12 Absolute accuracy and combination with path accuracy

This illustrates that the absolute accuracy of the robot is influenced by the transmission error of the microgears, as already identified for the path accuracy. Thus, for an optimization of such miniaturized robot structures, a calibration of the transmission errors of the drives is much more promising than a calibration of the robot structure.

15.4 Conclusion

In this chapter a methodological approach for the construction of two different microassembly robots has been presented. Theoretical considerations such as structural and dimensional tolerance sensitivity analysis have been discussed and then applied to the design of size adapted robot concepts. With knowledge of the general structure, other factors such as additional sensor integration as well as structural stiffness were then shown with the $Micabo^{f2}$. This basis has then been used to complete the construction of both robots and is followed by the different realized performances of the robot. A further discussion of the additional sensors found within the $Micabo^{f2}$ have been presented, along with how they are used to improve the assembly uncertainty. It has then been shown that the assembly uncertainty of an assembled demonstrative system is less then 10 μm.

For the *Parvus*, a further analysis of the robot specifications and adverse side effects of the microgears has been discussed. The robot analysis concluded that for the improvement of the precision characteristics of the robot, an optimization of the stiffness and a calibration of the transmission behavior of the microgears are necessary. It is foreseen that this transmission error can be determined and used to create a correction function. Based on the above presented analyses, enhancements of this functional model aimed to develop a new prototype of the *Parvus* for industrial use, described in Chap. 22.

Acknowledgements The authors gratefully acknowledge the funding of the reported work by the German Research Foundation within the Collaborative Research Center 516 "Design and Manufacturing of Active Micro Systems". They also thank the cooperation partner *Micromotion GmbH* for supporting the development of the *Parvus* robot.

References

[1] Burisch A, Wrege J, Raatz A, Hesselbach J, Degen R (2007) Parvus – miniaturised robot for improved flexibility in micro production. Assembly Automation 27(1):65–73
[2] Burisch A, Drewenings S, Ellwood RJ, Raatz A, Pisla D (2010) Analysis and inverse dynamic model of a miniaturized robot structure. In: Proc. of 3-rd European Conference on Mechanism Science, pp 117–124
[3] Burisch A, Raatz A, Hesselbach J (2010) Challenges of precision assembly with a miniaturized robot. In: Precision Assembly Technologies and Systems, Springer, pp 227–234
[4] Ellwood RJ, Raatz A, Hesselbach J (2010) Vision and force sensing to decrease assembly uncertainty. Springer, Berlin, vol 315, pp 123–130
[5] EN ISO 9283 (April 1998) Manipulating industrial robots – performance criteria and related test methods
[6] Heuer K, Pokar G, Hesselbach J (2003) Open architecture robot control based on Matlab/Simulink and a dSPACE real time system. In: Sensors and Controls for Intelligent Manufacturing III, Proc. of SPIE, Providence, USA, vol 5263, pp 1–9
[7] Merlet JP (2006) Parallel robots, Solid mechanics and its applications, vol 74, 2nd edn. Kluwer Academic Publishers, Dordrecht , Boston MA
[8] Micromotion GmbH (2009) Think Smaller. Catalogue
[9] Pokar G (2004) Untersuchung zum Einsatz von ebenen Parallelrobotern in der Mikromontage. Dissertation, TU Braunschweig
[10] Rathmann S, Wrege J, Schöttler K, Raatz A, Hesselbach J (2006) Sensor guided micro assembly by using laser-scanning technology. International Federation for Information Processing, Boston, MA, vol 198
[11] Sciavicco L, Siciliano B (2000) Modelling and Control of Robot Manipulators, 2nd edn. Springer-Verlag, London Ltd
[12] Simnofske M, Schöttler K, Hesselbach J (2005) micaboF2 – robot for micro assembly. In: WGP (ed) Production Engineering, vol XII/2, WGP, pp 215–218
[13] Slatter R, Degen I (2005) Miniature zero-backlash gears and actuators for precision positioning applications. In: Proc. of 11th European Space Mechanisms and Tribology Symposium ESMATS, pp 9–16
[14] Tsai LW (1999) Robot Analysis. John Wiley & Sons, Inc., New York
[15] Tuttle T (1992) Understanding and modeling the behavior of a harmonic drive gear transmission. Massachusetts Institute of Technology Cambridge, MA, USA

Chapter 16
Tools for Handling and Assembling of Microparts

B. Hoxhold, J. Wrege, S. Bütefisch, A. Burisch, A. Raatz[2], J. Hesselbach[2], S. Büttgenbach[1]

[1] *Institute for Microtechnology*
Technische Universität Braunschweig
s.buettgenbach@tu-braunschweig.de

[2] *Institute of Machine Tools and Production Technology*
Technische Universität Braunschweig
a.raatz@tu-braunschweig.de; j.hesselbach@tu-braunschweig.de

Abstract In the last several years many improvements of tools for handling and assembling of microparts (between 10 and 500 µm) have been made. This chapter will discuss the development and realization of electrostatic micro-grippers and magazine structures and mechanical microgrippers as well as their performance. The gripping forces and quick discharging techniques of the electrostatic devices were optimized. Concerning the mechanical gripping structures, several actuation principles, such as shape memory alloys, pneumatic drives and thermal expansion actuators, were optimized and integrated into flexible gripper gearings. These grippers and magazine structures, as well as pneumatically driven auxiliary microtools (such as centrifugal microfeeders and active clamping devices), can be used as supplement handling devices for desktop factories.

16.1 Introduction

Due to the continuous progress in the field of microelectronics and micromechanics, components of hybrid microsystems (e.g. microlenses, optical fibers and microtubes) decrease in size and often become highly fragile. The production of these systems sets high requirements and low tolerances to the microassembly process. These accuracies can hardly be achieved by conventional assembly equipment. Different assembling situations require a flexibility concerning size adapted gripper mechanics and gripping jaws, suitable material combinations and capable actuation mechanisms. Therefore, several kinds of microgrippers have been developed to allow handling and assembling of the tiny component parts. Figure 16.1 presents the general overview of the realized grippers and auxiliary microtools that are described in the following chapters.

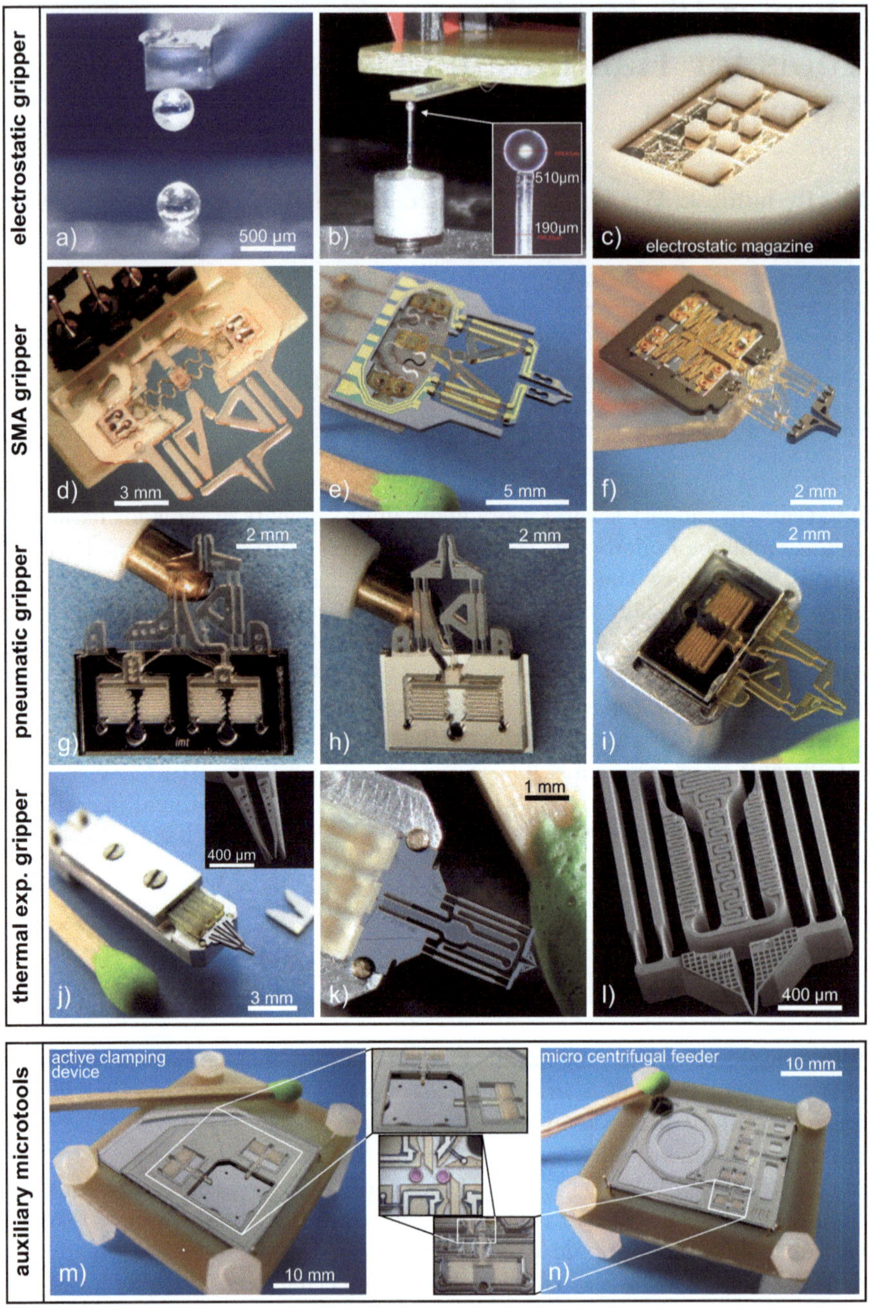

Fig. 16.1 General overview of the different grippers and auxiliary microtools.

16.2 Electrostatic Forces in Microhandling Processes

The development of microhandling devices requires a basic understanding of adhesion forces. Van-der-Waals, capillary or electrostatic forces are generally undesirable physical effects in a microassembly process. If these forces are not entirely controlled, microparts tend to stick to grippers, storage or chucking devices. Therefore, numerous researchers have suggested strategies to reduce adhesion forces during microhandling processes [6]. In addition to the proper selection of the mated materials for avoiding contact electrification effects, the contact areas should be minimized by using rough surfaces, hard materials and minimal grip forces. These measures are suitable to avoid unintentional Van-der-Waals forces, which act over very short atomic scale ranges. Capillary forces are comparably easy to reduce using dry atmospheric conditions. Therefore, a normal air conditioning system should be sufficient. Electrostatic forces seem to be manageable using conducting materials on grounded potential, but even very thin oxidized surfaces can carry significant surface charges. Such charged surfaces can cause forces which easily exceed the weight of microparts and act over relatively long distances [17]. Especially while working under normal atmospheric conditions, small parts can be observed jumping towards a tweezer tip more than several hundred micrometers.

Today, electrostatic forces are used in grasping tools, chucks or part feeders. The works described below focus on electrostatic gripper and magazine techniques.

The electrostatic microgrippers found in literature consist of simple electrode configurations, which generate electric fields. Homogeneous field distribution is sufficient to generate an electrostatic gripping force in the case that conducting materials have to be handled. Here, the forces are created by electrostatic induction, whereas inhomogenous fields are needed for handling insulating materials. The gradient of the electric field and the ratio between the dielectric constants of the material itself and the surrounding medium determine the dielectrophoretic forces acting on a dielectric body [16].

Electrostatic microgrippers described in literature [4, 5, 14, 15], can be classified by the electrode design, either unipolar or bipolar.

16.2.1 Centering Electrostatic Microgripper

The development of electrostatic handling devices at SFB 516 started with the design of a centering electrostatic microgripper. The design transfers the electrode configuration, recommended by Oh, into a planar design [8, 9, 17]. Concentric electrode designs were manufactured out of a sputter deposited gold layer on a 500 µm *Pyrex* substrate.

Altogether, five different shape variants were designed for further research. Variants (a), (b) and (c) in Fig. 16.2 were adapted to the shape of micro-

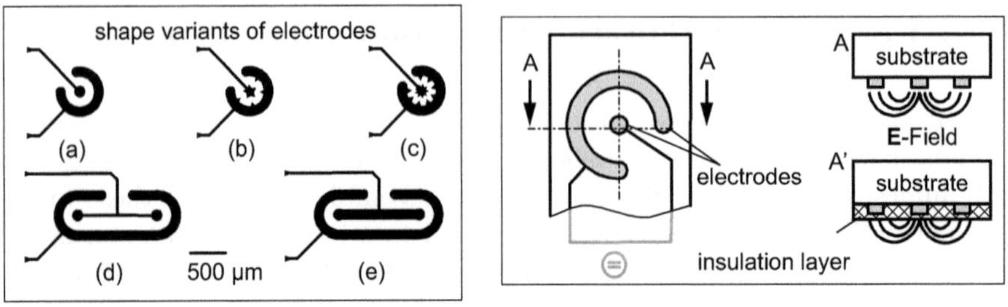

Fig. 16.2 Variants of planar concentric ring electrodes

spheres (diameter $d = 400\,\mu\text{m}$) where shape variants (b) and (c) had a different number of sharp edges. These electrode contours led to highly increased field strengths close to the edges. Generally, the inner electrode was dimensioned with 120 µm diameter and the distance between inner and outer electrodes was about 140 µm. Test grippers covered with 600 nm silicon oxide or 260 nm aluminum oxide as a dielectric insulation layer also were investigated.

Attainable electrostatic forces can be calculated with the finite element method (FEM). Fig. 16.3 shows the three-dimensional FEM Model of shape variant (c). A ruby sphere with relative permittivity $\epsilon_r = 11$ is placed in the center under the gripper electrodes. Boundary conditions of the model are the voltages applied to the gripper electrodes and, if needed, additional charge on the surface of the sphere. These models are used to calculate electrostatic forces acting on the ruby sphere in z-direction. Simulated forces can be plotted over the distance between the surfaces of the handling object and the gripper electrodes. For distances smaller than 26 µm, the curves presented in Fig. 16.3 exceed the calculated weight of the sphere. Therefore, it can be assumed that the ruby sphere can be picked up in all cases. The curves for the different shape variants differ only marginally, whereas only small amounts of surface charges on the handling object can increase the simulated forces significantly.

Test Rig for Electrostatic Force Measurement

Further research on electrostatic gripper forces was done with a test rig offering the capability to measure forces and distances in the micro-range. Three high accuracy translational axes were used as a 3 degrees-of-freedom (DOF) robot system for handling tests and additionally as measuring axes. The end effector of the robot carried either gripper electrodes, probes (dielectric spheres), charging bars or a camera. Gripper electrodes were driven by a DC high voltage (HV) source whose output polarity could be changed by a pole converter. An h-bridge circuit was developed and used to control the voltage of the gripper electrodes. This circuit achieved shorter discharge times compared to the simple shutdown of the HV source. In discharge mode, the

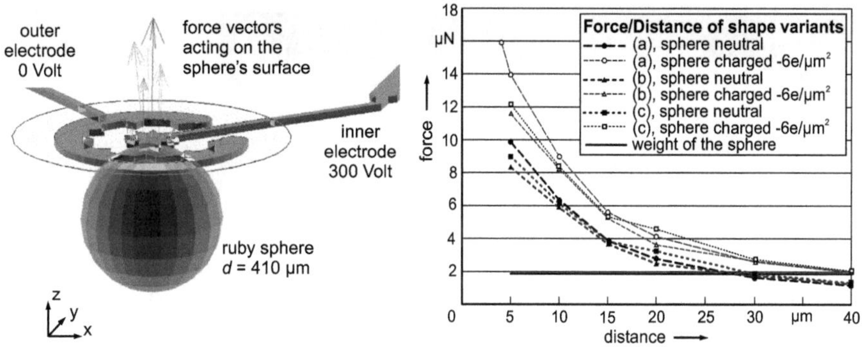

Fig. 16.3 FEM Model and simulated electrostatic forces

gripper electrodes were shorted to earth potential. Surface charges could be applied to specimens using customary charging generators with positive or negative DC polarity. For neutralization of surface charges, a 5 kV AC power unit with a single point electrode was installed.

A high accuracy OEM load cell was used to measure the electrostatic forces. They acted on the surface of a micropart when it was exposed to the electric field of the handling device. Lateral adjustment of probes and grippers fixed to the end effector can be observed with microscope cameras. An auto-touch routine determines the distance between a device fixed to the axes and a probe fixed to the load cell in vertical direction.

The repeat accuracy of the test rig maneuvers was demonstrated in an experiment in which the probe on the load cell was approached from the same arbitrarily chosen position thirty times. The variance of measured values did not exceed 0.2 μm, and the experimental standard deviation was 0.07 μm.

Experimental Results with Centering Electrostatic Microgrippers

The goal of the comprehensive experiments with the functional models of the electrostatic microgrippers was to determine the optimal operating conditions and to characterize the attainable electrostatic forces.

To set up an experiment, a gripper was mounted on the end effector of the axis and a sphere was fixed to the load cell as shown in Fig. 16.4 and Fig. 16.1.b. The sphere and the gripper electrodes had to be laterally aligned with the microscope camera and then the distance between the gripper and the sphere was measured by the auto-touch routine. Before a force measurement experiment started, the gripper was moved back to a predefined start position.

To measure the forces dependent on the distance, the operating voltage had to be switched on while the gripper was moved incrementally towards

Fig. 16.4 Setup of a force measurement experiment: (**a**) Top view; (**b**) Side view

the sphere. Both measured values, force and distance, were recorded simultaneously. Some basic results of the experiments are condensed in the diagram in Fig. 16.5. The measured curve progression for neutralized probes met the expectations. Experiments with charged probes illustrated the significant influence of surface charges on the probe. When the surface of the sphere was charged to 7.5 kV before the force measurement was started, significant discontinuous curves were observed. At first, the forces increased with decreasing distance as expected. Between the distance −25 and − 5 μm, the forces decreased in almost stochastic steps. In the diagram, this range is labeled "discharge area".

Depending on the polarity of the operating voltage, the step-like force decreases differed in number and loss of force. In the case of negative polarity, the curve progression fell in many small steps to the level of neutralized probes, whereas for positive polarity, the same effect took place in one or only few bigger steps. It was assumed that these phenomena were caused by gas discharges between the gripper electrodes. Similar effects were reported from experiments of contact electrification by Horn and Smith [10].

Experiments were also performed with functional models covered with the described isolation layers. With these grippers, a significant discharge area were observed for neutralized probes and a relative low operating voltage of 400 V, shown in Fig. 16.5. The measured curves showed many small steps of decreasing forces in the discharge area. At distances below 5 μm, the force increased very sharply. The maximum force with a 600 V positive operating voltage exceeded 1 mN, while the attainable force was about 0.7 mN for 600 V negative operating voltage. Such high forces can be explained if the surface of the thin insulating layer of the gripper is charged. Dielectric barrier surface discharges (DBD) were assumed to explain this phenomenon. Gibalov and Pietsch [7] described macroscopic experiments which showed extremely high surface charge densities on insulating surfaces after DBDs. The surface charge densities depend on the polarity of the discharge. These observations provided an explanation as to why thin dielectric layers on the electrodes are not useful to isolate high voltage electrodes. In fact, the accumulated surface charges on

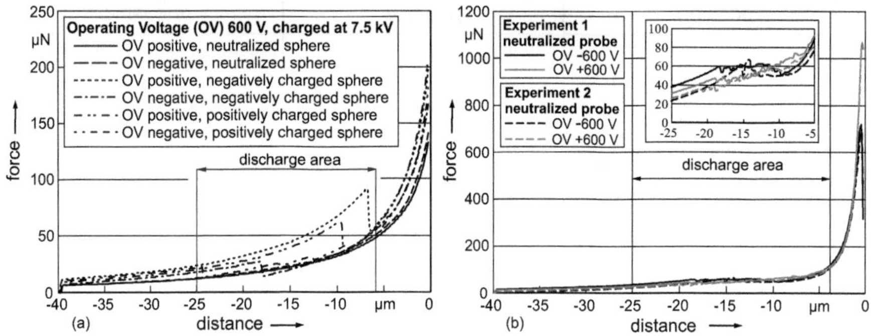

Fig. 16.5 Distance-force-diagrams: (**a**) Variant (a), ruby sphere $d = 405\,\mu\text{m}$, charged and neutralized probes; (**b**) Variant (a), ruby sphere $d = 405\,\mu\text{m}$, neutralized probes, Al_2O_3 insulation layer

the dielectric layer can cause a burnout of the electrode after several on/off switching processes.

Experiments using the electrodes with sharp edges (Fig. 16.2 (b), (c)) showed a different discharge phenomenon. At the beginning, the measured forces increased with decreasing distance as expected. The force started to decrease down to zero depending on the level of the operating voltage in a distance area between 3 and $0.8\,\mu\text{m}$. Such a short distance is less than the mean free path of electrons in air. It was assumed that in this case, the discharges were mainly caused by free electrons accelerated by the high electric field strength near the sharp edges of the electrodes.

Handling Tests with Centering Electrostatic Microgrippers

Grippers without insulating layers and smooth electrode contours were used to handle a wide variety of handling objects made of insulating materials. Round parts were centered and gripped by the inhomogeneous fields, but the centering effect failed for cubic parts. The variants (a) and (d) in Fig. 16.2 were used successfully for pick-and-place processes. To avoid malfunctions during object centering and releasing, surface charges had to be controlled by an active discharge of the handling object before the picking process was started. For the same reason, it was also important to limit the operating voltage to 400 V to avoid discharge phenomena. Reliable releasing of objects was achieved when the electrodes were grounded to earth potential. In these tests, microspheres as small as $400\,\mu\text{m}$ in diameter (as shown in Fig. 16.6 and Fig. 16.1.a) as well as $50\,\mu\text{m}$ thin hot melt foils ($2 \times 5\,\text{mm}^2$) were used as handling objects.

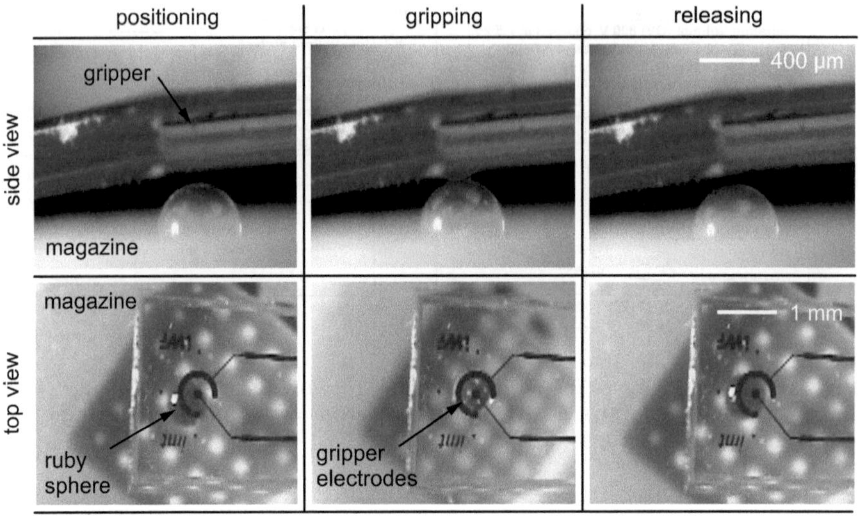

Fig. 16.6 Handling tests with centering electrostatic microgrippers (ruby sphere $d = 400\,\mu\text{m}$, mechanical centering magazine)

16.2.2 Handling Devices Generating Electrostatic Forces Without Electrodes

Surface charges on insulating bodies can be modified using active charging systems. In industrial applications, systems which consist of DC charging generators and unipolar pin electrodes are used to generate surface charges. In contrast, active discharging systems consisting of AC power units and bipolar electrodes (ionizer) are used to reduce surface charges. Surface charges on a body consisting of insulating material create an electric field in the surrounding medium.

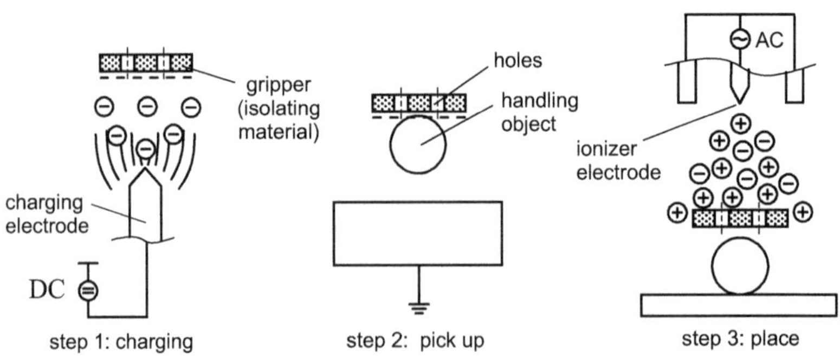

Fig. 16.7 Gripping with surface charges

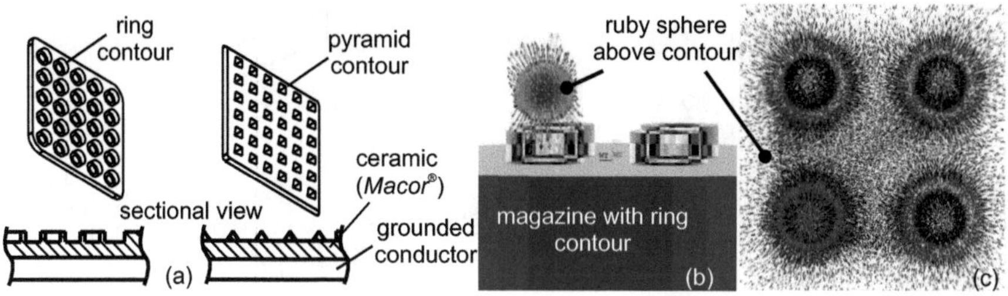

Fig. 16.8 (**a**) Layout of electrostatic magazines; (**b**) Force vectors acting on a surface of a ruby sphere; (**c**) Electric field caused by surface charges ((**b**, **c**): FEA simulations)

The initial idea was to use charged insulating surfaces to generate gripping or clamping forces. The first electrostatic microgripper without electrodes was manufactured out of a simple polymer plate. The handling process takes place in three steps as shown in Fig. 16.7. In the first step, the surface of the gripper is charged by a conventional active charging system. Afterwards, the gripper vertically approaches a handling object and picks it up. The object can then be detached by either joining forces acting in the assembly position or by actively discharging the gripper surfaces. The contact gap between the gripper and the object surfaces is shielded against the ions moving in the electric field of the ionizer if the ionizer is placed above the gripper. The diffusion length in the contact gap can be shortened by holes in the gripper body, which significantly reduce the time for the neutralization of the surface charges.

The same physical principal was used to manufacture electrostatic micromagazines made of machinable glass ceramic (*Macor*) without electrodes (Fig. 16.1.c). The surface of the isolating magazine body was structured as pyramid or ring contours by milling (Fig. 16.8). These contours cause an inhomogeneous electric field in the surrounding medium to generate dielectrophoretic clamping forces while providing an additional mechanical centering effect. The structures were dimensioned for ruby spheres with diameters from 400 to 600 μm. FEA-simulated chucking forces amounted to more than two times of the weight of the spheres when surface charge densities of about $3 \cdot 10^{-5}\,\mathrm{C\,m^{-2}}$ were assumed on the magazine surfaces.

Experimental Results by Modifying Surface Charges

For using modified surface charges on isolator surfaces, two major strategies were investigated. First, active charging was used to create electrostatic chucking and gripping forces. Second, active discharging was used to avoid unintentional sticking effects for pick-and-place tasks with mechanical microgrippers. These experiments are described later on. In the first experiment, the gripper consisted of a simple 360 μm thick SU-8 plate with two holes

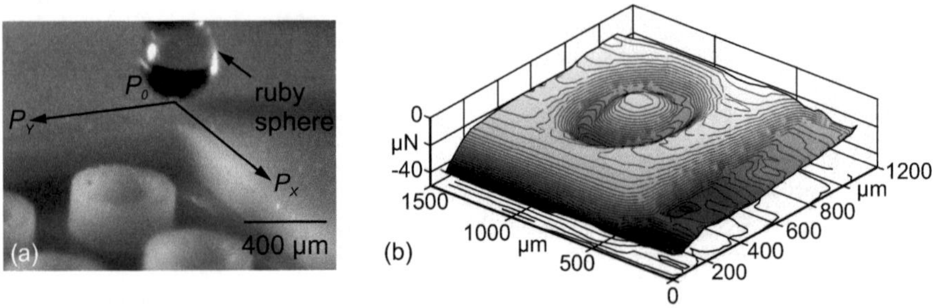

Fig. 16.9 Electrostatic magazines without electrodes: (**a**) Raster force measurement; (**b**) Measured force plane

(400 µm diameter) in the active surface, and the handling object was a ceramic die ($1 \times 1 \times 0.5\,\mathrm{mm}^3$). A single point charging bar with about 13 kV was used to charge the gripper surface at a distance of 1.5 cm. The die was picked up at a distance of approx. 0.5 mm while the gripper approached the ceramic die. To release the die, the gripper was actively discharged, which took about three seconds.

Several magazines were manufactured and investigated. They operate in the same way by charging and discharging the device. In Fig. 16.9.a, an experiment to measure the electrostatic force plane ($P0$-PX-PY) acting between an uncharged ruby sphere ($d = 400\,\mu\mathrm{m}$) and a charged magazine surface is shown. The sphere was moved incrementally with steps of 30 µm in x- and y-direction. The distance between the xy-plane and the top of the magazine surface was kept constant at 8 µm. For each step, the force was measured in z-direction and recorded together with the position data. The results of this experiment are shown in Fig. 16.9.b. Maximum measured forces were in the range of 40 µN.

This force was high enough to hold the ruby spheres against gravitational force for more than 24 hours.

16.2.3 Improvement of Process Reliability by Active Neutralization of Mechanical Microgrippers

Microparts such as spheres normally tend to stick to one of the gripper jaws while they were placed on a substrate (Fig. 16.10). Active neutralization of the direct environment in a handling process with a mechanical microgripper, such as described in Sect. 16.3, showed promising results. Pick and place experiments were performed with ruby spheres ($d = 200\,\mu\mathrm{m}$) and MEMS grippers made of silicon. In some extreme cases the spheres bounced between the two jaws while the gripper was opened. Hence a great influence of electrostatic forces on the placing process reliability was assumed. In 46 handling operations only 65% of the placing processes succeeded without sticking

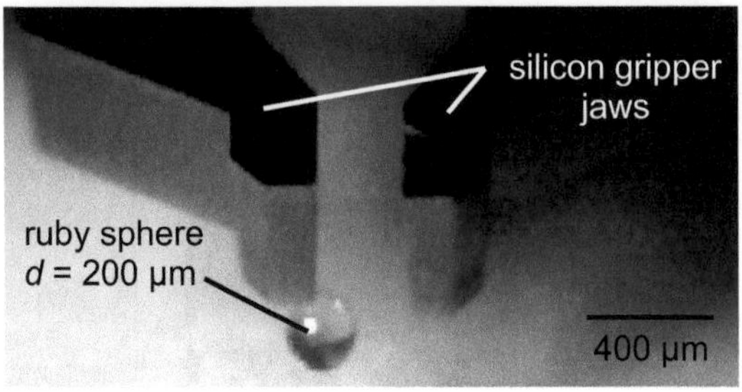

Fig. 16.10 Sphere sticking to gripper jaw

problems. In an ensuing experiment a customary static elimination gun was used to discharge the whole environment of the process (gripper, sphere and substrate). With this pre-processing out of 57 handling operations that were performed 94% of the placing operations succeeded. The first placing process that failed was observed after 48 successive handling operations. This could be caused by reappearing contact electrification between the dissimilar materials in the handling process. Finally this procedure gave promising results fo the use of mechanical microgrippers as presented in the following Section (Sect. 16.3).

16.3 Mechanical Microgrippers with Integrated Actuators

Even though the demand for gripping tools for precision assembling technology is increasing and the number of realized prototype microgrippers from research groups is quite high, the number of microgrippers on the commercial market is rather low [1]. One reason for this could be the fact that many of the working prototypes need several manual assembling steps in order to obtain a working device, which results in additional costs. One attempt to reduce this manual work is gripper fabrication via batch technology.

The following paragraphs describe several methods for the batch fabrication of SMA-, pneumatically and thermal expansion actuated microgrippers.

16.3.1 Basic Design of the Mechanical Microgrippers

All mechanical grippers that were developed in the SFB 516 are designed as parallel grippers with centering capability. To reach a synchronized and parallel gripping jaw movement, these grippers combine a four-bar-linkage mechanism with two additional linkages in a parallel-crank arrangement (Fig. 16.11.a). This guarantees a more reliable fixation of the clamped de-

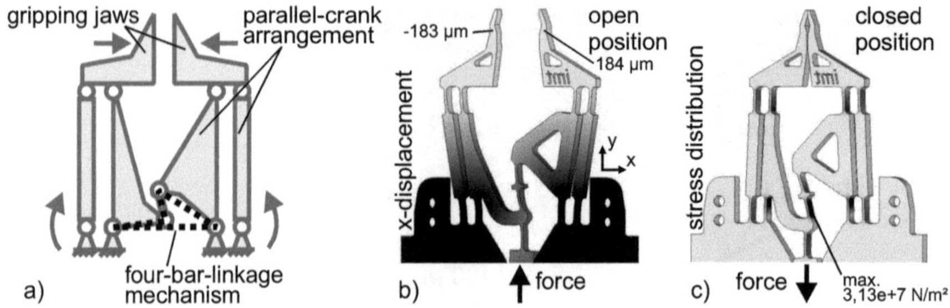

Fig. 16.11 Parallel microgripper gearing: (**a**) Schematic arrangement; (**b**) FEM calculated x-displacement; (**c**) Stress distribution of the *SU-8* gearing

vice and a more uniform gripping force distribution. To avoid internal play, the applied compliant mechanism uses flexural hinges for mechanical linkages. Figure 16.11.a depicts the described schematic arrangement. Figure 16.11.b,c show some results of FEM simulation of the later realized *SU-8* gripper gearing. FEM was used to optimize the synchronized opening motion and stress distribution of the gearings.

Depending on the application, the gearing structures can be fabricated from materials such as *Epon SU-8*, a photostructurable transparent epoxy (by *MicroChem Corporation*), monocrystalline silicon or a composite of both materials. These and some unusual materials like *FOTURAN*, a photostructurable glass (*Microglas Chemtech GmbH*), or copper *SU-8* compounds are also described in Sect. 16.3.5.

16.3.2 Microgrippers with SMA Actuators

The first of the actuation principles is based on the shape memory alloys (SMA) such as nickel-titanium-alloys, which are able to remember a pre-memorized shape when being heated from the martensitic to the austenitic phase. During this phase change, the actuator element is able to apply a force that causes actuator travel.

The SMA material is usually structured in the pre-memorized flat shape (Fig. 16.12.a,d). To enable an actuation movement, the SMA structure has to be stretched in the cold state (Fig. 16.12.b,e). During heating, the structure tends to return into the memorized shape. To reshape the actuator for another working step, two antagonistic working SMA actuators are heated alternately by electrical current (Fig. 16.12.c). In the early years of the SFB 516, Bütefisch assembled the first functional models of SMA-actuated gripper gearings [2]. These grippers build the basis for the batch fabricated SMA grippers in use today.

Wet Chemically Structured SMA/*SU-8* actuators. The first developed method to batch integrate SMA foils into mechanical structures is wet chemical etching [11]. Attached to *SU-8* sockets, a gold coated SMA foil (50 µm thick) is structured using lithographically patterned photoresist, gold etchant and NiTi etching solution. The remaining freestanding SMA actuators are embedded in a 360 µm-thick *SU-8* layer forming the gripper body (Fig. 16.1.d). Because of the isotropic etching process, the SMA actuator bars show a trapezoidal cross section area. This limits the design options dramatically.

Laser Micromachined SMA/*SU-8* Actuators. To reduce the minimum SMA actuator dimensions and to attain a more rectangular cross section of the structures, the second developed fabrication process uses laser micromachining with a Q-switched Nd:YAG laser [11]. The SMA structures are again attached to *SU-8* socket structures and finally embedded in a thick *SU-8* layer to form hook-shaped supporting structures around the actuator ends. These *SU-8* hook structures are used to fasten the actuator element in a separately fabricated gripper body (Fig. 16.1.e). Due to the anisotropic laser micromachining step, the minimum structure dimensions of the SMA actuators could be effectively reduced, enabling more compact and more complex actuators.

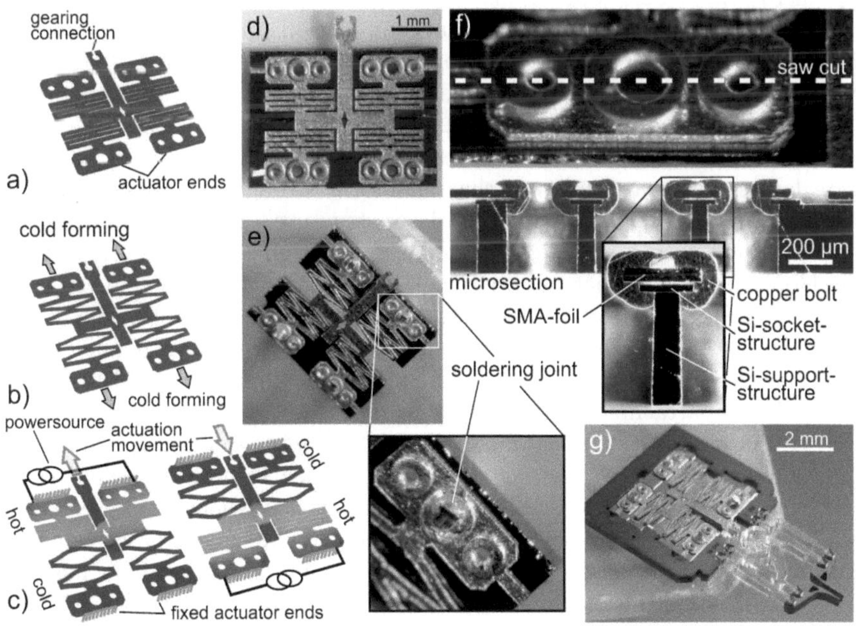

Fig. 16.12 SMA actuation principle: (**a**) Structuring of the SMA material; (**b**) Stretching the actuator and fixing the ends; (**c**) Alternately electrical heating of actuator; SMA/SOI actuator element: (**d**) After the dicing process; (**e**) After stretching and fixation; (**f**) Microsection of the material compound; (**g**) Attached gripper gearing

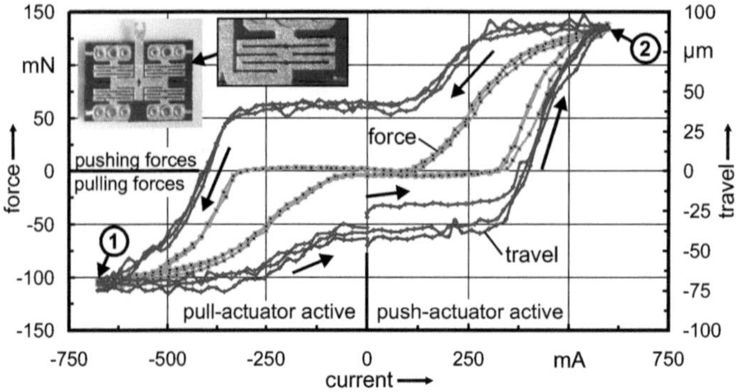

No.	P_{max}	T_{max}	d_{max}	F_{max}
1	416 mW	95 °C	74 µm	-110 mN
2	224 mW	75 °C	93 µm	152 mN

Fig. 16.13 Actuation force and travel versus heating current of the SMA/SOI drive

Laser Micromachined SMA/SOI Actuators. The third fabrication process was developed to reduce the overall dimensions of the actuator elements even more. The latest gripper generation, the microgripper construction kit (Sect. 16.3.5), requires a size reduction of more than 50%. To reach this and to strengthen the SMA fixation, the *SU-8* embedding process is substituted by micro copper bolt connections and silicon supports.

Fabrication of this atypical material combination starts with deep reactive ion etching (RIE) of a layered silicon-insulator-silicon substrate (silicon on insulator, SOI technology). The handle wafer (300 µm) forms the supporting structures and the device layer (20 µm) is used to create socket structures for the freestanding actuators. Holes with different diameters in both silicon layers create an undercut that enables a strong connection point for the electroplated copper bolts. The laser-structured SMA mesh is aligned and attached to the structured wafer using photoresist as adhesive. After electroplating the copper bolts through the SMA foil and the silicon holes (Fig. 16.12.f), the silicon-SMA actuator elements are separated with a dicing saw (Fig. 16.12.d). The mechanical and electrical connection is made from copper wires soldered on a printed circuit board adapter (Fig. 16.12.e) and ends with the attachment of a gripper gearing to the actuator (Fig. 16.12.g).

The characterization setup for the SMA/SOI actuators uses a laser triangulation sensor, a force sensor and a thermographical camera. A sample result of several actuation force and actuator travel measurements is presented in Fig. 16.13. The diagram shows the results for one alternating 3-cycle push-pull sequence. Concerning the actuation forces F_{max}, a maximum pushing force of 152 mN and a maximum pulling force of −110 mN was achieved. The actuation travel reached maxima d_{max} of 93 µm for the opening and −74 µm for the closing direction. The corresponding temperatures T_{max} and power values P_{max} for these points are given rightmost in the diagram.

16.3.3 Microgrippers with Pneumatic Actuators

The second actuation principle uses air pressure to power miniaturized cylinders for the microgripper drive. At the beginning of the SFB 516, Bütefisch developed micropneumatic actuators as a non-electrical alternative driving mechanism [3, 2]. Today, these drives have been improved, and additional auxiliary microtools for desktop factories use the same actuation principle for different tasks (Sect. 16.4). The latest single piston gripper (presented in Fig. 16.14.b) and Fig. 16.1.h,i) is driven and controlled by positive and negative pressures. It is particularly suitable for pick-and-place robots, which usually contain a suction nozzle with both types of pressures. Negative pressure closes the gripping tips, while positive pressure opens them [11].

The basic pneumatic actuator is made of a double-sided covered, flexible bellow structure, whose piston is pneumatically deflected. Recent micropneumatic systems are made of a perforated silicon bottom plate (for air in- and outlets), a structured *SU-8* polymer layer (including all movable and fixed piston and gearing structures) and a glass cover with a structured adhesive sealing. Actuation is achieved by sealing gaps on the top and bottom of the bellow structure leading to an inevitable but acceptable air leak, which is guided away from the gearing and drained off through small outlet holes (Fig. 16.14.a).

The fabrication of the pneumatic devices starts with lithographically structuring a sacrificial copper layer for the moveable structures onto a silicon wafer. The substrate material is structured via deep RIE, fabricating the air in- and outlets. An *SU-8* layer is spincoated and lithographically structured onto the wafer forming all piston and gearing structures. Finally, the sacrificial layer is removed and the grippers are separated (Fig. 16.1.g,h). The glass cover is made of a thin glass wafer. Structured gold conducting paths serve as targets for the electrodepositable photoresist *Intervia 3D-N*, which

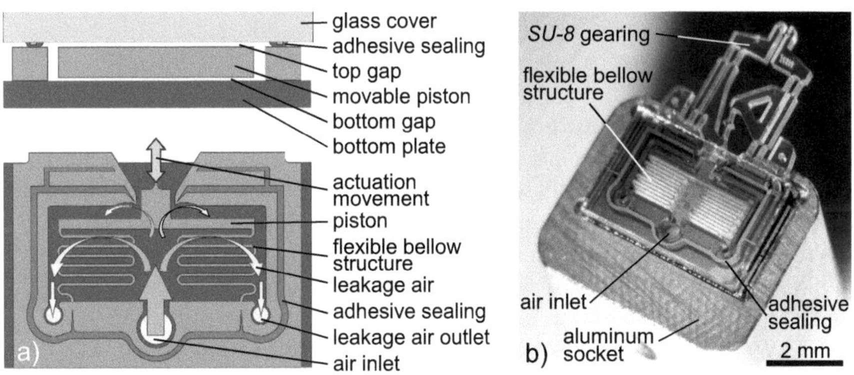

Fig. 16.14 (**a**) Schematic figure of the pneumatic actuator; (**b**) Fabricated single piston gripper

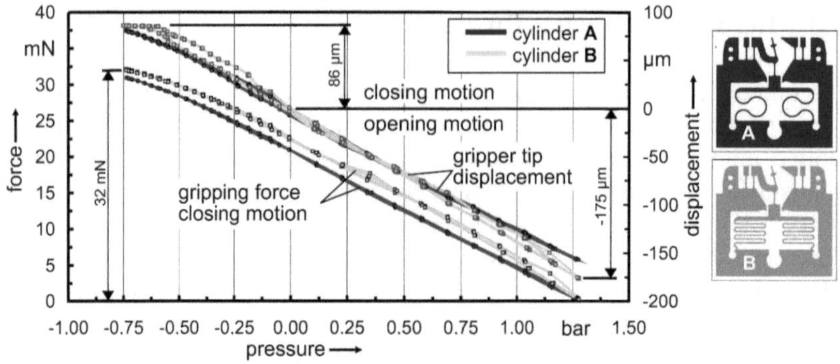

Fig. 16.15 Gripping force and gripper tip displacement versus operating pressure of the pneumatic gripper

is used as thermoplastic adhesive. This provides a method to batch apply lowest amounts of adhesive on predetermined areas. Finally, the saw-diced glass covers are aligned and fixed by heating. An aluminum socket connects the gripper to the handling robot (Fig. 16.14.b and Fig. 16.1.i).

To characterize the pneumatic grippers, an experimental setup containing a pneumatic control system, a laser triangulation sensor and a force sensor was used. Fig. 16.15 demonstrates the linear pressure-force-relation and the pressure-displacement-relation for two different cylinders shown rightmost. The maximum gripping force of 32 mN is reached at a pressure of −0.8 bar and the maximum opening width of 520 µm is achieved at 1.25 bar system pressure. The actuation frequency reaches up to 60 Hz with a small displacement fall-off starting at 10 Hz. Dynamic fatigue tests showed cycle counts between $5 \cdot 10^4$ and $1.1 \cdot 10^6$ before failure. Additionally, the pneumatic micro-grippers showed satisfying results in handling experiments with the desktop robot *Parvus* (Chap. 22).

16.3.4 Microgrippers with Thermal Expansion Actuators

A different principle uses thermal expansion of Joule heated silicon beams to drive the mechanics. The structure allows handling of objects smaller than 25 µm. More precisely, it is made for extracting single mirrors ($10 \times 10 \times 1\ \mu m^3$) from a digital micromirror device in video projectors (see Fig. 16.16.b). To ensure a high yield in mass production of these devices (containing up to 2 million mirrors), directed disassembling for inspection purposes is a helpful tool.

Fig. 16.16.a shows the schematic gripper gearing arrangement. The dark bars in the middle of the schematic figure demonstrate the inversely arranged expansion actuators. An opening motion is caused by heating the middle beam; a closing motion occurs when the outer beams are heated. To achieve

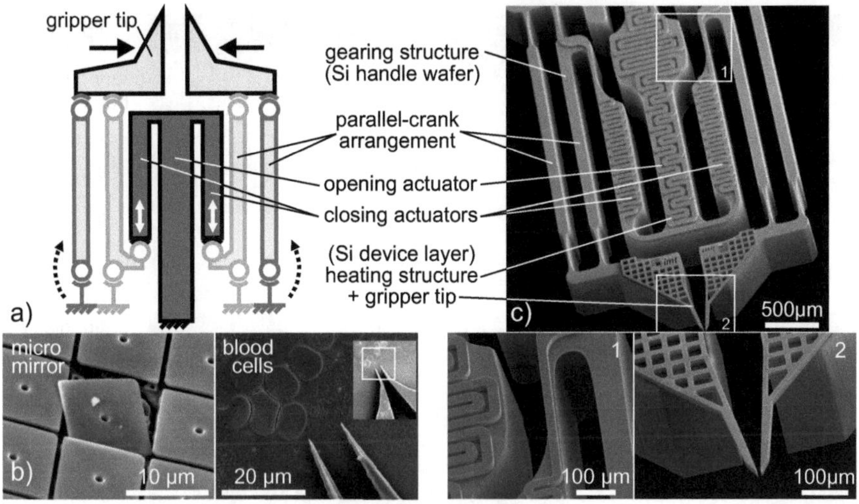

Fig. 16.16 (**a**) Schematic figure of the thermal expansion gripper; (**b**) Handling example: micromirror array; (**c**) SEM images of the gripper gearing

a robust gripper gearing with fine gripping tips, SOI technology is used. The robust gripper gearing and the expansion beams are fabricated from the silicon handling wafer (300 µm), and the heating resistor and the gripping tips are made of a highly doped silicon device layer (20 µm). Both silicon layers are structured using a deep RIE process. The gripper dice ($3 \times 5.5 \times 0.32\,\mathrm{mm}^3$) are separated by breaking from the RIE structured wafer matrix (Fig. 16.16.c). In addition to the gripper, a customized mounting mechanism (Fig. 16.1.j,k) was developed to ensures fast and simple gripper handling.

The tip displacement was measured by using a CCD laser triangulation sensor, and the actuator temperature was measured by using a thermographic camera. The measurements show a maximum opening displacement of 10.3 µm at 29 mA (1335 mW, 232°C) and a maximum closing displacement of 2.51 µm at 31 mA (1421 mW, 230°C) for each jaw. Considering the neutral opening width of 5.5 µm and the movement of the second gripping jaw, a total opening width of 26.1 µm can be achieved. The opening width of 10 µm for the DMD mirrors is achieved at a temperature of only 80°C. More information about this gripper is published in [13].

16.3.5 Gripper Construction Kit

Depending on the assembling situation, size- and material adapted gripping jaws, gripper mechanics and capable actuation mechanisms are required. A time saving way to adapt a gripper to a new handling situation presents the developed gripper construction kit. Due to a universal actuator-gearing in-

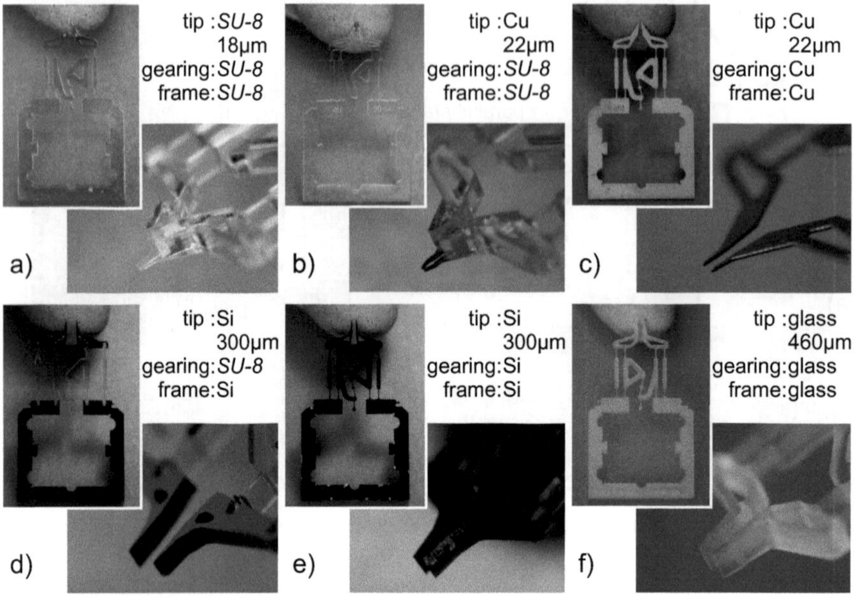

Fig. 16.17 Variations of the gripper gearing and tip material within the gripper construction kit

terface, different gripper gearings with varying gripping tips can be combined with different actuators and varying actuation principles.

Gripper Mechanics. Gearing structures were fabricated and tested with four different materials. Transparent *SU-8* gearings enable an observation of the handling process through the gripper and show the lowest gearing stiffness (ca. $164\,\mathrm{N\,m^{-1}}$). Structures made of single crystal silicon allow large mating forces and show the highest gearing stiffness (ca. $1820\,\mathrm{N\,m^{-1}}$). Apart from this, special gearings made of galvanic copper or the photostructurable glass *FOTURAN* (*Mikroglas Chemtech GmbH*) were also manufactured. The different gripper gearings are shown in Fig. 16.17.

Gripper Tips. To enable a large range of adaptability within the gripper construction kit, several gripping jaw materials and tip dimensions were created. A two-layer *SU-8* process results in thin gripping jaws with strong gearing structures (Fig. 16.17.a). A second gripper features very fine tips made of galvanic copper embedded in *SU-8* gearing structures (Fig. 16.17.b). Fig. 16.17.c illustrates a monolithically fabricated gripper consisting only of copper. Very hard silicon gripping tips are achieved by monolithic silicon gearing structures (Fig. 16.17.c) or combined with a flexible gearing made of *SU-8* (Fig. 16.17.d). The *FOTURAN* glass gripper features rough and hard gripping tips (Fig. 16.17.e). Material and tip design selection is dependent on the handling object and the assembling process itself.

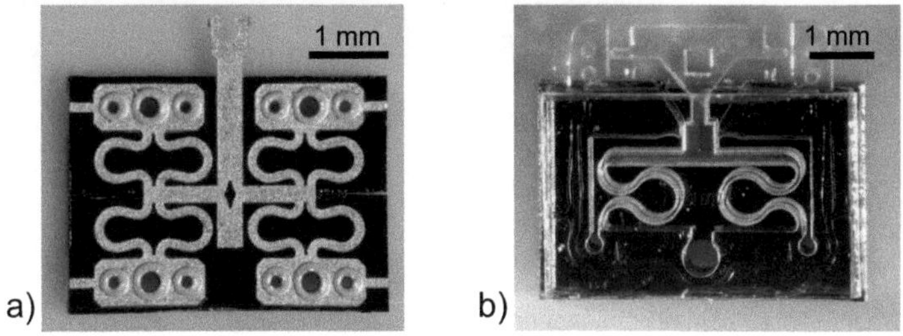

Fig. 16.18 Actuator elements from the gripper construction kit: (**a**) SMA/SOI actuator; (**b**) Pneumatic actuator

Actuator elements. Depending on the desired actuation force, the necessary tip displacement and the supplying energy source, three different actuation elements can be mounted. The previously described SMA/SOI actuator elements (Fig. 16.18.a and Sect. 16.3.2, Fig. 16.12.d) provide large forces, but they are hard to control. The pneumatic actuators Fig. 16.18.b are based on the pneumatic grippers from Sect. 16.3.3 (Fig. 16.14.b). They work without electric power and show a linear relation between the air pressure and the actuator force. The resulting forces are smaller than the SMA forces.

Fig. 16.1.f presents an example for a prepared gripper with *SU-8* gearing, silicon gripping tips and SMA actuator on a printed circuit board adapter.

16.4 Pneumatically Driven Auxiliary Microtools

Since size-adapted, modular production lines (desktop factories) are becoming more and more popular, handling tools of similar dimensions are essential. Glass balls for precision bearings or metrology styli usually come in tiny plastic bags or boxes and need to be manually prepared for further usage. Continuous machining with pick-and-place robots requires a well-known position for each ball, which can be achieved by camera-controlled robot systems or extensively prepared ball magazines. To improve these assembling preparations, the pneumatic microgripper batch process was adapted to fabricate active microassembling devices like centrifugal feeders and clamping units driven by micropneumatic actuators.

16.4.1 Centrifugal Feeder

The first device, a microcentrifugal feeder with integrated separation unit, was developed to feed an assembling robot with 200–300 μm glass balls (Fig. 16.19 and Fig. 16.1.n). The tiny bulk material is poured into the round storage chamber and circularly accelerated by air pressure from the chamber

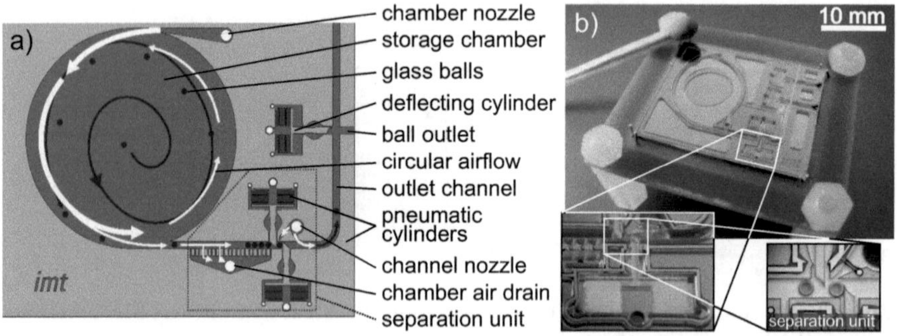

Fig. 16.19 (**a**) Schematic figure of the microcentrifugal feeder; (**b**) Prepared device

nozzle. A combination of centrifugal force and escaping air drives the glass balls into the outlet channel, where two pneumatic cylinders separate the lined-up balls and free single globes on demand. Additional channel nozzles push the balls through the channels, and deflecting cylinders guide the balls to different outlets. Deceleration of the glass ball at its final pick-up-position is accomplished via an aspiration port. Using these basic elements, different storage chamber variations and nozzle configurations were simulated, fabricated and tested.

All moving parts of the auxiliary microtools are based on the pneumatic piston actuator concept, which is described in Sect. 16.3.3. The required air flow for acceleration and transportation of the glass balls in the system is achieved via additional air nozzles and drains.

To characterize the fabricated systems, an experimental setup containing a pneumatic control system and a high speed camera is used. Repeatable tests with varying feeding systems prove that it is possible to release at least 6 balls in less than 840 ms on demand. After lining up the balls, releasing a single ball after another takes less than 90 ms per ball.

16.4.2 Active Clamping Device

The clamping unit (Fig. 16.20 and Fig. 16.1.m) is designed to align small components (such as mounting baseplates) automatically and fix them during the assembling process to avoid component slip. Pneumatic actuators are used to push and jiggle the handling object against multiple stoppers and finally clamp it in an aligned position. Vacuum holes underneath the object keep it in place and parallel to the base. An optional clamping cylinder holds the aligned component in position, even if all pneumatic supplies are disconnected.

The experimental testing of the clamping unit was performed by placing a silicon chip with distorted orientation in the device area, followed by a sequence of simultaneous or alternating piston strokes. Multiple tests showed that simultaneous piston strokes can align and fix the object with the first

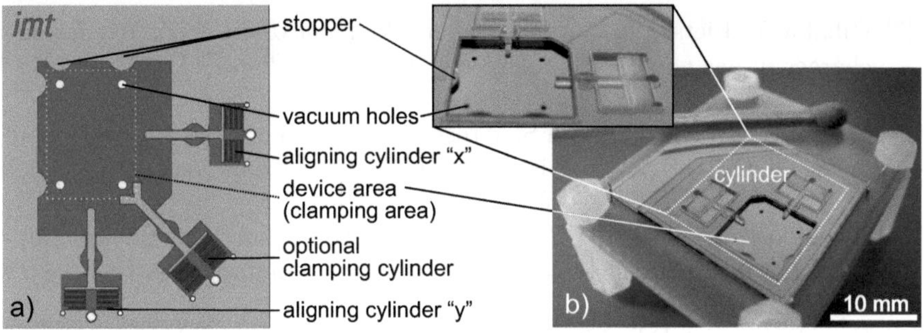

Fig. 16.20 (**a**) Schematic figure of the microclamping unit; (**b**) Prepared device

stroke (less than 80 ms). An alternating stroke method needs more than one stroke (120 ms). The object movement and the achieved final device position were analyzed by comparing the high speed camera pictures. Further details of both microtools are described in [12].

Acknowledgements The authors gratefully acknowledge the funding of the reported work by the German Research Foundation within the Collaborative Research Center 516 "Design and Manufacturing of Active Microsystems".

References

[1] Agnus J, Nectoux P, Chaillet N (2005) Overview of microgrippers and design of a micromanipulation station based on a MMOC microgripper. In: 2005 IEEE International Symposium on Computational Intelligence in Robotics and Automation, 2005. CIRA 2005. Proceedings, pp 117–123

[2] Bütefisch S (2002) Entwicklung von Greifern für die automatisierte Montage hybrider Mikrosysteme. Dissertation, Technische Universität Braunschweig

[3] Bütefisch S, Seidemann V, Büttgenbach S (2002) Novel micro-pneumatic actuator for MEMS. Sensors and Actuators A: Physical 97:638–645

[4] Enikov ET, Lazarov KV (2001) Optically transparent gripper for micro-assembly. In: Proceedings of Spie, vol 4568, pp 40–49

[5] Fantoni G, Biganzoli F (2004) Design of a novel electrostatic gripper. Int Journal of Manufacturing Science & Production 6(4):163–179

[6] Fearing RS (1995) Survey of sticking effects for micro parts handling. In: Proc. IEEE/RSJ Int. Conf. on Intelligent Robots & Systems (IROS), 'Human Robot Interaction and Cooperative Robots', Pittsburgh, PA, pp 212–217

[7] Gibalov V, Pietsch GJ (2000) The development of dielectric barrier discharges in gas gaps and on surfaces. Journal of Physics D 33(20):2618–2636

[8] Hesselbach J, Büttgenbach S, Wrege J, Bütefisch S, Graf C (2001) Centering electrostatic microgripper and magazines for microassembly tasks. In: Proceedings of Spie, vol 4568, pp 270–277

[9] Hesselbach J, Wrege J, Raatz A (2007) Micro handling devices supported by electrostatic forces. CIRP Annals – Manufacturing Technology 56(1):45–48

[10] Horn RG, Smith DT (1992) Contact electrification and adhesion between dissimilar materials. Science 256(5055):362–364

[11] Hoxhold B, Büttgenbach S (2008) Batch fabrication of micro grippers with integrated actuators. Microsystem Technologies 14(12):1917–1924

[12] Hoxhold B, Büttgenbach S (2009) Micro Tools with Pneumatic Actuators for Desktop Factories. Sensors and Transducers 7(10/09):160–169

[13] Hoxhold B, Büttgenbach S (2010) Easily Manageable, Electrothermally Actuated Silicon Micro Gripper. Microsystem Technologies published online:1–9

[14] Lang D, Tichem M (2006) Design and experimental evaluation of an electrostatic micro-gripping system. In: Proceedings of 3rd Int. Precision Assembly Seminar, pp 33–42

[15] Oh HS (1998) Elektrostatische Greifer für die Mikromontage. Fortschritt-Berichte VDI

[16] Van Brussels H, Peirs J, Reynaerts D, Delchambre A, Reinhart G, Roth N, Weck M, Zussmann E (2000) Assembly of microsystems. Annals of the CIRP 49(2):451–472

[17] Wrege J (2007) Elektrostatisch unterstützte Handhabungstechniken in der Mikromontage. Dissertation, Technische Universität Braunschweig

Chapter 17
Stereophotogrammetry in Microassembly

C. Keck, M. Berndt, R. Tutsch

Institute of Production Metrology
Technische Universität Braunschweig
r.tutsch@tu-braunschweig.de

Abstract Hybrid microsystems are preferably assembled serially by robots, which place the individual components successively in their intended locations. Microassembly processes are subject to various influences, which often consume a considerable part of the specified tolerance. Acceptable quality and economical yield can be achieved, if sensors are employed to detect and immediately correct deviations arising in the assembly process. This chapter describes a camera sensor for direct integration into robots for microassembly. The sensor uses beam-splitting mirror optics to record two images of the process from the top-left and the top-right with a single camera. It measures the spatial location of flat components in real-time by stereophotogrammetric evaluation of the two images. The parts are detected by means of circular marks or other features, which are deposited by electroplating or are made of fluorescent resist. Two arrays of light-emitting diodes on the sides of the mirror optics provide a constant dark-field illumination throughout the entire assembly process. The sensor uses fast methods for the evaluation of the images in order to meet the demanding real-time requirements of assembly. The compact design makes the sensor promising also for other applications such as the optical inspection of ball-grid array (BGA) packages on printed circuit boards (PCB).

17.1 Introduction

Active microsystems such as electromagnetic actuators are built as hybrid microsystems, allowing the designer to choose from a broad range of components manufactured in different technologies. For proper operation, these microsystems have to be assembled and joined with high accuracy. Typical tolerances for the positions of single components lie in the range of 1 µm and below.

The assembly of hybrid microsystems is preferably done by robots, which pick up the individual parts and place them successively in their intended locations (serial assembly). These robots have to cope not only with positioning inaccuracies, but also with the inacurracies of the assembly tools [8]. As microparts have small sizes and low mass, surface effects may also have a strong influence [10]. In the case of adhesive joints, the curing of the adhesive also has to be considered. The cumulative error of all these influences may reach the same magnitude as the specified tolerances, making it difficult to produce microassemblies with acceptable quality and economical yield.

Capable processes can be achieved, if the assembly steps are executed under sensor control [3, 8]. The sensors monitor relevant quality characteristics and provide direct feedback in order to compensate for deviations caused by internal and external influences, such as the handling of the parts, or the curing of adhesive bonds. This approach is often referred to as "sensor-guided assembly".

The most important quality characteristics in microassembly are geometrical quantities. In order to complete an assembly step successfully, the spatial (3D) locations (position and angular orientation) of all components involved in the step have to be under control. Optical sensors in combination with computer vision are well suited for such real-time measurement tasks, because they offer good accuracy and high measurement speed. At the same time, they do not require mechanical contact to the measured parts and can be quickly adapted to new assembly tasks.

Within the Collaborative Research Center SFB 516, the Institute of Production Metrology (IPROM), in collaboration with the Institute of Machine Tools and Production Technology (IWF), developed a compact camera sensor for direct integration into robots for microassembly [1]. The sensor records images of the process from two different viewing directions and measures the 3D location of flat microparts by stereophotogrammetric evaluation of the two images in real-time.

17.2 Photogrammetry

In photogrammetry, the 3D locations of objects are determined from previously recorded images of the region of interest [6]. Several aspects have to be considered not only in the design of the sensor, but also in the design of the whole measurement process:

Imaging System. The imaging system records the images of the assembly scenery. The field of view has to be large enough to register any part involved in an assembly step. At the same time, the integration of the sensor into the robot demands a compact, lightweight design.

Illumination. The sensor requires illumination to reliably detect the optical features on the surface of the parts. The illumination should not vary with the situation, as electronic image processing is heavily impeded by changing illumination.

Signalization. Optical features are needed for the reliable detection and precise registration of parts. These optical features may already be available on the parts, but may also be placed on the parts on purpose. The detection of the features should require only moderate effort in electronic image processing. The selection of features strongly depends on the characteristics of the parts of interest, especially on the complexity and the optical properties of the structures on the parts, as well as on the illumination.

Evaluation. After the images have been acquired and digitized, the optical features are extracted from the recorded images by electronic image processing. The 3D locations of the parts are then determined from the extracted features by photogrammetric analysis. The processing and evaluation of the images demand considerable computational resources. Therefore, the proper selection of methods and algorithms is critical for real-time operation in assembly.

Integration. In visually guided assembly, the sensor and the robot operate together closely to correct for deviations immediately. The sensor should not impede the robot's operation. At the same time, the robot must be prevented from disturbing the measurement.

17.2.1 Sensor Specification

The sensor was designed for the size-adapted *Micabo* ^{F2} prototype robot system (see Chap. 15). The handling tool and the camera sensor reside together inside the head (Fig. 17.1), which is shaped like a cylinder and is able to rotate around its vertical axis. The camera sensor can be moved up and down independently of the handling tool to set the optimum working distance for the sensor. The specifications for the sensor are mainly derived from the parts, the assembly process and the robot:

Parts. In the SFB 516, microsystems are mainly built from flat parts, which are made of silicon and bear electroplated structures. In order to avoid additional process steps, plated structures should also be used for part signalization. For parts without plated structures, different optical features have to be generated on the part. The lateral sizes of the parts range from 3 mm to 10 mm, while their thicknesses lie in the range from 0.5 mm to 2 mm.

Assembly process. The working volume of the sensor should have a minimum size of 5 mm by 10 mm. The height of the working volume was specified to be 5 mm in order to avoid frequent vertical adjustments of the sensor.

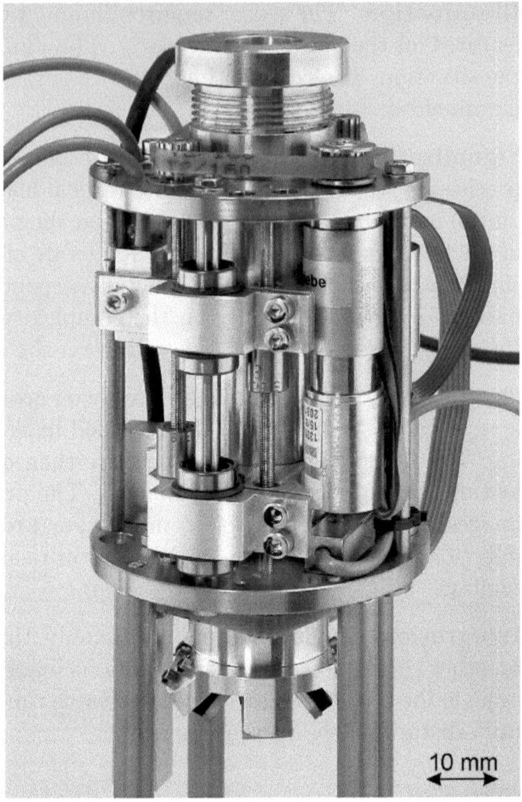

Fig. 17.1 The head of the *Micabo*^{F2} size-adapted assembly robot

Robot. The robot imposes strong design constraints on the sensor. The sensor has to be mounted in a borehole with a diameter of only 22 mm and a height of approximately 135 mm. Below the borehole, additional space with a diameter of 40 mm and a height of 30 mm is available.

The sensor should not impair the dynamic capabilities of the robot, and must not exert excessive forces or moments on the robot, neither in standstill nor in motion. Therefore, the sensor should have low mass, and all components of the sensor should be placed as closely as possible to the center of the robot's head.

17.2.2 Imaging System

The robot only permits visual monitoring of the assembly process downwards from its head, while the 3D measurement of positions requires at least two images taken from two different directions in space. Beam-splitting mirror optics such as that shown in Fig. 17.2 provide a solution for recording two images of the scene from two different points of view.

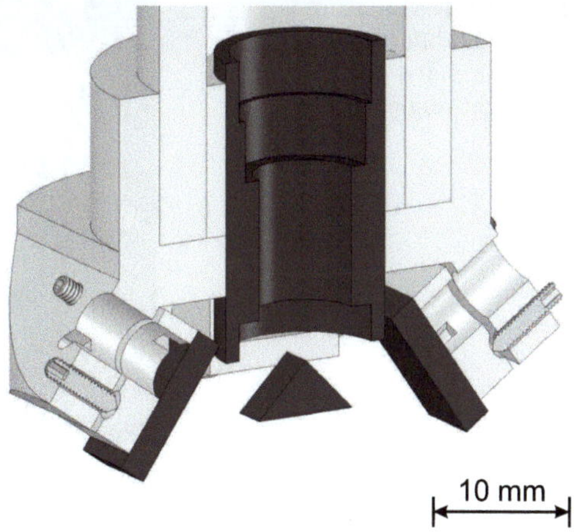

Fig. 17.2 Mirror optics (cross-sectional view) [1]

A ray tracing simulation (Fig. 17.3) illustrates the working principle. The two outer mirrors project the views onto an inner right-angled prism, which bears a reflective coating and in turn projects the views upwards into the

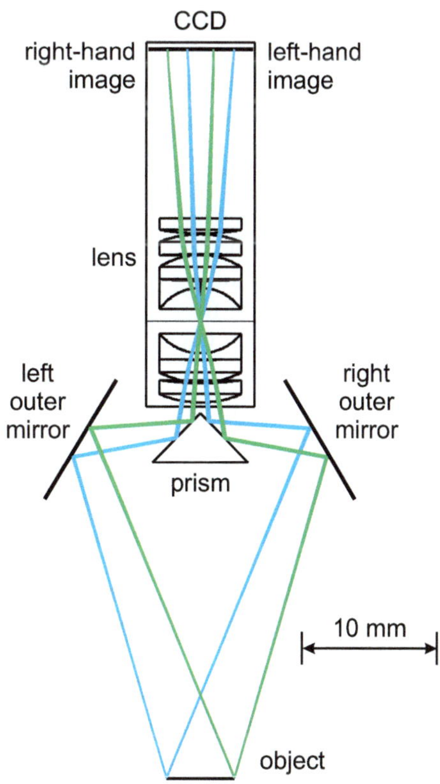

Fig. 17.3 Ray tracing simulation of sensor with mirror optics

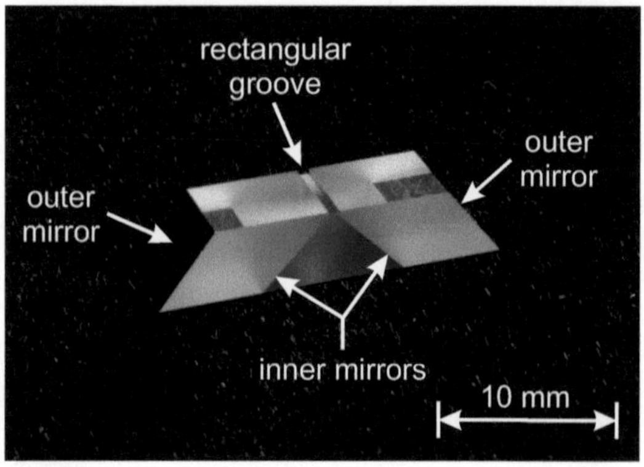

Fig. 17.4 Miniature prism module made of two four-sided prisms and one right-angled prism

camera. After passing the camera optics, the left-hand view is recorded by the right half of the CCD (charge-coupled device) image sensor, while the CCD's left half records the right-hand view.

This setup requires only one single camera, which can be placed directly above the working volume of the robot and makes optimum use of the available space. The compact setup has a convenient distribution of mass, in contrast to a standard convergent stereophotogrammetric setup, which would require two separate cameras to record the scenery from the top-left and the top-right.

The lateral sizes of the measuring volume are dissimilar, which may be a disadvantage in some applications. The images of the left and right views overlap partially in the sensor. To diminish this overlap, the aperture of the camera has to be small, reducing the sensitivity of the camera, but also increasing the depth of focus.

For operation in rough conditions (high accelerations, thermally unstable environments) prism modules provide a more robust solution. Fig. 17.4 shows a miniature module made from two four-sided prisms. Their lateral faces are coated with aluminum and serve as mirrors. The right-angled prism in the center provides mechanical support only. The rectangular groove between the prisms suppresses ghost images due to total internal reflection (TIR) at the upper faces of the prisms [11].

If prisms are used as mirror optics, the refraction at the interfaces between the prisms and the surrounding air (Fig. 17.5) causes additional distortions.The interfaces on the camera's side are orthogonal to the optical axis of the camera and cause radially-symmetric distortions, which are independent of the object distance. The interfaces on the object side cause distortions that increase with the object distance and are difficult to consider in measurements [6].

Two prototypes have been built and tested. The first prototype uses adjustable outer mirrors (Fig. 17.6), while the second prototype is based on the prism module (Fig. 17.4). Both prototypes use a miniature television cam-

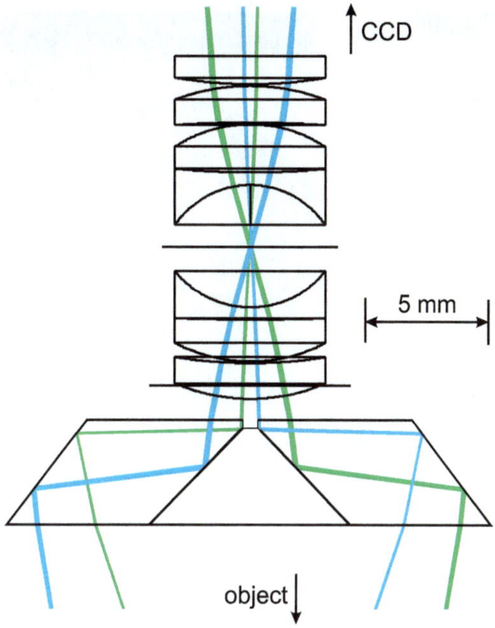

Fig. 17.5 Additional refraction at the glass-air interfaces of a prism module

era in a pen-type cylindrical housing (*Toshiba Teli CS4300BC* with a 12 mm lens). The CCD image sensor (*Sony ICX059*, 1/3" size) of the camera provides a resolution of 752 × 582 pixel. The aperture k of the camera was set to approximately 12, corresponding roughly to an entrance pupil of 1 mm.

The left- and right-view images have a resolution of approximately 350 × 580 pixel, as roughly 50 columns in the center of the image are lost due to the partial overlap between the left and right images. The observed area is 11 mm long, 6.5 mm wide, and approximately 5 mm high. The left and right lines of view enclose an angle of about 40°. The resolution is approximately 19 µm per pixel. The total weight of both prototypes is approximately 270 g. Both prototypes show similar performance in the measurement of circular marks. The repeatability (standard deviation) in position measurement is about 0.5 µm in the horizontal plane and 0.8 µm along the vertical axis.

After calibration, the prototype with outer discrete mirrors shows a maximum error of 1.5 µm in the measurement of horizontal distances between circular marks. The prototype with the prism module shows higher deviations of up to 5 µm, as the camera calibration cannot fully compensate for the additional radially-symmetric distortions due to the refraction at the interfaces between the prisms and the surrounding air.

The camera is an important factor for the performance of the sensor. The repeatablity of measurements can be considerably improved by using a camera with higher resolution. A high-resolution digital camera (*Pulnix TM1402CL*, 1/2" CCD, 1392 × 1040 pixel) improves the repeatability (stan-

Fig. 17.6 Sensor with discrete mirrors and LED panels for dark-field illumination [1]

dard deviation) in position measurement to 0.2 µm in the horizontal plane and to 0.3 µm along the vertical axis [1]. Unfortunately, there are no high-resolution cameras commercially available that meet the demanding space constraints of the robot's head.

17.2.3 Illumination

For the reliable and precise detection of optical features on the parts throughout the assembly process the sensor requires a dependable illumination in each step of the assembly process. The illumination has to emphasize the features of interest and should suppress unwanted influences from other sources.

In the case of the electroplated silicon chips, dark-field illumination is a good choice, as it achieves the best contrast between scattering and reflective surfaces [1]. It is provided by two arrays of green light-emitting diodes (LEDs), which are mounted at right angle to the mirror system (Fig. 17.6).

The light emitted by the LEDs reaches the horizontal part surface in a flat, slanting angle. Rough surfaces scatter a fraction of the incident light into the sensor and appear bright in the images. Flat polished surfaces appear dark, as they reflect the light to the side of the measurement volume, from where it cannot reach the sensor. As the LED arrays are mounted on the sensor itself, a constant illumination is provided that is independent of the robot's position in the working volume.

Green LEDs with a peak wavelength of 530 nm were chosen, as the sensitivity of the camera reaches its maximum at wavelengths between 500 nm and 540 nm. The camera requires a light intensity of approximately 20 cd for operation. The thermal losses of the LEDs are low enough not to influence the precision of the sensor.

17.2.4 Signalization

The sensor measures the location of a part by comparing the left- and right-hand images of the optical pattern on the part surface with the nominal pattern. Precise measurements require optical features that show a good reproducibility from part to part and can be detected precisely in the images. As assembly requires immediate response in real-time, the optical features should require only moderate computational resources for detection.

Signalization by Polygons

The electroplated patterns on the parts contain a large assortment of optical features that can also be employed for measurement. The structures for the linear electromagnetic actuators are made from electroplated nickel-iron (NiFe). They are patterned by photolithography and show very high geometrical accuracy.

In many cases, the pattern can be modeled by filled polygons [5]. The position of the corners can be extracted from the left- and right-hand images using standard image processing techniques. As the location of the part is generally not known beforehand, additional information is required to associate the corners in the images with the corresponding corners in the model. First, the straight edges connecting the corners are traced to assign the corners in the images to their corresponding polygons. Second, the polygons found in the images are associated with the polygons of the model, using the number of corners, the distance to neighboring polygons and the circumference as criteria.

The measurement of the part location using corners as optical features strongly depends on the measurement of the planar coordinates of the corners in the image. If the corner positions are determined directly from the image, the results show large systematic deviations [5]. Therefore, the corner coordinates should be determined from the neighboring edges, making better use of the image information. Although dedicated and efficient software for image processing is available, the measurement of 3D part locations from polygon corners still requires several 100 ms on standard computers, limiting its suitability for real-time operation.

Signalization by Scattering Circular Marks

Better performance in location measurement can be achieved with dedicated optical features that are only employed for measurement. They can be independently optimized in terms of precision and speed. On the other hand, such dedicated features cannot contribute to the intended function of a part and consume extra area, which is no longer available for functional structures.

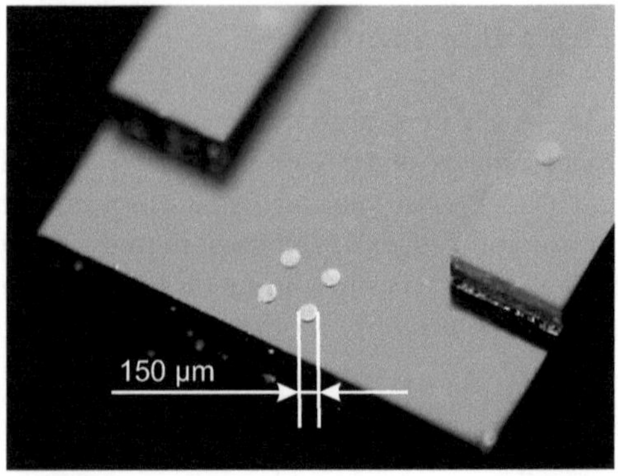

Fig. 17.7 Part with circular marks from electroplated NiFe [1]

The simplest dedicated optical features are circular marks, which can be easily distinguished from other geometrical forms. Circular marks always appear as elliptical disks in the images, and can be measured from all directions in space in the same way [1, 2]. The position of such circular marks can be measured in the images with high precision. First, the edges of the circular marks are detected. Second, the run of the edges is determined with subpixel precision using the method by Tabatabai and Mitchell [9], which performs linear interpolation between neighboring pixels. Finally, the position and the other geometrical parameters of the ellipses are determined by least-square fits.

Circular marks from electroplated NiFe (Fig. 17.7) can be easily produced in the same process as functional structures. Fig. 17.8 shows parts with such circular marks (diameter 150 µm, height 6 µm) under dark-field illumination during assembly. For a brighter appearance in the camera the electroplated marks were fabricated with an intentionally high roughness.

Electroplated circular marks from NiFe have been established as standard signalization within the SFB 516 because of their extraordinary compatibility with the existing electroplating process and their advantageous optical scattering behaviour. Experiments have also been done with electroplated marks from copper and gold, which tend to flake off and therefore cannot be relied on.

Signalization by Fluorescent Features

The disadvantage of using optical features made from reflecting or scattering materials is that the measurement is often disturbed by light, which is scattered or reflected by other objects in the surroundings. Also, the structures and the substrate may not show enough contrast under dark-field illumination.

Fluorescent optical features do not suffer from these disadvantages, as the emitted light has a different wavelength than the light required for ex-

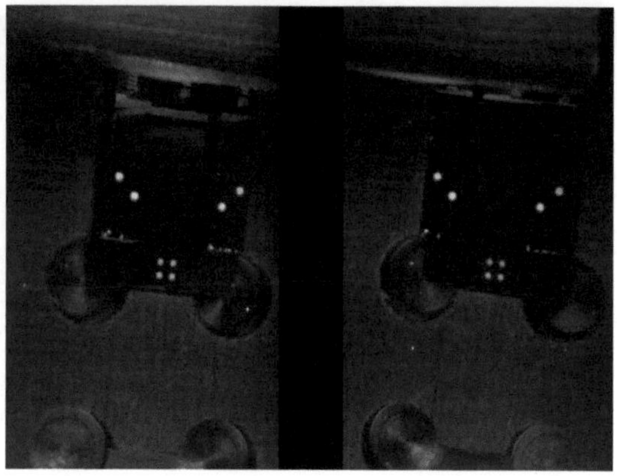

Fig. 17.8 Parts with circular marks from NiFe as seen by the sensor during assembly [1]

citation [2]. A color filter is sufficient to separate the light emitted by the fluorescent features from the other light, which is reflected or scattered elsewhere and would disturb the measurement. Features made from fluorescent resist permit a efficient signalization of parts, which is not deteriorated by reflection or scattering.

A fluorescent photoresist can be made from the resist *SU-8* by adding the fluorochrome *Lanxess Macrolex Fluorescent Yellow 10GN* [2]. *SU-8* is a negative, near-ultraviolet photoresist based on an epoxy resin. Butyro-1,4-lactone (γ-butyrolactone, GBL) is added to the *SU-8* to set the viscosity for coating. *Fluorescent Yellow* emits green light when excited with ultraviolet (UV) radiation and dissolves well in *SU-8* and GBL.

The fluorescent resist is prepared by first dissolving *Fluorescent Yellow* in GBL and then adding the solution to the *SU-8* resist. The *Fluorescent Yellow* molecules finely disperse in the resist film. Except for the setting of the viscosity, all other processing steps such as coating, softbake, exposure, post-exposure bake, development and hardbake require only minor modifications to reliably produce high-quality fluorescent optical marks.

Fig. 17.9 shows structures made of fluorescent *SU-8* resist under illumination with visible and ultraviolet light. It demonstrates the high improvement in contrast that can be achieved with a fluorescent resist. The emitted light of fluorescent *SU-8* reaches its maximum at 530 nm, perfectly matching the sensitivity of the CCD. The best stimulation of the fluorescent *SU-8* can be achieved with a wavelength of 405 nm, which is well-separated from the emitted green light.

For the measurement of fluorescent features, the green LEDs (Fig. 17.6) are replaced by LEDs emitting ultraviolet radiation. Unfortunately, the fluorescent *SU-8* requires a highly intensive irradiation in the UV-range to ensure a sufficient contrast in the camera images, which is difficult to achieve with current UV-LEDs [2].

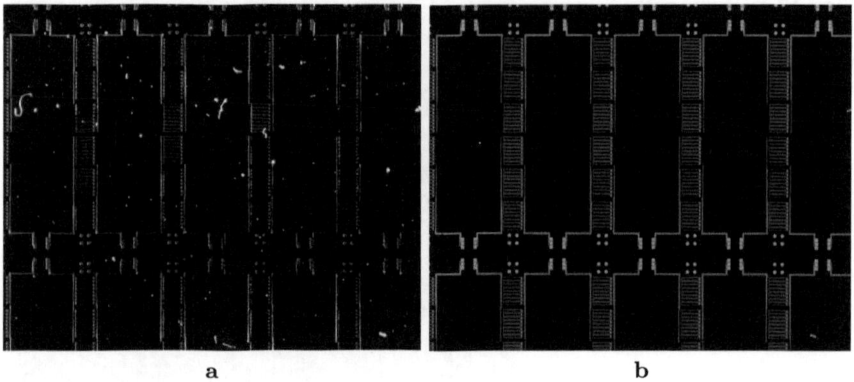

Fig. 17.9 Structures made of the fluorescent *SU-8* resist under illumination: (**a**) visible; (**b**) ultraviolet light [1]

17.2.5 Evaluation

After the images have been acquired and digitized, the optical features are extracted from the recorded images by electronic image processing. The key step is the determination of 3D positions of the optical features from their planar position in the left and right images of the scenery.

Camera Model

In order to determine the 3D object coordinates from the 2D coordinates of optical features in the left- and right-hand images, the photogrammetric evaluation requires a model, that describes the projection of object points to the camera images [6]. The so-called "collinearity equations" relate the coordinates of an object point $P = (X, Y, Z)^T$ in the 3D sensor coordinate system to the coordinates $P' = (x', y')^T$ of the point in the camera image.

$$x' = x'_0 - c \cdot \frac{r_{11} \cdot (X - X_0) + r_{21} \cdot (Y - Y_0) + r_{31} \cdot (Z - Z_0)}{r_{13} \cdot (X - X_0) + r_{23} \cdot (Y - Y_0) + r_{33} \cdot (Z - Z_0)} + \Delta x \quad (17.1)$$

$$y' = y'_0 - c \cdot \frac{r_{12} \cdot (X - X_0) + r_{22} \cdot (Y - Y_0) + r_{32} \cdot (Z - Z_0)}{r_{13} \cdot (X - X_0) + r_{23} \cdot (Y - Y_0) + r_{33} \cdot (Z - Z_0)} + \Delta y \quad (17.2)$$

The parameters of the collinearity equations can be divided into two subsets. The first subset with the position of the camera center $O = (X_0, Y_0, Z_0)^T$ and the direction angles $(\omega, \varphi, \kappa)^T$ defines the camera's line of vision, known as "outer orientation". The elements r_{ij} of the rotation matrix R are calculated from the direction angles $(\omega, \varphi, \kappa)^T$.

The rotation between the reference coordinate system and the line of vision of a camera can be described in different ways. The rotation matrix R below describes the transformation by three successive rotations about the moving y-, x- and z-axes of the sensor coordinate system [6]:

$$
\begin{aligned}
R &= \begin{bmatrix} r_{11} & r_{21} & r_{31} \\ r_{12} & r_{22} & r_{32} \\ r_{13} & r_{23} & r_{33} \end{bmatrix} \\
&= \begin{bmatrix} \mathrm{c}_\varphi \cdot \mathrm{c}_\kappa & -\mathrm{c}_\varphi \cdot \mathrm{s}_\kappa & \mathrm{s}_\varphi \\ \mathrm{c}_\omega \cdot \mathrm{s}_\kappa + \mathrm{s}_\omega \cdot \mathrm{s}_\varphi \cdot \mathrm{c}_\kappa & \mathrm{c}_\omega \cdot \mathrm{c}_\kappa - \mathrm{s}_\omega \cdot \mathrm{s}_\varphi \cdot \mathrm{s}_\kappa & -\mathrm{s}_\omega \cdot \mathrm{c}_\varphi \\ \mathrm{s}_\omega \cdot \mathrm{s}_\kappa - \mathrm{c}_\omega \cdot \mathrm{s}_\varphi \cdot \mathrm{c}_\kappa & \mathrm{s}_\omega \cdot \mathrm{c}_\kappa + \mathrm{c}_\omega \cdot \mathrm{s}_\varphi \cdot \mathrm{s}_\kappa & \mathrm{c}_\omega \cdot \mathrm{c}_\varphi \end{bmatrix}
\end{aligned}
\tag{17.3}
$$

using the following short-hand notation

$$
\begin{aligned}
\mathrm{s}_\omega &= \sin\omega & \mathrm{s}_\varphi &= \sin\varphi & \mathrm{s}_\kappa &= \sin\kappa \\
\mathrm{c}_\omega &= \cos\omega & \mathrm{c}_\varphi &= \cos\varphi & \mathrm{c}_\kappa &= \cos\kappa
\end{aligned}
$$

The offset $(x'_0, y'_0)^T$ and the camera constant c together with the optical distortions $\Delta x'$ and $\Delta y'$ form the second subset, which characterizes the imaging properties of the camera, also called "inner orientation".

The camera model considers distortions caused by the optical components of the sensor as well as distortions caused by the CCD sensor [1]. If the camera is looking at a flat object at a right angle, the refraction at the faces of lenses, optical filters or prisms leads to radially symmetric distortions.

Radially asymmetric and tangential distortions arise from the manufacturing of the camera objective, if single lenses are not centered on the optical axis of the camera objective, or are tilted to the optical axis. In the sensor, these distortions combine in a different way and usually cannot be separated from each other, as the camera is looking at the object through the mirror optics.

Electronic distortions originate from the geometry of the pixels and from the readout of the image out of the CCD sensor. Chromatic aberrations do not need to be considered, as long as the same monochromatic illumination is used throughout calibration and operation.

Calibration

The inner and outer orientations of the right and left parts of the sensor are determined by a calibration during the robot setup. The calibration procedure uses a flat glass slide as calibration artifact [1, 7]. The slide, shown in Fig. 17.10, bears a 9-by-11 pattern of vapor-deposited metallic circular marks. Each circular mark can be unambiguously identified in an image, as long as the five circular marks with larger diameter are visible in the image. The dis-

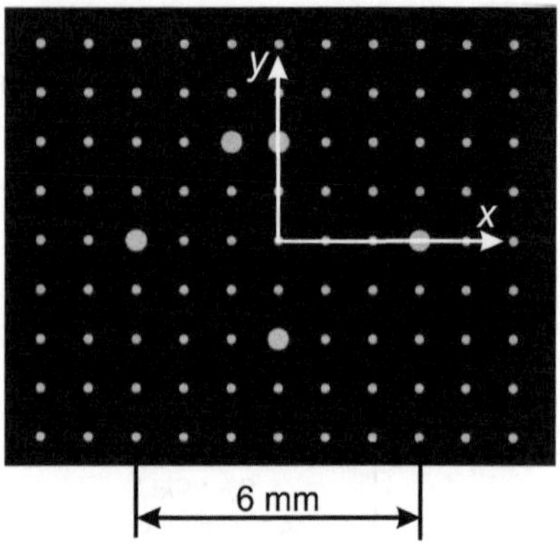

Fig. 17.10 Calibration artifact: glass slide with circular marks from chromium

tance between the outer left and right larger marks is calibrated to establish traceability back to a length standard.

For calibration, the slide is placed in various positions and orientations in front of the sensor, and left- and right-hand images are simultaneously recorded (Fig. 17.11) [1]. After the 2D coordinates of the circular marks have been measured in both images, the parameters of the inner and outer orientation of the left and right part of the sensor are determined by bundle adjustment, along with the positions and orientations of the slide in space as well as the planar positions of the circular marks on the slide.

The bundle adjustment assumes separate cameras for the left- and right-hand images. It neglects the fact that the sensor records the images using the left and right halves of the same camera, and that the distortions of the left- and right-hand images are closely correlated. Nevertheless, the calibration describes the optical sensor precisely enough that the maximum deviation of length measurement is limited to 1.5 μm for the prototype with discrete mirrors.

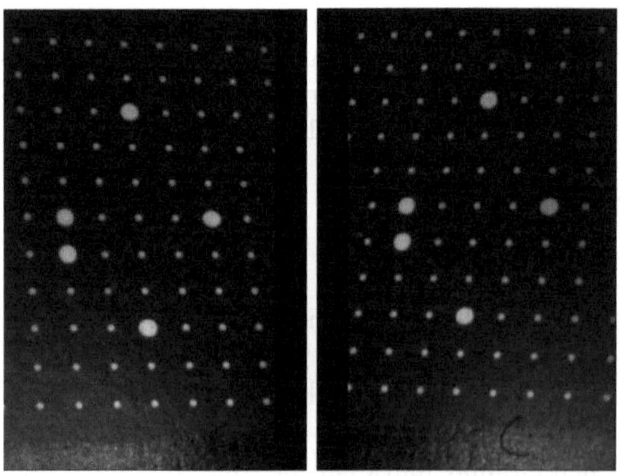

Fig. 17.11 Left and right view of the tilted and rotated artifact during calibration [1]

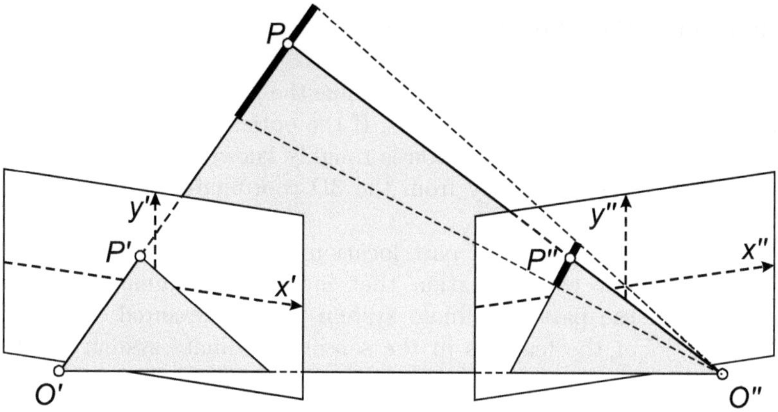

Fig. 17.12 Epipolar association

Epipolar Association

Before the 3D coordinates of an point on the object can be determined from its 2D image coordinates, the point has to be identified in both the left and the right image. This step requires additional information, which may come from the images themselves or from other sources. The sensor faces difficulties here, as the CCD sensor does not resolve enough details to permit the use of coded marks that could be easily identified in the recorded images.

Epipolar association provides an efficient method to search for associated points in a pair of left and right images. It relies on the fact that the left and right image points define a common plane with the object point (Fig. 17.12). The errors remaining after the calibration of the sensor and the repeatability lead only to minor violations of this rule. For a given point in the left-hand image, the point in the right-hand showing the least deviation from the common epipolar plane can be identified as the corresponding point.

If the upper and lower boundaries of the measuring volume are taken into account, the association reduces to a search for the point closest to a straight line in the right image, making it very efficient [1]. If two or more object points lie on a common epipolar line, the corresponding image points can be identified by their location within the images, as they always appear in the same left-to-right arrangement in both images.

In practice, the distance from the epipolar plane is also a good indicator for the condition of the sensor. The epipolar distances are close to zero if the sensor is perfectly calibrated. If the sensor leaves the calibrated state, the epipolar distances will immediately increase.

Determination of Part Location

The last step in the measurement is to determine the part locations from the measured positions of the optical features. If the optical features on a part form a regular pattern, and the location is roughly known already, the part location can be calculated directly from the 3D coordinates of the optical features [1].

In general, the measurement of part location corresponds to the determination of a Helmert transformation that maps the nominal positions $P = (X, Y, Z)^T$ in the part coordinate system to the measured positions $P' = (X', Y', Z')^T$ of the features in the sensor coordinate system [5]. It consists of a rotation about the fixed axes of the sensor coordinate system (rotation matrix R with elements r_{ij}) followed by a translation with an offset $P'_0 = (X'_0, Y'_0, Z'_0)^T$:

$$\begin{bmatrix} X' \\ Y' \\ Z' \end{bmatrix} = \begin{bmatrix} r_{11} & r_{21} & r_{31} \\ r_{12} & r_{22} & r_{32} \\ r_{13} & r_{23} & r_{33} \end{bmatrix} \begin{bmatrix} X \\ Y \\ Z \end{bmatrix} + \begin{bmatrix} X'_0 \\ Y'_0 \\ Z'_0 \end{bmatrix} \tag{17.4}$$

The determination of the rotation matrix poses ambiguity problems that lead to singularities, if the matrix elements are calculated using trigonometric functions [6]. In order to avoid these problems, the Helmert transformation is determined using the method by Drixler, which uses four quaternions to describe the rotation [4]. The four quaternions can be directly calculated from the 3D positions of the features, thus avoiding numerical methods.

17.2.6 Integration

The integration of the sensor into the assembly robot is an important factor for the robot's performance in assembly. The sensor should detect geometrical deviations reliably and enable the robot to initiate corrective steps immediately. At the same time, the processing of the images and the measurement of the 3D location of parts require considerable time for computation.

In photogrammetric measurement, accuracy can be lost due to motion blur if the sensor or the object move while images are recorded. As microassembly requires high measurement accuracies, the robot has to be brought to a standstill, before the sensor can perform a measurement. Therefore the *Micabo* ^{F2} employs a "look-&-move" strategy. In each step, a measurement is made first to detect deviations, before the robot is instructed to move, in order to correct for the deviations [8].

Different steps are taken to make efficient use of the computional resources of the control computer. First, the image processing is limited to those regions in the left and right images that actually contain information of interest. This saves much computational power, as the number of pixels to be processed grows quadratically with the size of the region of interest.

Second, the sensor and the position controller operate asynchronously. This enables the sensor to record and evaluate the images at the camera's full frame rate, without regard to the position controller. After having measured the location of the parts, the sensor first saves the position and orientation information in its own memory and only sends it to the position controller upon request [1, 8].

17.3 Conclusions

In this chapter, a camera sensor for the measurement of the spatial location (3D position and orientation) of parts in microassembly is presented. The sensor uses a single camera and beam-splitting mirror optics to record images of the microassembly process from two different viewing directions. For operation in rough conditions, prism modules may replace mirror optics built from discrete optical components.

Two arrays of green light-emitting diodes (LEDs) provide a dark-field illumination, which is especially suited for the detection of scattering features on reflective substrates. If no suitable features are available on the parts, dedicated features made of fluorescent resist can be employed, thus avoiding the bad influence of scattered or reflected light.

The 3D location of parts is measured using standard stereophotogrammetric techniques, which were selected and optimized for real-time operation on a standard personal computer. The preferred method of registering parts is to use circular marks, which can be easily distinguished from other features and demand only moderate computational capabilities. The identification of corresponding points in the right and left images is done by epipolar association.

The mirror optics allows for a very compact design of the sensor. In combination with suitable software for real-time operation, camera sensors with beam-splitting mirror optics provide a promising solution not only for microassembly, but also for other production and inspection tasks.

Acknowledgements The authors gratefully acknowledge the funding of the reported work by German Research Foundation (DFG) within the Collaborative Research Center SFB 516 "Design and Manufacturing of Active Micro Systems". The authors also thank their former and present colleagues for their contributions to the research project. Reinhold Ritter initiated the research project. The camera sensor was designed by Diana Ispas and Clemens Neumann. Marcus Petz developed the software for the photogrammetric evaluation of the images. Jan-Hinrich Eggers found new techniques to measure the location of parts from polygonal features on planar surfaces. Marie-Julie Artola evaluated the performance of the prototype based on the prism module.

References

[1] Berndt M (2007) Photogrammetrischer 3D-Bildsensor für die automatisierte Mikromontage. Dissertation, Technische Universität Braunschweig. Schriftenreihe des Instituts für Produktionsmesstechnik, Shaker Verlag, Aachen

[2] Berndt M, Tutsch R (2007) Accurate and robust optical 3D position control in microassembly using fluorescent fiducial marks. In: Optomechatronic Sensors and Instrumentation III, Proceedings of the SPIE, vol 6716, p 67160N

[3] Dittrich S (2004) Sensorgeführte Präzisionsmontage. Dissertation, Technische Universität Braunschweig. Schriftenreihe des Instituts für Werkzeugmaschinen und Fertigungstechnik der Technischen Universität Braunschweig, Vulkan-Verlag, Essen

[4] Drixler E (1993) Analyse der Form und der Lage von Objekten im Raum. Dissertation, Universität Karlsruhe, Deutsche Geodätische Kommission bei der Bayerischen Akademie der Wissenschaften, Reihe C (Dissertationen), vol 409. Beck-Verlag, München

[5] Keck C, Eggers JH, Petz M, Tutsch R (2009) Ein Verfahren zur Messung der räumlichen Position und Orientierung von Mikrobauteilen mit planaren Strukturen. In: 3D-NordOst 2009 – 12. Anwendungsbezogener Workshop zur Erfassung, Modellierung, Verarbeitung und Auswertung von 3D-Daten, Berlin, pp 81–90

[6] Luhmann T, Robson S, Kyle S, Harley I (2006) Close Range Photogrammetry: Principles, Techniques and Applications. Whittles Publishing, Dunbeath

[7] Reich C (1999) Vollständige optische Formerfassung durch photogrammetrische Verknüpfung von Teilflächen. Dissertation, Technische Universität Braunschweig. Cuvillier-Verlag, Göttingen

[8] Schöttler K (2008) Planung und Untersuchung automatisierter Mikromontageprozesse unter besonderer Berücksichtigung der Einflussgrößen. Dissertation, Technische Universität Braunschweig. Schriftenreihe des Instituts für Werkzeugmaschinen und Fertigungstechnik der TU Braunschweig, Vulkan-Verlag, Essen

[9] Tabatabai AJ, Mitchel OR (1984) Edge location to subpixel values in digital imagery. IEEE Transactions on Pattern Analysis and Machine Intelligence 6(2):188–201

[10] Wrege J (2007) Elektrostatisch unterstützte Handhabungstechnik in der Mikromontage. Dissertation, Technische Universität Braunschweig. Schriftenreihe des Instituts für Werkzeugmaschinen und Fertigungstechnik der TU Braunschweig, Vulkan-Verlag, Essen

[11] Yoder PR (2008) Mounting Optics in Optical Instruments. Second Edition. SPIE Press, Bellingham

Chapter 18
Use of Hot Melt Adhesives for the Assembly of Microsystems

G. Hemken, S. Böhm, K. Dilger

Institute of Joining and Welding
Technische Universität Braunschweig
k.dilger@tu-braunschweig.de

Abstract Adhesive bonding is gaining importance as a key technology for precision engineering and microsystem technology. New packaging and bonding processes for microsystems frequently use bonding methods requiring adhesives with suitable properties. Permanently increasing specifications for micro assembly processes limit the use of conventional liquid adhesives. In contrast, hot melt adhesives (HMA) offer reproducible dosage sizes in a sub-nanoliter range. The extremely short setting times of HMAs allow to cut process time. Hybrid bonding systems combining hot melts with e.g. epoxies provide faster process times, higher performance and thermal stability of the joints. Focus of this paper is to present basics and methodology of a joining technology offering good initial strength, shorter setting times and batchwise pre-application of minimal adhesives volumes to join even miniscule bond surfaces.

18.1 Adhesive Bonding as Micro Joining Technology

The high degree of miniaturization of MEMS and MOEMS and the need to reduce production costs require batchwise joining processes, which allow short process cycles and are capable of joining even smallest bond areas [10]. That is why adhesive bonding processes have gained significant importance [11]. Since 80% of the costs result from joining and assembly, micro adhesive bonding offers both enormous technical innovation potential and value-added solutions. When joining and assembling micro-parts of dissimilar materials, adhesive bonding often is the only feasible solution. Main objective is multi-functionality, i.e. the joint must offer mechanical fixing plus conductivity even under dynamic stress. Other goals are more efficient and faster processes providing new freedom of design. Lower process temperatures, good aging stability in terms of temperature and humidity, and even reduced shrinkage are also important objectives.

Here, it must be mentioned that the properties of adhesives applied on a micro-scale are different from those applied on a macro-scale. Adhesives which may join two substrates well on a macroscopic scale may be completely unsuitable in the microscopic world, e.g. because of their high viscosity. Tiniest amounts have to be dispensed and positioned accurately. In order to achieve high joint quality and consistent microsystem performance, adhesive dispension must be reproducible. To meet all those requirements, a new joining technology using hot melt adhesives has been developed by the Institute of Joining and Welding (Technische Universität Braunschweig), combining the benefits of adhesive bonding with fast process cycles. Currently, viscous adhesives are the commonly used adhesive for MST applications. The problem areas of viscous adhesives in an assembly and joining process are identified below:

1. Holding times providing structural strength required for component handling are longer ($\geq 10\,\mathrm{s}$),
2. Adhesive application (squeezed out adhesive),
3. Inaccuracies in the assembly process due to insufficient adhesive setting, ultimately causing components to tilt,
4. Doses of adhesives with minimum viscosity do not allow to join miniscule geometries ($\leq 250\,\mu\mathrm{m}$),
5. Pot life depends on adhesive,
6. No batchwise mass production [6].

Using hot melt adhesives avoid the aforementioned limitations to a great extent [3].

18.2 Properties of Hot Melt Adhesives

Hot melt adhesives are an important group in the category of physically setting adhesives, commonly used for macroscopic assembly processes. Many commercially available thermoplastic hot melt adhesives are polyamides, saturated polyesters, polyolefines, ethylene–vinylacetate–copolymers, block polymers like styrene–butadiene–styrene and polyimides [4, 7].

The category of reactive hot melts combines the benefits of both hot melts and chemically curing adhesives. Through crosslinking, reactive hot melts become infusible and insoluble. Major constituents for prepolymer reaction are polyurethanes with solid polyesterpolyols. Heating the adhesive induces crosslinking. A new technology are polyurethane hot melt films containing solid isocyanate crosslinkers which are applied prior to the joining process and are activated by heat above melting temperature [5]. Another reactive HMA-systems are indicated by the reaction between thermoplastic polyamide and isocyanates crosslinkes to humidity resistant networks.

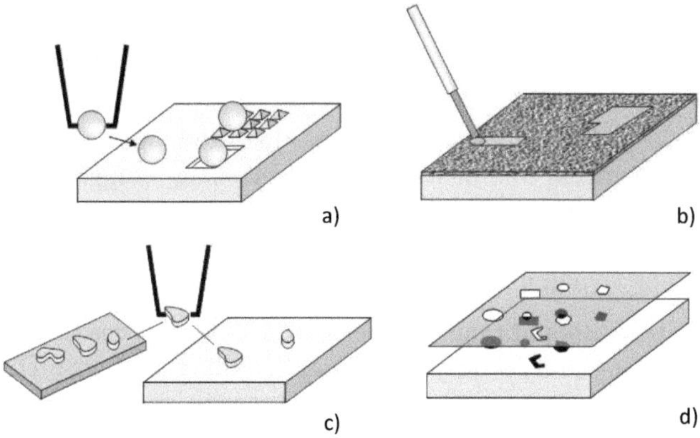

Fig. 18.1 Various hot melt pre-application methods: (**a**) Discrete application of adhesive balls/spheres; (**b**) Application of powder adhesives and selective melting using laser systems; (**c**) Application of adhesive films; (**d**) Stencil printing of hot melt dispersions

The melting temperature of most HMAs ranges between 60°C up to 220°C. That is why substrates sensitive to temperature can be bonded at low temperatures. When assembling microsystems, a major benefit of hot melts over viscous adhesive systems is the possibility to apply hot melt systems as powder, spheres, balls and films, and as dispersion or solution prior to the joining process, see Fig. 18.1. The actual adhesive bonding process does not necessarily have to take place directly after applying the adhesive to the substrate but can be initiated any time during the joining process. Hot melts have no pot life [1]. The adhesive melts through thermal impulse only during the bonding process and thus wets the surface of the other substrate. The adhesive bonding process can be started by either directly warming the adhesive or indirectly heating substrate and/or adhesive and substrate. The adhesive sets during cooling. Experiments have proven that hot melts set very quickly, i.e. joints achieve the required handling strength (usually the ultimate strength) in less than three seconds, if suitable heat management is used [2].

18.3 Adhesive System Selection Criteria

This section compiles a requirement profile for the use of HMAs in micro system assembly processes.

18.3.1 Hot Melt Adhesives

The adhesive properties below served as assessment criteria in a pre-selection of hot melt adhesives.

- Heat-sensitive substrates require for minimum distortion and residual stress in the joining process low melting- and processing-temperatures. Melting temperatures of hot melt adhesives should range between 60–120°C. Melting temperature ranges are evaluated by Differential Scanning Calorimetry (DSC) or Dynamic Mechanical Analysis (DMA).
- Processing temperatures should range between 60–160°C in order to provide good device wetting and to quickly achieve handling strength. The higher the processing temperatures the shorter it takes to reach handling strength because of the heat dissipation required by the system. Assessment criteria, such as the surface tension of adhesives and substrates and melting viscosities are measured by tensiometers and rheometers.
- Setting times of hot melts must be in a very short range of 3 seconds or less which is the time period necessary to bond to acceptable strength within short process times and product cycles. Assessment criteria for the setting characteristics are glass transition temperature and time-dependent viscosity measured by DSC and DMA.
- The selected hot melt adhesives must not be tacky at room temperature because it would make handling of pre-coated parts with grippers not feasible. Tackiness was assessed by test series on a micro-tensile test rig.
- Low melt viscosity is an important criterion for good wetting of the joining partners and even for adhesive application. Low viscosities are also required for specific applications, such as piezo dosage or micro balls. The viscosities are determined by rheometer measurements.
- The pot life is the degree of stability in molten state and the inclination to decompose and char. In automated application processes, oxidative reduction generates contamination which could disrupt the hot melt application machine. The skin forming on the adhesive through thermal reduction was verified by visual inspection. The thermal stability – the resistance against oxidative reduction – was examined in a DSC–OIT (oxidative induction time) test.
- The hot melt's suitability for processing and handling micro-balls, films and powder must be verified on an experimental basis. Additional informations provide the molecular weight and the glass transition temperature by DSC of the tested adhesive systems.
- When assembling complex hybrid micro systems with difficult joining geometries, a shear strength of 5 MPa is required. Assessment criterion is the compressive shear strength measured on a modified micro-tensile test rig and with the shear head of a wire bonder.

18.3.2 Adhesive Systems with Enhanced Thermal Stability

In addition to adhesives with the aforementioned requirement profile, there is a demand for hot melts with higher thermal stability but fast processing times. These adhesives crosslink through humidity or elevated temperatures. Reactive hot melt systems crosslinking under the influence of humidity are not suitable for batchwise pre-application because crosslinking starts directly after adhesive application making a downstream joining processes impossible. Due to these constraints, it was focused on two different systems with enhanced thermal stability.

- Reactive hot melts with thermal crosslinking: different chemistries like 2K–PUR, IPDI and reactive phenolic resin with nitrile rubber were used for the experiments.
- "Hybrid-bonding": In a first step of a hybrid bonding process, micro parts are quickly fastened by hot melts. In the second step, a delayed and locally separated bonding (e.g. underfill) process with a thermosetting adhesive of low viscosity (e.g epoxy) achieves maximum bond strength.

The following assessment criterion completes the requirement profile:

- Increase of the thermal stability by modifying adhesive systems. These systems must be creep- and heat-resistant when exposed to high temperatures and/or under extended periods of high static load. Heat and load resistance are measured and verified by DMA and shear tests.

Based on the aforementioned requirement profile for the selection of suitable adhesives, nine adhesive systems were chosen and further analyzed.

18.4 Different Particle Shapes of Micro-Scale Hot Melt Adhesives

Extensive tests regarding production, handling and processing of micro-scale hot melts were conducted to optimize the application of the hot melt adhesives. The shape and its minimal deposition volume of the suitable adhesive is an important aspect for the pre-application. The following shapes were assessed:

- Hot melt balls and spheres,
- Hot melt films,
- Hot melt particles plus dispersed particles.

Hot melt powder were tested on PTFE substrates in convection furnaces. During the tests, individual particles were applied to the PTFE carrier and heated up to melting temperature. Due to the adhesive surface tension and

Fig. 18.2 SEM of a hot melt ball, diameter about 500 μm, volume of 65 nl

the small surface energy of PTFE, a contact angle of approximately 180° between adhesive and PTFE surface must be maintained for the hot melt particles to become ball-shaped with adhesive volumes in picoliter-range. Yet, insufficient wetting caused the hot melt particles to have a convex-shaped lens geometry. Another attempt was made to produce balls by remelting free-falling hot melts in the air. By using a piezo dispensing unit, reproducible ball-shaped hot melt systems could be dosed appropriately (Fig.18.2).

Adhesive films (thickness approx. 50 μm and less) were generated through hot pressing hot melt powder, flat film extrusion (chill roll procedure), coating by squeegee application of polymer dispersions and spin-coating of soluble hot melt systems. Table 18.1 shows the results of different film processing procedures. Mechanical cutting, laser beam cutting or die cutting allows to shape different contours. Commercially available hot melt powders were classified in different grain fractions by means of filters or wind sifters. Fine-milling hot melt powder to produce grain fractions smaller than 40 μm was not effective. To join micro systems in μm-range smallest joints must be adhesively bonded. Pilot tests show that the particle size of hot melt powders must be less than 40 μm to achieve miniscule bonds.

Table 18.2 shows different methods to produce particles of hot melts in μm-dimensions. To avoid agglomerations caused by moisture or electrostatic influence, different additives were added to the hot melt powders. DMA-measurements show no effect on the mechanical characteristics of the modified material. Long term tests verified that only one additive (*Aerosil 200*) was capable to avoid powder agglomerations. The modified powders are pourable for 14 days at a temperature of 23°C and 50% relative humidity.

Table 18.1 Overview: HMA film processing

Method	Film Thickness	Remarks
Coating of polymer dispersions	30 µm	Inhomogenities in the film thickness
Flat film extrusion (chill roll)	50 µm	Inclusion of air bubbles
Hot pressing	25–50 µm	Reproducible
Spin-coating with soluble HMAs	1–2.5 µm	Highly reproducible

Table 18.2 Overview: HMA particles processing

Method	Process	HMA	Particle-Size Distribution	Remarks
Bottom-Up	Spray Drying	*Platamid*	d (90) $\leq$ 14 µm	Agglomeration; Particle-separation by additive addition
Top-Down	Cryo Mill	*Technomelt*	d (90) $\leq$ 63 µm	Agglomeration
Top-Down	Cryo Mill and Air Jet Sieving	*Vestamelt*	d (90) $\leq$ 53 µm	Agglomeration

18.5 Application Methods for Micro-Scale Hot Melt Adhesives

Dry application of hot melts can be accomplished by

- Discrete application of adhesive balls and spheres,
- Application of film adhesives,
- Application of powder adhesives,
- Printing of hot melt dispersions.

In order to use hot melts in microsystem assembly processes, HMA particle shapes must be smaller than 500 µm. Handling tiny particles, however, causes problems because compared to their behavior on the macroscopic scale the weight forces are not dominant. With smaller part sizes, the influence of surface forces, such as Van-der-Waals forces, electrostatic forces and adhesion forces increase (see also Chap. 16). This influence becomes apparent by the different part behavior [2]. Several tests with different handling methods were carried out. Hot melt adhesive **balls/spheres** and **films** were gripped by vacuum and electrostatic handling devices. A capillary vacuum gripper is used to grip hot melt balls and deposit them on the substrate. The balls can only be applied on the substrate's surface by heating. The hot melt's surface becomes tacky and the adhesive sets on the substrate. In some cases, hot melt balls stick to the capillary which is caused by electrostatic forces and/or warming of the vacuum tool. Another problem is the force-free placing of the balls. If the deposition force is too high, balls are pushed into the capillary tool. With a modified vacuum gripping tool it is possible to pick up and deposit contoured films of up to $150 \times 150\,\mu m^2$. The electrostatic application of films can only be accomplished if the substrate is heated, as described. Gripping tests using charged surfaces show no positive results. Film application by spin-coated soluble hot melts is very easy to handle and reproducible.

Summary of the application tests using these kinds of hot melts:

- HMA balls and films can only be applied in series processes,
- Spin-coating and the application of HMA powder and dispersions can be executed batchwise.

Due to their suitability for mass production some application methods are described in greater detail below. The feasibility of **pre-applying powder** both on large-surfaced substrates, such as wafers, and also on smaller devices is one reason for the good suitablility of hot melts in batch processes. Another reason is that the actual adhesive process (wetting and physical setting) can be controlled by exact temperature increases and/or decreases. This process ensures a change in viscosity. Another benefit is that heat can be input directly or indirectly, laminar or locally. Thermal energy can be applied for example by

- Local heating of the substrate by laser (1064 nm, 532 nm, 355 nm),
- Local heating of the component by laser (1064 nm, 532 nm, 355 nm),
- Laminar heating by a heating plate or infrared radiation.

Local heating of hot melts by means of laser radiation can take place directly in the hot melt or indirectly through the hot melt on the substrate surface. It depends on the laser wavelength and the hot melt's absorption properties. A precision heating plate can be used as an alternative to laser radiation. This plate is particularly suited for large-surfaced substrates (e.g. wafers). In comparison to laser heating, it takes longer to cool down the plate. Using hot melt powder, joining surfaces can be shaped by laser-sintering. These modified powders are applied homogeneously by electrostatically charging the substrates creating a monolayer of powder particles. In the process of laser sintering, the surface is scanned by a high-focus laser beam. A Nd:YAG disk laser (1064 nm, focal diameter 30 µm, galvo scanner and CAD-CAM-module) melts the adhesive system in cw-modus at 5.5 W indirectly through the hot melt on the silicon substrate. The thermal impulse locally melts and fuses the adhesive and contours the joining surface, see Fig. 18.3 and 18.4. A subsequent cleaning process (e.g. compressed air plus electrostatic powder discharge) removes the remaining non-adhesive powder.

Stencil or screen-printing of hot melt dispersions is one of the newest and interesting processes when pre-applying hot-melts. Stencil printing of

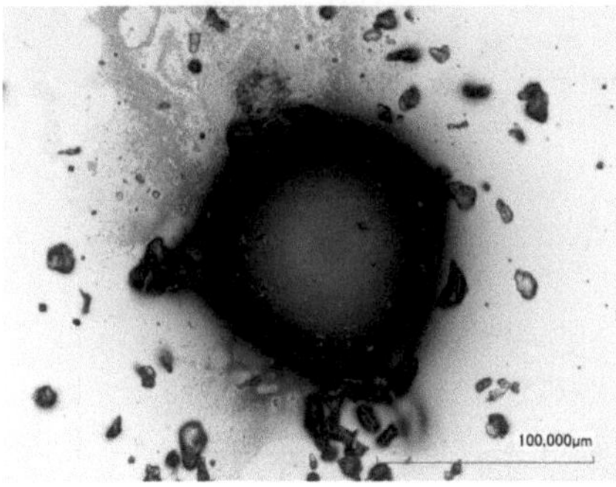

Fig. 18.3 Locally melted powder, diameter of structure about 100 µm, dot volume 630 pl

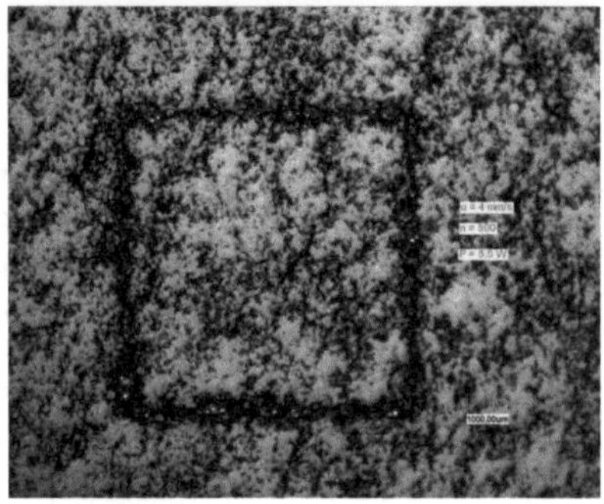

Fig. 18.4 Contoured melted powder, width of structure about 250 µm

soldering paste is an established procedure for which widely recognized tests have been carried out. Correlations and results of these tests cannot be directly related to the printing process of hot melt dispersions. In terms of physical and rheological behavior, dispersions are different to soldering pastes. A typical stencil printing process is described in the next section. As the squeegee blade moves across the stencil during the print cycle, dispersion or paste fill the stencil apertures. The paste is released to the pads on the substrate during the substrate/stencil separation cycle. Ideally, 100% of the paste filling the aperture during the printing process is released from the aperture walls and attaches to the pads on the board, forming a complete adhesive deposition. To optimize stencil printing of hot melt dispersions with regard to minimum volume application, with high reproducibility and reliable processes, the following influencing factors, Fig. 18.5, must be verified.

During these tests several influencing factors were tested which could be altered by the operator. To minimize testing effort only one stencil – a nano-coated stainless steel of about 100 µm thickness with circular apertures – was examined. The finest apertures of this stencil are about 150 µm in diameter with an aperture volume of 1.767 nl. The aperture volume is the theoretical possible printing volume (= 100% printing efficiency). Printing results can be evaluated by the following methods: The qualitative optical evaluation examines whether the substrate surface is clean of particles, and whether all the dispersion was released from the stencil. The quantitative evaluation is derived from the measurement of deposition volume, diameter and thickness of the printed spots. This measurement data was generated automatically by a laser scanning microscope. In order to interpret the measurement results, the printing efficiency was defined:

Printing efficiency = deposition volume / aperture volume × 100%

For a better understanding of the cause-effect relationship of the influencing factors in a printing process a series of experiments was created.
The following criteria were defined as "good" printing result:

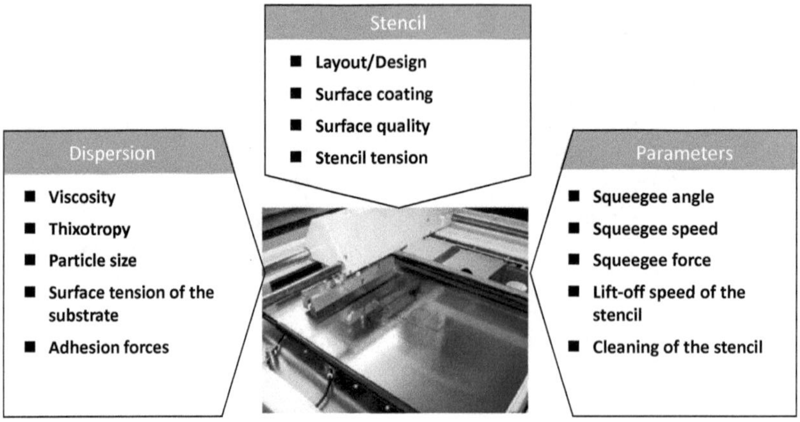

Fig. 18.5 Main factors of influence in a stencil printing process

- No visual defects – the adhesive deposited was completely printed,
- Height of the depot is about 60% of the foil thickness,
- Deposition diameter equals aperture diameter, permissible tolerance $\pm$ 2%,
- Deposition volume is about 70% of the aperture size,
- Standard deviation of a test series must be under 10% of printed and measured deposition volumes,
- Substrate surface must not be contaminated.

In preliminary tests various stencil printing parameters were optimized. After these tests, the three control parameters squeegee-force, -speed and stencil lift-off speed were verified with a factorial DoE (design of experiments). The following summary shows the results. The optimum squeegee speed should range between 60 and 70 $\mathrm{mm\,s^{-1}}$. If the squeegee is too fast, the paste will not be properly dispersed within the apertures. If it is too slow the dispersion does not have enough kinetic energy to fill the apertures and air bubbles form in the printed depositions. The squeegee force should be set in a range that allows thorough stencil cleaning. Increasing the squeegee force bends the squeegee blades resulting in a decreased squeegee angle ultimately improving the printing quality. A sharp angle ranging between 60° and 35° is the best solution. However, if a specific squeegee force is required, an optimal squeegee angle must be set. Various different HMA dispersion formulations have been examined to characterize and optimize the application process. In a first step, formulations of three dispersions with different hot melt powder particle sizes of $\leq$ 20 µm, $\leq$ 40 µm and 20–40 µm were optimized. The basic constituents of hot melt dispersions are a solvent (destilled water), a thickening agent and powder particles of different sizes. After the printing process, hot melt dispersion and substrate are heated in a convection oven. In a gradual heating process the water evaporates in most cases at 80°C, below the melting temperature of the hot melt. In a second step, the dry deposited adhesive is gradually heated to a temperature of 150°C. The adhesive melts and the liquid hot melt forms a hemispherical structure under the influence of surface tension. During the remelting process the printed hot melt dispersion volume shrinks about 45–55% depending on the particle size.

The stronger the dispersion is sheared the lower is its viscosity. This characteristic favors easy aperture filling during the printing process. Measurements show that the viscosity is decreasing proportionally with increasing shear rates.

1. Viscosity at a shear rate of $0.1\,s^{-1}$ amounts in most cases to about 400 Pa s. It is near the – for a good printing result – required 450 Pa s published by Kolbe [8].
2. Viscosity at a shear rate of $100\,s^{-1}$ is in range of 10–15 Pa s, the dispersion fills the apertures with ease.

Hot melt dispersions of all particle sizes show a shear thinning characteristic. Another rheological characteristic is the relaxed viscosity after a printing cycle. Measurements with a 3-phase-rheometer were carried out to test these thixotropic properties These 3-phases describe the typical shear thinning characteristics of a stencil printing process. In a first step, a shear rate of $0.01\,s^{-1}$ is set. This shear rate simulates the passive state of the dispersion. In a second step, the shear rate is increased to $100\,s^{-1}$. This simulates the dispersion moving across the stencil during the print cycle and the dispersion flow into the stencil apertures. Eventually, the dispersion was adjusted to a shear rate of $0.01\,s^{-1}$. The measurement was carried out by means of a rotary rheometer in a plate-plate test arrangement with a measuring gap of 300 μm. Table 18.3 shows the results of the 3-phase measurement.

Table 18.3 Results of 3-phase measurement

Particle size in dispersion	Viscosity at shear rate $0.01\,s^{-1}$	Viscosity at shear rate $100\,s^{-1}$	Viscosity directly after forced impact	Relaxation-time to original viscosity
$\leq$ 20 μm	120.000 Pa s	15 Pa s	15.000 Pa s	27 min
20–40 μm	100.000 Pa s	15 Pa s	50.000 Pa s	25 min
$\leq$ 40 μm	50.000 Pa s	10 Pa s	8.000 Pa s	38 min

The dispersion with a particle size of $\leq$ 20 μm shows the highest viscosity. Dispersions with particles of 20–40 μm are in about the same viscosity range. Dispersions of $\leq$ 40 μm have a lower viscosity. All dispersions have thixotropic

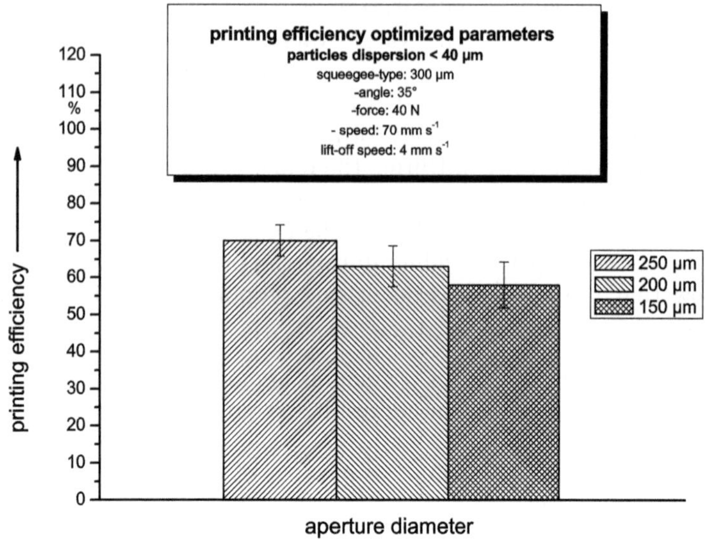

Fig. 18.6 Printing effiency optimized parameters

characteristics and show no bleeding effects after printing. Hot melt dispersion can change their rheological properties during longer storage periods in closed storage cups. Possible reasons could be particle sedimentation, solvent evaporation or particle agglomerations. The long-term dispersion stability is good because dispersions show no tendency to sedimentation. Rheology measurement prove, that after 4 days of storage in open cups the viscosity does not change. After 10 days a substantial decrease in viscosity was measured. As a result, dispersions should not be used in stencil printing processes after 4 days of open storage.

After the most important influencing factors in a hot melt dispersion printing process were optimized, formulations could be established for printable dispersions and printing parameters applicable to minimal reproducible adhesives volumes in a sub-nl-range. Hot melt dispersions of $\leq 40\,\mu m$ yielded reproducible prints, see Fig. 18.6. Aperture diameters of 200 µm showed a high printing efficiency in a range of 65–70%, see Fig. 18.7. Tests with dispersions types $\leq 20\,\mu m$ yielded even better results in a range up to nearly 78–80%. However, dispersions with particles of 20 µm show many inconsistent printing results leading to the realization that particles were too small and many of the fine particles stuck to the aperture walls. Up to an aperture diameter of 150 µm, it is more useful to use dispersion with particles of 40 µm and smaller. Small hot melt dispersion depositions can be reproducibly printed in volumes of about 0.40–0.43 nl by a diameter of 150 µm.

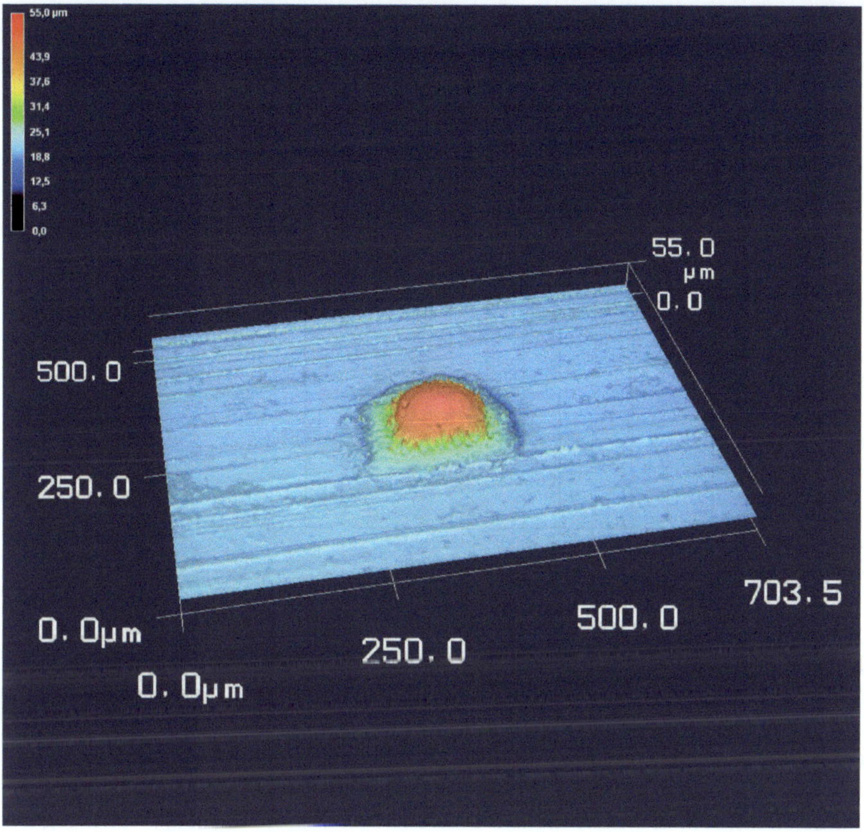

Fig. 18.7 Test structure of printed hot melt dispersion, diameter 200 µm, volume 0.75 nl

18.6 Properties of Hot Melt Adhesive Bonds

It is the main objective of this research project to evaluate the selected adhesives, develop application methods and different types of process controls by testing the adhesive bonds. Destructive testing methods were used to examine the bonds. The compression shear strength of bonded Si-Chips ($1.4 \times 1.4\,\text{mm}^2$) on Si-substrates ($4 \times 10\,\text{mm}^2$) was detected with two different test setups. Some of the specimens underwent destructive tests using a modified micro tensile test rig and others were tested by means of the shear head of a wire bonder. A visual inspection of the fracture surfaces completed the tests. The joining gap of the tested structures amounted to 25 µm. Silicon (Si) was cleaned with isopropanol and has a surface energy of $57\,\text{mN}\,\text{m}^{-1}$. Fig. 18.8 shows the compression shear test results.

The adhesive systems *Technomelt*, *CaPa* and *Jowat* had the lowest compression shear strength of all tested systems in the range of 4.5 to 5.5 MPa. A

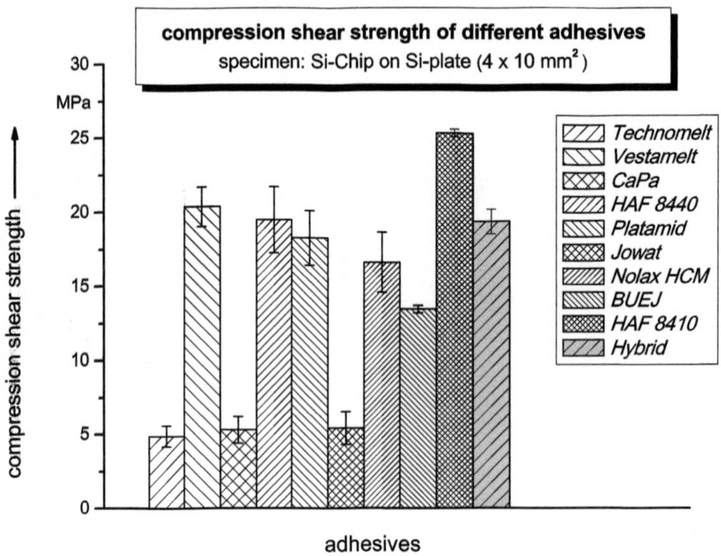

Fig. 18.8 Results of compression shear strength

compression shear strength of 5 MPa is sufficient for most micro technology applications. An example is a micro grinding process of hot melt bonded ceramic substrates in which shear forces up to 10 N were induced. With regard to the joining geometry, a maximum compression shear strength of 2.5 MPa is required. In most tests, adhesive systems with a low melting viscosity of (≤ 1.000 Pa s), *Technomelt* and *Jowat* show cohesive failure behavior after the visual fracture analysis. These types of adhesives feature a flexible matrix. Best system of adhesives with compression shear strengths higher than 10 MPa, is the *HAF 8410* with 25.3 MPa. In most failure analyses, this adhesive shows very good adhesion to silicon. By EDX (Energy Dispersive X-ray Spectroscopy) silicon fragments broken out of the Si-substrate were detected on the adhesive surface.

The compression shear strengths of different adhesive systems with enhanced thermal stability were measured in extensive tests. Reactive hot melts crosslinking under heat show very good results. These systems feature both compression shear strengths ranging from 13 to 25 MPa and a thermal stability of 160°C and higher at a (DMA) storage modulus of 1 MPa. These results are verified by compression shear tests under thermal load. Hybrid-bonding, the second approach to achieve thermal stability, is an alternative to currently available systems. A low viscosity epoxy system provides thermal hot melt stability which is essential for fast fixing. Typical shear compression strengths range between 15 and 22 MPa depending on the adhesive system.

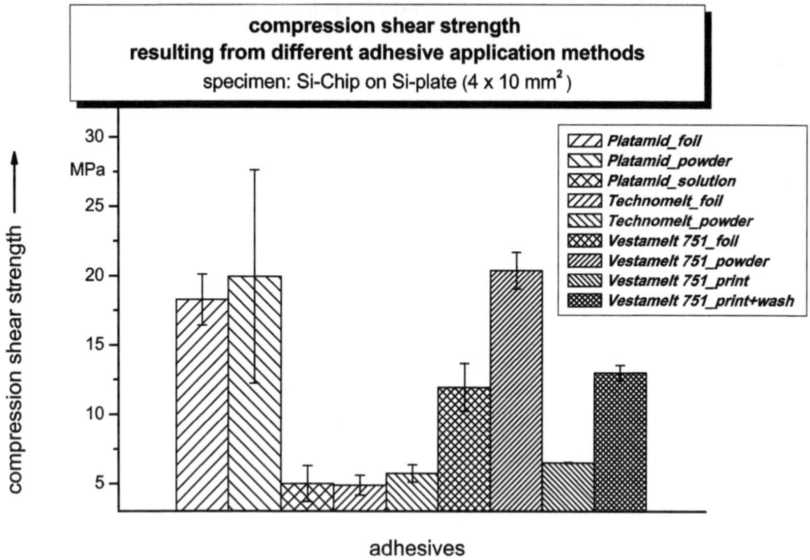

Fig. 18.9 Compression shear strength application methods

Different shear compression strengths relative to the application methods are shown in Fig. 18.9. Powder deposition of hot melt particles shows high compression shear strength with various adhesives. Layers of particles not applied over the entire surface are usually characterized by a cohesive failure pattern. The high deviation is caused by an uneven particles deposition, i.e. fluctuations in the thickness of the adhesive layer. In contrast, film application shows less compression shear strength, the corresponding fracture analyses show adhesive failure patterns. This is because films do not bond tightly to the substrate. Dispersion stencil print yields different results: Compression shear strength varies greatly which is due to the fact that the printed and remelted dispersion contaminates the adhesion surface. The thickening agent diffuses up to the surface. Washing with distilled water removes contaminations and yields better compression shear strengths. Eventually, dispersion stencil printing produces very accurate adhesive volumes of and thus high compression shear strengths with low standard deviations. In comparison to other application methods, soluble hot melts show constant strengths.

Variable process parameters, such as joining -force, -time and -temperature had the following effects: If the joining temperature is at the tail end of the process mechanical properties of many adhesive systems begin to fluctuate. That translates into reduced compression shear strength and thus compromised adhesion to the substrate. Compression shear tests yielded good results in processing temperatures ranges of 150 to 180°C. Minimal joining forces should be in a range of 3–5 N. For small Si-devices (thickness $\leq 250\,\mu m$), a process time of 3 s is adequate.

18.7 Conclusion

This paper presents a joining technology using HMAs for micro assembly processes. Major benefit of HMA systems is that they allow to join even smallest geometries with minimum doses of adhesives. The short setting time of about 1–2 seconds which is required until the joint reaches handling strength is another advantage compared to presently used viscous adhesive systems. HMAs feature handling strengths (usually the ultimate strength) in a range from 5 to 25 MPa related to the compression shear strength. Various thermally stable systems were tested for special applications where a higher thermal stability of the adhesive was required. These systems are heat-resistant up to 160°C. "Hybrid-bonding" fixes microparts with thermally stable hot melts, and uses an underfill with a viscous adhesive in a second process step to achieve maximum bond strength. Here, hot melts enable faster process times, higher performance and thermally stable joints. The extremely fast setting times of HMAs reduce cycle time. If all process parameters, such as joining -force, -time and -temperature are optimized and the heat management is adapted to the assembly process, process times of 3 seconds are possible. The use of high-viscous hot melts prevents inaccuracies in the assembly process, such as tilted components and inaccurately squeezed out adhesive. In addition, hot melts offer the benefit of batchwise pre-application of various different micro parts. Such a pre-application does not require joining directly after the application of adhesive to substrate. The joining process can take place subsequently or even in a different location. Various different application methods have been developed: Discrete application of adhesive balls and spheres gripped by vacuum and electrostatic handling devices were described. Films can be applied with different gripping systems or by spin-coated soluble hot melts. Batchwise application methods are deposition of powder particles and stencil printing of hot melt dispersions. Both methods allow to apply large-surfaced substrates, such as wafers, but also smallest devices in batch processes. The previously described methods offer reproducible application processes using the following minimum volumes of adhesive:

- Balls with a diameter of 500 µm and a volume of 65 nl, undefined spheres in pl-range possible,
- Films to a thickness of 25 µm,
- Spin-coated soluble hot melts up to 1 µm layer thickness,
- Powder particles to 15 µm, melted powder particles to volumes of 1 nl,
- Stencil printing of hot melt dispersions with good reproducibility: spot diameter of 150 µm with 0,5 nl/spot.

Thanks to these specifications even smallest volumes of adhesive can be applied to join miniscule bond areas. In order to connect micro parts electrically, hot melts dispersions are modified with different conductive fillers. First tests show volume resistivities as of standard isotropic conductive adhesives [9].

Acknowledgements The authors gratefully acknowledge the funding of the reported work by the German Research Foundation within the Collaborative Research Center 516 “Design and Manufacturing of Active Micro Systems”.

References

[1] Böhm S, Dilger K, Hesselbach J, Hemken G, Rathman S (2005) Pick and stick – Automatisiertes Fügen klebstoffbeschichteter Bauteile. Mikrosystemtechnik Kongress 2005 pp 799–802
[2] Böhm S, Dilger K, Hesselbach J, Stammen E, Wrege J (2005) Adhesive joints for MEMS using hotmelts. MST News 5(1):34–35
[3] Böhm S, Hemken G, Rathmann S, Wrege J, Hesselbach J (2007) A hot melt based interconnecting technology for MEMS and MOEMS. Smart System Integration 2007 pp 435–438
[4] Brockmann G, Geiss PL, Klingen J, Schröder B (2005) Klebtechnik, vol 2. Wiley, Weinheim
[5] Büchner J, Henning W, Stepanski H, Raffel B (2005) Lagerstabile latentreaktive Klebfolien und ihre Einsatzchancen. Adhäsion – Kleben und Dichten 56(7/8):23–28
[6] Götz F, Gesang T, Hennemann OD (2003) Gefüllte Klebstoffe: Mikrodispensen ohne Aussetzer. Adhäsion – Kleben und Dichten 47(7/8):38–42
[7] Habenicht G (2006) Kleben: Grundlagen, Technologien, Anwendungen, vol 5. Springer, Berlin
[8] Kolbe J (2008) Screen or stencil printing of accurate adhesive structures. Adhesion – Adhesives and Sealants 9(1):24–28
[9] von Schilling A, Hemken G, Böhm S (2009) Kontaktierung von Mikrosystemen auf Textilien durch elektrisch leitfähige Schmelzklebstoffdispersionen. Proceedings Mikrosystemtechnik Kongress 2009 pp 773–776
[10] Van Brussel H, et al (2000) Assembly of microsystems. Annals of the CIRP 49(2):451–472
[11] Zäh M, Schilp M, Jacob D (2002) Kapsel und Tropfen – Fluidauftrag für Mikrosysteme. wt-Werkstattstechnik online 92(9):428–431

Chapter 19
Design of a Microassembly Process Based on Hot Melt Adhesives

S. Rathmann, J. Ellwood, A. Raatz, J. Hesselbach

Institute of Machine Tools and Production Technology
Technische Universität Braunschweig
a.raatz@tu-braunschweig.de; j.hesselbach@tu-braunschweig.de

Abstract Within a microassembly process, four independent variables can be identified, which influence the overall accuracy of the assembled part in a high dimension. These variables are the uncertainty of the moving parts of the handling system, the component fixtures, the gripping elements and the joining technology. Due to the complexity of these variables, an all-embracing conclusion of the resulting joining uncertainty, primarily taking into consideration the different effects of liquid adhesives, cannot be made. In this context, the curing process, the surface properties, and the design of the assembly process are discussed in detail, with special emphasis on a new approach that takes advantage of the favorable characteristics of physical setting hot melt adhesives instead of chemically reactive liquid adhesives. The integration of the newly developed assembly processes using hot melt adhesives results in increasingly accurately assembled microcomponents.

19.1 Introduction

The production of hybrid microsystems places high demands on the interaction of different production processes: not only the fabrication of MST-components and the control of the complex thin and thick film processes, but also the design and assembly processes. In the production of hybrid microsystems, the assembly method has a drastic influence on the functionality of the product. Therefore, it is very important in the assembly to comply with tolerances given in the design process. Within MST assembly processes, these tolerances are in the range of a few micrometers to several hundred nanometers.

Recent investigations have shown that not only the development of high-precision assembly systems is necessary, but also the presence of high-precision handling and joining technologies as well as adapted assembly strategies. This is mainly because micro-specific characteristics must be considered by reason of the small size of components. Therefore, the miniaturization of production machines is not only sufficient [3]. The use of new technologies, which sometimes use micro-specific component properties e. g. for gripping (Chap. 16) and the use of new joining technologies, their use does not appear practical in macroscopic applications, can lead to a solution of these problems. In particular, the optimization of existing joining technologies and the development of new joining technologies play a crucial role in this context. Schttler [12] enumerates a few influence variables which affects a assembly process. The influence of these variables are are not even known partially, or are not completely controllable. The largest and least controllable influence were the parameters of the joining process. The new joining technology, presented in Chap. 18, describes the possibility of making the influences of joining more controllable, because hot melt adhesives use a less complex setting mechanism, which can be simply controlled. In Chap. 18, the technical requirements for creating suitable hot melt volumes and application techniques were presented. In this chapter, the development of the technical requirements that enable the semi-automated and automated joining of micro parts with this new joining technology will be presented.

Before the various aspects of implementation of a hot melt based joining technology are discussed, the influence variables of a microassembly process will be explained. In addition, assembly strategies will be described that can exploit or avoid the above described micro-specific component behavior.

19.1.1 Variables Influencing the Assembly Process

The acknowledgment of possible influence variables is important for a highly accurate automated assembly of hybrid microsystems. The possibility to control the influences allows to increase the accuracy and leads to a safe and reproducible assembly process. The influence variables identified from macroscopic assembly processes can be adopted for the most part, but have to adapt into the microworld. Mainly the influence of the special characteristics of the micro world have to account. The most important influence variables to the micro assembly are listed below. A detailed description of the influences is given in [13].

1. Tolerance of the component (dimension, surface)
2. Environmental influences
3. Error influences of the positioning system
4. Uncertainties of the measurement systems
5. Error influences of the gripping technology
6. Setup of the fixture systems
7. Influences of the joining technology
8. Influences of the assembly strategy

19.1.2 Assembly Strategies

There are two methods to assembling microparts: serial and parallel assembly. The serial assembly strategy is used if single parts, which mostly come from different manufacturing processes, have to be assembled. An advantage of this strategy is the possibility to adjust every part individually so that a high positioning accuracy can be reached.

The idea of the parallel assembly stems from the idea that parts that are produced on batch processes have a high degree of reproducibility. This result can be used to assemble many parts at the same time with an accuracy which relates from the previous manufacturing process. This assembly strategy shortens the assembly time, especially for batch-produced components [8].

Due to the special properties of microparts a third assembly strategy is possible. The relation between surface and volume forces allows in some cases an adjustment of the parts to each other by itself. This strategy is called self assembly [4]. At this strategy, the microparts position and orientate themselves due to liquid forces, electrostatic forces or the use of ultrasound [2, 7].

The choice of a assembly strategies depends on different parameters. At this point a detailed conclusion can not been given. Important parameters, which have to consider at the choice are the quantity, the precision and the complexity of the assembly process.

19.1.3 Joining Technologies

MST-joining technologies are especially based on the technologies used in the packaging of integrated circuits. These methods can be categorized in joining by soldering, joining by welding and joining by gluing. In the MST in each group specially adapted or new joining technologies are used. Many of these technologies which can be used in microassembly are still in development. Soldering technologies are for example the Active soldering, the Transient Liquid Phase Bonding and the Laser beam soldering. The joining technologies that use welding are Laser beam welding, Electron beam welding and Ultra sound welding. A good overview is over these joining technologies is given in [6]. Most often, however, the gluing is applied, since this technology can bond different materials simply and quickly. Traditionally, viscous adhesives are used when gluing. These adhesives crosslink by chemical reactions. A distinction is made between the moisture-curing, heat-curing and radiation-curing adhesives. The main disadvantage of these adhesives is that their setting time is extremely long (10 s or longer). This can move or tilt the joined components. In contrast, physical bonding adhesives set very quickly. When using appropriate heat management, setting is done within a few seconds to below one second [1]. For this reason, in this work, the bonding with hot melt adhesives has been used as joining technology.

19.2 Process Design for the Joining with Hot Melt Adhesives

When using hat melt adhesives for the joining of micro components the heat management is a crucial point. Mainly the heat introduction and dissipation is a very important object. in this chapter the fundamentals for suitable heat management concepts will be developed. Finally two heat management concepts will be discussed which uses different approaches for the heat introduction and dissipation.

19.2.1 Tasks of a Heat Management

Due to the influence of viscosity to the processability of the adhesive and the assembly process, the use of a suitable heat management is required. in order to maintain optimal adhesive properties of the hot melt adhesive. The fundamental functions of heat management in the process are to introduce heat for melting the adhesive, to keep the heat for optimal assembly process and to dissipate the heat in order to quickly cool off the adhesive for a stable bond. The heat management is tightly coupled with the handling system and especially with the gripper or fixture system. The design of these components has to be considered separately and will be discussed later in this chapter.

Heat Introduction

To heat a hot melt adhesive from room temperature to the processing temperature, a heat $Q_{Adh,\,all}$ is needed. As the condition of aggregation is changed during the process of heating, an additional amount of energy $Q_{S,\,Adh}$, the so called latent heat of fusion, is needed for melting. In most cases it is not only the adhesive that has to be heated, but also the coated component (see Chap. 16 and the handling equipment. Therefore additional energies Q_{Comp} and Q_{Hand} are required. The outcome is the entire amount of heat:

$$Q_{all} = Q_{Adh,\,all} + Q_{Comp} + Q_{Hand} \tag{19.1}$$

whereas

$$Q_{Adh,\,all} = Q_{Adh} + Q_{S,\,Adh} \tag{19.2}$$

Heat Dissipation

The adhesive has to cool down after the assembly process. The temperature of the adhesive, the component and the handling equipment has to fall below the melting temperature T_S so that the adhesive can set and a stable bond can be achieved. This can be done with two mechanisms: running off the heat into a cold joining partner or actively bringing down the temperature by cooling the complete system. In the first mechanism, the joining partner has to be a sufficient heat sink or the heat transport has to be provided by the conduction of heat into the surroundings, (handling equipment of the joining partner). In the second mechanism, the heat is drawn from the joining patch and transported to the surroundings via a connected cooling system. This mechanism needs extra active components that enhance the complexity of the components responsible for the heat management.

As the assembly process does not take place in infinitesimally short time, the adhesive has contact with the joining partner before the final joining position is reached. For the first mechanism, it has to be taken into account that the heat begins to conduct into the colder joining partner at the first contact of adhesive and component. In this case dimensioning the heat management to provide sufficient heat to finish the assembly process with optimal adhesive properties is important. This also applies for the active cooling system, where the appropriate time for connecting it has to be defined.

Gripper

The connectors between the components and the source of heat are the handling components, in particular those for gripping and storing (magazines). The requirements to these components are:

- Holding the components under the influence of variable temperatures. This includes the fact that the position and orientation of the component are not to be changed during the assembly process by the heat or forces.
- Integrating the source of heat with slight loss of the conduction of heat to the component.
- Integrating active or passive heat sinks that absorb excessive heat after the assembly process and conduct it into the surrounding.
- Isolating the heat sensitive components of the handling system and potentially heat sensitive measuring and monitoring systems from the source of heat. In this context, particular attention has to be paid to the heat expansion of the handling components.

Implementation

The implementation of the tasks described above into an assembly process can be done in multiple ways. Fig. 19.1 gives an overview of solutions of the tasks: heat introduction and gripping. Another important function for the heating of components is the heat transfer. It is necessary for heat introduction and heat dissipation. In the figure also solutions for heat transfer are shown.

Aspect	Solution		
heat generation / introduction	electrical resistance	electro-thermal principle	chemical reaction
heat transfer	conduction	convection	radiation
gripping force	mechanic	electrostatic	vakuum

Fig. 19.1 Solutions for the aspects of heat management

From this figure, different approaches for the implementation of a heat management for assembly processes are possible. In the following two heat management concepts will be introduced and there advantages and disadvantages will be discussed.

19.2.2 Heat Management

Heat management is the integral part of the selection and the design of an assembly process using hot melt adhesives, as the volume of the hot melt is quite small and the thermal capacity is rather low. This results in the hot melt is not being able to store much thermal energy and thus heating the hot melt adhesive directly is not recommended. Contrarily the thermal capacity of the component and the gripping system is much higher. Hence the characteristics of the grippers and the components, such as thermal capacity and thermal conductivity, influence heat management much more. For this reason, special considerations have to be taken when designing a specific assembly process, as the choice of a suitable heating source is crucial for process design.

In general the concept of heat management can be split into two different variants. On the one hand, a heat storage component, which is loaded before the joining process, is used to introduce the heat energy into the component and adhesive. This concept is called passive heat management. On the other hand, a heat source is integrated in the handling devices, whereby heat can be introduced continuously. This is called active heat management. The functionality as well as the advantages and disadvantages of these different concepts will be described and discussed in the following sections. The principle temperature profile is displayed in Fig. 19.2.

Passive Heat Management

Passive heat management uses the heat accumulating properties of adhesives, components and grippers for adjusting the temperature profile that has to be passed through. Negative effects are caused by the size of the source of heat as well as by the distance to the assembly site. As a result, there are additional

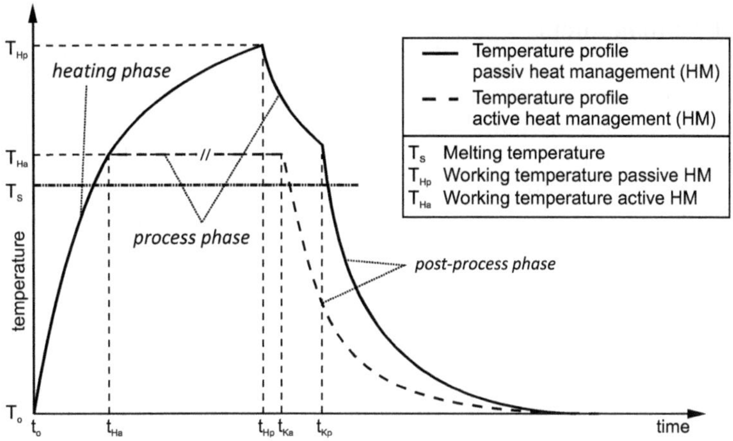

Fig. 19.2 Principle runs of passive and active heat management

paths during the assembly process. Within an assembly process, three phases have to be differentiated: the heating phase, the process phase and the post-process phase. During the heating phase, gripper, component and adhesive are brought to working temperature T_{Hp}. The working temperature is an important process parameter. It is mainly chosen due to the range of the melting temperatures respectively the processing temperature of the hot melt adhesive. In combination with the heat capacity of the gripper, the component and the adhesive the length of the process phase will be influenced. The time to heat the system is called t_{Hp}. During the process phase the component will be moved to the assembly place. In this phase, possible measurement and post- positioning of the components is included t_{Kp}. As the gripper is removed from the source of heat, the temperature declines. As soon as a component makes contact with the substrate, the temperature of the adhesive and the gripper and the component declines much faster, due to the heat being conducted into the substrate. This phase is called the post-process phase. At the time of the contact between adhesive and substrate there has to be sufficient heat energy to wet the substrate. If this is not the case, the parts will not be joined. From this, the correlation between adjusted working temperature during the heating phase and the needed heat energy for the assembly process is evident. When applied to passive heat management, this correlation is critical and has to be adapted to every particular assembly task. The advantage of this concept is the simplicity of the gripper system. If the heat capacity of the component is not sufficient, the gripper can be equipped very esay with an additional heat reservoir.

Active Heat Management

Active heat management, in contrast to passive heat management, actively carries the source of heat within the gripper system. As an additional element a cooling system within the gripper system would be possible, in order to achieve a faster heat dissipation after the assembly process. The process flow of an assembly process using active heat management can also be divided into the three phases that were identified in the section about passive concept. The difference from the passive concept are the values of the process and post-process phase. During the process phase there is no temperature decline in component and adhesive taking place, due to the continuous heat input supplied by the carried source of heat. Based on this fact, the working temperature T_{Ha} can be kept lower. This is mostly done in the range of the processing temperature. As a result, the component has to bear a smaller thermal load. When component and substrate comes into contact the heat source has to provide enough heat energy to cover the substrate with the melted adhesive. During the subsequent post-process phase (t_{Ka}) the temperature of the adhesive has to decline rapidly below the melting point range, just as in the passive concept. The heat input is then stopped by switching off the source of heat and switching on the cooling system.

The implementation of these two heat management concepts is different because of the different design of the process phases. In the following sections an example of implementation for passive and active heat management will be described.

19.3 Implementation of a Passive Heat Managment

19.3.1 Design of Passive Heat Management

Process Design

Prior designing the components which implement the assembly process and the heat management, the assembly process has to be planned taking the heat management into consideration. When planning the assembly process, the following steps have to be taken into account:

1. Measuring and gripping the component
2. Positioning the component
3. Adjusting the component
4. Carrying out the joining movement
5. Releasing the component
6. Examining the joining process (optional)

In addition, certain heat management tasks must be considered:

a) Heating the adhesive, the component and the gripper (heating phase)
b) Keeping the heat until the joining process is started (process phase)
c) Cooling the component (post-process phase)

In the process planning it is important at which moment during the assembly process (1–6) the steps of the heat management (a–c) are integrated. With passive heat management steps b) and c) cannot be influenced directly. The heating has to take place before the joining process starts. While planning the process it should be considered whether the heating should take place before or after adjusting the component. Alternatively, the heating of the component or the gripper could take place before gripping the component. Fig. 19.3 displays three different possible variants of the process. The steps of the heat management method are illustrated with dashed lines.

For the three variants of the process, the different length of the process phases is remarkable. It does not necessarily mean that they take more time, but that a different number of process steps is included. This can lead to a longer process phase. The advantage of process variant 1 is that the process phase needs the shortest time to be completed. Using Fig. 19.2 one can conclude that the working temperature T_{Hp} can be chosen lower than on the other variants. A disadvantage is that adjusting the components takes place before heating and thus errors occurring through the heating cannot be compensated. These errors can be taken into account in the adjustment in variant 2. However, this extends the process phase with the time needed for the adjustment and leads to a higher working temperature T_W. Variant 3 makes sense when the heat capacities of the component and the gripper are high enough to keep the temperature of the adhesive close to the processing temperature T_P between gripping and joining. The advantage of this variant is that the heating requires no time during the process phase and thus the time needed for the assembly cycle is shortened. Using these process models, a detailed design for the assembly using passive heat management can be done.

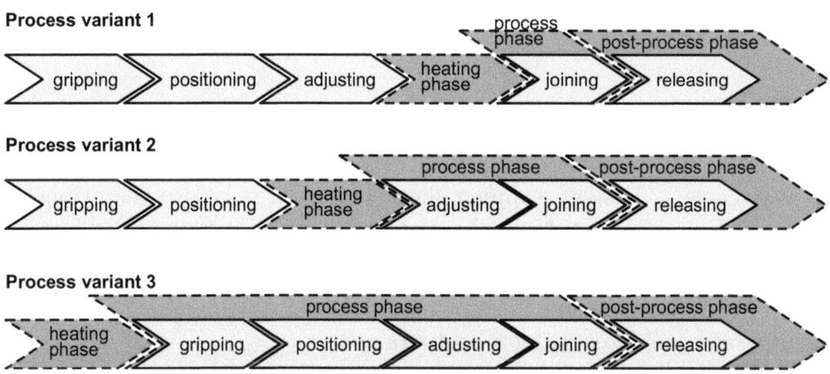

Fig. 19.3 Process variants by using passive heat management

Fig. 19.4 Heat flow through the use of heat radiation

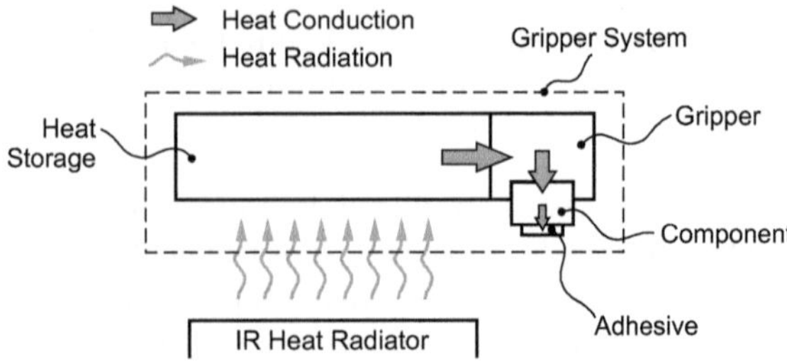

Thermal Design

As shown in Fig. 19.1, heat production and induction can be done using three principles: lost heat via electric resistances, the electro-thermal principal (peltier effect) and exothermic chemical reactions. Using chemical heat seems to be infeasible for this application. The electro-thermal principal can be used but initially seems to be ineffective due to the low power density. Use of lost heat at electric resistances a high power density can be achieved. Moreover, a wide range of developments is available (e.g. heating plates, heating cartridges, heating film, etc.). The heat transmission from the generator to the adhesive can take place via heat conduction or thermal radiation. Convective heat transfer is possible as well, but due to the enormous loss during the transmission it is not as effective as the other two transmission methods. For process variant 1 and 2 in Fig. 19.3 heat conduction cannot be used for the heat transmission between source and adhesive, because a direct contact between component and heat source would be needed for this method. This could lead to shifting the component within the gripper and thus influence the accuracy of the assembly process. In process variant 1 and 2 heat transmission via thermal radiation can be used. Possible wavelength ranges are those from the range of infrared to hyperfrequency waves. As generating hyperfrequency waves is a lot more complicated compared to generating infrared waves, using infrared waves is recommended. Fig. 19.4 schematically depicts the heat flow for process variant 1 and 2.

Gripper Design

Based on Fig. 19.3, an adequate gripper concept has to be chosen for passive heat management. Besides gripping the component, it is also the gripper's task to function as port between heat storage and component. As a result, the gripper has to fulfill the following functions and requirements:

- Gripping the component
- Incorporating or integrating a heat storage
- Thermal stability
- High thermal conductivity into the gripped component

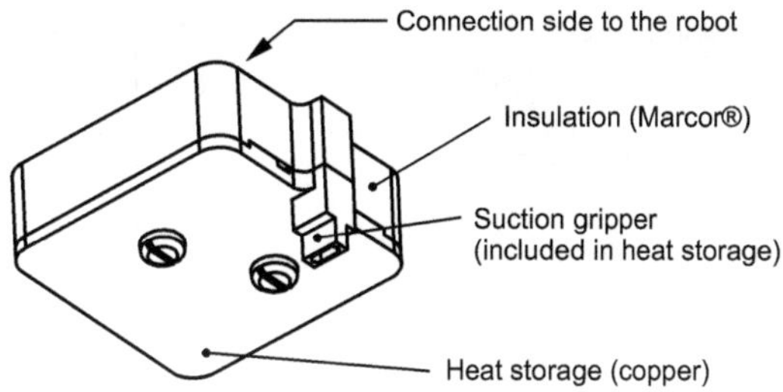

Fig. 19.5 Draft of the passive gripper

From these requirements, specific properties can be derived. To fulfill the last three issues, the gripper should be designed in the following way. Metals are particularly suitable to store heat. This is due to the high heat capacity and good heat conduction. These characteristics result in a high thermal diffusivity, a measure of the time change of the spatial temperature distribution in objects. This can guarantee a good heat conduction into the component. Due to the good machinability of metals, they can be easily integrated into existing gripper systems. Thermal stability in this context means that gripper's functionality must also be guaranteed for the large variations in temperature (room temperature – temperature processing). This is achieved with a simple construction and few moving parts. The considerations mentioned above lead to the use of a vacuum gripper, which can produce high holding forces. The contact space can be designed in any size. Only the impermeability of the vacuum system has to considered. This kind of gripper has no movable parts, and the parts can be produced of any material. Besides the listed functions, a good thermal insulation to the environment and especially to the handling system must of course also be guaranteed.

In Fig. 19.5 a blue print for a gripper for passive heat management and process variant 1 and 2 is displayed. The essential part of the gripper consists of insulation and the heat storage which at the same time is responsible for the gripping function. The gripper is constructed in such a way that it can be installed at the handling system described in Chap. 15. During construction, it was noted that the component can be recorded almost entirely by image processing. The vacuum supply is carried out through the insulation, so that there are no additional parts coming in contact with the heat.

19.3.2 Simulation of Assembly Processes

The gripper system in Fig. 19.5 was developed using simulative thermal analysis while it was constructed. This was mostly done to design the heat storage. Heating the gripper is done with the help of infrared waves according to process variant 1 and 2. The analysis of the heat impact to the gripper through infrared waves and the heat conduction into the component is therefore important for the lay-up of the heat storage. The insulation was made

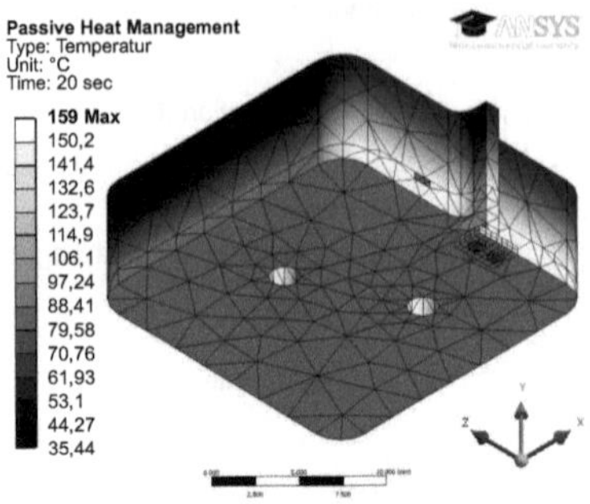

(a) Model

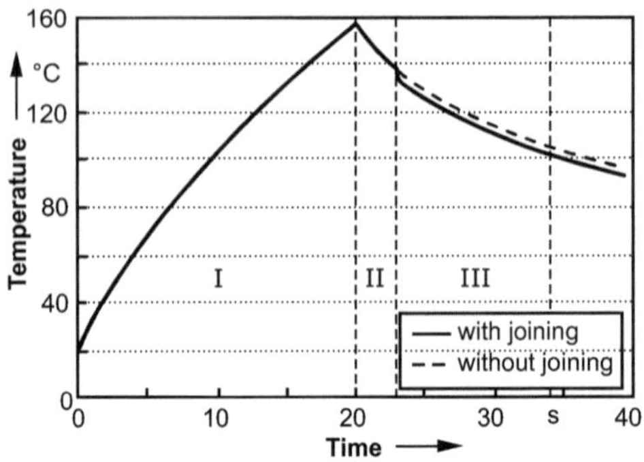

(b) Results

Fig. 19.6 Simulation model and results for an assembly process with the passive gripper system

from *Marcor*. This is a ceramic with a high thermal resistance and a low thermal expansion. Additionally, the machinability of this material is very good. The material chosen for the heat storage at first was copper due to its outstanding properties in heat conduction. As the copper's absorbance factor for infrared waves is very low ($\alpha \approx 0.2$), the irradiated area was coated with baking enamel. This enabled increasing the absorbing factor up to $\alpha = 0.97$. To calculate the heat introduction the dispensed radiation power of the IR radiator P_1 was measured:

$$P_1 = \epsilon \, \sigma \, T_1^4 \, A_1 \tag{19.3}$$

Where ϵ is the radiator's emission ratio, σ is the Stefan-Boltzmann constant, T_1 is the temperature of the radiator and A_1 is the radiator area. Index 1 stands for the radiating body. Index 2 thus stands for the irradiated body. Taking the sight coefficient F_{12} and the absorbance factor of the heat storage into account, the heat introduction Q_2 can be calculated.

$$Q_{in} = \epsilon_1 \, \epsilon_2 \, F_{12} \, \sigma \, T_1^4 \, A_1 \cdot \tag{19.4}$$

With these determined parameters, a simulation of the process phases (heating phase, process phase and post-process phase) can be conducted. For the post-process phase it was assumed that the forming of the adhesive due to the joining speed happens faster than the general heat outlet. This means that for the simulation, the changing surface of the adhesive during the joining is not considered. In Fig. 19.6.a, the simulation model of the adequate passive gripper is displayed. Fig. 19.6.b depicts the temperature profile of the bottom side of the component during the three process phases, once with and once without joining process. The diagram shows the apparent faster decrease of temperature in the component when a joining process is performed.

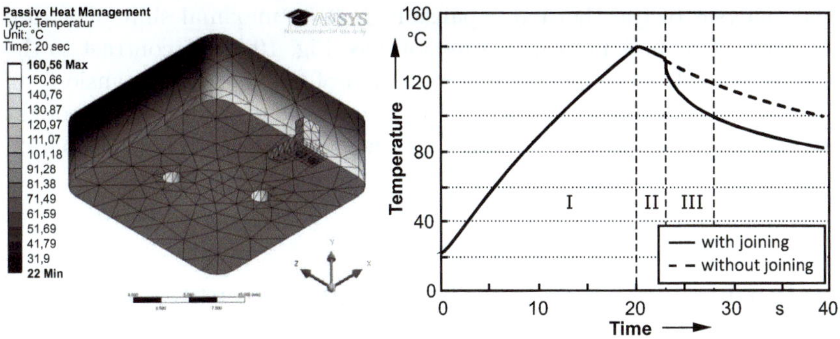

(a) Model (b) Results

Fig. 19.7 Redesign of the passive gripper, simulation model and results

Analysis of the properties of the thermal expansion shows that using copper as heat storage material is impractical for high accuracy. Due to large dimension which is needed for heating the gripper, the gripper is moved about 50 µm to the front and about 18 µm up at the moment of maximal temperature (end of the heating period). As a result, a new design of the heat storage is required. The new material used for the heat storage is *Invar*, which has about the same heat capacity as copper but has only one tenth of the length. However, it has a clearly smaller heat conductance than copper. In order to conduct enough heat to the compound and the adhesive, *Invar* was used for the heat storage and copper for the gripper. The redesigned passive gripper is depicted in Fig. 19.7.a. The chart in Fig. 19.7.b illustrates the results of the process simulation.

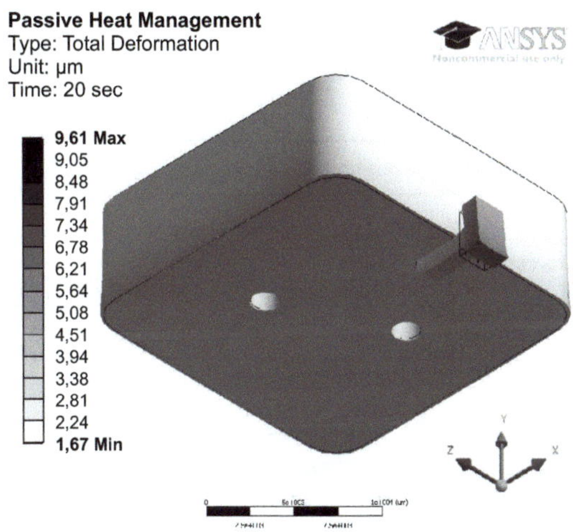

Fig. 19.8 Results of the thermal deformation simulation

The analyses of the thermal expansion shows a maximal shift of 9.6 µm in the x-direction and 1 µm in z-direction (see Fig. 19.8). In contrast to the heat storage mode of copper a large reduction of the thermal expansion can be reached by using *Invar* as storing material. The lower expansion is bought by an also low thermal conductivity. This leads to a faster heat dissipation during the post-process phase.

19.4 Implementation of Active Heat Managment

Active heat management can be approchached in the same way as passive heat management as described in Sect. 19.4. The considerations made for process planning and heat management in Chapt. 19.3.1 have to be rethought and adapted to the new conditions for the active heat management.

19.4.1 Design of Active Heat Management

Process Design

When planning an assembly process with active heat management, one has to consider the same steps as with passive heat management, but the priorities in considering the process steps for the heat management change. While keeping the heat plays an important role in passive heat management it is rather unimportant for the active heat management as there is a constant heat input. This process step is more or less limited to the moment of joining. In contrast to passive heat management, the cooling of the component is an important process step, as active engagement is needed to stop the heat

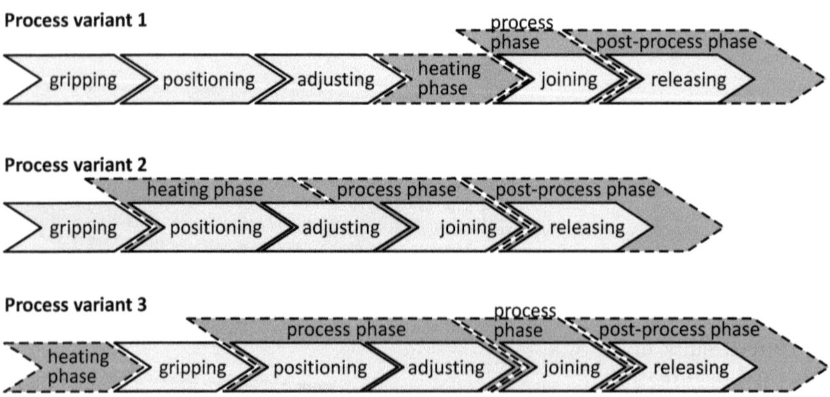

Fig. 19.9 Process variants by using active heat managament

input. Fig. 19.9 depicts possible assembly process variants with active heat management.

Variant 1 can be changed so that the heating phase starts in the adjustment phase. Variant 3 is an alternative where the energy application into the gripper can be reduced. If the component is heated up to working temperature before the start of the assembly, a lower energy input is needed to keep the component at that temperature. As a result, simplifications can be made at the constructive design for the insulation.

Thermal Design

As with passive heat management, heat introduction and transfer can be done with the possible methods shown in Fig. 19.1. Likewise, using chemical energy is not recommended with active heat management. Electrical heat can be used for active heat management just as for passive heat management. Unlike in passive heat management, it is possible in active heat management to use the electro-thermal principal (peltier effect). This is mostly due to the fact that there is small power consumption as there is no heat storage that has to be heated. Another advantage is to reverse the effect. Peltier elements pump heat from one side of the element to the other when voltage is applied. Reversing the voltage direction also reverses the direction of the flow. It is therefore not only possible to heat the component, but also to cool it. Fig. 19.10.a depicts the thermal flow with active heat management while 19.10.b depicts the thermal flow for active cooling.

Gripper Design

To the design of the gripper system for the active heat management the same requirements as to the gripper for the passive heat management have to applied. Instead of integrating a heat storage, a heat source has to be integrated. Due to the similarity between active and passive heat management, a vacuum gripper system has also been chosen for the active system.

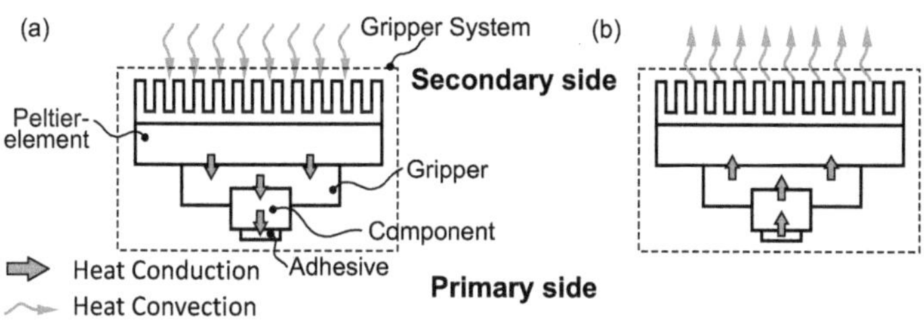

Fig. 19.10 Heat flow through the use of the peltiereffect: (**a**) Heating; (**b**) Cooling

Fig. 19.11 Draft of the active gripper

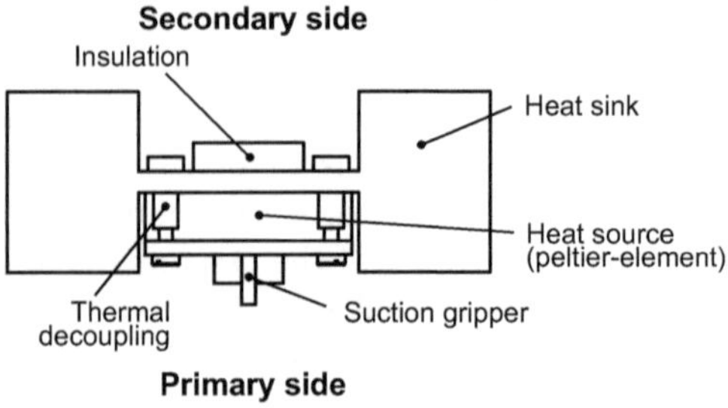

Fig. 19.11 shows the active gripper system. The most important part of the gripper are heat source (peltier element), heat sink, vacuum gripper plate, the insulation above the handling system and thermal decoupling between primary and secondary side.

The underside of the peltier element is called the primary side since it is important for the heat exchange with the component. Thus the top side of the peltier element is called secondary side. The task of the cooling element is to keep the temperature of the secondary side constant. A cooling element is convenient for this task, since it has a large surface and can achieve an improved convective flow of heat. Fig. 19.12 shows the buoyed active gripper system.

19.4.2 Implementation and Experiments

The size adapted assembly system which was described in Chap. 15 was used for the assembly experiments and the validation of the active gripper system. The standard control of this assembly system was enhanced for the control of the active gripper. Thus a fully automatic assembly process with hot melt adhesives and active heat management is possible. A continuous detection of the components with the integrated 3D-vision system was used to detect possible shifts of the components during the joining process. The assembly was done with a $2 \times 3\,\mathrm{mm}^2$ silicon chip. The process variant 2 was implemented to achieve a short process cycle time. For the experiments, the

Fig. 19.12 Implementation of the active gripper system

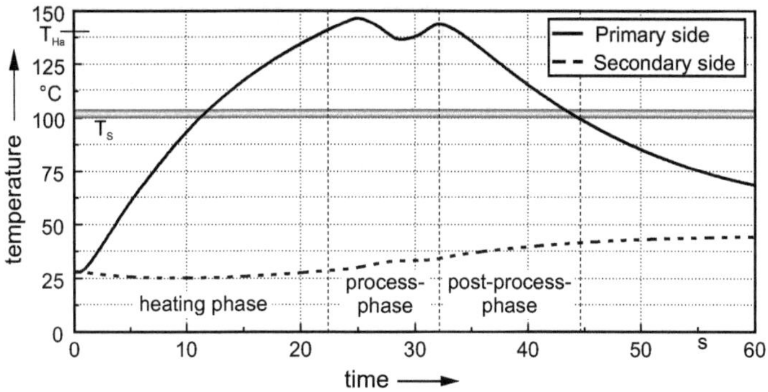

Fig. 19.13 Temperature profile of the active gripper in a typical assembly process

hot melt adhesives chosen in Chap. 16 were used. Experiences with these adhesives show that a working temperature T_{Hp} of about 140°C is suitable for joining.

Figure 19.13 shows the temperature profile of an assembly process. The temperature curves of the primary and secondary side are displayed. The three phases of heat management are marked.

The curves show an assembly process starting at room temperature at all components, in which the duration of the heating phase is about 21 s. The working temperature was set to 140 °C an is controlled by a bang-bang control. Cycle test with the gripper system have shown, if the gripper is in a stable working state, the time for heating can be reduced to about 12 s. During heating phase and process phase, the components are positioned and adjusted. The components are aligned with a maximum error of 1 μm. The positioning correction is done with a lock and move algorithm. The components will then be joined. During the joining process, the assembly system adjusts the components with a gap of about 50 μm between them. The joining force was nearly constant during the test assemblies and was in the range of about 1 N. During the post-process phase, the temperature of the primary side drops below 100°C within about 12 s and the hot melt adhesive sets. Then the part can be released. The total assembly process takes less than 50 s. In this time also releasing and moving to start position is included. With the executed assembly examination using the active gripper concept an assembly uncertainty of 11.9 μm was reached (Fig. 19.14).

The hight of the deviation of the assembly uncertainty can be explained by the influence of the viscosity of the adhesive. This viscosity depends extremely on the temperature of the adhesive. The fast change of the temperature at the skin zone of the adhesive at the time of the first contact of component and substrate results in a inhomogeneous allocation of different areas of viscosity. This seems to produce strains in the adhesive layer and leads to a displacement of the components during and after the joining process. Increasing the working temperature can affect this process positively. However, this increas-

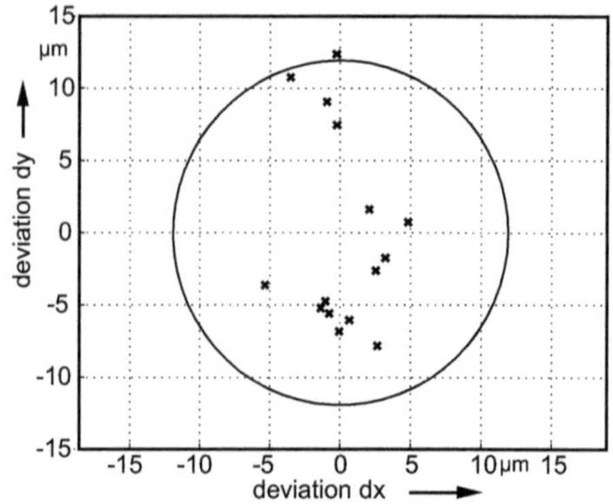

Fig. 19.14 Uncertainty of the assembly process with active heat management

ing leads to a higher thermal stress to the component and higher demands to the gripping system related to the thermal insulation. The use of a hot melt adhesive with a lower viscosity can also enhance the assembly uncertainty. This fact should be considered at the design of the joining process.

19.5 Conclusion

The requirements of hybrid microsystems have been constantly challenging the limits of available assembly technology. This applies particularly to the required accuracy and reproducibility, which are constantly being pushed to their limits. The technology presented in this chapter for joining with hot melt adhesives is an important step towards achieving this goal. Hot melt adhesives, in contrast to chemically reactive adhesives have several advantages: e. g. fast setting time and re-meltability. Joining with hot melt adhesives is therefore an excellent alternative in MST. As has been shown, the use of hot melt adhesives requires adapted thermal management concepts.

In this chapter two different heat management concepts were introduced. The passive heat management is a concept which can be implemented very easy into existing microassembly systems. It use an internal heat storage, where the important challenge is the design of this storage. The described approach to use a simulative design for the gripping systems proved to be successful. The simulative examinations and the comparison to experiments which was described in [5, 9, 10, 11] show the utilizability of the passive heat management in the micro assembly.

The alternative concept, which uses integrated heat sources, results in more complex handling components. However, advantages of this concepts are the huge flexibility at the process design. In Chap. 19.4 a approach using a peltier element as integrated heat source was introduced. In addition to the possibility to warm up the adhesive it is possible to change the polarity of the element and to cool the handling parts to reach faster process times. The results, which were presented in the Chap. 19.4, show, that the approach to join micro components using hot melt adhesives is also suitable. However, to reach high accuracies in the assembly of micro components an enhancement of the adhesives relating to the adhesives viscosity are necessary.

Summarized the results show a practicable alternative joining concept which uses hot melt adhesives. The introduced heat management concepts have different advantages and disadvantages making them suitable for different applications in the MST. The use of sensor guided assembly concepts is also applicable as the assembly of parts with a heat capacity which is high enough. The described problems concerning the assembly uncertainty are solvable by optimizing the gripper components, the handling system and the adhesive viscosity.

Acknowledgements The authors gratefully acknowledge the funding of the reported work by the German Research Foundation within the Collaborative Research Center 516 "Design and Manufacturing of Active Micro Systems".

References

[1] Böhm S, Hesselbach J, Wrege J, Rathmann S, Dilger K, Stammen E, Hemken G, Ma W (2006) Micro bonding with non-viscous adhesives. Microsystem Technology p 1432

[2] Böhringer KF, Goldberg K, Cohn MB, Howe R, Pisano A (1998) Parallel Microassembly with Electrostatic Force Fileds. In: International Conference on Robotics and Automation, pp 1204–1211

[3] Brussel Hv, J Peirs, D Reynaerts, A Delchambre, G Reinhart, N Roth, M Weck, E Zussman (2000) Assembly of Microsystems. CIRP Annals - Manufacturing Technology 49(2):451–472

[4] Cohn MB, Kim CJ, Pisano A (1991) Self-Assembling Electrical Networks as Application of Micromachining Technology. In: International Conference on Solid-State Sensors and Actuators, pp 490–493

[5] Dietrich F, Rathmann S, Repenning A, Raatz A (2008) Aktiv beheiztes Wärmeführungskonzept für die Montage von SMD mit Schmelzklebstoffen. wt Werkstattstechnik online 11/12/2008:969–973

[6] Dilthey U, Brandenburg A (eds) (2005) Montage hybrider Mikrosysteme: Handhabungs- und Fügetechniken für die Klein- und Mittelserienfertigung. VDI-Buch, Springer, Berlin

[7] Hesselbach J, Wrege J, Raatz A (2007) Micro Handling Devices Supported by Electrostatic Forces. CIRP Annals - Manufacturing Technology 56(1):45–48

[8] Pokar G (2004) Untersuchung zum Einsatz von ebenen Parallelrobotern in der Mikromontage. Vulkan-Verl., Essen

[9] Rathmann S, Blumenthal P, Hemken G, Raatz A (2009) Aktives Greifkonzop zur Montage hybrider Mikrosysteme mit Schmelzklebstoffen. In: Seidel H (ed) Mikrosystemtechnik Kongress 2009, VDE Verlag, Berlin Offenbach, pp 808–811

[10] Rathmann S, Ellwood J, Hemken G, Keck C, Böhm S, Tutsch R, Raatz A (2009) Präzisionsmontage schmelzklebstoffbeschichteter Bauteile mit Hilfe eines 3D-Bildsensors. In: Vollertsen F, Büttgenbach S, Kraft O, Michaeli W (eds) 4. Kolloquium Mikroproduktion, BIAS Verlag, pp 151–156

[11] Rathmann S, Raatz A, Hesselbach J (2010) Active Gripper for Hot Melt Joining of Micro Components. In: Ratchev S (ed) Precision assembly technologies and systems, Springer, Ifip advances in information and communication technology, pp 191–198

[12] Schöttler K (2008) Planung und Untersuchung automatisierter Mikromontageprozesse unter besonderer Berücksichtigung der Einflussgrößen. Vulkan-Verl., Essen

[13] Schöttler K, Raatz A, Hesselbach J (2008) Precision Assembly of Active Microsystems with a Size-Adapted Assembly System. In: Ratchev S, Koelemeijer S (eds) Micro-Assembly Technologies and Applications, Springer US, IFIP International Federation for Information Processing, vol 260, pp 199–206

Chapter 20
Design of an Automated Assembly for Micro and Nano Actuators

S. Rathmann, S. Hansen, A. Raatz[1], H. H. Gatzen[2]

[1] *Institute of Machine Tools and Production Technology*
Technische Universität Braunschweig
a.raatz@tu-braunschweig.de

[2] *Institute for Microtechnology (now: IMPT)*
Leibniz Universität Hannover
gatzen@impt.uni-hannover.de

Abstract In addition to the development of new technologies for the manufacturing of electromagnetic micromotors, the development and production of a micromotor was in the main focus of this Collaborative Research Center. This chapter presents an approach for the assembly of the variable reluctance microstep motors and the automation of the assembly. The results have been achieved by the interdisciplinary cooperation of different subprojects in the domains of design, manufacturing, and assembly technologies. After a short introduction on the importance of an automated assembly of the high-tech product microactuator and its components, the assembly concepts for the linear microactuator and the xy-microactuator are explained.

20.1 Introduction

20.1.1 Motivation

The developed microstep motor consists of various components, which were fabricated in different processes on separate substrates. Due to a decreasing yield, proportional to the chip size, the components cannot be fabricated in a monolithic integrated process on one substrate. Thus, a concluding assembly of the xy-micro and nanopositioner is necessary. In this case, the assembly is comprised of all operations to join, connect, and house the components. Here, special considerations of the electrical and thermal connections need to be made. The required assembly tolerances of the xy-micro and nanopositioner demand a very high precision alignment of the micromotor components to each other. The gap between traveler and stator is 8 µm over a range of 20 mm. Furthermore, the angles between traveler and stator as well as to other components are very critical. These requirements make an assembly necessary which achieves an accuracy of about 1 µm. A manual assembly of the microactuators, even if a mechanical bedstop is used, is only suitable for a prototype assembly. For this reason, an automated or at least a semi-automated assembly is advisable. For the assembly process commercial assembly systems as well as an assembly system developed in the Collaborative Research Center can be used. Advantages of the developed assembly system compared to the commercial ones are described in detail in Chap. 15. Based on this knowledge, a semi-automated assembly of the microactuators was developed. The high accuracy assembly of a microactuator with the requirements defined above puts constraints on the design and manufacturing of the components. For example, to guarantee the visibility of the components for a vision system, marks on the components are advantageous. The use of marks, which will be deposited in combination with the functional structures, will result in a high alignment resolution. This technology is described in detail in Chap. 17. This assembly concept requires that a great deal of attention is being paid on the planarity and thickness of each part. In this context, the thickness is a very important property due to the fact that all the components will be assembled in one layer. Therefore, a variation of thickness can lead to a tilt or even to a deadlock of the traveler in the assembled state. These requirements were discussed during the development phase and as well as possible included in the design process. The assembly of the high-tech product microactuator has not only high demands on the components' design, but also on the assembly process and the systems' components. The requirements are for example:

- Tolerance of the assembly
- Reproducibility
- Assembly process cycle time
- Product flexibility
- Quantity flexibility
- Flexibility of the usable joining technology
- Possibility to monitor the assembly process
- In-Process product testing
- Dust-free environment considering the micro-specific behavior of the parts
- Low load of the components (thermal, electro-static, mechanical)

Additionally, it is important to know external and internal influences which disturb the assembly process and its accuracy. These influences are discussed in Sect. 19.1.1. The requirements and special characteristics are the motivation for specialized automated microassembly processes.

20.1.2 Components of the Microstep Motors

Before discussing the assembly of the microactuators, the standard components of the micromotors are described shortly. Stator and traveler are the key components of the microactuator for generating the driving force. Their design is described in Sect. 13.6. In this microatuator system, the stator generates the magnetic flux for driving the traveler. For a nanopositioner, a frictionless motion of the traveler is desirable, it is achieved by choosing a magnetic levitation guide. The active component is principally the same one as the stator of the pole-based micromotor, while the passive part is a flux guide (the traveler without poles) which is attracted by the field generated of the active part.

The air gap sensor consists of two components, an electrode without any electrical connections which is located on the moving part, and a stationary electrode which generates the electrostatic field used for the measurement of the displacement. The principle and the design of the air gap sensor is described in Chap. 7.

20.2 Assembly Concept for a Linear Microactuator with Levitation System

In Fig. 20.1 the assembly of the linear microactuator is shown. In addition to the components mentioned so far, which are described in Sect. 20.1.2, special mechanical parts for housing and connecting the components are also required. These parts are the bottom and top plate, which carry the active components of the motor, sensor, and guide. Between these plates the traveler plate, carrying the passive components, is located. Bottom plate, top plate, and traveler plate are optimized regarding to their weight. The assembly blocks on the right and left side are needed to connect the bottom and top plate. For the connection of the active components, printed circuit boards (PCB) at the bottom plate as well as the top plate are needed. These PCBs hold the circuit and special marks which allow the adjustment of the components during the assembly process. In the following the assembly concept and the assembly sequence of this microactuator are described.

The system integration of the microactuator starts with the assembly of the three carriers, bottom plate, traveler plate, and top plate. Each plate is assembled separately. All plates have to be aligned with a planarity less than 1 µm. First of all, the carrier plates have to be fabricated and adapted to the actuator. The bottom plate and all other plates are made of Aluminum-Titanium-Carbide (AlTiC), a very hard material machined by grinding and dicing. Out of the raw material, a plate with a thickness of 1.2 mm and

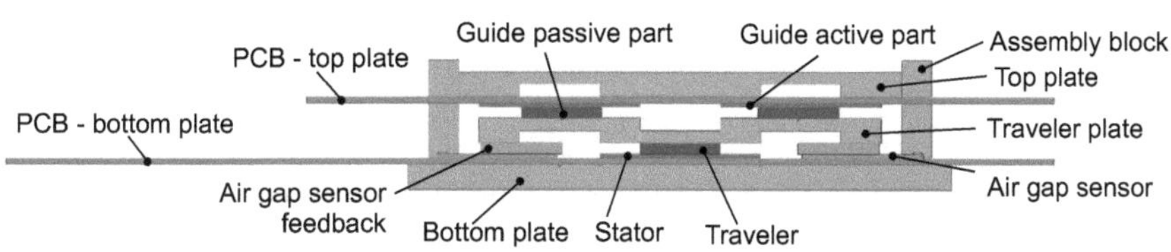

Fig. 20.1 Cross section of the assembled linear microactuator

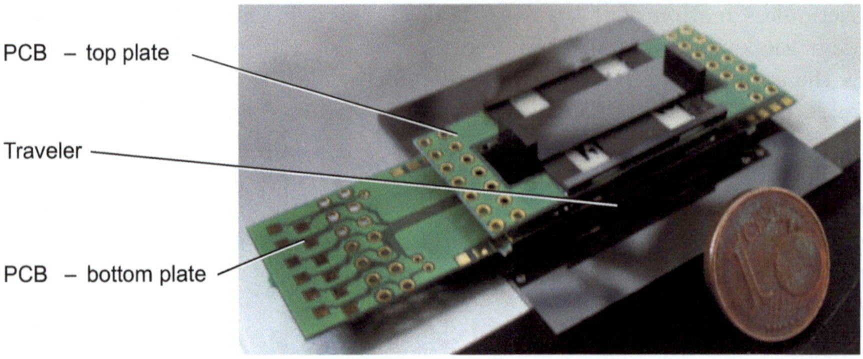

Fig. 20.2 Assembled linear microactuator with levitation system

a planarity of less than 1 μm is grinded. Channels are integrated into this plate to allow the alignment of components as well as for adhesive bonding. During bonding, the channels are filled with the adhesive which, while curing, establishes a connection between component and plate. This allows a very planar connection of the components with the carrier plate. At the bottom plate, the PCB and the air gap sensors are mounted. The second step is the assembly of the traveler plate. This assembly is divided into two phases. First, the high accuracy assembly of the traveler and the air gap sensor feedback is accomplished and secondly, the assembly of the passive parts of the guide takes place. Due to the fact that for the levitation of the traveler plate only the normal force of the motor is needed, the alignment of these components can be done with less accuracy. The third step is the assembly of the top plate, at which the active guidance components are mounted. Since it is important to consider the weight of the traveler plate, the plate is thinned. After the assembly of the three plates, the actuator is completely assembled. During this actuator assembly, a spacer foil with a thickness of 16 μm is used to adjust the air gap between stator, traveler, and the guide components. First, the spacer foil is attached to the stator. After that, the assembled traveler plate and the top plate are placed on top of the spacer. Then, the assembly blocks are added, properly adjusted to the bottom and top plate, and bonded with adhesive. In a final step, the spacer foil is removed. To show the feasibility of the system assembly in a design study the linear microactuator it is built up manually. This setup is presented in Fig. 20.2.

Fig. 20.3 Top view of the xy-micro and nanopositioner

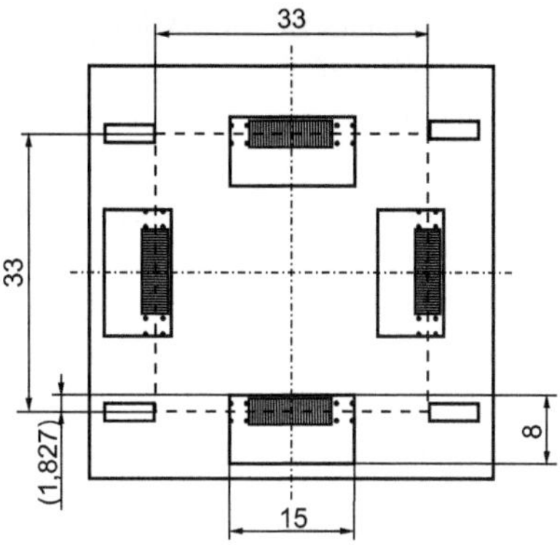

20.3 Assembly Concept for a xy-Micro- and Nanopositioner

The assembly of the xy-micro and nanopositioner is based on the same principles as the linear variant described above but is much more complex. By using tooth based VR microstep motors as drive systems and using microstepping as a drive scheme (see Chap. 13), the actuator may be used as a nanopositioner. The design of the xy-micro and nanopositioner consists of four single micromotors which are arranged in a square, located in the same plane. Figure 20.3 shows the arrangement of the stators of the linear microactuator. The build-up in vertical direction is the same as for the linear actuator. Figure 20.4 shows an exploded view of the xy-micro and nanopositioner. The

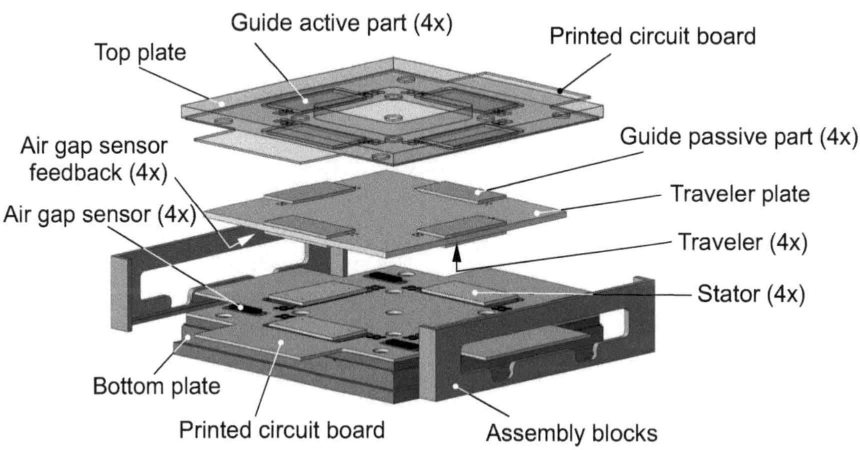

Fig. 20.4 Assembly view of the xy-micro and nanopositioner

device consists of a bottom plate, a traveler plate, a top plate, plus application specific PCBs.

During the assembly of the xy-micro and nanopositioner, it is even more important to guarantee a high planarity of the carrier plates. The accuracy requirement is the same as for the linear microactuator variant (1 μm), but the area where it has to be maintained is much larger. To achieve this accuracy, an appropriate wafer polishing process was applied. As previously, the bottom plate features integrated channels, which are used to bond the components. The lower PCB is also bonded to the bottom plate. For the alignment of the components, circular marks as described in Chap. 17 are used, which are applied to the PCB. For the assembly, the system described in Chap. 15 is utilized. This system allows to achieve a precision of 1 μm. The requirement on the assembly of the air gap sensors is not as high as the one on the motor components, therefore the air gap sensor components can be placed without any alignment marks. For the assembly of the traveler plate, the same requirements as on the bottom plate assembly exist. Since the traveler does not require a PCB, the marks for the assembly have to be applied to the traveler plate directly. Using the marks, the assembly of the four travelers and the passive components of the guide is done the same way as the assembly of the bottom plate. The air gap sensor is placed with a simple teach-in assembly step at a predefined position. For the mounting of the active parts of the guide system to the top plate, the marks on the upper PCB can be used. The alignment of the top plate to the base plate is done with help of the lateral assembly blocks. The air gap between stator and traveler as well as the active and passive components of the guide is adjusted with a spacer foil having a thickness of 16 μm. The assembly of the three modules is done manually. Figure 20.5 shows the assembled xy-micro and nanopositioner.

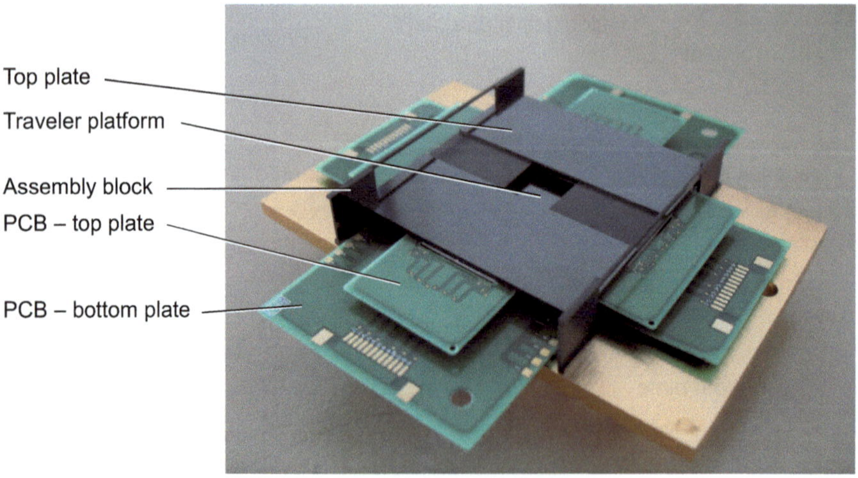

Fig. 20.5 Assembled xy-micro and nanopositioner

20.4 Conclusion

The approaches and results presented in this chapter reflect the work of an interdisciplinary group of researchers from the Collaborative Research Center. These are not only the authors but also members of the working group "Production of the demonstrators". The goal was to define and implement a strategy to assemble the high-tech product xy-micro and nanopositioner, which poses high demands on the assembly strategy and accuracy. The presented strategy to assemble the linear microactuator, which was designed at the *imt* and the *IAL*, shows a feasible approach for the semiautomated assembly of the linear microactuator and the xy-micro and nanopositioner. The analyses of the assembly task have shown, that besides the components of the micromotor, additional parts are necessary for the adjustment and fixation of these components. The requirements to the additional parts are comparable to the components of the microactuator. A particular challenge were the demands on the planarity of the mounting surfaces. In cooperation with production engineers of the Collaborative Research Center, a strategy for the manufacturing of the important additional parts and the automated assembly process was developed. The assembly of the microactuators could be tested with sample parts of the linear VR microstep motor. In this project it was shown, that the interdisciplinary cooperation is necessary for a successful development of technical microsystem products.

Acknowledgements The authors gratefully acknowledge the funding of the reported work by the German Research Foundation within the Collaborative Research Center 516 "Design and Manufacturing of Active Micro Systems". The authors would also thank S. Cvetković (*imt*) for supporting the system assembly.

Part V
Industrial Applications

Chapter 21
Bistable Microvalve for Biomedical Usage

A. Balck, C. Kirsch, U. Schmid, H. Seidel[2], M. Leester-Schädel[1], S. Büttgenbach[1]

[1] *Institute for Microtechnology*
Technische Universität Braunschweig
m.leester@tu-braunschweig.de; s.buettgenbach@tu-braunschweig.de

[2] *Chair of Micromechanics, Microfluidics/Microactuators*
Universität des Saarlandes
seidel@lmm.uni-saarland.de

Abstract One example of the SFB 516 transfer work is a bistable microvalve which is planned to be integrated in implantable high precision dose infusion pumps typically used for patients with chronic pain or spasticity. The drug pumps are designed to have a 35 year lifespan within the patients body and work without an external energy supply. To adjust the drug rate to the individual needs of the patient, it is of utmost importance to provide a variable drug output rate. To achieve this goal, an innovative low-energy microvalve has been developed. Development and manufacturing of the microvalve are based on research results of the SFB 516, in particular those applicable to the drive unit, the silicon bulk micromachining, and the optimization of friction and wear.

21.1 Introduction

In this section, the motivation for the project as well as the state of the art of bistable microvalves are outlined.

21.1.1 Implantable Infusion Pump and Requirements for the Valve

The implantable high precision dose infusion pumps produced by the industrial partner consist of a metal body with a drug reservoir in the shape of a bellow. A pressure chamber filled with liquid gas applies constant pressure to the drug reservoir. This pressure produces a continuous drug stream through a choke path into a catheter. The flow rate is predominantly determined by the length and the cross-section of the choke path channel. The drug pumps are designed to have a 35 year lifespan and work without an external energy

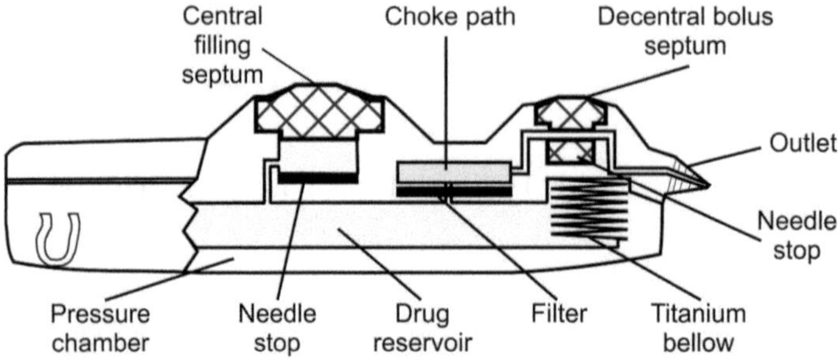

Fig. 21.1 Cross section of the infusion pump by the industrial partner

supply. A physician is able to refill the device with a special cannula through a self-locking septum (see Fig. 21.1).

For many reasons which may occur daily, such as a substantial change in the patient's health or an adjusted day and night dose, it is desirable to modify the drug output according to specific needs of the patient. For this purpose, a microvalve is developed to open and close the entrance to a second choke path so that at least two different flow rates can be realized. Further improvements will also be implemented to make finer dose steps possible.

Conventional 2–port/2–way valves need a permanent energy supply to hold at least one of both positions. To extend the lifespan of the pump, the required energy, which could be provided in a first approach by an integrated battery, needs to be minimized. Therefore it is essential to develop a bistable valve, which is able to hold both possible positions without energy supply. The switching of valve positions has to be technically simple. Also, the risk of unintentional activation through daily activities of the patient has to be avoided. Another major obstacle for this project is the selection of a suitable energy supply. Other tasks focus on the reduced selection of available materials due to the chemical properties of the drug and the overall request for biocompatibility. The valve also has to be sterilizable and the in- and outlet have to fit to the adapter of the pump. Last but not least, the valve has to close against a pressure of 350 kPa.

21.1.2 State of the Art

Since the first publication of microvalves in 1979, many research activities stimulated the realization of various concepts [6]. Apart from the common types, typically known as normally open or normally closed, different approaches for the realization of bistable microvalves have been reported. A true bistabile performance, however, has to be challenged in some cases.

Goll et al. already published a microvalve with a bistable polymer-diaphragm in 1996. This diaphragm is curved and can be turned to either position by pressure pulses generated by a resistively heated coil [3]. In 2001, Qui et al. presented a similar bistable actuator principle that was also based on a turning membrane [7]. In 2000, Böhm et al. published a bistable microvalve with an electromagnetic actuator. Using a silicon valve, the switching operation is activated electromagnetically so that one state is held by a spring, the other by a permanent magnet [1]. However, a valve based on this principle was claimed in the United States Patent No. 5497135 already in 1994 [10]. Capanu et al. also used an electromagnetic actuator, in which both states are locked by two separate magnets [2].

In the description of the US patent 6565526B2 a bistable microvalve that is based on a shape memory polymer [9] is described. It was designed for application in a catheter. For a switching operation, the polymer is heated above the glass transition temperature by a laser so that the pressure is increased or lowered to either close or open the catheter, respectively. To fix the position, the polymer has to cool down below glass transition temperature. This system is bistable, but the switching operation requires a reversible pressure generation, which can not be realized in some cases.

In the German patent 10037251A1, both switching states are fixed by a centrally located permanent magnet, which closes a magnetic circuit in each position. The switching operation is induced by an electromagnetic field [4].

A silicon valve based on an electrostatic actuator principle is described in the US patent 6168395B1 [8]. It consists of two cavities, which are linked to each other by a channel, two diaphragms, and two pairs of electrodes. If one pair of these electrodes is activated, the corresponding diaphragm is pulled down by electrostatic attraction. The other diaphragm is moved up by the resulting overpressure and closes the valve seat located above.

None of the above mentioned valve systems seem to be suitable for the planned application on account of insufficient pressure, the selected approach for triggering the actuator elements, and the large energy consumption.

21.2 Concept

The bistable valve presented in this study is specially designed to fulfill the demands of the implantable pump. In this paragraph, a short overview on the design and the functionality of key components is given.

21.2.1 Basic Design Issues

The valve consists of four layers, as shown in Fig. 21.2. The bottom layer contains the in- and outlet, as well as the valve seat. It is covalently bonded to the first membrane layer. As either an anodic or a silicon direct bonding process is used for joining both components, the resulting chamber is per-

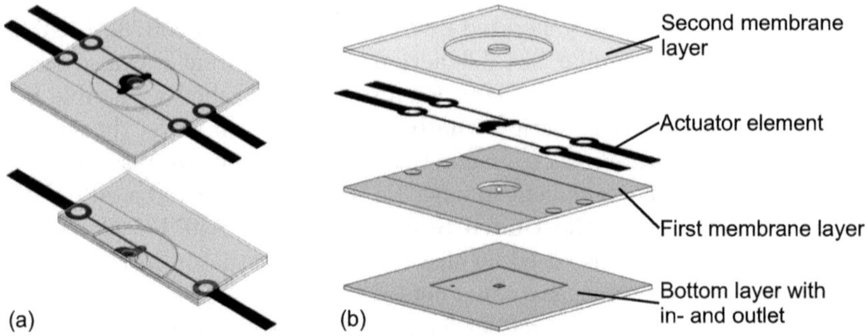

Fig. 21.2 Schematic view of the valve: (**a**) Assembled valve; (**b**) Exploded view

fectly applicable for medical use. No adhesive or any other non biocompatible substance comes in contact with the drug that is handled in the chamber.

Above this chamber, the actuator is located. It is made of shape memory alloy (SMA) and is able to move laterally. The fourth layer serves as a cover and generates a counteracting force against the valve to close and against the SMA element to stay in its second position. It is bonded to the frame of the first membrane layer, so that the SMA actuator element acts like a wedge between the boss of the first membrane layer and the second membrane layer. In an alternative design of the second membrane layer a pre-deflected membrane is used as a micromechanical spring. This pre-deflection originates from residual stress of a sputtered tungsten thin film deposited on the wet etched membrane.

21.2.2 Functionality of Key Components

The SMA actuator element has two stable rest positions: The first one is quite close to the boss of the first membrane layer. In this position, the inactive SMA element does not transfer any force on the first membrane layer and the valve is open. When electrically activated, the middle part of the SMA element is laterally pulled onto the boss. The distance between the boss and the second membrane layer is less than the thickness of the SMA element. Therefore, the boss is pressed downwards and the valve closes. The position of the SMA element on top of the boss is the second stable position. The SMA element, being inactive again, is held by friction forces and is positively stabilized. When the SMA actuator element moves back into its first position, the stress stored in the membrane lifts it up and the valve opens (see Fig. 21.3).

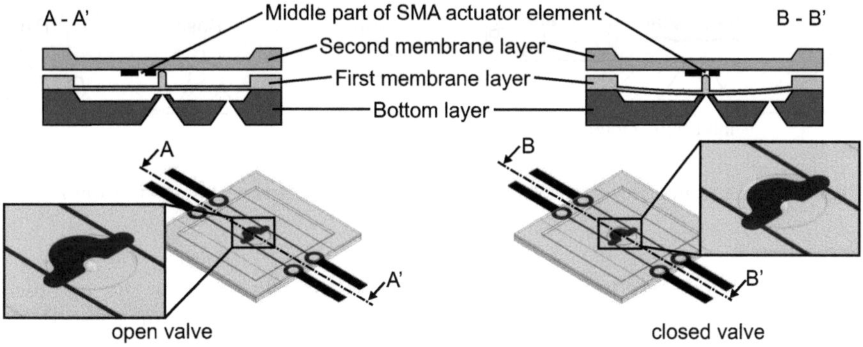

Fig. 21.3 Schematic view of the valve in open and closed position

21.3 Transfer of SFB Knowledge

In one subproject of the SFB 516 described in Chap. 16, microgrippers that are driven by SMAs, pneumatic forces or by using thermal expansion are designed and fabricated. The collected data and knowledge were of great importance to the choice and design of the actuator in this project.

Because of the know-how gained from two other subprojects (see Chaps. 5 and 6) on tribologically optimized surfaces and alignment structures, the first valve version was completely redesigned to reduce friction forces, abrasion and tilting problems. As a consequence, in the new bistable valve system, only the SMA actuator is sliding between the tailored boss structure and the membrane.

Futhermore, a self developed modular computer-aided design environment (see Chap. 4) has been used to assist the entire development process of the microvalve.

21.4 Realization

The design variants, eligibility criteria, and fabrication processes of the different layers are described in the following paragraph. Also, details on how these layers were connected with each other are explained.

21.4.1 Design and Simulation

Bottom Layer with In- and Outlet. The bottom layer features one inlet- and one outlet-port (see Fig. 21.4.a). The port diameter on the peripheral side and the distance between both ports are predetermined with 0.8–1 mm and 4 mm, respectively. The overall dimensions of the bottom layer are $20 \times 20 \times$

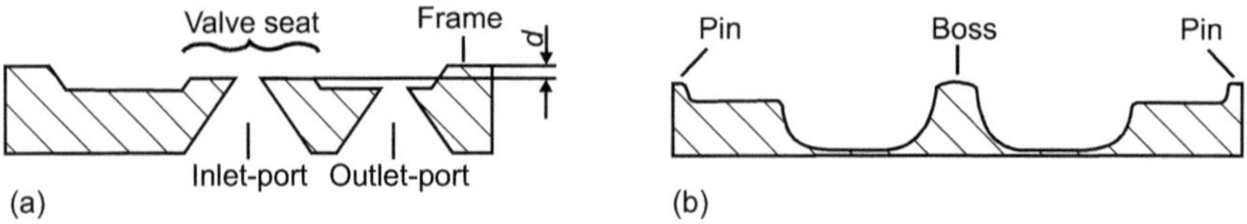

Fig. 21.4 Sectional view: (**a**) of the bottom layer; (**b**) of the first membrane layer

0.36 mm^3. The top side structure, which is facing the valve inside, includes the valve seat. The bottom layer is preferably made of silicon or glass. Both materials are medically compatible and can be patterned by wet chemical etching techniques.

First Membrane Layer. The first membrane layer is the second layer within the complete device. The key element is a membrane with a boss in the middle (see Fig. 21.4 b) that deflects vertically, causing the closing and opening of the inlet-port. The displacement is defined by the height of the valve seat related to the frame thickness of the bottom layer. In the actual design, d is about 25 µm. Suitable materials are, again, silicon or glass. As a consequence of the available etching techniques, the membrane is either square with a square boss in the middle, or round with a round boss geometry. The average membrane thickness is 30 µm. The lateral membrane dimensions are optimized with respect to a maximum deflection while simultaneously requiring minimum forces.

To close the inlet-port, the first membrane layer can be placed with the boss towards the valve seat. The challenges are to line up the diameters exactly and to obtain defect free surface characteristics. Alternatively, the membrane layer can be placed the opposite way, so that the inlet-port is closed by the flat membrane surface. Although the resulting surface seal involves the risk of being leaky because of adhering particles, this design was determined to be more suitable. In the future, the inside of the valve will also be covered with parylene.

The parylene layer will improve the sealing quality by means of softer material performance compared to silicon or glass. In addition, particulates that would otherwise block the closing of the valve will be embedded permanently in the parylene layer. The top face of the first membrane layer includes four pins for mounting the SMA actuator elements (see Fig. 21.4.b). The key elements of the various membrane designs in combination with the bottom layer are summarized in Fig. 21.5.

One of the main results of the theoretical and experimental investigations is that isotropically etched membranes provide the best ratio with respect to the deflection and the applied force.

Actuator Element. According to the SMA actuator element of the microgripper (see Chap. 16), the valve actuator element is operated on basis of a differential working principle. It consists of two identically designed single

membrane design	square silicon	round silicon	round silicon	round glass
fabricated by	wet chemical, anisotropic etching	dry etching	wet chemical, isotropic etching	wet chemical, isotropic etching
schematic view of the membrane	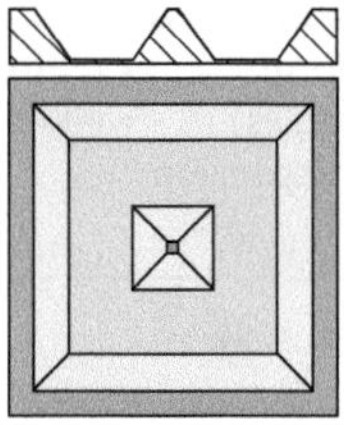	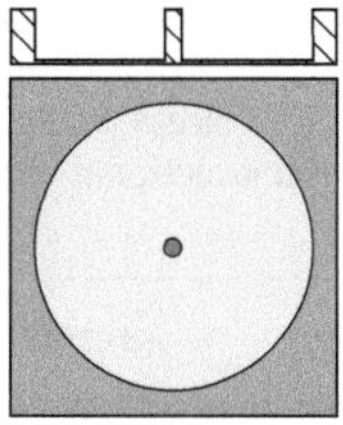	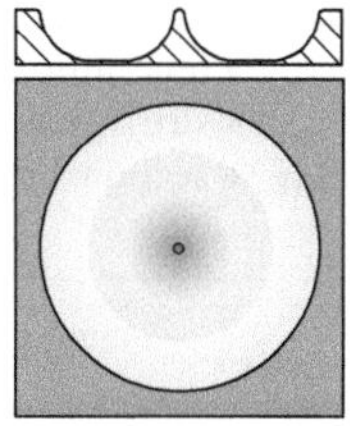	
general properties	• areas with large mechanical stress • even at maximum membrane dimensions displacement too small	• mechanical stress distribution more homogeneous • sufficient displacement and forces	• mechanical stress distribution most homogeneous • sufficient displacement and forces • optimal pin design • pin with rounded dome possible	
connection to bottom layer with in- and outlet made of silicon	• silicon direct bonding challenging • anodic bonding with additional glass layer promising			• anodic bonding advantageous
connection to bottom layer with in- and outlet made of glass	• anodic bonding advantageous			• gluing may cause medical incompatibility • glass direct bond challenging

Fig. 21.5 Different designs of the first membrane layer

elements connected in series. The outer ends are fixed to the pins of the first membrane layer. Being electrically heated above the austenitic finish temperature, each element contracts about 300 μm. An alternate heating of the single elements generates a 300 μm lateral movement of the middle part.

The design of the SMA actuator element is optimized with respect to the movement of the middle part, an acceptable force to close the valve and a high cooling rate. Three different designs have been analyzed up to now, as shown in Fig. 21.6. The design labeled as "straight structure" promises excellent properties and will be further investigated.

To be heated above the austenitic finish temperature (nearly 100°C), the SMA single element needs a current pulse of 200 mA for a time period of 3 s. For a rough estimate, it can be assumed that the valve will be opened and closed two times a day. Two types of batteries have been found that could serve as an energy source and that could be integrated inside the existing housing of the infusion pump: CR14250 Li/MnO_2 and SL-770 $Li/SOCl_2$. The first one has the advantage of being very small (i.e. 14.5 mm in diameter, 25 mm thick) and lightweight (11 g), but its operation lifetime of about 6.4 years is rather short. The second alternative has larger geometrical dimensions (i.e. 26.2 mm in diameter, 49.8 mm thick) and more weight (50 g), but it would provide energy for more than 18 years.

	looped structure	railing-like structure	straight structure
actuator design	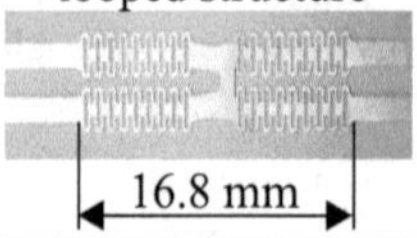	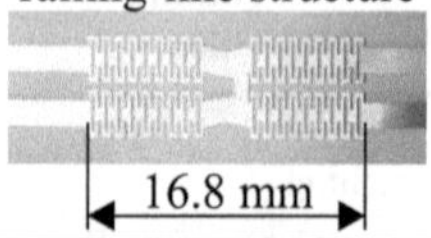	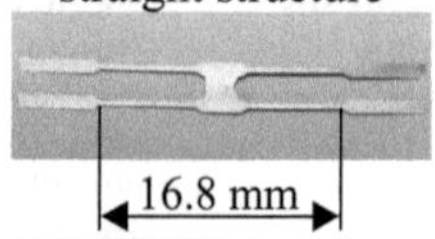
stress distribution	• non-uniform • stress peaks (loops) • sectional overexpansion −	• most non-uniform • stress peaks (edges) • sectional overexpansion −	• uniform • no stress peaks +
actuator displacement	several mm +	several mm +	350 μm O
actuator force	200 mN −	200 mN −	6000-8000 mN +
general properties	• out-of-plane deflection	• out-of-plane deflection • larger surface than looped structure →faster cooling →faster switch frequency	• no deflection • simple mech. principle • no stress peaks (active areas) • design limitations
fabrication	• long laser process time • the lasered structures are difficult to unhinge		• short laser process time • unhinging easier

Fig. 21.6 Different designs of the SMA actuator element

Second Membrane Layer. The maximum closing force is reached when an unstructured wafer is used as a second membrane layer. To increase its elasticity and hence to lower the force perpendicular to the SMA actuator element, the thickness of the center section of the second membrane layer is reduced. Preferably, materials for realization are again silicon or glass.
In an alternative design a membrane offering the possibility for larger deflection characteristics by a sufficient closing force has been investigated. Initially, finite element methods (FEM)-based simulations of the membrane performance have been carried out modelling a silicon membrane that was anisotropically wet etched to a thickness of 20 μm and covered with a 1 μm thick tungsten thin film. Previous experiments showed that sputtered deposited tungsten layers on silicon substrates can have compressive stress values of up to 1100 MPa. Taking into account the design of the other valve layers, a membrane size of $5000 \times 5000\,\mu m^2$ was chosen. To reduce the simulation time to an acceptable level, the suspension of the membrane can be removed and replaced by a "fixed-patch" boundary condition. In pre-investigations, it was demonstrated that this assumption simplifying the model leads to a substantially lower maximum deflection error of about 5% in the range of up to 70 μm. Initial investigations revealed that without any further boundary conditions, the direction of the membrane deflection is arbitrary. This is due to the buckling effect. The intrinsic film stress of the tungsten layer acts as a load for the membrane element which is above the critical load for buck-

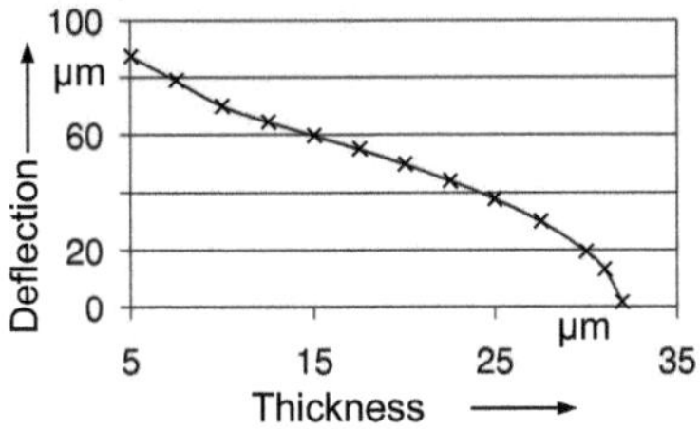

Fig. 21.7 Deflection depending on the membrane thickness

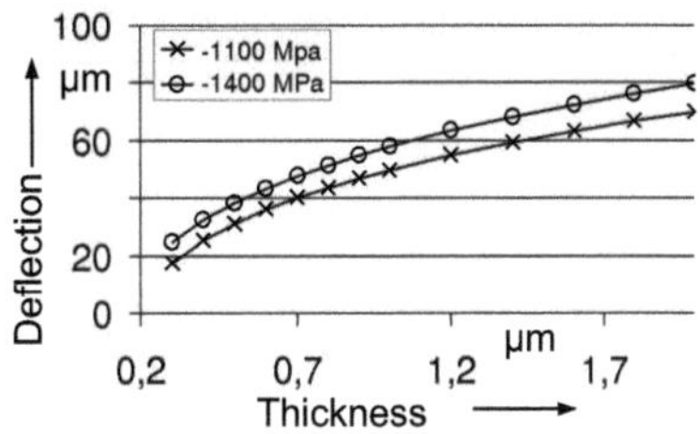

Fig. 21.8 Deflection depending on the tunsten thickness

ling according to the Euler principle. Thus, one more boundary condition is necessary by applying an additional pressure load of 10^{-9} Pa perpendicular to the membrane surface, otherwise this situation represents a singularity. This pressure load does not have any influence on the amplitude. It serves only as direction vector. Fig. 21.7 shows the theoretically predicted deflection amplitude of a $5000 \times 5000\,\mu m^2$ square-shaped membrane associated with the membrane thickness including a tungsten layer of 1 µm. The closing path of the valve depends on the gap (d in Fig. 21.4) between the first membrane layer and the valve seat. To avoid a serious throttling effect in the opened state, the gap has to be at least 20 µm. For a conservative dimensioning, the thickness of the silicon membrane should not exceed a value of 25 µm so that the deflection amplitude is large enough. Assuming a 20 µm thick membrane, Fig. 21.8 shows the deflection relative to the thickness of the tungsten layer for two intrinsic stress values (i.e. 1100 MPa and 1400 MPa). If, however, the displacement of the membrane is supposed to be greater than 30 µm, a minimum metal layer thickness of 0.5 µm is required. It is worth mentioning that in these simulations the residual stress was assumed to be independent of film thickness, which is only a rough estimate for real devices.

Pressure Resistance. Not only the deflection amplitude is of interest, but also the resistance against counter pressures. The counter force is applied by the boss of the first membrane layer which is here assumed to have a square surface of about 250 µm edge length. Therefore, a corresponding area in the middle of the membrane is defined as the active surface where the pressure is applied. Fig. 21.9 shows the results. Also, in this simulation, the buckling effect can be verified. Starting at a deflection of about 50 µm, the amplitude decreases until a transition point is reached. A further increasing of the counter pressure forces the membrane to switch to negative amplitudes. This behavior should be avoided, as in this situation the membrane has reached a stable point and will not switch back when the counter force is switched off. This means that the valve would keep its open position. There are two ways to prevent this situation: Either the counter force is limited to values below the switching point, or the membrane is redesigned. The second graph in Fig. 21.9 represents a new design in the form of an additional boss structure

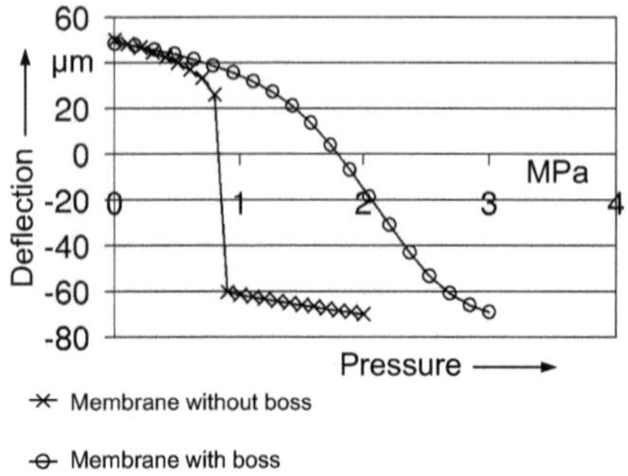

Fig. 21.9 Deflection amplitude with applied counter pressure

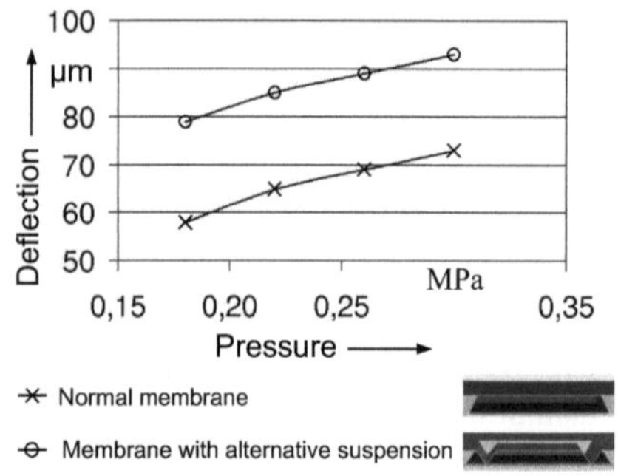

Fig. 21.10 Comparison between normal and special suspension

on the backside of the membrane. This increases the stiffness of the whole element and preserves a normal deflection behavior. At the bottom of Fig. 21.10, the cross-section of an alternative design for the closing membrane is shown. In contrast to the previous approach, the mounting of the membrane is different. Two double-sided anisotropically wet etched V–trenches offer a more flexible suspension. In [5], this type of structure is used for decoupling topside mounted sensors from disturbing vibrations. The intention is to use the additional flexibility provided by this design to get larger deflection values. As simulations show, the average maximum displacement as a function of hydrostatic pressure increases about 30% (see Fig. 21.10). However, there are also disadvantages associated with this design. Due to the square design, areas of high stress values are generated in the silicon substrate, mainly at corners. A high fracture probability is associated with stress levels being in the range of 600 MPa for silicon. Depending on the arrangement of the etched structures, a much higher amount of stress is even possible. Another disadvantage is the complicated manufacturing process. Because of the anisotropic wet etch process, the convex corners have to be protected against undercutting.

21.4.2 Fabrication Processes

Bottom Layer with In- and Outlet. Both the in- and outlet and the valve seat are fabricated applying a multistep wet chemical etching process using ⟨100⟩ silicon wafers. Low pressure chemical vapor deposition (LPCVD) silicon nitride layers are used to mask during wet etching of silicon with 40% potassium hydroxide (KOH) at 80°C. Plasma enhanced chemical vapor deposition (PECVD) silicon oxide layers are used as mask layers during the wet etching of silicon nitride in 180°C concentrated phosphoric acid. The silicon oxide layers, in turn, are masked with conventional photo resist (ma–P 1215, *Micro resist technology*, Germany) and etched in hydrofluoric acid.

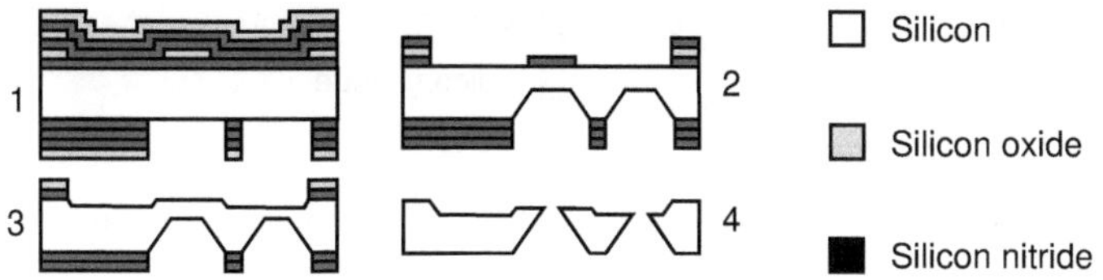

Fig. 21.11 Process steps of the bottom layer with in- and outlet and valve seat

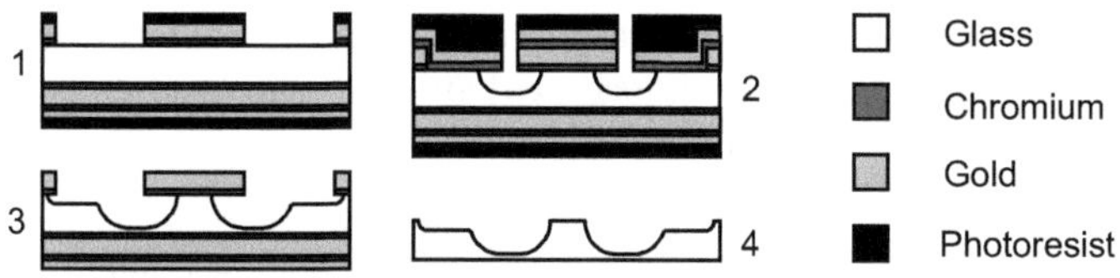

Fig. 21.12 Process steps of the first membrane layer

The layer composition can be seen in Fig. 21.11 step 1. After the processing and patterning of the mask layers, two funnel-through holes are etched from the backside to a local wafer thickness of 60 µm (see Fig. 21.11 step 2). In the next step the top silicon oxide and nitride layer are removed and the silicon is etched for another 10 µm to form the valve seat, as shown in Fig. 21.11 step 3. Finally, the second silicon oxide and nitride layer are removed and the funnel-through holes are etched through. As the valve seat is not masked anymore, its resulting height is reduced by 25 µm. To finish the in- and outlet wafer, the remaining mask layers are removed (see Fig. 21.11 step 4).

First Membrane Layer. The first membrane layer is fabricated either of silicon or pyrex glass. The silicon membrane layer is structured using deep reactive ion etching (DRIE) technique to generate round membrane shapes. Square designs were fabricated by means of an anisotropic wet chemical etching process similar to the one explained above.
The pyrex glass membrane layer is fabricated in a two-step wet chemical etching process. Two different mask layers are positioned right over each other. After having completed the first glass etching process step, the upper mask layer is removed to bare the bottom mask layer followed by the second etching step. This is shown in Fig. 21.12. Because the first wet etching process step in hydrofluoric acid requires a longer etching periode, a photo resist (AZ9260 *Mallinckrodt Baker Inc.*, Germany) is used in combination with a bi-layer of chromium and gold (30 nm/300 nm). The chromium layer serves as an adhesion promoter for the top metallization. In the second shorter-lasting etching step, a thicker gold layer of 800 nm (90 nm chromium) is applied and no additional photo resist layer is used.

SMA Actuator Element. To fabricate the SMA actuator elements, a 50 µm thick NiTi foil is cut by direct laser writing. Therefore a Q–switched Nd:YAG–laser is used with a basic wavelength of 1064 nm, an average power

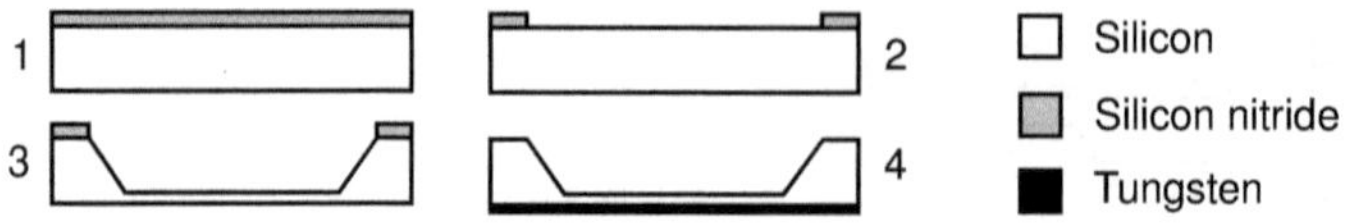

Fig. 21.13 Process steps of the alternative design of the second membrane layer

of 6 W, a pulse frequency of 1 kHz, and a cutting speed of 0.4 mm s^{-1}. To minimize the heat affected zone and its negative impact on the actuator performance, a mixture of nitric acid, hydrofluoric acid and water in combination with an ultrasonic bath is used for removal.

Second Membrane Layer. The second membrane layer made of silicon or pyrex glass can be fabricated applying a single wet chemical etching step. The process flow chart to fabricate the alternative pre-deflected membrane design is shown in Fig. 21.13. To start, silicon nitride is deposited with a PECVD process on a silicon wafer with ⟨100⟩ orientation. This layer serves as a hard mask for KOH wet etching and is structured by a 6% hydrofluoric acid solution using conventional photo resist (AZ–1518, *Microchemicals GmbH*, Germany) as mask material. Next, the KOH etch process can be applied to the wafer, which leads to the typical 54.7° etch slopes. This step is followed by another hydrofluoric acid bath to remove the residual parts of the hard mask. To get a stress induced deflection, the membrane is coated on the backside with a tungsten thin film, which is applied by a sputter process. Depending on the sputter parameters, especially power and pressure, the residual stress can be adjusted. Fig. 21.14 shows the mean biaxial stress values for a series of measurements as a function of the plasma power and back pressure in the deposition chamber during film synthetization. It can be concluded that a high degree of freedom is available to tailor the intrinsic stress in tungsten films within certain limits (i.e. tensile to compressive up to about 1000 MPa). Thus, it is possible to a large extent to adjust nearly every pre-defined stress value. For large thicknesses and high stress values, delamination effects occur. The adhesion of the tungsten film on the silicon substrate limits the achievable stress values.

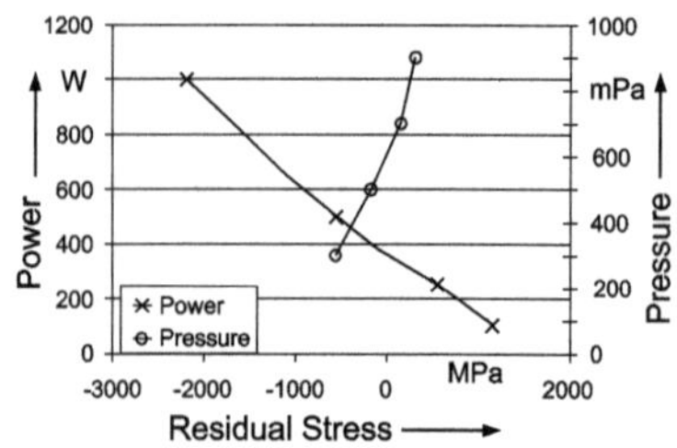

Fig. 21.14 Stress depending sputter parameter

For deflection structures with fixed borders, compressive stressed films are much more effective than those having tensile stress. This is again the result of the buckling effect according to Euler.

21.5 Combination of the Valve Layers

Three joining techniques are used for the realization of the presented valve: Anodic bonding, silicon direct bonding and bonding with adhesives.

If the bottom layer, as well as the first membrane layer, are made of silicon, silicon direct bonding is preferred. Anodic bonding is applied if the first membrane layer is made of pyrex glass. It can also be used if both the bottom layer and the first membrane layer are made of silicon. With this approach, an additional 1.7 µm pyrex glass layer is deposited onto the upper side of the bottom layer.

In the next step, the SMA actuator element is adhered to the upper side of the first membrane layer. Dissimilar working SMA elements have to be elongated by applying an external force before their first actuator cycle. Therefore, the SMA element is accurately fixed between two clamps and elongated 300 µm by means of a micrometer screw. The appropriate experimental setup is displayed in Fig. 21.15. Then, the SMA actuator element is electrically activated for the first time. Thus, the middle part is pulled to the open valve position (beside boss) and the SMA element is adjusted to the first membrane layer. The second membrane layer is adhered to the frame of the first membrane layer.

Fig. 21.15 Setup to combine the valve layers

21.6 Pre-Evaluation of the Intermediates

Although the presented development of the bistable microvalve is in the early stage, testing and evaluation of the intermediate products and first prototypes is of utmost importance to get a feedback for further development and optimization steps in the near future. A large variety of methods were used to characterize different membrane designs and to evaluate the tightness of the valve as well as the SMA behavior. Some of these test methods are explained in more detail in the following paragraph.

Experimental Setup for Membrane Characterization. An essential part of the membrane characterization was to determine which forces are needed to deflect the membrane and how far the membrane can be deflected before failure. Therefore, the first membrane layer is clamped between two aluminum plates which are fixed to a step motor. The motor traverses the membrane against the probe of a force sensor. An optical interferometer is arranged above the membrane to detect the membrane deflection at different mechanical loads.

Experimental Setup for Force-dependent Leak Tightness Characterization. With the experimental setup, shown in Fig. 21.16, the leakage behavior of the valve seat as a function of the closing force can be measured. In this case, the valve (i.e. bottom and first membrane layer) is clamped between two aluminum plates which realize the fluidic connection to the macro world. The hole in the bottom aluminum plate above the membrane layer

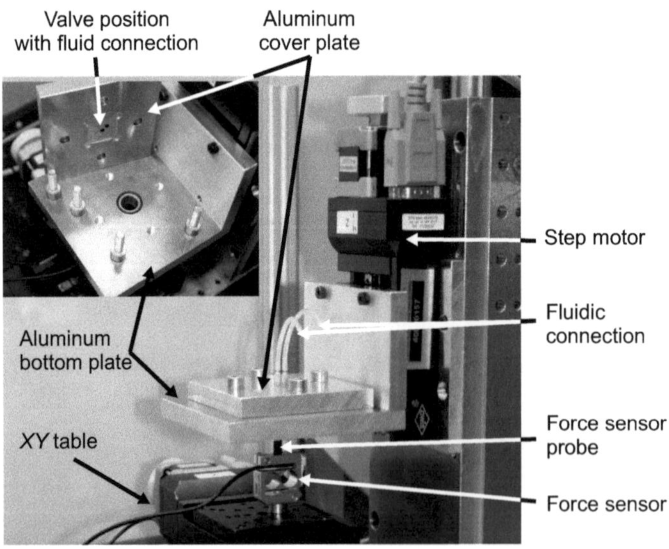

Fig. 21.16 Experimental setup to measure leak tightness against force

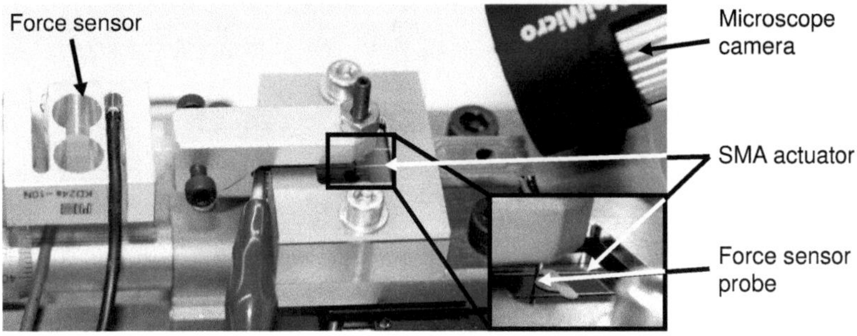

Fig. 21.17 Setup to measure force and displacement of the SMA actuator element

enables the transfer of different forces to the boss, which are measured by a force sensor as shown below. With pressure sensors implemented behind and in front of the valve, the leakage behavior can be detected simultaneously.

Experimental Setup to Measure Leak Tightness against Membrane Deflection. Another aspect which has to be investigated is the degree of tightness which can be achieved by the deflected membrane. For characterization, a stylus is placed in a slide bearing installed in an adjusting knob. In order to detect the contact point, an electric circuit is closed as soon as the stylus is in contact with the boss. In the following tests, the membrane can be deflected continuously by turning the adjusting knob to defined angles.

Experimental setup for SMA characterization. The experimental set-up to investigate the displacement and force characteristics of the SMA elements is presented in Fig. 21.17. The SMA element is already bonded to the first membrane layer. To measure the force of the SMA element, the probe of a commercial force sensor is connected into the middle part. The displacement of the middle part is measured optically. The information of interest is whether the actuator force is large enough to pull the middle part between boss and second membrane layer.

Verification of Membrane Amplitudes. Pre-investigations show that the direction of the membrane amplitude is arbitrary, as theoretically predicted by the FEM-simulation. The test structures consist of a single silicon chip with 9 wet etched membranes. After the sputter process, some membranes show a positive displacement while the others show a negative one. Figs. 21.18 and 21.19 show the deflection amplitudes according to the sputter time for two different membrane thicknesses. Both figures show some unexpected behavior. The amplitude should increase with the thickness of the tungsten layer and sputter time. But in nearly every graph, a region of constant deflection with increasing tungsten film thickness or even a local minimum is measured. It is assumed that varying conditions in the sputter deposition chamber when synthesizing tungsten films with different thicknesses may lead to these results. Stress coupled behavior mainly depends on

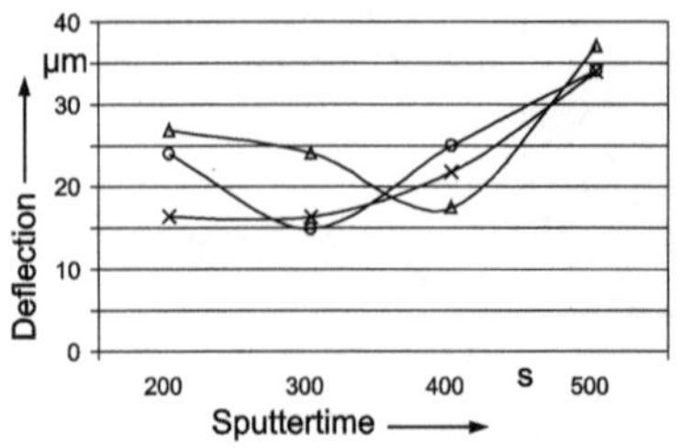

Fig. 21.18 Deflection depending on the sputter time for 20 µm membranes

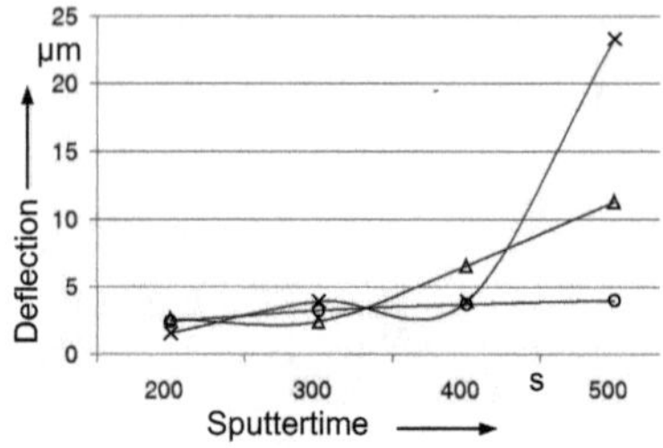

Fig. 21.19 Deflection depending on the sputter time for 26 µm membranes

the parameters of the growth by means of the presence of different imperfections, such as a varying density of defects and grain boundaries. The impact of these parameters by varying systematically the sputter conditions has to be investigated in the near future to achieve reliable deflection amplitudes.

21.7 Summary and Prototype

In the scope of this industrial application project an innovative 2–port/2–way microvalve that fulfills the predefined requirements and is integrable into the implantable infusion pump of the industrial partner has successfully been developed and fabricated. Due to the differential SMA actuator principle and the interaction of three mechanically connected silicon or glass layers, the microvalve is truly bistable. This means that only the switching operation between open and closed state requires electric power, whereas the two rest positions can be kept without consuming any power. An adequate energy source is e.g. a Li/MnO2 or a Li/SOCl2 battery, which can be integrated into the pump body.

The tasks to develop the functional valve principle, to design, fabricate and test the single layers as well as the SMA actuator element and to combine them to a first prototype (see Fig. 21.20) have already been finished. For testing and evaluation purposes, different complex experimental setups have been developed and built up. The functionality of the four valve layers could successfully be proven by means of FEM simulations and practical tests. Force, displacement, and power requirements that need to be fulfilled by the different SMA actuator elements have been measured. A further optimization of the individual fabrication steps and a detailed testing of the prototypes will be done in the near future.

Fig. 21.20 Prototype of the bistable microvalve

21.8 Outlook

Further research activities should focus on the interaction of all valve parts. In addition, the corresponding joining techniques on the wafer level as well as the mechanical bond connections need to be characterized in more detail with completely assembled prototypes. Furthermore, the sputter process has to be optimized to get more consistent results concerning the residual stress values of the tungsten thin layers. The simulations concerning the buckling effect and the counter forces have to be verified. For a more advanced layout, investigations on alternative membrane designs have to be considered. On the one hand, the effects of the stressed tungsten thin film layer, compared to the normal suspended membranes, should be analyzed. On the other hand, a second active membrane layer that assists the SMA actuator element by lifting-up during switching is planned and would substantially increase the overall device performance. Therefore, the influence of varying temperature levels on the membrane deflection using multilayer systems will be investigated.

Finally, a coating of the chamber between the valve seat and first membrane layer with e.g. parylene is targeted. Despite this broad range of technology related aspects that need further research efforts, it could be demonstrated within this project that the proper operation of key components for the realization of a truly bistable microvalve is feasible, indicating the opportunity for a successful testing of the complete device.

Acknowledgements The authors gratefully acknowledge the funding of the reported work by the German Research Foundation within the Collaborative Research Center 516 "Design and Manufacturing of Active Micro Systems". The authors also would like to thank *Tricumed Medizintechnik GmbH*, especially Dr. Volker Zacharias for the cooperation and support.

References

[1] Böhm S, Burger GJ, Korthorst MT, Roseboom F (2000) A micromachined silicon valve driven by a miniature bi-stable electro-magnetic actuator. Sensors and Actuators, A: Physical 80(1):77–83

[2] Capanu M, Boyd IV JG, Hesketh PJ (2000) Design, fabrication, and testing of a bistable electromagnetically actuated microvalve. Journal of Microelectromechanical Systems 9(2):181–189

[3] Goll C, Bacher W, Büstgens B, Maas D, Menz W, Schomburg WK (1996) Microvalves with bistable buckled polymer diaphragms. Journal of Micromechanics and Microengineering 6(1):77–79

[4] Kordon R (31.07.2000) Bistabiles Magnetventil. BSH Bosch und Siemens Hausgeräte GmbH, Munich (DE), patent DE10037251A1

[5] Offereins HL, Sandmaier H, Folkmer B, Steger U, Lang W (1991) Stress free assembly technique for a silicon based pressure sensor. pp 986–989

[6] Oh KW, Ahn CH (2006) A review of microvalves. Journal of Micromechanics and Microengineering 16(5):R13–R39

[7] Qiu J, Lang JH, Slocum AH (2001) A centrally-clamped parallel-beam bistable mems mechanism. pp 353–356

[8] Quenzer HJ, Wagner B (10.02.1997) Bistable microactuator with coupled membranes. Fraunhofer-Gesellschaft zur Förderung der Angewandten Forschung e.V., Munich (DE), patent US6168395B1

[9] Seward KP (08.03.2001) Bistable microvalve and microcatheter system. The Regents of the University of California, Oakland, CA (US), patent US6565526B2

[10] Wisskirchen M, Geser B (30.03.1994) Bistable electromagnet, in particular magnetic valve. Schrott, Harald, patent WO9423435

Chapter 22
Microassembly Following the Desktop Factory Concept

A. Burisch, S. Deumlich, R. Degen[2], A. Raatz[1], J. Hesselbach[1]

[1] *Institute of Machine Tools and Production Technology*
Technische Universität Braunschweig
a.raatz@tu-braunschweig.de; j.hesselbach@tu-braunschweig.de

[2] *Micromotion GmbH*
degen@micromotion-gmbh.de

Abstract Assembly lines and clean rooms for millimeter-sized products often measure some tens of meters and are mostly too expensive for small- to medium-sized businesses. This motivates the development of new miniaturized production lines (desktop factories) adapted to micro systems technology products. These equipment following the Desktop Factory concept underlies several promising assumptions that will be discussed in detail. The state of the art for Desktop Factory devices points out the need for new miniaturized machine components and handling equipment. The optimization and the new design concept of such a miniaturized handling device, the *Parvus* robot, will be discussed concerning its industrial use in a visionary Desktop Factory. The experimental verification of the use of this robot within a miniaturized assembly setup points out the challenges of miniaturized devices for micro assembly. This Desktop Factory assembly setup enclosed in a localized clean room cell, consists of the size adapted handling device *Parvus*, microgrippers, feeders, as well as standard microscope cameras. On this platform the 3D-assembly of an exemplary industrial application, a micro-CMM probe consisting of three main parts, is demonstrated.

22.1 Introduction

During the last few years, a trend of miniaturization with regard to product development in several industrial sectors has been observed. Millimeter-sized Micro Electro Mechanical Systems (MEMS) are now expected to grow by 12% within the coming years [2]. With this in mind, most machines for precision assembly are many orders of magnitude larger than the workpieces to be handled as well as the necessary workspace. The dimensions and costs of the production systems compared to the products become more and more disproportional. Assembly lines and clean rooms for millimeter-sized products often measure some tens of meters and are mostly too expensive for small-to medium-sized businesses. Therefore, many micro-products are assembled by hand. This results in high assembly costs that amount to 20% to 80% of the total production costs [16].

22.1.1 Concepts for Miniaturization

Promising solutions for the aforementioned situation are miniaturized flexible handling systems. Most of these concepts relate to one of the two general groups described below.

The first group covers small, autonomous, moving microrobots and handling machines that are mostly piezo-driven [8]. The second group deals with miniaturized robots, better described as size-adapted handling devices, which fill the gap between small, walking microrobots and conventional robots. Such robots can be developed using innovative, miniaturized machine parts for realizing the highest degree of miniaturization of conventional robot technology. These size-adapted handling devices can range from several centimeters to a few decimeters in size. They are easily scalable and can be used in highly flexible production systems. General assumptions for the potentials of such systems are discussed in the following.

22.1.2 Assumptions Concerning Desktop Factories

The primary variables that determine the performance of size-adapted production machines and desktop factories are flexibility and cost efficiency. Flexibility can be described in four ways: flexibility of function, flexibility of production volume, flexibility of placement and flexibility of property.

Flexibility of function. Modern miniaturized robots can be equipped with a range of functionalities similar to those of conventional industrial robots (e.g. degrees of freedom, sensors and tools). A microproduction system with a wide range of functions can be achieved by combining these miniaturized machines with freely programmable control systems, miniaturized drive systems and microgrippers.

Flexibility of production volume. Even though a smaller space is required for miniaturized production machines to operate, the productivity related to space of such microsystems can be higher compared to that of conventional machines. These so-called "desktop factories" thus increase the availability of unused space for other applications. Hence, the manufacturer is more flexible in the number and variety of produced pieces.

Flexibility of placement. Because of their size, miniaturized production systems can be placed in conventional clean rooms as well as in local clean room cells. These systems are also easier to move from location to location;

thus companies profit from a high flexibility of placement with regard to the development of processes as well as the expansion and relocation of the company.

Flexibility of property. It is difficult to adapt most conventional production systems for additional applications. For companies, a complete substitution of these production systems is prohibitively expensive. Thanks to miniaturized production machines, already existing larger production systems can be equipped with extended functionalities. Miniaturized systems are small enough to be integrated within already existing systems. This symbiosis between larger and smaller systems yields improved flexibility.

Based on these flexibilities, size-adapted robots and desktop factories allow further assumptions to be made that make them very promising solutions for microproduction:

- **Higher density of functionality:** Combining the flexibilities of function and production volume, it can be seen that the productivity of a microsystem per unit space is large compared to conventional machines.
- **Increased accuracies:** Smaller machine parts can be manufactured in the same clamping, which improves the tolerance. In addition, phenomena such as expansion and contraction due to change in temperature are less prominent at smaller scales.
- **Dynamic properties:** Because of reduced mass of machine components, smaller torques are required during fast movements. Reduced moved mass and smaller necessary torques result in acceptably velocities, although the miniaturized drives have less driving torque.
- **Operation in a local clean room cell:** Since clean rooms are expensive to maintain, it is preferred to reduce the amount of space that a given system uses. Smaller systems are therefore very cost efficient.
- **Lower maintenance / operation costs:** Small and easy-to-manage systems allow direct access to all parts when maintenance is required. Small systems consume less energy and resources than larger systems. They are thus well suited for concepts of *Green Manufacturing.*
- **Lower manufacturing / initial costs:** Smaller machine parts require fewer materials. This directly lowers the initial cost of the system.

Based on the aforementioned assumptions, several of the current problems in microassembly and the semiconductor industry can be solved by integrating miniaturized robots into production facilities, pursuing two different strategies:

Strategy 1. As a component for miniaturized production systems such as visionary desktop factories, e.g. in microproduction or the microassembly industry.

Strategy 2. As a miniaturized production machine integrated into a conventional bigger machine, e.g. in testing machines for conductor boards.

Most of the size-adapted systems in research and industry are based on the first strategy and follow the desktop factory concept. The aforementioned concepts in flexibility and other assumptions will allow microproduction in small- and medium-sized businesses to be more productive and cost efficient.

22.1.3 State of the Art for Desktop Factories

The first impulse for desktop factories was given in 1990, when a research group of the Mechanical Engineering Laboratory (MEL) in Japan (Tsukuba) estimated that a 1/10 size-reduction of production machines leads to a decrease of energy consumption about 1/100 compared to a conventional factory [20]. Nowadays the MEL, with its Desktop Machining Factory, is deemed to be a pioneer in the field of microfactories [20, 25]. Basic ideas and questions related to desktop factories are discussed by Breguet in his paper "Toward the Personal Factory" [3]. Deliberating about advantages and disadvantages, he highlights the vision of conventional factories and desktop factories coexisting in the future. The aim is not the all-purpose desktop factory, but rather a highly modular system adapted in size to MEMS. In conventional automation technology, examples of modular production cells such as the system of *IMSTec* [14] can be found. However, there are only a few examples in research and industry of concepts for modular desktop factories. The concept of assembly modules mounted around a fixed platform is followed up in research projects by Gaugel et al. [9] and Dembélé et al. [6] and by the industrial manufacturer *MiLaSys* [18]. Many other concepts follow up the idea of a fixed production cell equipped with a main handling device and several subsystems, e.g. Uusitalo et al. [26] in research and *Klocke Nanotechnik* [15] as a manufacturer. In particular, these concepts and the aforementioned system of *IMSTec* point out that the size of the whole production system is limited by the size of the conventional precision robot used. These miniaturized production systems obviously require highly miniaturized conventional precision robots and a miniaturized environment. For this reason, miniaturized precision robots such as the *PocketDelta* of *EPFL* and *HTI-Biel*, Switzerland [5, 27] and the *Parvus* [4] (described in Chap. 15 and 22) of *IWF* and *Micromotion GmbH* have been developed. Inspired by the small size of these robots, Tampere University of Finland developed modular miniaturized factory cells [24].

Finally it becomes clear that all of these introduced concepts underlie the same challenging requirement: the dependency on miniaturized high precision machine components such as drives and sensors. For this reason a few examples of such components are discussed below.

22.2 Miniaturized Components

In recent years, commercially miniaturized electric drives and high-resolution encoders have been noted as an emerging technology. A few examples for such components will be discussed in the following.

22.2.1 Miniaturized Drive Components

The main driving principles for miniaturized drives with larger strokes are piezoelectric actuators and electromagnetic motors. Piezoelectric drives use the stick slip, ultrasonic or inchworm effect to convert the small deflection (a few micrometers) of the piezoceramics into larger strokes of a few millimeters. Therefore, a high driving frequency is necessary, which in some cases produce disturbing noise and vibrations. In the case of the stick slip effect, there are unknown states of friction that make a control difficult in some instances. Nevertheless, piezoelectric drives enable very small, high resolution movements, but these depend on the measuring system that is the limiting factor for the drive's precision. Due to the fact that miniaturized sensors in high resolution are not available in the size of only a few millimeters, high resolution rotational piezoelectric motors still measure 20 mm in diameter or more.

Another strategy is using miniaturized electromagnetic motors. They can be used as a direct drive that also requires a diameter of more than 20 mm for generating enough rotational moment. In this case, the measuring system is also important for the resolution of the entire driving system.

The third strategy to achieve adequate drives with high resolution for microassembly systems with diameters of a few millimeters is using miniaturized electrical motors in combination with microgears, as developed by *Micromotion GmbH*. The *Micro Harmonic Drive* offers the same positioning accuracy as the large-scale *Harmonic Drive* gears used in industrial robots. Fig. 22.1.a shows a *Micro Harmonic Drive* gear that only uses six components to achieve reduction ratios between 160:1 and 1000:1. These ratios are required to create necessary torques from currently available micromotors, which are capable of rotational speeds of up to 100,000 rpm with torques of only a few μNm. Electroplating techniques, related to the LIGA-technique (Lithography, Electroplating, and Molding), are used to manufacture the single gear wheels of the *Micro Harmonic Drive*, which consist of a nickel-iron alloy. Because of the high yield point of $1500\,\mathrm{N\,mm^{-2}}$, the low elastic modulus of $165000\,\mathrm{N\,mm^{-2}}$ and its good fatigue endurance, this electroplated alloy is well-suited to operate in this microgear system. By providing an angular repeatability of 10 arc seconds, the *Micro Harmonic Drive* gear is the only microgear currently available that provides sufficient accuracy for a miniaturized robot meeting the requirements given in Chap. 15. During the operation of the gears within the robot it was noted that the driving torque

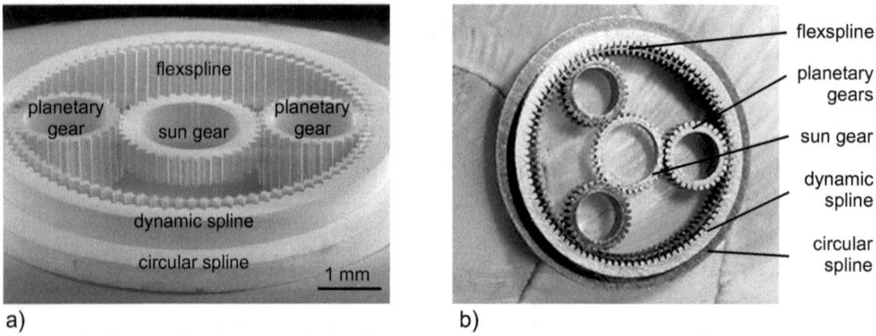

Fig. 22.1 *Micro Harmonic Drive*: (**a**) Components; (**b**) Gear with three planetary wheels

of the gear is very important, and therefore some improvements on the teeth were made. These improvements were focused on a higher overlap rate of the gears that leads to better meshing. The new P-tooth profile achieved a 200% increase in both torque capacity and torsional stiffness [17]. An increase in both stiffness and dynamics represents the new *Micro Harmonic Drive* gear with a transmission ratio of 120:1 (see Fig. 22.1.b). The special features of this gearbox are the three rotating planet gear wheels inside the flexspline. This leads to increased stiffness and better transmission behavior compared to the previously mentioned *Micro Harmonic Drive* gears. In addition to the gear series based on the *Harmonic Drive* principle, conventional planetary gearboxes with five elastic pre-stressed planetary gear wheels [17] for high speed applications are also available. This set of diverse types of miniaturized gearboxes makes it possible to choose the ideal gearbox for actuating a robot with requirements of several different applications.

The capabilities and challenges of the *Micro Harmonic Drive* gears can be evaluated by the use within the *Parvus* robot, as presented in Chap. 15. This investigation demonstrated that *Micro Harmonic Drive* gears are well-suited for use in a miniaturized robot with a good repeatability. However, it became clear that the special transmission behavior of the *Micro Harmonic Drive* gears cause some side effects. An approach is discussed in Chap. 15 to compensate for these effects. For this compensation e.g. the angular movement of the robot arms has to be roughly detected by miniaturized sensors. The optimization of the *Parvus* is thus dependent on the availability of such miniaturized devices as described below.

22.2.2 Miniaturized Sensors

The use of miniaturized motors as servo drives requires miniaturized precision angular sensors. In general, such sensors can be divided into inductive, capacitive, magnetic, magneto- and opto-resistive sensors [11, 21]. Due to the low torque of miniaturized motors, frictionless and contact-free angular sensors are preferred in contrast to potentiometers that have high frictional moments. Frictionless angular sensors always consist of two mechanically separated components. One is connected to the stator and the other to the rotor. A representative example give magnetic angular sensors, based on the Hall principle (*RLS*, Type: *AM8192B* [23], *Austrian Micro Systems*, type: *AS5045* [1]). The magnetic sensor principle allows absolute angular measurement over a complete rotation (single-turn or continuous-turn). Miniaturized angle sensors achieve typical resolutions of 12 bits (0.088°) to 14 bits (0.022°). Their accuracies vary, depending on the sensor type, between 0.5° and 0.1°. Contact-free sensors are based on a non-contact coupling, whereupon the moved components are mounted in a housing (*Numerik Jena*, Type: *Ulthemius* [19]). Due to the bearing and the seal in the housing frictional moments occur during operation. Magnetic sensors of this type achieve resolutions up to 15 bit (0.011°) and optical types up to 0.0036° (after quadruple interpolation), whereas optical sensors achieve accuracies better than 0.1°.

In the search for angular sensors, it was found that although there are many miniature angular sensors, they do not have sufficient resolution and accuracy for direct position control in precision handling equipment. Currently the use of a motor encoder in combination with a high transmission microgear still achieves a better theoretical resolution. It is therefore of great interest to develop solutions for the use of microgears and handle the challenges of transmission behavior of these components.

22.3 New Prototype of the *Parvus* Robot

The first functional model of the *Parvus* robot served in the investigation of the interaction between the miniaturized robot structure and the miniaturized servo drives using *Micro Harmonic Drive* gearboxes. During this research the robot specifications, such as the workspace, several types of accuracies and the transmission behavior of the microgears, have been observed, as presented in Chap. 15. The investigation pointed out some major aspects for the optimization of the robot design. Thus a new robot concept was developed meeting the following requirements:

- **Frictional moments:** The microgears are highly influenced by frictional moments inside the robot structure. The aim is to reduce the friction inside the robot joints.
- **Stiffness of microgears:** As friction influences the gear, a high stiffness of the gear is important. New concepts for a stiffer gear and better transmission behavior must be found.
- **Coupling of gears with the robot structure:** The mechanical connection between the gears and the robot structure may produce tensions. Misalignments must be prevented or compensated by a coupler.

- **Transmission behavior:** To optimize the transmission behavior of the microgears, a new control strategy has been found. This approach requires a direct angular sensor at the gear output to be integrated into the robot.
- **Functional density:** The first version of *Parvus* has external servo amplifiers and vacuum control. New miniaturized components enable a much denser design of the robot with more integrated functions.
- **Housing:** The first functional model was designed without housing so that internal functions could be observed for research purposes. The new prototype needs to have a housing to protect the inside components and give a product-like appearance.
- **Safety functions:** The first model of the *Parvus* is only equipped with basic safety functions using a single sensor per axis. The new prototype needs more sensors for safety functions and a user friendly homing procedure without any assistance of the user.

22.3.1 New Design Concept

For the new prototype of the *Parvus* robot, user friendly functions and a product-like design was desired. Thus the robot needed to have body housing for protecting the interior and giving a professional appearance. As for professional products, the body housing and the robot's interior were developed at the same time (see Fig. 22.2). This resulted in an increased functional density of the robot. The new concept now covers the robot arm structure with the hand axis, the drives, several types of sensors, the z-axis, a vacuum generator, the servo amplifiers, heats sinks, and several signal converter electronics. The most important points of the design process will now be described in detail:

- Four small power amplifiers are now integrated into the robot body. Due to their small size and extensive features they are ideally suited. In contrast to the previous servo amplifiers, they support higher sampling rates and more complex control algorithms. This will also improve the robot's specifications, such as speed and safety functions.
- The z-axis could have been miniaturized by using a new small precision ball screw directly connected (without belt) to the servo drive. In addition, the direct coupling of the spindle (ball screw) with the motor leads to a better accuracy of the whole axis.
- To reduce friction torques inside the robot structure, special low friction bearings have been selected for the new design.
- For the vacuum supply of the gripper, a vacuum generator can be integrated into the robot body.
- The platform of the main drives for the parallel arm structure is housed to protect the motors and delicate cables. It also enables a better integration of the air tube of the vacuum gripper.

- A miniaturized coupler has been integrated into the output shaft of the gearbox. It compensates possible misalignments between the drives and the robot arms to avoid tensions and torques on the sensitive microgears.
- The whole robot is covered with a housing that protects the electromechanical components inside the robot body and furthermore gives the robot a product-like appearance.
- Two low-cost absolute angular sensors have been integrated under the robot arms. They are used for the new control strategy to compensate the transmission behavior of the microgears. Furthermore, they can be used to optimize the home procedure and other safety functions.

The main aspects for the major improvement of this new robot concept were given by the experimental use of the first prototype of the *Parvus*. These challenges and experimental verifications for microassembly are described in the following.

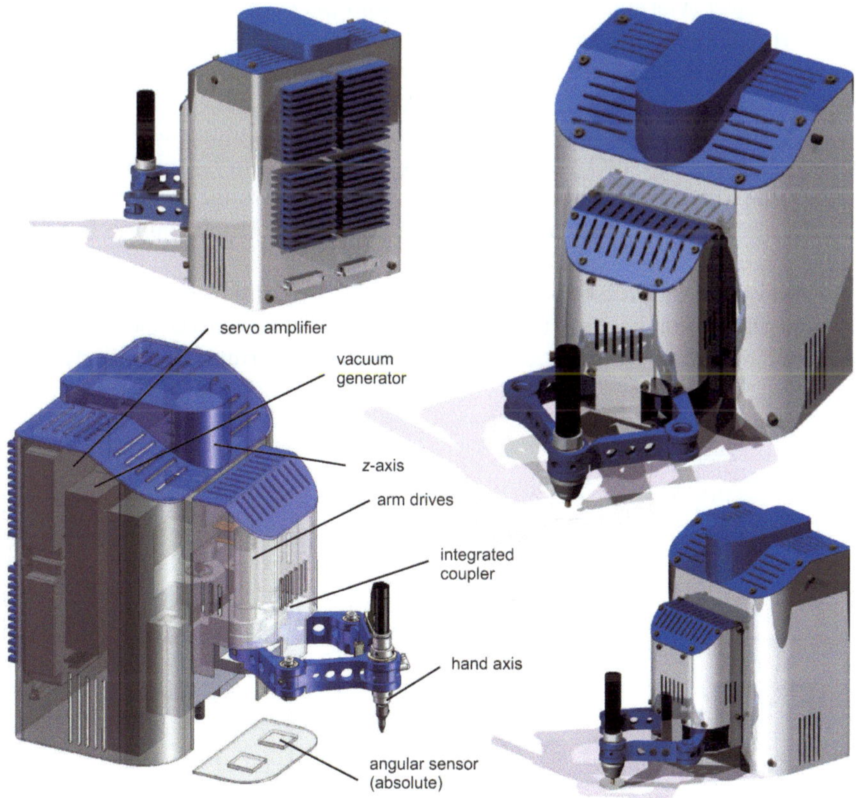

Fig. 22.2 New prototype of the *Parvus* robot

22.4 Experimental Verification

Many experimental measurements described in Chap. 15 and experimental handling operations have been done with the *Parvus* robot. An example assembly process of a micro-CMM probe gave a lot of information about the applicability and challenges of the robot concerning microassembly.

22.4.1 Challenges of Precision Assembly

Precision assembly of MEMS needs highly precise pick-and-place applications of the related parts. Typical examples for such tasks are the placement of optical parts such as prisms and photodiodes or optical ball lenses in a micro-optical LIGA bench [10] to be assembled with accuracy in the range of 1 µm. The demonstrators of microelectric actuators, described within this book and developed within in the Collaborative Research Center 516 in Braunschweig and Hannover (Germany) also need a pick-and-place application of small ruby balls with a diameter of a few hundred micrometers [12].

For all these pick-and-place applications, it is essential that the device always reaches the same assembly position multiple times. This is well described and characterized by the repeatability of the handling device following the standard of EN ISO 9283 [7]. The repeatability can be taken as a reference value for the maximum position accuracy of the robot end-effector. It is hereby assumed that the robot always reaches the position coming from the same direction (Fig. 22.3.a).

However, in the case of picking several different parts from magazine trays around the assembly area, the handling device will reach the position from different directions (Fig. 22.3.b). To get further information about the be-

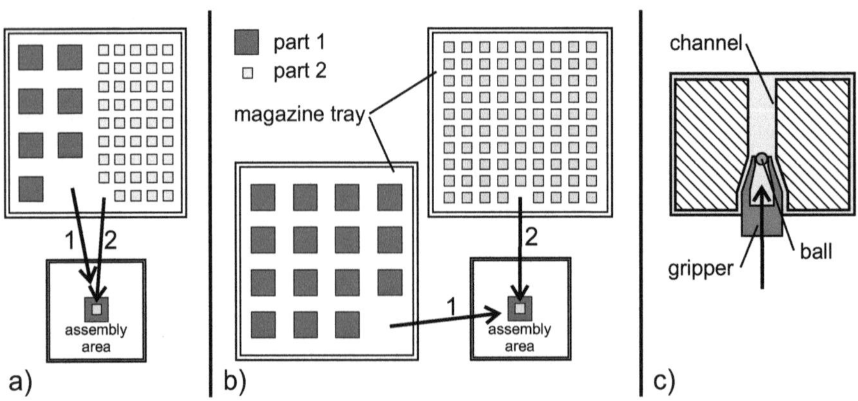

Fig. 22.3 Challenging assembly situations for: (**a**) Repeatability; (**b**) Multiple direction accuracy; (**c**) Path accuracy

havior of the robot in this case, the multiple direction position accuracy [7] has to be taken into account.

A further possibility for the evaluation of a robot is the path accuracy. In most cases it is only important for machine tools and not for pick-and-place robots. However, the example scenario of picking a glass ball out of a channel and moving the end-effector precisely along this channel shows that path accuracy can also be very important in microassembly (Fig. 22.3.c). In this case, good path accuracy [7] of the robot is necessary to move a gripper precisely inside the channel without collision, as later demonstrated in the experimental assembly process Sect. 22.4.5.

With these challenges in mind, an example assembly has been set up, as described below.

22.4.2 Product to be Assembled

As an example product, a three-dimensional microprobe has been chosen for the assembly process. Similar microprobes are used in coordinate measuring machines (CMM) such as *Zeiss F25.* The microprobe (see Fig. 22.4.a) is a tactile force sensor, which has overall dimensions of $6.5 \times 6.5\,\mathrm{mm}^2$. Its development and application are described in Chap. 9 and Chap. 6. The sensing element (boss membrane) is integrated into the base frame, which has a square opening of $3 \times 3\,\mathrm{mm}^2$ and a vertically projected area of $1 \times 1\,\mathrm{mm}^2$ the same as the center boss. The tactile stylus with a length of 5 mm has to be assembled on this center boss. A ruby ball with a diameter of 300 µm is placed as a probing sphere on the tip of the stylus [22].

22.4.3 Process Chain of Example Assembly Process

The above-described microprobe consists of three main parts made of different materials. This makes complex precision assembly necessary. To identify the necessary assembly equipment, a process plan with the main steps of the process chain has been investigated (see Fig. 22.4.b).

First of all, the bounding conditions and process preparations for the assembly process have to be considered. This includes the allocation of parts as well as additives such as glue, the UV-light supply and the arrangement of cameras as well as light sources. Other perturbations and disturbing influences, such as electrostatic charges of the assembly periphery and the parts to be assembled, must be reduced.

The challenges of each assembly step from 1 to 10 (see Fig. 22.4.b) are given here:

1. Low precision; boss membrane must not be damaged; grip, move and release the part.

2. Move base frame into predetermined position; fix the part.
3. High precision; grip the delicate microstylus, move part to glue reservoir.
4. Dip microstylus into glue; prevent hard contact and large vertical forces while still ensuring contact with glue.
5. High precision; position shaft on the center of membrane; release shaft; harden glue.
6. Low precision; grip, move and dip stamp into glue; prevent large vertical forces.
7. Low precision; apply glue on the upper side of the microstylus; prevent hard contact force while ensuring application of glue on stylus.
8. Separate one ball, move ball into a definite gripping position.
9. High precision; move gripper into channel of ball feeder while preventing collision; grip and move ball.
10. High precision; bring ball into precise assembly position; harden glue; release ball.

22.4.4 The Assembly Setup

Several other handling devices have been developed apart from the robot *Parvus* within the Collaborative Research Center 516 at Technische Universität Braunschweig. Such miniaturized grippers [13], feeders and active magazines manufactured with UV-depth-lithography are described in Chap. 16.

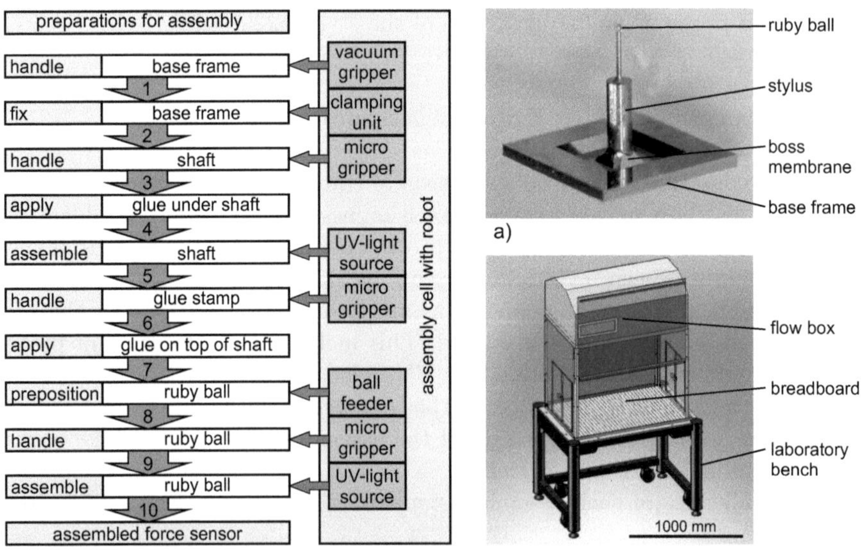

b) c)

Fig. 22.4 Example process: (**a**) Microprobe [22]; (**b**) Process chain; (**c**) Table sized clean room

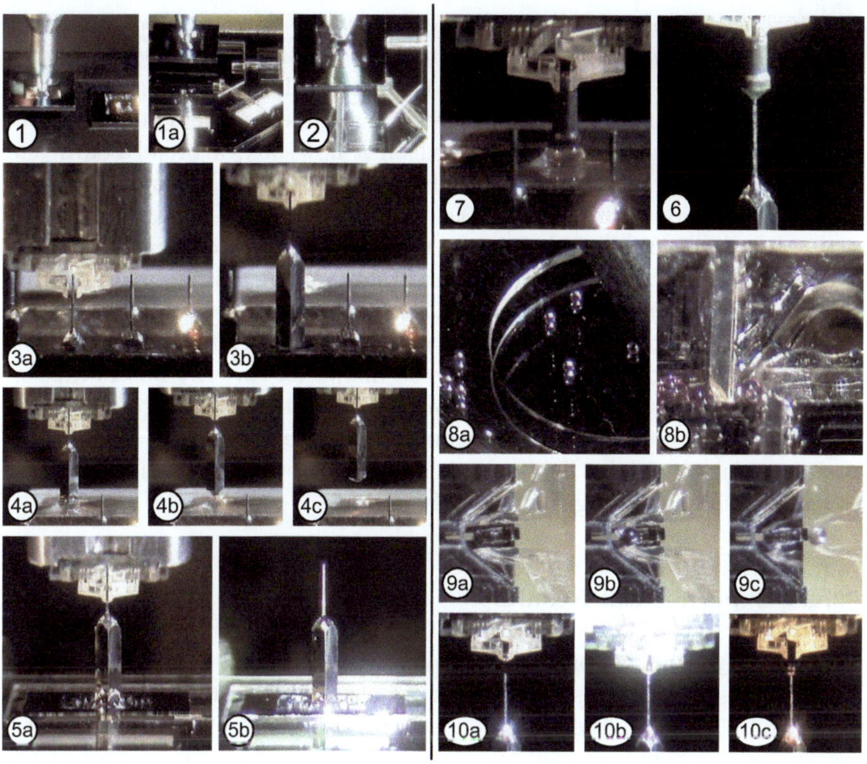

Fig. 22.5 Process steps of the assembly process of the microprobe

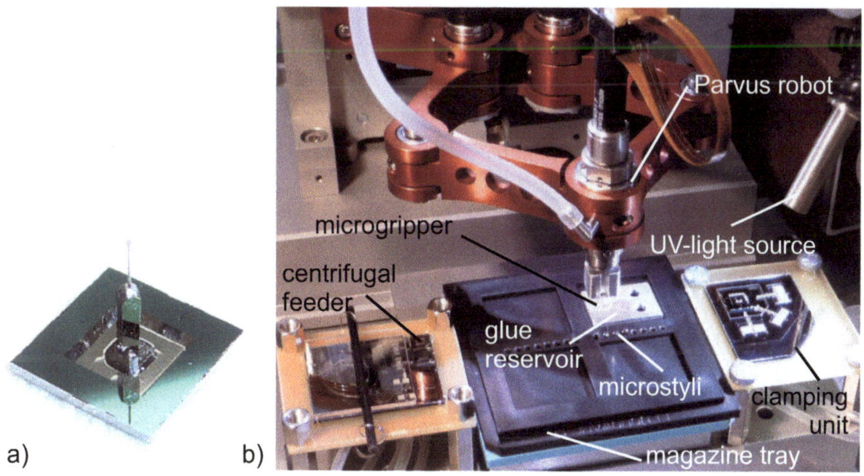

Fig. 22.6 (**a**) Assembled microprobe; (**b**) Experimental Setup

Owing to the cooperation between the *Institute of Machine Tools and Production Technology (IWF)* and the *Institute for Microtechnology (IMT)*, a miniaturized setup for microassembly has been realized.

As a platform for the assembly setup, a table-sized environment (see Fig. 22.4.c) has been developed based on a flow box, a bread board and an easily movable laboratory bench. Inside the flow box a maximum clean room classification up to ISO 2 (100) can be achieved. The devices for the assembly setup can be easily fixed on the integrated breadboard.

The magazine tray, the ball feeder and the clamping unit are placed within the workspace of the *Parvus* (see Fig. 22.6.b). Microscope cameras that are arranged around the workspace can monitor the assembly positions. The horizontal robot structure of the *Parvus* allows for an easy observation of the assembly situation using two cameras. Having two images from different points of view, it is possible to achieve a precise teach-in of the process positions. An ultraviolet light-source is placed above the assembly position to harden the glue during the assembly process.

22.4.5 Experimental Results

The assembly setup was prepared with a magazine tray filled with base frames (sensor chips) and microstyli, which were already vertically positioned (see Fig. 22.6.b). The ball feeder was loaded with ruby balls of 300 µm diameter. An ionized air flow was used to neutralize the electrostatic charges of the microparts and the environment. A reservoir with a thin layer of ultraviolet glue was prepared on the magazine tray. Figure 22.5 illustrates the assembly of the microstylus. The handling (1) of the base frame with the vacuum gripper requires only low precision because of the clamping device. With this device, the roughly positioned base frame is aligned and clamped (2) to a defined assembly position. The stylus is gripped at the thinner end with the microgripper (3a) in order to prevent a further change in the gripper. The gripping force is high enough to lift the stylus (3b). After dipping the stylus into the glue (4a,b), a noticeable glue drip is visible underneath (4c). The stylus can be positioned (5a) on the center boss with a repeatability of 5.7 µm (not including gripping uncertainty). The stylus does not move after it is released from the gripper. By applying ultraviolet light (5b) the glue is hardened within a few seconds. Steps (7) to (10) in Fig. 22.5 show the assembly of the ruby ball on top of the stylus. The stamp, dipped into the glue (6), allows for the application of the glue on the upper side of the stylus (7). Within the chamber of the ball feeder, the ruby balls are pushed into a channel by an air flow (8a). Within this channel, a single ball is separated (8b) out and blown into the gripping position. The ball can be precisely picked from this position (9a-c), thereby the path accuracy of the robot has to be taken into account to prevent a collision between the gripper and the surrounding equipment. The ball is finally positioned (10a) on the top of the stylus with a repeatability of 5.7 µm (not including gripping uncertainty). While the glue is being hardened with UV-light (10c), the gripper must hold the ball in its final assembly position. After releasing the gripper (10b) the assembly process is finished. This experiment demonstrates the plausibility of assembling the three-dimensional microprobe (see Fig. 22.6.a) with the given equipment in desktop factory size.

22.5 Conclusion

For visionary desktop factories, it is necessary to develop highly modular systems by playing on the potentials of size adapted flexible handling systems, as presented in the previous Sect. 22.1.2. Therefore, miniaturized components and handling devices with high flexibility of placement have to be developed. The realization of such a highly miniaturized prototype was possible due to the availability of miniaturized machine components such as micromotors, microencoders and highly precise *Micro Harmonic Drive* gears. Motivated by the availability of such components in the last few years, the miniaturized precision robot *Parvus* has been developed. The design and analysis of the first functional model is presented in Chap. 15. Intensive analyses and applications with this first functional model gave motivation for designing an optimized prototype of this robot concept. This new concept presented in Sect. 22.3 will cover more functions and a new control strategy for optimized accuracy and transmission behavior. Finally, an example application of the robot has shown its potentials and challenges concerning its use in a miniaturized assembly setup.

Within this project the future focus will be on the implementation of new control strategies, such as the compensation of the gear transmission errors and general safety functions. The application of this robot and the overall setup of a miniaturized assembly setup will be evaluated by further example tasks and analysis of the assembled products.

Acknowledgements The authors gratefully acknowledge the funding of the reported work by the German Research Foundation within the Collaborative Research Center 516 "Design and Manufacturing of Active Micro Systems". Many thanks go to the *Institute for Microtechnology (IMT)* and especially to Björn Hoxhold who supported the experimental assembly process with miniaturized peripheral handling devices.

References

[1] Austria Microsystems (2008) 12 Bit programmable magnetic rotary encoder (AS5045). www.austriamicrosystems.com
[2] Bouchaud J, Dixon R (2010) MEMS Back to Double Digit Growth in 2010. MEMS H1 2010 Market Tracker, iSuppli Corporation
[3] Breguet JM, Bergander A (2001) Toward the personal factory? In: Proceedings of SPIE, vol 4568, pp 293–303
[4] Burisch A, Wrege J, Raatz A, Hesselbach J, Degen R (2007) PARVUS–Miniaturised Robot for Improved Flexibility in Micro Production. Assembly Automation 27(1):65–73
[5] Codourey A, Perroud S, Mussard Y (2006) Miniature Reconfigurable Assembly Line for Small Products. Precision Assembly Technologies for Mini and Micro Products pp 193–200
[6] Dembélé S, Rochdi K (2006) A three DOF Linear Ultrasonic Motor for Transport and Micropositioning. Sensors and Actuators A: Physical 125(2):486–493
[7] EN ISO 9283 (1999) Industrieroboter, Leistungskenngrößen und zugehörige Prüfmethoden. Beuth-Verlag, Berlin
[8] Fatikow S (2007) Automated Nanohandling by Microrobots. Springer Verlag
[9] Gaugel T, Bengel M, Malthan D (2004) Building a Mini-Assembly System from a Technology Construction Kit. Assembly Automation 24(1):43–48
[10] Gerlach A, Ziegler P, Mohr J (2001) Assembly of hybrid integrated micro-optical modules using passive alignment with LIGA mounting elements and adhesive bonding techniques. Microsystem Technologies 7(1):27–31
[11] Hesse S, Schnell G (2009) Sensoren für die Prozess- und Fabrikautomation, 4th edn. Vieweg
[12] Hesselbach J, Pokar G, Wrege J, Heuer K (2004) Some Aspects on the Assembly of Active Micro Systems. Production Engineering 11(1):159–164
[13] Hoxhold B, Büttgenbach S (2008) Batch Fabrication of Micro Grippers with Integrated Actuators. Microsystem Technologies 14(12):1917–1924
[14] IMSTec (2010) Automationsplattform. www.imstec.de
[15] Klocke V, Gesang T (2003) Nanorobotics for Micro Production Technology. In: Proceedings of SPIE, vol 4943, pp 132–141
[16] Koelemeijer S, Jacot J (1999) Cost Efficient Assembly of Microsystems. MST-news, Verlag VDI/VDE Innovation+Technik GmbH 1:30–32
[17] Micromotion GmbH (2009) Think Smaller. www.micromotion-gmbh.de
[18] MiLaSys technologies GmbH (2010) Systembaukasten. www.milasys.de
[19] Numerik Jena GmbH & OPTOLAB Microsystems GmbH (2010) Ulthemius Micro Encoder. www.optolab.com

[20] Okazaki Y, Mishima N, Ashida K (2002) Microfactory and Micro Machine Tools. In: Proc. of Korean-Japan Conference on Positioning Technology, Daejeon, Korea
[21] Parthier R (2006) Messtechnik, 3rd edn. Vieweg
[22] Phataralaoha A, Büttgenbach S (2005) A Novel Design and Characterization of Micro Probe Based on a Silicon Membrane for Dimensional Metrology. In: Proc of Eurosensors XIX, Barcelona, vol 2, p WPb31
[23] RLS doo (2009) Angular magnetic encoder IC (AM8192B). www.rls.si
[24] Siltala N, Heikkilä R, Vuola A, Tuokko R (2010) Architectures and Interfaces for a Micro Factory Concept. Precision Assembly Technologies and Systems pp 293–300
[25] Tanaka M (2001) Development of desktop machining microfactory. Riken Review 34:46–49
[26] Uusitalo JJ, Viinikainen H, Heikkilä R (2004) Mini Assembly Cell for the Assembly of Mini-Sized Planetary Gearheads. Assembly Automation 24(1):94–101
[27] Verettas I, Clavel R, Codourey A (2005) Pocket Factory: Concept of Miniaturized Modular Cleanrooms. In: 1st Topical Meeting of Desktop MEMS and Nanofactories (TMMF2005), Tsukuba, JP

Chapter 23
Automated Optical BGA-Inspection – AUTOBIN

D. Gnieser, R. Tutsch

Institute of Production Metrology
Technische Universität Braunschweig
r.tutsch@tu-braunschweig.de

Abstract Today's highly complex microelectronic integrated circuits require large numbers of electrical connections for power supply and signal input/ouput. Ball Grid Arrays (BGA) have found a wide-spread use in microelectronic packaging, as they offer a large number of connections with advantageous electrical and thermal parameters. The quality inspection of printed circuit boards faces difficulties, because the soldering connections of the BGA are hidden under the component and are neither accessible for electronic probing nor for ordinary visual inspection of the quality of the soldered joints. Only quite expensive X-ray systems are applicable, but even these do not reveal all possible types of defect. Visual inspection of soldered joints can be done by applying microendoscopes that can be used to look into the small gap between the component and the electronic board. Such systems are commercially available, but the adjustment of the fiberoptic endoscope and the illumination fibres requires precise manual adjustments. The interpretation of the images is left to the user. The aim of this transfer project is to transfer this approach to an automatic testing equipment. In the SFB 516, a miniaturized optical 3D-sensor has been developed to control microassembly processes. This sensor will be modified and adapted to the task of adjustment of the endoscope relative to the BGA component. A miniaturized pattern projector has to be developed and integrated into the compact housing. The industrial partner will integrate this sensor into an automated optical inspection system and will develop image-processing software to evaluate the soldered joints.

23.1 Introduction

Electronic assemblies are key components of nearly all technical products today. In many applications, e.g. in automotive industry, electronic assemblies have to work reliably in a harsh environment with extreme climatic conditions and mechanical vibrations. Solder joints of high quality are mandatory.

In 2006 the Directive 2002/95/EG of the European Union, better known as RoHS Directive, banned all soldering alloys containing lead. Lead-free soldering processes have been introduced, but the process windows are much smaller than those of traditional soldering processes. They require higher temperatures and longer process times. The soldering flux evaporates faster during soldering leading to wetting problems between soldering alloy and pads. As a consequence the importance of test procedures in electronic production has increased.

Quality control in electronic production in general can be divided in electrical tests on one hand as well as optical and X-ray inspection on the other hand. In electrical testing either the correct function of the electronic assembly is checked or an In-Circuit Test (ICT) is performed by contacting discrete components of the assembly using a bed-of-nails adapter or a flying probe machine. A special case is the Boundary Scan Test, that uses standardized test functions implemented in the integrated circuits to check the connections between the different ICs on the board.

Electrical tests are very powerful and can be performed automatically, but they cannot detect all relevant defects on a board. Components that protect the assembly against overvoltage, e.g., might be defect or even missing without impairing the function of the assembly under regular conditions. Ceramic capacitors with wrong polarity might survive a short test procedure but will soon fail under working conditions. A deformed IC pin might have no connection with the pad on the board but is pressed against this pad during the ICT by the test needle. A very important shortcoming of electrical tests is that they give no information about the quality of the soldering joint.

Visual inspection as a complementary test procedure to electrical testing is essential and is performed either manually by trained personnel or using digital image processing, in this case known as automatic optical inspection (AOI) . Important criteria for visual inspection are the completeness of the assembly, the correctness of the type, the position and the orientation of all components and the judgement of the quality of the solder joints. The latter can be done by inspection of the shape and the surface appearance of the solder [8].

The main components of AOI systems are electronic cameras, imaging optics, illumination systems, fixtures and positioners for the assembly under test, computers and special test software. Two different types of AOI systems are available: 2D-systems evaluate single images and are usually sufficient for checking the completeness, the position and the orientation of components. In 3D-systems either several images are combined using photogrammetric algorithms or a structured illumination is applied. The resulting 3D-data can be used for additional checks, e.g. the determination of the volume of solder deposits, the measurement of the deviation of the board from planarity and the check of parallelism between the surface of an IC and the board.

Quality control of solder joints is a very demanding task in AOI. The strong curvature of the surface and the variation between specularly reflecting and diffusely scattering regions often make the application of flexible illumination systems and the combination of different viewing directions necessary to remove ambiguity.

According to the empirically stated Moore's law the number n_G of gates in integrated circuits doubles approximately every two years. Another empirical formula, Rent's rule [3] states that the number n_P of necessary connections of an integrated circuit increases with n_G according to $n_P \propto n_G^p$ with an exponent p of $0.3 < p < 0.7$, depending on the type of IC. Therefore the demand of a large number of connection pins for power supply and signal input/output is a consequence of the increasing complexity of microelectronic integrated circuits.

This led to the development of different packages starting from Dual In Line (DIL) packages introduced 1965 with 14 to 16 pins in two parallel rows with a pitch of 2.54 mm to a variety of Quad Flat Packs (QFP) with four rows of pins with pin distance between 0.4 mm and 1.0 mm and up to 304 pins altogether in the 1990s. A further increase in pin density would cause several severe problems:

- The sensitivity to positioning and adjustment errors would become extremely high.
- Due to the small mechanical stiffness of the pins they could be easily bent or damaged during transportation and handling.
- The soldering process would become unstable, especially concerning shorts between adjacent pins.

In recent years microprocessors and gate arrays with a pin count of more than 1000 appeared on the market. For these components Ball Grid Array (BGA) packages are advantageous because the contacts are distributed across the area, not along the sides of the package. While the distance between neighbouring pins of fine-pitch QFPs is 0.5 mm or even 0.4 mm, the distance between the balls of BGAs is typically larger than 0.75 mm. Robust and reliable positioning and soldering processes are therefore much easier to implement for BGAs. Furthermore because of the large cross section and the low height of the balls the electrical and thermal resistance of BGA connections is much lower than that of QFPs. The inductance is also lower, an advantage for high frequency applications. Figure 23.1 shows an example of each of the three chip-types: a dual in-line package represented by a SOIC, an QFP and an BGA.

(a) (b) (c)

Fig. 23.1 Three types of packages: (**a**) shows the typical two parallel connection-rows of a SOIC; (**b**) an QFP with pins on each side; (**c**) an BGA-chip, the first row of solder balls is barely visible beneath the body

Instead of classical leads the BGA components have small solder balls that are arranged in a 2D array on the bottom side of the package. The printed board has small metallic lands in a corresponding arrangement. Solder solid deposits are distributed to the lands by screen printing. The BGA component is precisely placed on the board and fixed with a small amount of adhesive that has been applied to the board before. In a reflow oven with carefully controlled temperature profile the solder deposits melt and form a connection to the solder balls of the BGA. On cooling down again the solder solidifies, forming a mechanically stable connection between component and board.

Several testing tasks have to be solved:

- The volume and the position of the solder deposits and the glue drop on the board have to be checked before the BGA component is attached.
- The position and alignment of the BGA component relative to the board has to be checked before and after the reflow process.
- The quality of the solder joints has to be checked after soldering.

According to a survey [1] about 1/3 of all problems in electronic board manufacturing are caused by faulty solder joints, making this the most important source of errors. A problem in manufacturing electronic boards with BGA components is that the contact balls are hidden under the component and are neither accessible for electronic probing nor for visual inspection of the quality of the joints. The following strategies are common for testing BGA solder joints:

- Electrical testing is rendered possible by designing special testing lands in the printed board that are connected to pins of interest of the BGA component and are accessible to test needles. This requires careful consideration in the design procedure of the board and is refered to "design for testability".
- Even pins that are not accessible can be tested electrically if a boundary scan structure is implemented in the integrated circuit. The test, however, is limited to the check of connections, the function of the circuit cannot be observed this way.
- X-ray inspection can be used to see "through" the board and the chip. Plastics, ceramics and silicon are only weak absorbers for X-rays, while the absorption by metals is significant. By digitizing the shadow that is thrown when illuminating the board with X-rays the balls and the structures of the metallization layers are visible with good contrast.
- Optical inspection is possible applying microendoscopes with 90° beam-bending prisms that can be used to look into the small gap between the component and the electronic board.

X-ray systems are able to detect shorts between adjacent balls and show even defects within the balls like voids and vertical cracks. Horizontal cracks, however, might not be visible. Because non-metallic materials are nearly transparent to X-rays, remaining flux deposits that may cause corrosion will not appear in the shadow images, either. It is also not possible to judge the quality of the solder joint (e.g. surface-check) in x-ray images. X-ray systems are very expensive, need shielding and require specially trained personnel.

Optical testing with microendoscopes on the other hand allows for inspection of the surface quality of the solder balls, flux deposits are visible as well as shorts between neighboring balls. Horizontal cracks might also be detectable. Commercially available optical BGA inspection systems are considerably less expensive than X-ray systems. On the other hand testing takes much more time because the adjustment of the fiberoptic endoscope and the illumination fibres is very tedious and has to be done manually. The interpretation of the images is also left to the user. The aim of the transfer project T3 of the Collaborative Research Center 516 was to transfer this approach to an automatic testing equipment.

In the Collaborative Research Center 516 a miniaturized optical 3D-sensor has been developed to control microassembly processes (Chap 17). This sensor will be modified and adapted to the task of adjustment of the endoscope relative to the BGA component. On the one hand the measurement volume has to be increased considerably, on the other hand a miniaturized pattern projector has to be developed and integrated into the compact housing. The industrial partner, the company *e-cube* [4] will integrate this sensor into an AOI system and will develop image-processing software to evaluate the soldering joints.

23.2 Specification of the Target System

23.2.1 The Inspection System

The inspection system will be used to record images of the outer row of solder balls that bond the BGA-chip to a circuit board, respectively of the free space between two neighboring columns of these balls. These images will then be sent to an image-processing software for further analysis.

The system consists mainly of four parts. A camera for recording the images, endoscopic optics for reaching down to the small gap between board and chip, a deflecting prism to divert the optical path by 90° towards the side of the chip and the illumination. Depending on the actual inspection task two different illumination techniques are possible: front side illumination shows the surface of the solder beadlets while backside illumination makes shorts between solder balls as well as remaining flux visible. Fig. 23.2 shows the layout of the system.

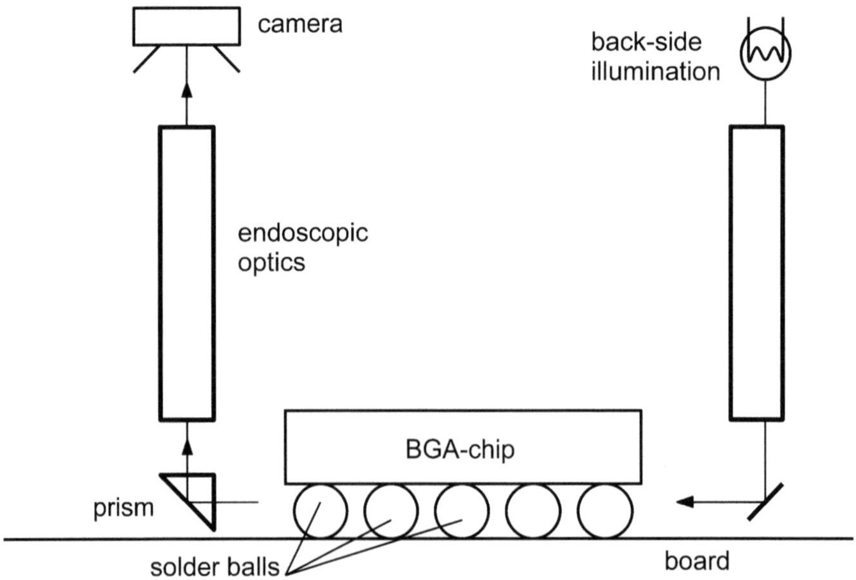

Fig. 23.2 Layout of the inspection system: camera, endoscopic optics, prism and illumination. Front side illumination parallel to the optic path is not shown in this figure

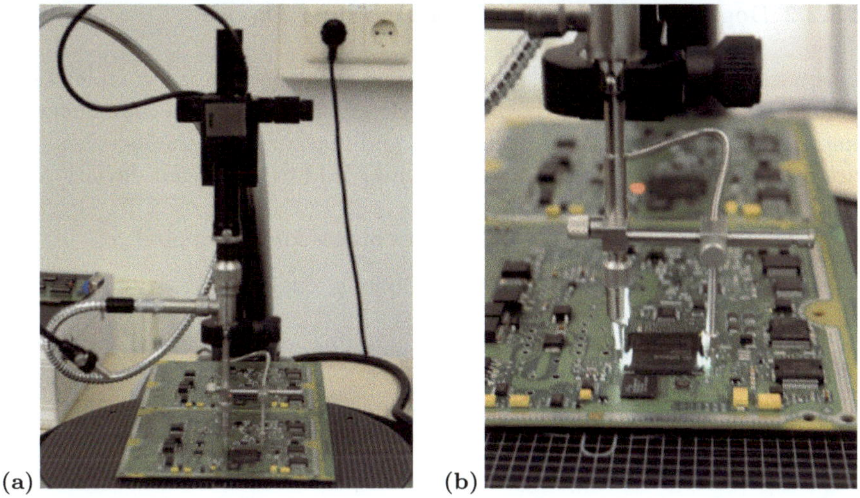

Fig. 23.3 (**a**) is a view over the whole obtained inspection-system; (**b**) shows more closely how prism and back side illumination are positioned at a chip under test

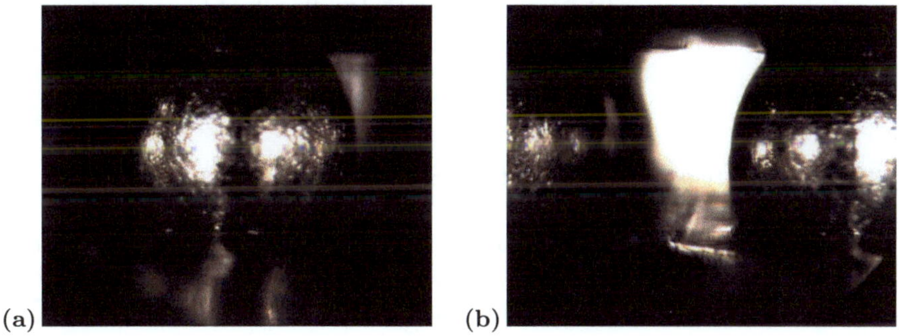

Fig. 23.4 (**a**) A single solder-ball of the front row of an BGA-chip observed by the obtained inspection-system; (**b**) A view between two collumns of the solder balls, no blockage for the back side illumination, so no short can be detected

A commercially available system was obtained for analyzing the performance of the different components and for using some of these components in a later test assembly. It was chosen to purchase the *ERSASCOPE* by *Ersa* [5] that can be seen in figure 23.3. Other suppliers of similar devices are e.g. *Optilia* [6], *Olympus* [7] or *TechnoLab* [2].

An example of images that can be taken by the inspection-system is shown in Fig. 23.4.

23.2.2 Degrees of Freedom of the Handling Device

The positioning of the inspection system and the 3D-sensor is a complex task. Though the nominal geometry of the electronic board and the components is known in advance, warping of the board and inaccuracy of the its edges that are used as reference within the fixtures may lead to significant deviations in reality.

A handling device with several degrees of freedom is required:

- A planar motion in the XY-plane (x-axis, y-axis) is necessary to reach all edges of the BGA-components.
- In addition a rotation in the XY-plane (angle ϕ) is required to adjust the inspection system and the 3D-sensor to the edges of the BGA-component in case of misalignment. This rotational axis is also necessary to access all four sides of the BGA-package. In addition, components might be arranged in a 45° orientation relative to the edges of the board.
- The distance d between the endoscope and the back side illumination has to be adjusted to the dimensions of the component under test.
- The endoscope and the back side illumination need individual z-axes (z_1-axis, z_2-axis) because the board might be locally tilted because of warp.
- The focus of the inspection system has to be adjusted to the row of solder balls under test.

23.2.3 3D-Sensor and Projection System

The inspection unit has to be positioned close to the surface of the board and close to the BGA-component, but must not touch either of them. While the nominal geometry of the board and the components is known, a fine adjustment in most cases cannot be avoided.

In order to do this adjustment quickly and automatically a 3D-sensor will be employed. This sensor is derived from the 3D-sensor used for the control loop of the microassembly robot *micabo*^{F2} that is described in Chap. 17. Summarized, it is a micro-stereophotogrammetric sensor. Two mirrors monitor the scene from two different directions. Both views are mapped onto different regions of the same camera sensor for analysis. The sensor is applied horizontally, to get a left and right view of the edge of the device under test.

In addition a pattern projector is needed, within the cramped confines of the setup. A projector displaying a line-pattern perpendicular to the component's edge was chosen. A phase-shift algorithm would be too time consuming, so a simple static pattern was favored: a laser creating a pattern via a diffractive optic element (DOE) is also space-saving. With a slight tilt of the projector (and also the sensor to register it) it actually is possible to project parts of the pattern on the side flanks of the chip to gain even more information of the height of these flanks. Interpretation of this pattern can

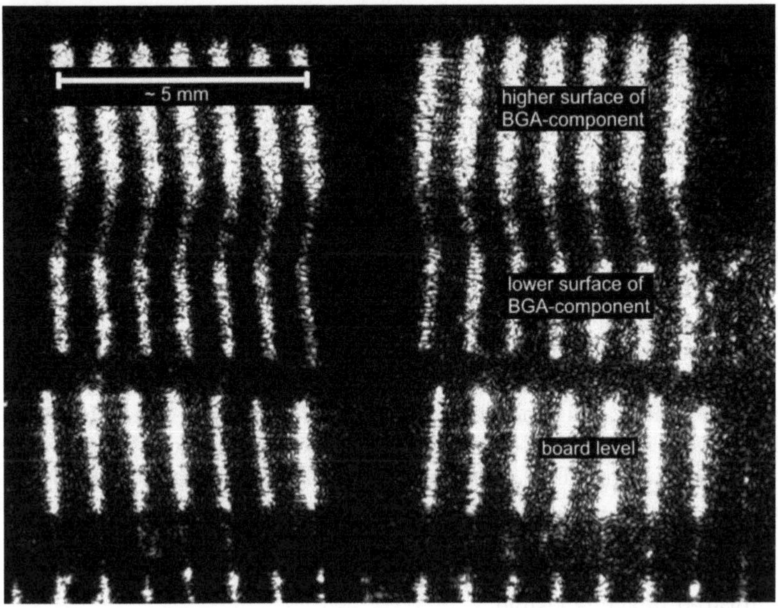

Fig. 23.5 Image of the two views of the 3D sensor. The 7-line-pattern is clearly visible. The BGA-chip has two levels, a high one and a low one, the step about at the middle of the figure is the difference from the lower part of the package to the board. The dark part in the lower third of the image is the intersection to the higher orders of the DOE

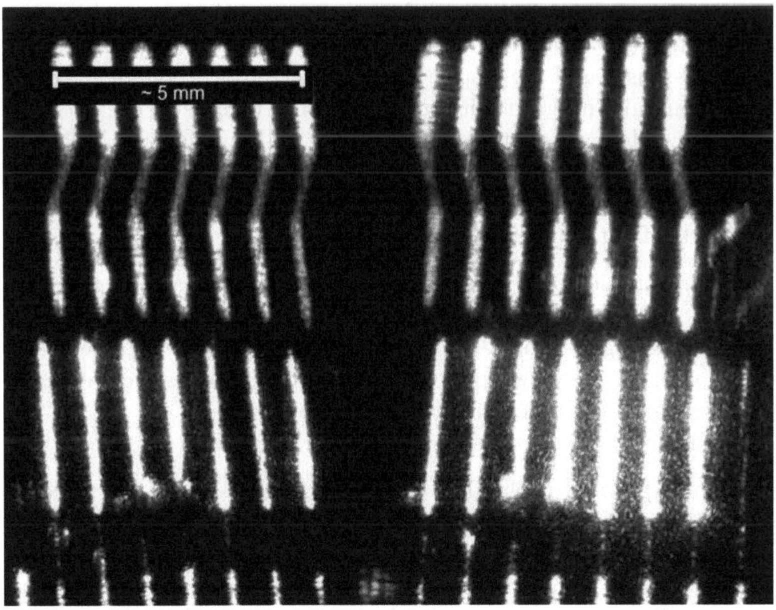

Fig. 23.6 Image of the two views of the 3D sensor. Same view as Fig. 23.5 but with a small vibration applied to reduce speckles

cause problems, if the correlation of the same lines in the two views of the 3D sensor does not match correctly. Two different DOEs were tested to support unambiguous matching. First approach was having a central line thicker and brighter than the others, but the image of the central line was not outstanding enough and in the case, that this special line gets shrouded, e.g. by a ridge perpendicular to the component's edge the analysis could fail. The second idea was to have a low number of lines (seven in the test build-up) that are entirely imaged in each of the two stereo-views. This approach proved to be realizable. Shrouding a single line can be adjusted for with the previous knowledge of the actual number of lines.

A big problem when using a laser for patterning are the speckles as can be seen in Fig. 23.5. A line-identification algorithm running about perpendicularly to the lines (along the pixels in x-axis of the image) will encounter difficulties. Speckle-reduced lasers are expensive, sometimes not very effective or more space consuming. E.g. using vibrating mirrors to despeckle the beam of the laser. But small vibrations induced at the target system reduce speckles, too – potentially at the cost of reduced accuracy. In Fig. 23.6 a small vibration parallel to the lines was applied. A clear improvement compared to the non-vibrated image is visible.

23.2.4 Inspection Procedure

In principle there have been two possible approaches to perform a complete inspection procedure. First, a so called "online" procedure, at which the 3D measuring and the endoscopic inspection are performed quasi-simultaneously. The 3D-sensor and the inspection tools have to be mounted close to each other. Second an "intermittent" procedure that completes the 3D measurment of the whole component and calculates a 3D model before the inspection system starts up. The second approach was chosen, so some of the descriptions in former parts of this chapter already were related to it. Both procedures start off with:

- Placing the electronic board at the inspection-table and fastening it.
- Using the knowledge of the nominal dimensions of the board, applying the 3D-sensor to the register marks of the board to get a first alignment and an estimation of the global warping of the board. For the distinctive marks no pattern generator is necessary.
- Approaching the component under test.

In the "online"-procedure, the field of view of the 3D-sensor has to cover the entire length fo the edge of the package, because positioning of endoscope and back side illumination occurs at the same time as measuring the geometry of the component – so only two 3D-measurements (one each in x and y) are necessary. But a pattern is required, that covers the frontside and the backside of the chip and still is capable of precisely exposing edge-deviations.

It is no simple task to determine the dimensions of the field of view of the sensor, because dimensions of BGA-components vary between 2 mm and 50 mm. Beneficial is a higher speed of the syncronous measurement and inspection and it may be possible for the sensor to monitor the endoscope and back side illumination. A tilted projection (see Sect. 23.2.3) is not possible, because of analysing front- and backside simultaneously. Some BGA-chips have a heat conducting zone centered at their surface, that may cause optical difficulties for the sensor or pattern projector.

For the "intermittent" procedure only a small field of view at the edges of the component is measured, so all four sides have to be passed over, but it is independent of the size of the component and different materials at the center of the surface of the chip, additionally the projected pattern is easier to generate. A 3D model is computed from this measurement and afterwards the inspecting procedure is applied based on the former measurements. This approach is supposed to be more time-consuming. A tilted pattern projection is possible.

The inspection tool then records images of the beadlets. Either single images based on the knowledge of the distances and dimensions of the solder balls will be recorded, or an overlapping series of images – the software will then have to be able to select appropriately. An image-processing will be applied afterwards to evaluate the solder joints.

The research work will be completed after the termination of Collaborative Research Center 516.

Acknowledgements The authors gratefully acknowledge the funding of the reported work by the German Research Foundation within the Collaborative Research Center 516 "Design and Manufacturing of Active Micro Systems".

References

[1] Berger M (2005) Prüfmethoden im Vergleich. Mechatronik F&M pp 14–16
[2] BGA.Inspector, http://www.technolab.de (2010)
[3] Christie P, Stroobrandt D (2000) The interpretation and application of Rent's rule. IEEE Trans on VLSI Systems, Special Issue on System-Level Interconnect Prediction 8(6):639–648
[4] e-cube GmbH, http://www.automation-cube.de (2010)
[5] ERSASCOPE 1, http://www.ersa.de/smt-bga-inspektion.html (2010)
[6] Flexia Digital BGA Inspection System, http://www.optilia.eu (2010)
[7] http://www.olympus-ims.com (2010)
[8] IPC (2010) Acceptibility of Electronic Assemblies. Association Connecting Electronics Industries, A-610E

Chapter 24
Slider with Integrated Microactuator (SLIM) for Second Stage Actuation in Hard Disk Drives

C. Ruffert, H. Saalfeld, H. H. Gatzen

Institute for Microtechnology (now: IMPT)
Leibniz Universität Hannover
ruffert@impt.uni-hannover.de; gatzen@impt.uni-hannover.de

Abstract To achieve a perfect track registration in Hard Disk Drives (HDD), a second stage actuation is desirable for compensating the frequency limitations of the main actuator. For addressing these requirements, Micro Electromechanical System (MEMS) technology was applied to develop a Slider with an Integrated Microactuator (SLIM). A pair of electromagnets actuates a mounting block to which a chiplet is attached which contains the read/write element. By appropriately energizing the electromagnets, a second stage actuation for ultra precision track following of the chiplet takes place. This paper discusses the fabrication process of the SLIM device and presents experimental results.

24.1 Introduction:

In Hard Disk Drives (HDD), an approach for achieving a perfect track registration, i.e. guiding the recording head's read/write element perfectly above the desired data track, is the use of dual-stage actuators (DSA), which has been suggested for quite a while [4]. A DSA servo system consists of a conventional electromagnetic actuator, formed by a voice coil-motor (VCM), plus a microactuator. While the VCM acts as a coarse, low-speed, but large-stroke actuator, the microactuator serves as a high-precision and high-speed, but small-stroke positioning system. A combination of both types of systems in a DSA allows to compensating respective deficiencies of either one.

There are various designs for DSA systems. In a moving suspension-type actuator typically a pair of piezo elements is located between the suspension's mounting plate and the flexure body [19], making it rather easy to fabricate compared to other DSA alternatives. However, its servo bandwidth is limited to about 3 kHz due to suspension resonances. In a moving slider-type actuator, the second stage microactuator is mostly a Micro Electro-mechanical System (MEMS) based rotary system located between the suspension's gimbal and the slider itself [12, 10]. Its servo bandwidth is 5 kHz or higher. Yet, both types of actuators result in an increase in costs of the read/write head, one due to the more expensive flexure, the other due to extra costs for the MEMS microactuator. While there is a growing need for a DSA in HDD, for this highly cost driven field extra costs required for its implementation so far turned into the main stumbling block for gaining an industry acceptance.

The Leibniz Universität Hannover (LUH) teamed up with the Berlin University of Technology in Germany, the University of Cambridge in the United Kingdom, and INESC-MN in Portugal to develop a cost competitive approach DSA solution for HDD [5]. Within the program, the LUH was chartered to develop the micromagnetics, conduct most of the system integration, perform final testing, and execute the program co-ordination.

24.2 Concept

Fig. 24.1 depicts the approach taken for accomplishing a second stage actuation. Fig. 24.1.a presents a regular "pico" size slider (length × width × height: 1.240 mm × 0.990 mm × 0.300 mm) [16]. It has the read/write element positioned at its trailing edge. The read/write element is fabricated in a (rather complex) thin-film process on a thick wafer. The trailing edge area of the slider represents the real estate required on the wafer, the slider length equals the wafer thickness. For the new DSA, the following approach was taken: instead of placing the read/write element at the trailing edge of the slider itself, it will be located on a chiplet only a third the size of the slider. This way, there is a potential for compensating the costs for the DSA by substantially reducing the fabrication costs of the read/write element. This approach also allows for actuating the chiplet by integrating a microactuator inside the slider resulting in a Slider with an Integrated Microactuator (SLIM) [8, 9]. Fig. 24.1.b depicts a schematic representation of the SLIM de-

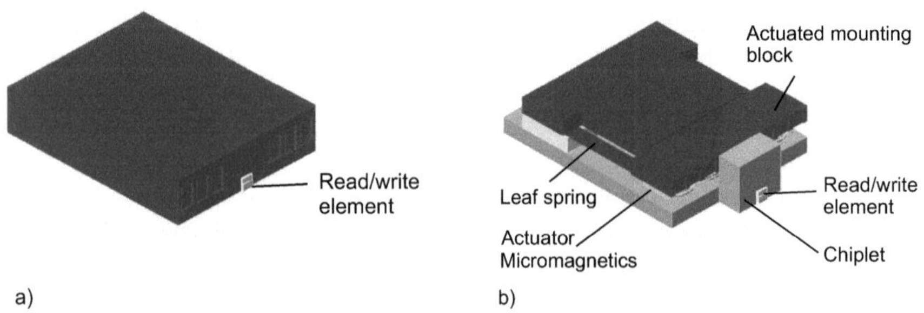

Fig. 24.1 Comparison between the sliders: (**a**) Standard HDD "pico" format slider; (**b**) Slider with an Integrated Microactuator (SLIM) [16]

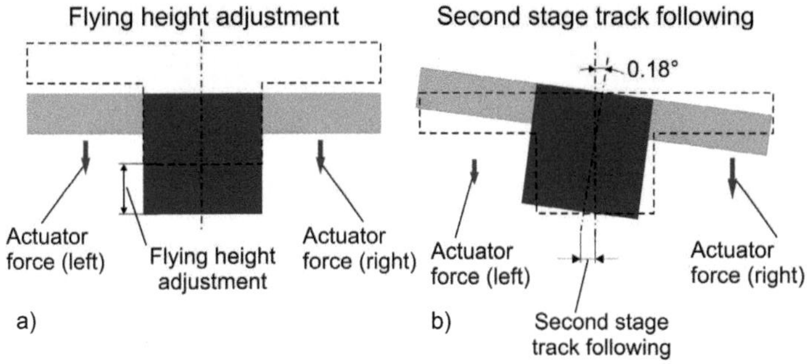

Fig. 24.2 Intended SLIM functions: (**a**) Height adjustment; (**b**) Second stage track following [8]

sign. An integrated magnetic system activates a mounting block, the chiplet containing the read/write element is attached to it.

The actuator allows two motions: a vertical displacement for lowering the read/write element to the quasi-static flying height and a very slight rotation executing a lateral displacement for track following (Fig. 24.2) [8]. Fig. 24.2.a schematically shows the flying height adjustment function, accomplished by energizing both halves of the microactuator. As it turned out, this flying height adjustment function became obsolete with the implementation of thermal flying height control (TFC) in HDD heads, which controls the head-to-disk spacing by a thermally actuated protrusion of the read/write element [13]. Although a vertical movability of the chiplet is still required to support an optimal flying height, no more dynamic flying height adjustment by the microactuator is intended.

Fig. 24.2.b depicts a schematic representation of the track following function. Alternatively exciting one half of the microactuator results in a minute positive or negative rotation of the chiplet. However, due to the extremely small angles involved, the motion causes a lateral displacement of the read/write element. At the program's start, a maximal lateral displacement of 625 nm was expected to be necessary, resulting in a rotation of 0.18°. Due to the small rotational angle, the resulting change in the flying height at the read/write gap is only about one nanometer. The SLIM device is fabricated on two wafers, one for the micromagnetics (ultimately forming the bottom of a slider) and one for the micromechanics (ultimately forming the top). Sandwiched in between the top and bottom part is a spacer; it provides a compensation for the magnetic microactuator's building height. The microactuator's active part consists of a pair of variable reluctance (VR) microactuators [8]. The actuator mechanics are composed of a mounting block suspended by a pair of leaf springs. Attached to the mounting block is the chiplet, on which the read/write element resides. A cost competitive solution will be achieved if the costs for fabricating the SLIM components plus the costs for the SLIM system integration stay below two-thirds of the cost of a present-day slider.

24.3 Micromagnetics Design and Fabrication

24.3.1 Micromagnetics Design

For finding an optimal design, the microactuator was modeled and simulated by a Finite Element Method (FEM) analysis using the software tool *ANSYS* Multiphysics [3]. During this design phase, not only performance requirements, but also technological aspects of the thin-film fabrication were taken into account. This way, important technological parameters defined by the photolithography processes like the components aspect ratio and flank angle, were considered appropriately. The lateral widening occurring during the exposure is considered in the design of the photolithography masks. To facilitate its desired function, a microactuator consisting of two subsystems was required. Each subsystem features a U-shaped, soft magnetic core carrying a pair of two double-layer spiral excitation coils (Fig. 24.3). In its final design, each magnetic pole has a length of 100 µm, a width of 20 µm, and a total height (excluding the flux guide height) of 40 µm, while the flux guide itself has a length of 100 µm, a width of 234 µm, and a height of 20 µm. A magnetics microactuator subsystem has (seen in HDD recording direction) a length of 284 µm, a width of 440 µm, and a height of 61 µm. Each pole carries a double-layer spiral coil, each of its conductors is 10 µm wide and 20 µm thick. The total number of turns per subsystem is 20. The insulation thickness between conductors is approx. 8 µm, between a conductor and a pole 10 µm, and between coil layers 250 nm. Compared to most MEMS, the microactuator magnetics features a rather great building height, as well as aspect ratios of the microactuator components. This is due to the fact, that the forces this microactuator is able to exert depends on the system's volume.

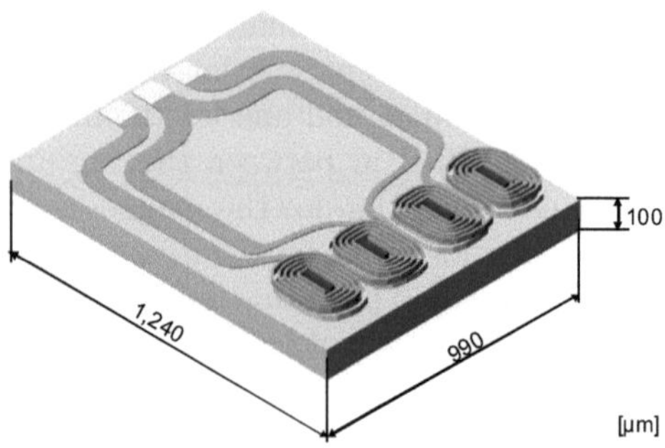

Fig. 24.3 Schematic representation of the microactuator magnetics [8]

Fig. 24.4 Micromagnetic simulation results [3]

For calculating the magnetic forces, both 2D and 3D simulations were conducted. At the beginning, an operating point at a rather low air gap length was envisioned. Fig. 24.4 depicts the simulation results for a single sided system. A 2D simulation predicted a magnetic force of about 355 µN at an air gap of 2.5 µm between the poles and the upper flux guide while exciting the coils with a current of 175 mA. A 3D simulation resulted in a magnetic force of about 340 µN for the same air gap. Ultimately, a much greater air gap had to be chosen. This was due to the fact, that a magnetic force pulling down one side of the mounting block also resulted in a vertical displacement of its other side. Fig. 24.5 depicts the force curves of the microactuator and the working point chosen. The optimal air gap range for the non-excited state is 24.5 µm to 29.5 µm.

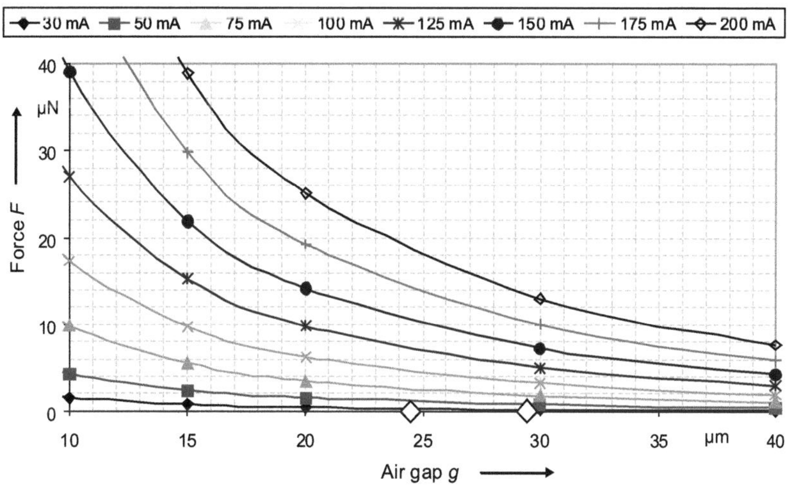

Fig. 24.5 Magnetic force curve of a single system. Air gap tolerance for the non-excited state: 24.5 µm to 29.5 µm (diamonds)

24.3.2 Micromagnetics Technology Basics

For fabricating this actuator micromagnetics with its substantial building height, High Aspect Ratio Micro Structure Technology (HARMST) was applied [17]. By using UV depth lithography, high aspect ratio patterns were created, serving as micromolds for electroplating. The materials employed in the electroplating processes were Cu for the coils, as well as NiFe45/55 for the magnetic flux guides and magnetic poles. NiFe45/55 was favoured over NiFe81/19, since it has a saturation flux density B_s of 1.6 T compared to 1.0 T for NiFe81/19. It also typically has a higher relative permeability μ_r [18]. However, NiFe45/55 is highly magnetostrictive, which is no disadvantage for actuator materials.

For executing electroplating on non-conductive substrates, a conductive seed layer is required. Such a seed layer was sputter deposited before the electroplating process was executed and removed by Ion Beam Etching (IBE) afterwards, if required. As embedding material, photosensitive epoxy (*SU-8*) was used, which also served as an insulation material between the coils as well as the coils and the poles. For achieving both flat surfaces and a constant layer thickness, Chemical-mechanical Polishing (CMP) was applied [2]. As vertical insulation between the coil layers, low stress Si_3N_4 deposited by Plasma Enhanced Chemical Vapor Deposition (PECVD) was used. By choosing Si_3N_4, the total height of the microactuator could be reduced from 90 μm to 61 μm compared to using *SU-8* also for the vertical insulation. Furthermore, the Si_3N_4 deposited by PECVD provides an excellent edge coverage, exhibits low stress, and offers a good dissipation for the JOULE's heat created by the coil. The same material was used for completely coating the finished system with a passivation layer.

24.3.3 Micromagnetics Fabrication Steps

For fabricating the microactuator magnetics, a 525 μm thick oxidized Si wafer (with a 100 nm thermal oxide layer on top) was used [16]. The first fabrication step was creating a conductive seed layer by sputter depositing a 50 nm thick Cr adhesion layer, followed by a 200 nm thick Au layer. As mentioned before, the existence of a conductive bottom material is a precondition for an electroplating process.

After depositing the Cr/Au seed layer, Ni fiducials were fabricated, followed by NiFe45/55 lower flux guides, and Cu electrical leads as well as the vias serving as contacts between electrical leads and coils. For all processes, a positive tone *AZ* photoresist was used to create micromolds for the respective electroplating process. At the end of the process sequence, the seed layer was removed by an IBE step to avoid a short circuit between the conductors.

During the following process sequence, the structure fabricated so far was embedded in *SU-8*, followed by a CMP planarization. By depositing a 250 nm thick low stress Si_3N_4 film using a PECVD process, a vertical insulation layer was created. For opening windows at the location of poles and vias, the Si_3N_4 layer was patterned. To do so, a positive tone *AZ* photoresist etch mask was created, followed by IBE and resist stripping.

The purpose of the next process steps was creating the lower coil layer, which also includes the bottom part of the poles. As previously, the process started with depositing a Cr/Au seed layer. In the following step, a photolithography process created *AZ* resist patterns. Using them as micromolds in an electroplating step created the first layer of the Cu coils. Afterwards, a stripping process removed the photoresist and an IBE step etched away the seed layer. After completing the lower coil layer, the bottom of the magnetic pole was fabricated. To do so, first a 400 nm thick NiFe45/55 seed layer with a 50 nm Cr adhesive layer underneath was fabricated by sputter deposition. Next, an *AZ* micromold was created, followed by filling it with electroplated NiFe45/55. Afterwards, the photoresist was stripped, and the seed layer was removed by means of IBE. Subsequently, the whole coil and pole layer was embedded in *SU-8*, which also served as a lateral insulation material. To conclude the first coil layer fabrication, the structure was planarized using CMP.

The fabrication sequence for the second coil layer (which includes the top poles) started with fabricating a Si_3N_4 layer by PECVD. It served as vertical insulation between the bottom and top coil layer. The next step was patterning the Si_3N_4 layer to create windows for poles and vias. As before, the patterning was done by a combination of photolithography and IBE. For the remaining fabrication steps, the same sequence was applied as for the first coil layer.

To complete the micromagnetics fabrication, there were two remaining tasks: passivating the whole system and creating the bond pads necessary for the electrical contact. For passivation, the whole system was coated with Si_3N_4 deposited by PECVD. This step is rather important, since the device contains an organic material (cured epoxy), which is hygroscopic and thus prone to moisture expansion. For creating windows for the bond pads in the Si_3N_4 layer, photolithography in conjunction with IBE was applied. For the bond pad fabrication, another Cr/Au seed layer was sputter deposited. Next, an *AZ* micromold was fabricated and filled with electroplated Cu. Finally, the Cu bond pads were covered by electroplated Ni serving as a diffusion barrier and an Au layer for corrosion resistance as required for the bonding process. After stripping the photoresist and removing the seed layer by IBE, the actuator micromagnetics were completed.

Fig. 24.6 depicts an entire micromagnetic system. Fig. 1.6.a shows an optical micrograph, Fig. 24.6.b an SEM image; Fig. 24.6.c presents a cross section through a microactuator subsystem, consisting of one C-core and its two poles surrounded by two double-layer coils. The system integration of the micromagnetics and micromechanics parts is done on a double rowbar level. Each double rowbar consists of two slider rows, with ten sliders (i.e. SLIM

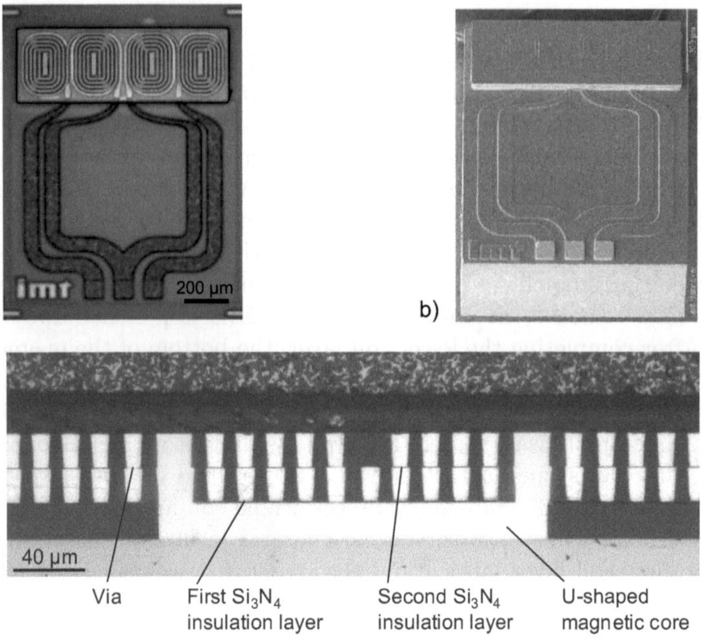

Fig. 24.6 Completed micromagnetic system of the microactuator: (**a**) Optical micrograph; (**b**) SEM micrograph (**c**) Cross section of a magnetic microactuator subsystem [16]

devices) each. The leading edges of the sliders are facing each other. The magnetics double rowbars were created by dicing the wafer appropriately.

24.4 Micromechanics Design and Fabrication

Besides the actuator micromagnetics, the micromechanics is another key part of the SLIM device. The actuator micromechanics was designed and fabricated by the *Microsensor & Actuator Technology Center* at the Berlin University of Technology in Berlin, Germany.

24.4.1 Micromechanics

Fig. 24.7 presents a schematic view of the actuator micromechanics, consisting of a base (forming the front part), a pair of leaf springs, and a mounting block, to which the chiplet is attached (chiplet not shown). Rather than using a sacrificial layer between the leaf springs and the micromechanics body, the whole material above the leaf springs is notched off. This approach avoids

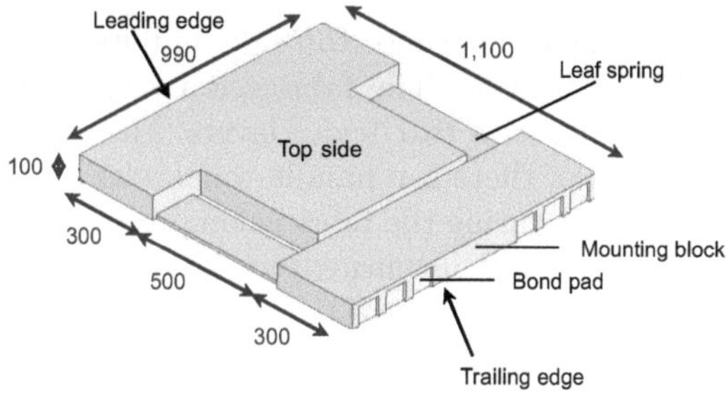

Fig. 24.7 Schematic representation of the SLIM micromechanics

the risk of stiction between the leaf springs and the slider body induced by a cleaning liquid. In the basic design, the leaf springs have a thickness of 5 µm, a length of 500 µm, and a width of 150 µm. For test purposes, parts with other leaf spring thicknesses were fabricated. Fig. 24.8 depicts two leaf spring and mounting block alternatives. Fig. 24.8.a shows the standard SLIM configuration with two 5 µm thick, 500 µm long, and 150 µm wide leaf springs and a solid mounting block. Fig. 24.8.b presents a version with 10 µm thick, 500 µm long, and 100 µm wide leaf springs and mass reduced mounting block. Due to its stiffer leaf springs and reduced mounting bar mass, it promises an improved frequency behavior.

At the mounting block's bottom, there is a pair of flux guides. Exciting the stator coils on an assembled system results in a force being created between stator and flux guide, actuating the mounting block. Six bond pads each are located on the rear (trailing edge) and front (leading edge) part of the micromechanics, providing electrical contacts to the chiplet mounted on the platform. They supply the write head, the read head, and the heater for thermal flying height control (two bond pads needed for each). Every bond pad has a size of $80 \times 80\,\mu m^2$. Six Al interconnections on the bottom side of the actuator micromechanics are joining the bond pads on the front side with those on the rear side. Every interconnection has a width of 20 µm. An Al

a)

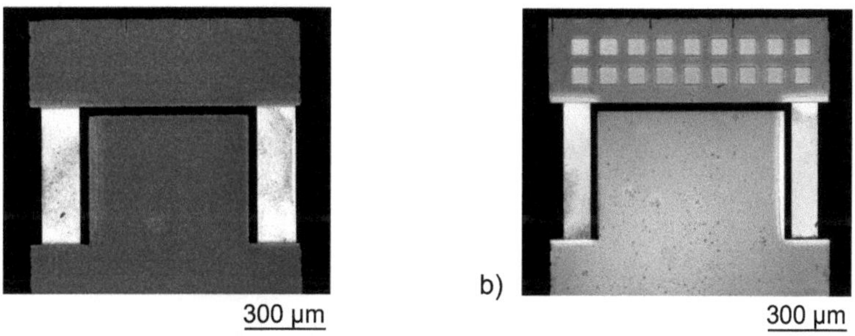

b)

Fig. 24.8 Flexure and mounting block alternatives: (**a**) 5 µm thick, 500 µm long, and 150 µm wide leaf spring and solid mounting block; (**b**) 10 µm thick, 500 µm long, and 100 µm wide leaf spring and mass reduced block

coat at the leaf springs' top side inhibits a warping of the leaf springs due to JOULE's heating. The fabrication was conducted on a Silicon-on-Insulator (SOI) wafer with two device layers. The overall wafer thickness is 300 µm. At the bottom, there is a handle wafer with a thickness of 200 µm, which serves as a carrier during the fabrication process. The two device layers with a total thickness of 100 µm ultimately form the micromechanics body and the leaf springs. The thickness of the top device layer defines the leaf spring thickness. Two buried oxide layers, each one with a thickness of 300 nm, are located in between and may serve as an etch stop during the fabrication process.

In preparation for system integration, the completed wafer was sliced in double rowbars. As in the case of the actuator micromagnetics, a double rowbar contains two rows of ten sliders each, with their leading edges facing each other. On both sides of each row, there is an anchor, immobilizing the delicate mounting block and leaf springs, to minimize damage to them during processing.

24.5 SLIM System Integration

24.5.1 Double Rowbar Stacking

The first step of the system integration of the SLIM components was stacking the double rowbars, sandwiching a spacer in between, and attaching the spacer to both sides by adhesive bonding. This work was executed by the Berlin University of Technology. Fig. 24.9 shows the stacked double rowbar on top of a one Eurocent coin [11]. While the micromechanics double rowbars already had their desired thickness, the size of the micromagnetics double rowbars was still the one of the original wafer, 525 µm. This provided rigidity during the stacking process. However, it also required that the stacked double rowbar had to be thinned to a thickness of 300 µm. For the prototypes, the height reduction was performed by a "thinning-by-dicing" process [1], followed by a nanogrinding [7] and CMP process. In a thinning-

Fig. 24.9 Stacked SLIM double rowbar a one Eurocent coin

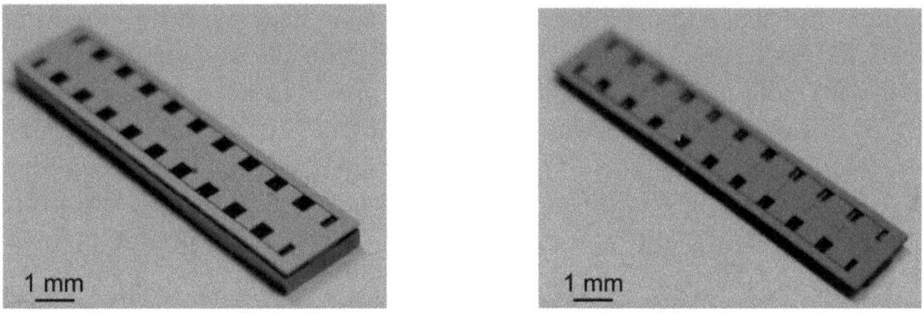

Fig. 24.10 Double rowbar: (**a**) Before thinning; (**b**) After thinning

by-dicing process, the wafer was first grooved by outside diameter grinding. Afterwards, the land areas created between the grooves was ground of. This step was followed by nanogrinding and CMP to flatten and smoothen the surface [1]. Fig. 24.10 depicts a double rowbar before and after thinning.

24.5.2 Slider Air Bearing Surface

At the bottom of a slider, facing the disk, there is an Air Bearing Surface (ABS), which controls the spacing ("flying height") between the recording disk and the slider trailing edge (on standard HDD sliders the location of the read/write element). The SLIM device also requires such an ABS. Its design was subcontracted to the *Center for Magnetic Recording Research (CMRR)* at the University of California, San Diego (UCSD), who came up with a sub-ambient pressure ABS design (Fig. 24.11). Simulations were conducted for a rotational disk speed of 7,200 rpm, aiming at a flying height between 2 nm to 3 nm. Fig. 24.11.a depicts a schematic representation of the sub-ambient pressure ABS profile. It features two different levels. The first (lower) level has a depth of 1 µm, the second (upper) level a depth of 260 nm. For creating the recessions, IBE was applied. Fig. 24.11.b shows a section of a double rowbar

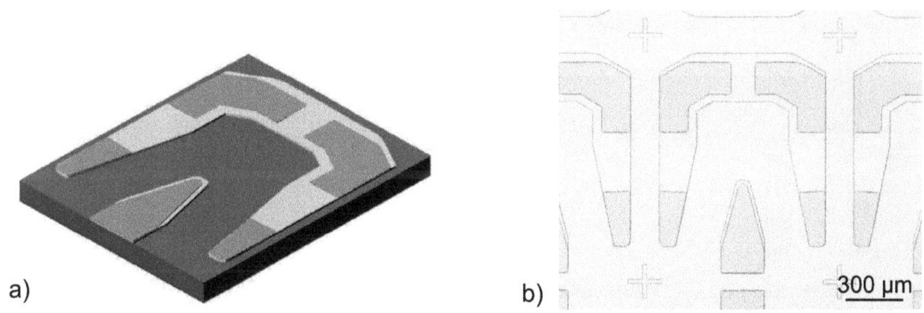

Fig. 24.11 Sub-ambient pressure ABS: (**a**) Schematic representation of the ABS design; (**b**) Section of the ABS on a double rowbar

after completing the ABS etching process [11]. For improving its corrosion resistance and its wear protection, a slider ABS typically is coated with approx. 1 nm of Diamond-like Carbon (DLC). To provide such a protection for the ABS of SLIM, the Department of Engineering of the University of Cambridge developed a hydrogen-free type of DLC called "tetrahedral amorphous carbon" or ta-C, deposited by filtered cathodic vacuum arc [14, 15]. However, except on dummy parts, no such coat was used on the SLIM prototypes.

24.5.3 Chiplet Mounting and Slider Dicing

After completing the ABS, the double rowbars were separated into single rowbars by a slicing process. This process also opened up the leading edge bond pads, allowing a static electrical test on the actuator coils. This allows to verifying which SLIM devices are electrically functional. Furthermore, the actuators were inspected for the gap length of the microactuator.

The next step was attaching the chiplet to the mounting block. On a fully functional SLIM device, the chiplet will carry the read/write element. However, the prototype chiplet didn't feature such elements. The dummy chiplets used in the prototype SLIM devices were made of Si. Since the chiplets are rotating slightly during operation when engaged in track following, they do require a cross crown radius chamferred at their bottom. This contour was created by a nanogrinding process [6].

For the microactuator to function properly, the chiplet has to be mounted with very tight tolerances at the appropriate height (in plane with the slider ABS) and at the desired location (exactly at the slider's center). To accomplish the task, an application specific assembly tool was developed (Fig. 24.12). The bottom of the fixture contains a slot connected to vacuum, acting as a vacuum chuck to hold down the chiplet's radiused bottom by suction. The first step in the process was to insert the rowbar in the assembly tool. A micrometer screw registered the rowbar appropriately against a reference surface. Then, the chiplet was picked up; due to its small size (its

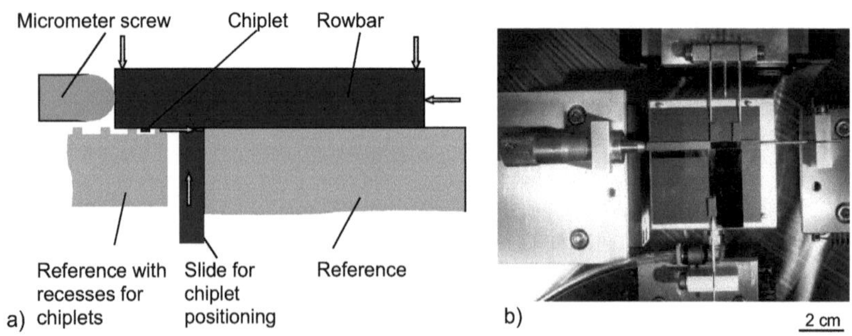

Fig. 24.12 SLIM chiplet mounting tool: (**a**) Schematic representation; (**b**) Picture

length and width are 150 μm, its height is 280 μm), a vacuum needle was used to manipulate it. Next, a drop of adhesive (cyanoacrylate) was administered to the chiplet. Then, the chiplet was placed on the vacuum slot and pushed against the mounting block using a slide. At the same time, it was pushed sideways against a reference surface using the vacuum needle for an appropriate lateral alignment and held in place until the adhesive was cured [11]. As soon as every electrically functional slider with an appropriate air gap length was equipped with a chiplet, the rowbar was diced into single sliders. This process is very delicate, because it also releases the mounting block, putting the leaf springs they are suspended from in action. The process was executed with the whole rowbar bonded upside down to a thermal release tape, which particularly immobilized the mounting blocks after their release [11].

24.5.4 Head Assembly

For assembling a slider into a recording head (also called a "head gimbal assembly" or HGA), two things have to happen: the slider has to be electrically connected (in this case, this can be limited to the microactuator) and mounted on a flexure. On existing sliders, the contacts are on the trailing edge of the slider. Due to the fact, that the SLIM device has a moving chiplet at its trailing edge, its electrical connections were moved to the leading edge. As flexure, a commercial part recommended by *Seagate Technology* and supplied by *Magnecomp* was used. First, Au wires were connected to the three bond pads. Next, these three wires were spliced into the flexure's existing flex cables. Then, the head assembly was completed by bonding the SLIM device to the flexture. For this process, a bonding tool developed and built by *AheadTek* was used. Fig. 24.13 presents a comparison between a classic HDD head (*Seagate Momentus 5.400.3*) and a SLIM head assembly.

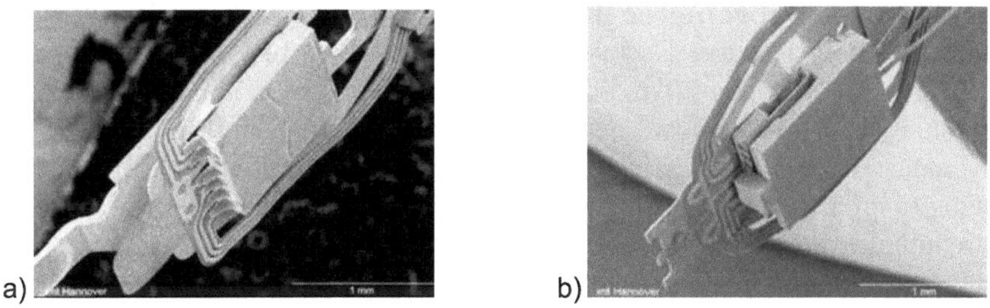

Fig. 24.13 HDD Slider: (**a**) Head assembly for the *HDD Seagate Momentus 5.400.3*; (**b**) SLIM head assembly

24.6 Experimental Investigations

24.6.1 Component Level Tests

Magnetic Force Verification. To verify the microactuator's magnetic force curves, the Berlin University of Technology developed a piezoresistive cantilever, which features the same flux guides as used in the SLIM system. This way, the magnetic force curves for various actuator excitations could be measured and compared to the FEM simulations. Fig. 24.14 depicts the results. A good correlation between simulation and measurement could be found for air gaps greater than 15 µm. For smaller gap lengths, there were discrepancies. Rather than being a proof for simulation errors, they were more likely caused by measurement difficulties on small gap lengths.

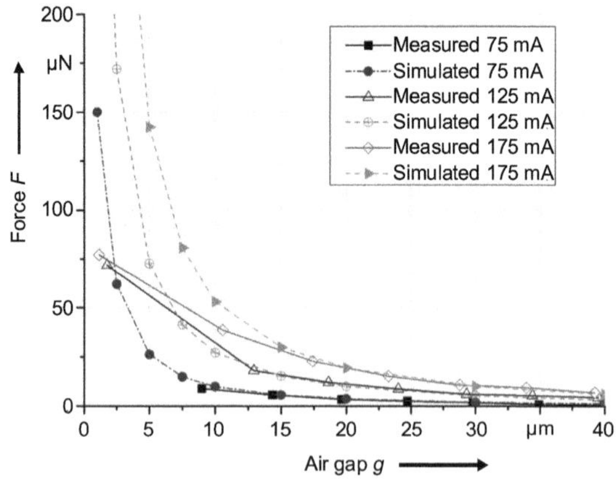

Fig. 24.14 Magnetic force curve: comparison between simulation and measurements results

Laser Doppler Vibrometer Tests. For analyzing the dynamic behavior of the SLIM device, Laser Doppler Vibrometer (LDV) tests were conducted (Fig. 24.15). Measurements can be either made for bodies moving out-of-plane or in-plane. Test specimen were separate micromagnetics and micromechanics systems. By rotating one by 180°, they could be mounted separately, still allowing to creating the desired air gap length. This way, a number of mechanical configurations could be subjected to tests rather easily. For the tests, no chiplets were mounted. Fig. 24.15.a presents a schematic representation of the test setup, while Fig. 24.15.b depicts the device in its second resonance mode, responsible for track following. Table 24.1 presents the results of LDV measurements carried out with various leaf spring/mounting bar systems featuring no mass reduction and a mass reduction of 25% or 50%, respectively.

Table 24.1 Resonant frequency of various leaf spring/mounting bar systems

Thickness [μm]	Width [μm]	Mass reduction [%]	1st harmonic [kHz]	2nd harmonic [kHz]
5	100	0	1.2	5.4
5	150	0	1.4	5.9
10	100	25	3.6	16.3
10	100	50	4.3	19.7
10	150	25	4.5	18.5
10	150	50	5	21.3

a) 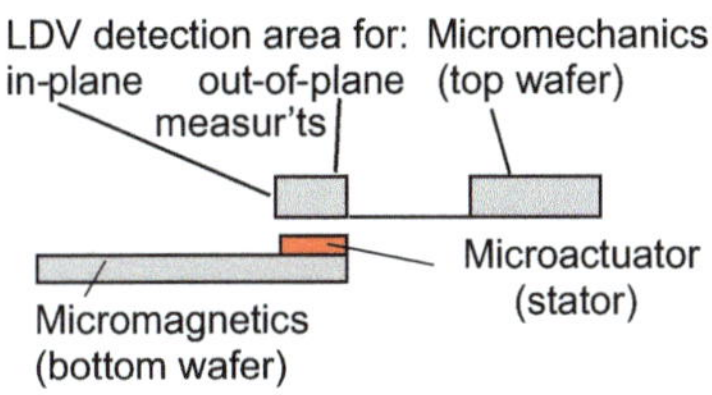

b) 

Fig. 24.15 LDV tests of SLIM components: (**a**) Test setup; (**b**) Second resonance mode responsible for track following

24.6.2 Device Level Tests

Flying Height Tests. Since there were no flying height measurement capabilities at LUH or any of the other of the consortium members, *AheadTek*, a company fabricating burnishing heads for disk vendors generously allowed the flying height tests to be conducted at their facility in San Jose, CA. The tests showed that no vertical chiplet actuation is necessary to adjust the flying height. Instead, the chiplet takes up a stable value, which is an equilibrium between BERNOULLI forces pulling the chiplet down and lift forces caused by the chiplet's cross-crown. Since the load bearing capability of an

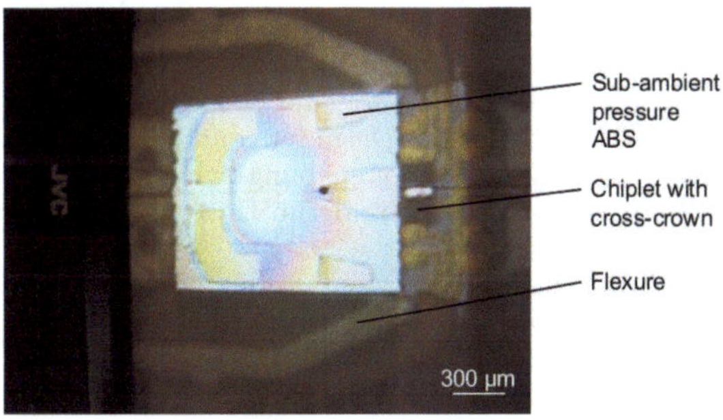

Fig. 24.16 SLIM head flying on the glass disk of a flying height tester

air bearing surface with a cross-crown is minimal, the resulting flying height at the chiplet is in the nanometer range, which is desirable. Fig. 24.16 depicts a SLIM head flying on the glass disk of a flying height tester. The rectangular bright area at the chiplet's bottom indicates a proper chiplet position.

Actuator Excitation. For these tests, the actuators were powered by an actuator control box developed by *INESC-MN*. The SLIM head assembly was mounted in a static tester. Then, one side of the actuator was excited and the displacement of the mounting bar was determined. By measuring at two different locations during two consecutive tests, the actual rotation of the mounting bar (and thus the rotation of the chiplet) could be determined. The resulting rotation was 0.106°, corresponding to a lateral displacement of the chiplet of 370 nm.

24.7 Conclusion and Outlook

The work conducted demonstrates the great potential of magnetic MEMS, in this specific case in creating a dual-stage actuation approach for hard disk drives. Most of all, the Slider with an Integrated Microactuator (SLIM) presents a cost competitive solution for dual-stage actuation. Both in the magnetics area, as well as in solving the system integration challenges, the program work greatly benefited from the research results gained within the Collaborative Research Center 516. Particularly the HARMST developed within this research programme was the key enabling technology required for the SLIM development.

Due to the inability of the LUH for conducting flying height tests, dynamic testing of SLIM devices in Hannover is too risky. Therefore, it is intended to conduct dynamic tests with flying heads at a HDD company in the U.S.

Acknowledgements This work was sponsored in part by the European Commission within the Framework 6 Information Society Technologies Programmes Specific Target Research Project (STREP) PARMA (Performance Advances in Recording through Micro Actuation). Furthermore, the authors are indebted to *Seagate Technology*, Bloomington, MN, for supporting the program with parts, test equipment and advice, F. Talke, UC San Diego for conducting the ABS simulations, and T. Higgins, *AheadTek*, San Jose, CA for enabling flying height tests. They also would like to thank the following persons at the Berlin University of Technology, the University of Cambridge, *INESC-MN*, and the Leibniz Universität Hannover, who contributed to the work: J. Borme, S. Cvetkovic, D. Dinulovic, P. Freitas, W. Heumann, W. Kurniawan, E. Obermeier, J. Robertson, and H. Yu.

References

[1] Cvetkovic S, Gatzen HH (2009) Technology development for the thinning of a slider with an integrated microactuator (slim). In: Proc. EUSPEN 9th Int. Conf. 2009, San Sebastian, Spain, vol 2, pp 455–458

[2] Cvetkovic S, Miletic D, Gatzen HH (2009) Relationship between removal rate and friction force of a CMP process for MEMS. In: Proc. EUSPEN 9th Int. Conf. 2009, San Sebastian, Spain, vol 2, pp 499–502

[3] Dinulovic D, Saalfeld H, Celinski Z, Field S, Gatzen HH (2008) Integrated electromagnetic second stage microactuator for a hard disk recording head. IEEE Trans on Magn 44(11):3730–3733

[4] Fan LS, Ottesen HH, Reiley TC, Wood RW (1995) Magnetic recording head positioning at very high track densities using a microactuator-based, two-stage servo system. IEEE Trans Ind Electron 42(3):222–233

[5] Gatzen HH (2004) Read-write head with integrated microactuator. German Patent 10260009B4 / U.S. Patent 20060238924, (2006)

[6] Gatzen HH, Cvetkovic S (2009) Cross crown nanogrinding of a chiplet for a slider with an integrated microactuator (slim). In: Proc. ASPE, 24th Ann. Meet. 2009, Monterey, CA, USA, p 95

[7] Gatzen HH, Mätzig JC (1996) Nanogrinding. Journal of the American Society for Precision Engineering 21:134–139

[8] Gatzen HH, Freitas PJ, Obermeier E, Robertson J (2008) A slider with an integrated microactuator (SLIM) for second stage actuation in hard disc drives. IEEE Trans on Magn 44(11):3726–3729

[9] Gatzen HH, Freitas P, Obermeier E, J R (2010) Second stage actuation for hard disc drives through MEMS resonant frequency of various leaf spring / mounting bar systems technology. IEEE Trans on Magn 42(3):782–789

[10] Hirano T, White M, Yang H, Scott K, Pattanaik S, Arya S, Huang FY (2004) A moving-slider MEMS actuator for high-bandwidth HDD tracking. IEEE Trans on Magn 40(4(2)):3171–3173

[11] Hoheisel D, Cvetkovic S, Kurniawan W, Obermeier E, Gatzen HH (2010) System integration challenges for a slider with an integrated microactuator. IEEE Trans on Magn 46(6):2163–2166

[12] Imamura T, Koshikawa T, Katayama M (1996) Transverse mode electrostatic microactuator for mems-based hdd slider. In: Proc. IEEE MEMS Workshop, pp 216–221

[13] Meyer DW, Kupinski PE, Liu JC (1999) Slider with temperature responsive transducer positioning. US Patent 5991113

[14] Robertson J (2002) Diamond-like amorphous carbon. Materials Science and Engineering:Reports 37(4):129–281(153)

[15] Robertson J (2003) Requirements of ultrathin carbon coatings for magnetic storage technology. Tribology International 36(4–6):405–415

[16] Saalfeld H, Dinulovic D, Gatzen HH (2009) Improved fabrication of an integrated electromagnetic second stage microactuator for a hard disk

recording head. In: Proc. ASME IMECE 2009, Lake Buena Vista, FL, USA, IMECE2009-11076, pp 1–8

[17] Seidemann V, Kohlmeier T, Föhse M, Gatzen HH, Büttgenbach S (2004) High aspect ratio spiral and vertical meander micro coils for actuator applications. Microsystem Technologies 10(6-7):564–570

[18] Wurz MC, Dinulovic D, Gatzen HH (2004) Investigations on the permeability of electroplated and sputtered permalloy. In: Processes and Devices, Proc. 8th International Symposium on Magnetic Materials, pp 525–536

[19] Zhong ZW, Sun J (2004) A suspension integrated with a piezoelectric microactuator for dual stage actuation. The International Journal of Advanced Manufacturing Technology 24(9-10):686

Index

3D Microprobes, 146
3D-model synthesis, 56

a-C, 91, 92, 99–102
absolute accuracy, 272, 284
aerostatic guide, 110
Air Bearing Surface (ABS), 437
air hammer effect, 110
Al_2O_3, 113
anodic bonding, 391
artefact, 160, 161
assembly process, 406
assembly strategies, 351
assembly system, 370
AUTOBIN, 415
automatic optical inspection (AOI), 416

Ball Grid Array (BGA), 417
batch processes, 331
beam propagation method, 129
beam splitter, 136
bearing, 12, 13, 19, 21, 22
bearing force, 110
$BICEP^3S$, 51
bistable microvalve, 379
bonding processes for microsystems, 331
boreholes, 260
borosilicate glass, 112
boss membrane, 147
buck converter, 32
building blocks, 51, 55

capacitive displacement sensor, 121
centrifugal feeder, 289, 307
Chemical-mechanical Polishing (CMP), 432
clamping device, 289, 308
closed-loop gap control, 111
CMM, 151, 153
coefficient of friction, 101–103
CoFe, 173, 175, 176, 186
coil, 12, 14, 16, 17, 20–22, 26, 27
collimator, 135
Computational Fluid Dynamics, 113
Computed Torque Feed Forward, 284
construction kit, 302, 305
contact area, 156
copper losses, 26
CVD coated grinding pins, 260
CVD coated grinding wheels, 247

dead volume, 111
desktop factories, 399
DFB laser diode, 137
diffraction simulation, 62
digital control, 111
dimensional standards, 160
displacement sensors, 111
DLC, 75–84, 86, 98, 101, 438
drive systems, 29
dry lubricating film, 90–92, 101, 104
dual-stage actuators (DSA), 427

Eagle2000, 129
ED-resist, 213
eddy current, 25, 26
efficiency, 32, 33, 37
electro contact discharge dressing (ECDD), 252
electro-depositable photoresist, 190, 193, 203
electro-depositable resist, 205
electromagnetic guide, 118

electroplating, 169–171, 174–176, 181, 182, 186

fast process cycles, 332
Finite Element Method, 13
flexibility, 398
force, 12–15, 18–20, 22–25
FOTURAN, 300, 306
friction, 71, 74, 76, 86
friction force, 100–106
field-programmable array (FPGA), 139
flow resistance, 110
flying height, 441

gas flow sputtering, 186
GRIN lens, 136
gripper, 353
  active heat management, 363
  passive heat management, 359
guidance, 12, 13, 19, 25

Hard Disk Drives (HDD), 427
hard magnetic materials, 167, 168, 179, 181, 183
Harmonic Drive, 280
head gimbal assembly, 439
heat management, 354
  active, 354, 362
  passive, 354, 356
helical coil, 189, 190, 198, 203–205, 215
High Aspect Ratio Micro Structure Technology (HARMST), 191, 205, 226, 237, 432
HMA film processing, 337
HMA particles processing, 337
hot melt adhesives (HMA), 331
hybrid step motor, 11, 17, 18, 225, 227
hybrid-bonding, 335

In-Circuit Test (ICT), 416
incremental sensor, 128
indentation hardness, 157
indentation measurements, 155
indentation modulus, 156
infusion pump, 380
inspection
  photogrammetric sensor, 312
InterVia 3D-N, 216
inverse dynamic model, 284

joining technologies, 351

kinematic synthesis, 271

LabVIEW, 138
laser, 132
Laser Doppler Vibrometer (LDV), 440
laser micromachining, 301
laser tracker, 283
linear series regulator, 31
linear stepper motor, 209
look-&-move, 277, 326
low process temperatures, 331

magnetic anisotropy, 13
magnetic flux, 189, 190, 201
magnetic material, 11–13, 21
meander coil, 189, 190, 198, 199, 201–205, 212
metal bonded grinding pins, 261
metal bonded grinding wheels, 252
Micabo F2
  photogrammetric sensor, 313
  robot, 273
micro adhesive bonding, 331
Micro Electro-mechanical System (MEMS), 427
micro force sensor, 95
Micro Harmonic Drive, 280, 401
microactuator, 369, 427
microassembly, 369, 406
  automated, 370
  challenges, 406
  inspection, 312
  look-&-move, 277, 326
  photogrammetric sensor, 312
  requirements, 370
  sensor-guided assembly, 312
microcoil, 196, 197, 205
microendoscopes, 419
microfriction, 89, 94
microgears, 401
microgrinding, 246
microgrinding pins, 260
microgripper, 289, 299
microguide, 90, 91, 93, 97, 101
microlens array, 141
micromotor, 225, 370
micronozzles, 112
micropattern, 89, 90, 92, 101–103, 106
microstructure, 99, 100
microtribology, 69, 70, 77
miniaturized machine components, 401
miniaturized robot, 277, 403
modified sliding mode controller, 42
modular software environment, 51
multiple direction accuracy, 272, 282

NiFe45/55, 168, 173–176, 186, 229, 236

NiFe81/19, 168, 173–175, 178, 226, 228
nozzle, 110

optical coupler, 129
optical grating, 134, 135
optical interferometer, 127

parallel robot structure, 278
Parvus robot, 280, 281
path accuracy, 272, 283
pattern projection, 422
PECVD, 78
permanent magnet, 11, 16–22
permeability, 13, 23, 24, 27
phase of preliminary design, 51
photo detector, 137
photogrammetric sensor, 311
  calibration, 323
  illumination, 318
  imaging, 314
  microassembly, 312
  mirror optics, 314
  prism module, 316
photogrammetry
  camera model, 322
  epipolar association, 324
  evaluation, 322
  part localization, 325
  signalization, 318
PID control, 117
piezoresistive elements, 149
piezoresistors, 149
planar waveguide, 129
Plasma Enhanced Chemical Vapor Deposition (PECVD), 432
pneumatic, 289, 299, 303, 307, 308
pocket bearing, 110
polymer magnet, 219
polymer magnet rotor, 222
position control, 40
pre-deflected membrane, 386, 390
probing forces, 157
probing repeatability, 151
process chain, 407
process sequence synthesis, 58
process variants
  active heat management, 356, 363
PVD, 69, 77, 78
pyrex glass, 389, 390

Q-Model, 51
quadrant photo diode, 137

reactive hot melts, 335
refractive index modification, 129
relative permeability, 168, 172, 174, 177, 178, 185
relative position, 275
relative positioning, 273
repeatability, 272, 279, 282
resist
  fluorescent, 321
robot design, 278
rotating stepper motor, 215
rotor, 217

saturation flux density, 168, 172, 174, 175, 178, 186
SC DC-DC converter, 33
SCARA, 278
Schönflies, 269, 278
scratch test, 72, 80
second stage actuation, 427
sensitivity analysis, 271, 278
serial robot structure, 278
shape memory alloy, 300, 307, 382, 384, 389
shielding electrodes, 122
Si_3N_4, 189, 197, 201, 205
signalization, 318
  by circular marks, 319
  by fluorescent features, 320
  by polygons, 318
  by scattering features, 319
silicon, 388–390
silicon direct bonding, 391
silicon on insulator, 302
simulation, 13, 23, 27
size-adapted robot, 272
Slider with an Integrated Microactuator (SLIM), 428
Sliding Mode Controller, 41
soft magnetic materials, 167, 168, 171, 173, 174, 178
speckle reduction, 424
spiral coil, 189, 190, 198–202, 205
state control, 117
stepping mode, 39
stick-slip, 93, 100, 103, 105, 106
stiffness of membrane, 153
SU-8, 189, 190, 194–198, 202–205, 236
  fluorescent, 321
synchronous micromotor, 225, 230
synchronous motor, 11, 16, 17, 20–22, 219
system identification, 44
system integration, 436

thermal expansion, 289, 304
thinning-by-dicing, 436
tip rounding, 155
tolerance analysis, 56
tool development, 248
torque, 14, 18, 21
track following, 429
transmission error, 283
tungsten thin film, 382

ultrasonic assisted holegrinding, 261
UV adhesive, 127
UV depth lithography, 201, 202, 432

variable reluctance micromotor, 225, 226
variable reluctance step motor, 11, 15–19, 22, 24–26

waveguide coupler, 128
waveguide writing, 129
wear, 69, 70, 72–76, 79, 82, 84–86
wet-etch simulation, 60
workspace, 281

X-ray inspection, 419
xy-microactuator, 225, 234, 369

FSC
www.fsc.org
MIX
Papier aus verantwortungsvollen Quellen
Paper from responsible sources
FSC® C105338

If you have any concerns about our products,
you can contact us on
ProductSafety@springernature.com

In case Publisher is established outside the EU,
the EU authorized representative is:
Springer Nature Customer Service Center GmbH
Europaplatz 3, 69115 Heidelberg, Germany

Printed by Libri Plureos GmbH
in Hamburg, Germany